PROBABILITY, STATISTICS, AND DECISION FOR CIVIL ENGINEERS

PROBABILITY, STATISTICS, AND DECISION FOR CIVIL ENGINEERS

Jack R. Benjamin

and

C. Allin Cornell

DOVER PUBLICATIONS, INC.
Mineola, New York

Bibliographical Note

This Dover edition, first published in 2014, is an unabridged republication of the work originally published by McGraw-Hill, Inc., New York, in 1970.

Library of Congress Cataloging-in-Publication Data

Benjamin, Jack R. (Jack Ralph), 1917–1998, author.
  Probability, statistics, and decision for civil engineers / Jack R. Benjamin, C. Allin Cornell. — Dover edition.
      pages cm
  Reprint of: New York : McGraw-Hill, 1970.
  Summary: "This text covers the development of decision theory and related applications of probability. Extensive examples and illustrations cultivate students' appreciation for applications, including strength of materials, soil mechanics, construction planning, and water-resource design. Emphasis on fundamentals makes the material accessible to students trained in classical statistics and provides a brief introduction to probability. 1970 edition"— Provided by publisher.
  ISBN-13: 978-0-486-78072-6 (paperback)
  ISBN-10: 0-486-78072-4 (paperback)
  1. Probabilities. 2. Mathematical statistics. 3. Bayesian statistical decision theory. I. Cornell, C. Allin, author. II. Title.

TA340.B45 2014
519.5—dc23

2014012854

Manufactured in the United States by Courier Corporation
78072402   2014
www.doverpublications.com

To Sarah and Elizabeth

Preface

This book is designed for use by students, practitioners, teachers, and researchers in civil engineering. Most books on probability and statistics are intended for readers from many fields, and therefore the question might be asked, "Why is it necessary to have a book on applied probability and statistics for civil engineers alone?" Further, within civil engineering itself, we might ask, "Why is it reasonable to attempt to write to an audience with such widely varying needs?" The need for writing such a book is a result of the unusual status of probability and statistics in civil engineering.

Although virtually all forward-looking civil engineers see the rationality and utility of probabilistic models of phenomena of interest to the profession, the number of civil engineers trained in probability theory has been limited. As a result there has not yet been widespread adoption of such models in practice or even in university courses below the advanced graduate level. Indeed, there has not yet been sufficient development of such models to permit a unified probabilistic approach to the many aspects of the strength of materials, soil mechanics, construction planning, water-resource design, and many other subjects where the methods could clearly be useful.

Many departments of civil engineering now require that their students study probability or statistics. Subsequent undergraduate technical courses, however, usually lack material that draws upon that study, and this often leaves the student without an appreciation for possible implications of the theory. Through the use of illustrations and problems taken solely from the civil-engineering field this book is designed to develop this appreciation for applications. It can serve as a primary text for a course within a civil-engineering department or as a supplementary text for courses taught in other departments. The major compromise here was the acceptance of the risk of obscuring or confusing the basic theory in describing the illustrative physical problem. We could have avoided this risk by using only examples dealing with coins, dice, red and black balls, etc., but the desire to develop the ability to construct mathematical models of physical phenomena in civil-engineering applications weighed heavily in favor of professional illustrations.

Students finishing such a professionally oriented probability course may or may not go on to advanced courses in the theory. This consideration dictated at least an introductory coverage of a wide variety of material coupled with a thorough treatment of the fundamentals. The kinds of compromises involved in trying to decide between breadth in subject matter and depth in fundamentals are obvious. It is unlikely that our many decisions of this kind will be judged optimal by all, or, indeed, by ourselves in time.

This emphasis on a thorough understanding of fundamentals, on a broad coverage of subject matter, and on illustrations drawn from familiar practice are appropriate for a book aimed at a growing number of practicing civil engineers, researchers, and students who have an interest in teaching themselves the elements of applied probability and statistics. Here the continued exposure to professionally based illustrations, as simplified as they may be, can help to sustain that interest and stimulate ideas for developments of direct personal concern. (Although it is not a basic aim, the usefulness of the text as a reference book is enhanced by a number of convenient tables.)

A book designed for a broad audience of civil engineers is also important because the training of civil engineers familiar with probability and statistics is presently widely varied in depth, breadth, and source. Many misuses and misinterpretations of probability and statistics exist in practice and in the profession's literature. The emphasis in this book on depth in fundamentals is again critical. In a number of places in the book a significant effort has been made to place additional emphasis on commonly misunderstood points. To the beginning reader the effort and words may seem excessive, but it is hoped that these discourses will serve as warnings to the reader to be particularly alert when similar situations are encountered. Readers with some experience in applications of probability and statistics will recognize these subjects and perhaps benefit by the additional discussion. It is hoped in particular that readers who have studied only statistics in the past will appreciate the opportunity to study the probability theory upon which statistics is based. A common reaction to such a study is, on one hand, surprise at the degree of underlying unity in statistical methods and, on the other, a new appreciation for the limitations of classical statistics.

The most significant consideration in our decision to develop a book for civil engineers of diverse backgrounds is the result of a major change within the theory of applied probability and statistics. A new, still somewhat controversial theory is being developed around a framework of economic decision making. The implications to civil engineers are many. In place of earlier emphasis on obtaining proper, objective *descriptions* of repetitive physical phenomena, the new concern is with making *decisions* involving economic gains and losses when uncertainty exists in the decision maker's mind regarding the state of nature. This new emphasis, with its new interpretations and its new methods, is far more appropriate and natural for civil engineers, whose profession is more closely involved than any other in the economical design of one-of-a-kind systems subject to the uncertain demands of natural and man-made environmental factors.

The new development (associated variously with such phrases as Bayesian statistical decision theory, subjective probability, and utility theory) is, in the authors' minds, of such central importance that convenient access to it must be developed for civil engineers. Even researchers and teachers well versed in traditional probability and classical statistics can benefit from exposure to decision theory. They will find a number of very reasonable new ideas and a more liberal interpretations of some old ideas (including, in a sense, justification for some natural "mistakes" they may very well have been making in applying classical statistics).

The necessity of making this new material accessible to engineers who have already been trained in classical statistics was the source of what the authors feel to be their major compromise. The authors are frankly Bayesianists, but because these classically trained engineers are making the present developments in the field and because they are the engineering educators of the future, we decided to *present classical as well as modern statistics*. It is hoped that a reader can use the book either way, if not both, and that this approach will accelerate the development and spread of probabilistic methods within our profession. Inclusion of both approaches will also aid those with some previous experience to interpret and judge the methods he is more familiar with.   Finally, knowledge of both approaches is desirable, if only to understand the meaning and the limitations of the widely used classical methods.   We apologize for the fact that the inclusion of both methods has led to a certain lack of uniformity in position and in treatment that the reader may find puzzling. We wish to thank the many students and colleagues who have contributed over the past seven years to the development of this text.

The book can be used in a variety of ways.   The reader seeking no more than a brief elementary introduction to probability might read as little as Chap. 1, Secs. 2.1, 2.2.1, 2.4.1, 3.1.1, 3.1.2, 3.2.1, 3.2.2, 3.3, and 4.1.1.   Minimum course coverage should probably include, in addition, the material in Secs. 2.2.2, 2.3.1, 2.4.2, 3.1.3, 3.6.2, 4.1.2, 4.1.3, 4.2, 4.4, and Chap. 5.   (Much of the material in Secs. 3.1, 3.2, and 3.4 can be covered in lecture courses as illustrations of the material in Chap. 2 rather than separately.)   At Stanford and M.I.T. we teach several quite different courses, using the same text and relying on it to supplement the material not emphasized in lectures.

JACK R. BENJAMIN
C. ALLIN CORNELL

Contents

PROBABILITY, STATISTICS, AND DECISION FOR CIVIL ENGINEERS

Introduction

There is some degree of uncertainty connected with all phenomena with which civil engineers must work. Peak traffic demands, total annual rainfalls, and steel yield strengths, for example, will never have exactly the same observed values, even under seemingly identical conditions. Since the performance of constructed facilities will depend upon the future values of such factors, the design engineer must recognize and deal with this uncertainty in a realistic and economical manner.

How the engineer chooses to treat the uncertainty in a given phenomenon depends upon the situation. If the degree of variability is small, *and* if the consequences of any variability will not be significant, the engineer may choose to ignore it by simply assuming that the variable will be equal to the best available estimate. This estimate might be the average of a number of past observations. This is typically done, for example, with the elastic constants of materials and the physical dimensions of many objects.

If, on the other hand, uncertainty is significant, the engineer may choose to use a "conservative estimate" of the factor. This has often

1

been done, for example, in setting "specified minimum" strength properties of materials and when selecting design demands for transportation or water resource facilities. Many questions arise in the practice of using conservative estimates. For example:

How can engineers maintain consistency in their conservatism from one situation to another? For instance, separate professional committees set the specified minimum concrete compressive strength and the specified minimum bending strength of wood.
Is it always possible to find a value that is conservative in all respects? A conservative estimate of the friction factor of a pipe should be on the "high side" of the best estimate in order to produce a conservative (low) estimate of flow in that pipe, but this estimate may produce unconservative (high) estimates of the flows in other parallel pipes in the same network.
Is the design of the resulting facility unnecessarily expensive? The consequences of occasional flows in excess of the capacity of a city storm drainage system may be small, but the initial cost of a system capable of processing a conservative estimate of the peak flow may be excessive.
Can the behavior of the facility be adequately predicted with only a conservative estimate? For example, the ability of an automatic traffic control system to increase the capacity of an artery depends intimately upon the degree and character of the *variability* of the traffic flow as well as upon the total volume.

In short, only if the situation permits can the engineer successfully treat uncertainty and variability simply through best estimates or conservative estimates. *In general, if the decision he must make* (i.e., the choice of design or course of action) *is insensitive to the uncertainty, the engineer can ignore it in his analysis.* If it cannot be ignored, uncertainty must be dealt with explicitly during the engineering process.

Often, particularly in testing and in laboratory, research, and development work, the individual involved may not know how the observed variability will affect subsequent engineering decisions. It becomes important, then, that he have available efficient ways to transmit his information about the uncertainty to the decision makers.

Although the distinction between these two needs—decision making and information transmission in the presence of uncertainty—is not always clear, both require special methods of analysis, methods which lie in the fields of probability and statistics.

In Chap. 1 we will review quickly the conventional means for reducing observed data to a form in which it can be interpreted, evaluated, and efficiently transmitted.   Chapter 2 develops a background in probability theory, which is necessary to treat uncertainty explicitly in engineering problems.   Chapter 3 makes use of probability theory to develop a number of the more frequently used mathematical models of random processes.   In Chap. 4 statistical methods that comprise a bridge between observed data and mathematical models are introduced.   The data is used to estimate parameters of the model and to evaluate the validity of the model.   Chapters 5 and 6 develop in detail the methods for analyzing engineering economic decisions in the face of uncertainty.

1
Data Reduction

A necessary first step in any engineering situation is an investigation of available data to assess the nature and the degree of the uncertainty. An unorganized list of numbers representing the outcomes of tests is not easily assimilated. There are several methods of organization, presentation, and reduction of observed data which facilitate its interpretation and evaluation.

It should be pointed out explicitly that the treatment of data described in this chapter is in no way dependent on the assumptions that there is randomness involved or that the data constitute a random sample of some mathematical probabilistic model. These are terms which we shall come to know and which some readers may have encountered previously. The methods here are simply convenient ways to reduce raw data to manageable forms.

In studying the following examples and in doing the suggested problems, it is important that the reader appreciate that the data are "real" and that the variability or scatter is representative of the magnitudes of variation to be expected in civil-engineering problems.

1.1 GRAPHICAL DISPLAYS

Histograms A useful first step in the representation of observed data is to reduce it to a type of bar chart. Consider, for example, the data presented in Table 1.1.1. These numbers represent the live loads observed in a New York warehouse. To anticipate the typical, the extreme, and the long-term behavior of structural members and footings in such structures, the engineer must understand the nature of load distribution. Load variability will, for example, influence relative settlements of the column footings. The values vary from 0 to 229.5 pounds per square foot (psf). Let us divide this range into 20-psf intervals, 0 to 19.9, 20.0 to 39.9, etc., and tally the number of occurrences in each interval.

Plotting the frequency of occurrences in each interval as a bar

Table 1.1.1   Floor-load data*

Bay	Base-ment	1st	2d	3d	4th	5th	6th	7th	8th	9th
A	0	7.8	36.2	60.6	64.0	64.2	79.2	88.4	38.0	72.7
B	72.2	72.6	74.4	21.8	17.1	48.5	16.8	105.9	57.2	75.7
C	225.7	42.5	59.8	41.7	39.9	55.5	67.2	122.8	45.2	62.9
D	55.1	55.9	87.7	59.2	63.1	58.8	67.7	90.4	43.3	55.2
E	36.6	26.0	90.5	23.0	43.5	52.1	102.1	71.7	4.1	37.3
F	129.4	66.4	138.7	127.9	90.9	46.9	197.5	151.1	157.3	197.0
G	134.6	73.4	80.9	53.3	80.1	62.9	150.8	102.2	6.4	45.4
H	121.0	106.2	94.4	139.6	152.5	70.2	111.8	174.1	85.4	83.0
I	178.8	30.2	44.1	157.0	105.3	87.0	50.1	198.0	86.7	64.6
J	78.6	37.0	70.7	83.0	179.7	180.2	60.6	212.4	72.2	86.0
K	94.5	24.1	87.3	80.6	74.8	72.4	131.1	116.1	53.6	99.1
L	40.2	23.4	8.4	42.6	43.4	27.4	63.8	18.4	16.2	58.7
M	92.2	49.8	50.9	116.4	122.9	132.3	105.2	160.3	28.7	46.8
N	99.5	106.9	55.9	136.8	110.4	123.5	92.4	160.9	45.4	96.3
O	88.5	48.4	62.3	71.3	133.2	92.1	111.7	67.9	53.1	39.7
P	93.2	55.0	80.8	143.5	122.3	184.2	150.0	57.6	6.8	53.3
Q	96.1	54.8	63.0	228.3	139.3	59.1	112.1	50.9	158.6	139.1
R	213.7	65.7	90.3	198.4	97.5	155.1	163.4	155.3	229.5	75.0
S	137.6	62.5	156.5	154.1	134.3	81.6	194.4	155.1	89.3	73.4
T	79.8	68.7	85.6	141.6	100.7	106.0	131.1	157.4	80.2	65.0
U	78.5	118.2	126.4	33.8	124.6	78.9	146.0	100.3	97.8	75.3
V	24.8	55.6	135.6	56.3	66.9	72.2	105.4	98.9	101.7	58.2

* Observed live loads (in pounds per square foot); bay size: 400 ± 6 ft².
Source: J. W. Dunham, G. N. Brekke, and G. N. Thompson [1952], "Live Loads on Floors in Buildings," Building Materials and Structures Report 133, National Bureau of Standards, p. 22.

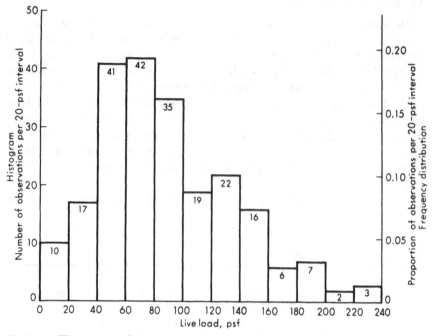

Fig. 1.1.1   Histogram and frequency distribution of floor-load data.

yields a *histogram*, as shown in Fig. 1.1.1.   The height, and more usefully, the area, of each bar are proportional to the number of occurrences in that interval.   The plot, unlike the array of numbers, gives the investigator an immediate impression of the range of the data, its most frequently occurring values, and the degree to which it is scattered about the central or typical values.   We shall learn in Chap. 2 how the engineer can predict analytically from this shape the corresponding curve for the *total* load on a column supporting, say, 20 such bays.

If the scale of the ordinate of the histogram is divided by the total number of data entries, an alternate form, called the *frequency distribution*, results.   In Fig. 1.1.1, the numbers on the right-hand scale were obtained by dividing the left-hand scale values by 220, the total number of observations.   One can say, for example, that the proportion of loads observed to lie between 120 and 139.9 psf was 0.10.   If this scale were divided by the interval length (20 psf), a *frequency density distribution* would result, with ordinate units of "frequency per psf."   The area under this histogram would be unity.   This form is preferred when different sets of data, perhaps with different interval lengths, are to be compared with one another.

The *cumulative frequency distribution*, another useful graphical representation of data, is obtained from the frequency distribution by calculating the successive partial sums of frequencies up to each interval division point. These points are then plotted and connected by straight lines to form a nondecreasing (or monotonic) function from zero to unity.

In Fig. 1.1.2, the cumulative frequency distribution of the floor-load data, the values of the function at 20, 40, and 60 psf were found by forming the partial sums $0 + 0.0455 = 0.0455$, $0.0455 + 0.0775 = 0.1230$, and $0.1230 + 0.1860 = 0.3090$.† From this plot, one can read that the proportion of the loads observed to be equal to or less than 139.9 psf was 0.847. After a proper balancing of initial costs, consequences of poor performance, and these frequencies, the designer might conclude that a beam supporting one of these bays must be stiff enough to avoid deflections in excess of 1 in. in 99 percent of all bays. Thus the design should be checked for deflections under a load of 220 psf.

Some care should be taken in choosing the width of each interval

† When constructing the cumulative frequency distribution, one can avoid the arbitrariness of the intervals by plotting one point per observation, that is, by plotting i/n versus $x^{(i)}$, where $x^{(i)}$ is the ith in ordered list of data (see Fig. 1.2.1).

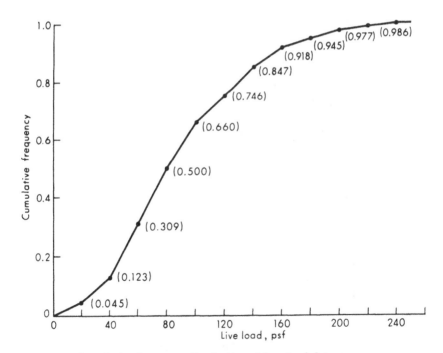

Fig. 1.1.2 Cumulative frequency distribution of floor-load data.

in these diagrams.† A little experimentation with typical sets of data will convince the reader that the choice of the number of class intervals can alter one's impression of the data's behavior a great deal. Figure 1.1.3 contains two histograms of the data of Table 1.1.1, illustrating the influence of interval size. An empirical practical guide has been suggested by Sturges [1926]. If the number of data values is n, the number of intervals k between the minimum and maximum value observed should be about

$$k = 1 + 3.3 \log n \tag{1.1.1}$$

in which logarithms to the base 10 should be employed. Unfortunately, if the number of values is small, the choice of the precise point at which the interval divisions are to occur also may alter significantly the appearance of the histogram. Examples can be found in Sec. 1.2 and in the problems at the end of this chapter. Such variations in shape may at

† If advantageous, unequal interval widths may be preferable (see Sec. 4.4).

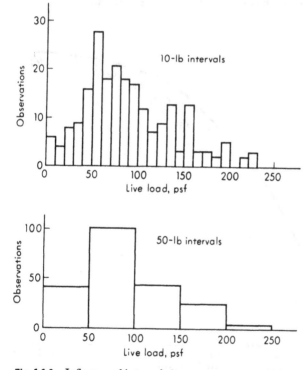

Fig. 1.1.3 Influence of interval size on appearance of histogram (data of Table 1.1.1).

first be disconcerting, but they are indicative of a failure of the set of data to display any sharply defined features, a piece of information which is in itself valuable to the engineer. This failure may be because of the inadequate size of the data set or because of the nature of the phenomenon being observed.

1.2 NUMERICAL SUMMARIES

Central value measures   The single most helpful number associated with a set of data is its average value, or arithmetic mean. If the sequence of observed values is denoted x_1, x_2, . . . , x_n, the *sample mean* $\bar{x}$ is simply

$$\bar{x} = \frac{1}{n} \sum_{i=1}^{n} x_i \qquad\qquad (1.2.1)$$

Fifteen reinforced-concrete beams built by engineering students to the same specifications and fabricated from the same batch of concrete were tested in flexure. The observed results of first-crack and ultimate loads, recorded to the nearest 50 lb, are presented in Table 1.2.1. Their cumulative frequency distributions are shown in Fig. 1.2.1. (They were

Table 1.2.1   Tests of identical reinforced concrete beams

Beam number	Load at which the first crack was observed, lb	Failure load, lb
1	10,350	10,350
2	8,450	9,300
3	7,200	9,600
4	5,100	10,300
5	6,500	9,400
6	10,600	10,600
7	6,000	10,100
8	6,000	9,900
9	9,500	9,500
10	6,500	10,200
11	9,300	9,300
12	6,000	9,550
13	6,000	9,550
14	5,800	10,500
15	6,500	10,200

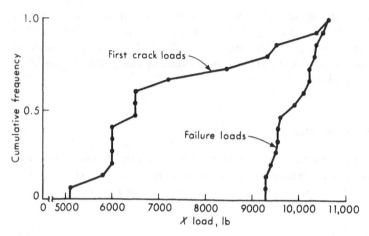

Fig. 1.2.1 Cumulative frequency distributions; beam data.

plotted by the method suggested in the footnote on page 7.)† The
scatter of the data might be attributed to unrecorded construction and
testing differences, inconsistent workmanship, human errors, and inherent
material variability as well as observation and measurement errors. The
mean value of the failure loads is computed to be

$$\bar{x} = \frac{1}{n} \sum_{i=1}^{n} x_i = \frac{148,350}{15} = 9890 \text{ lb}$$

The sample mean is frequently interpreted as a typical value or a
central value of the data. If required to give only a single number, one
would probably use this sample mean as his "best prediction" of the load
at which a nominally identical beam would fail.

Other measures of the central tendency of a data set include *the
mode*, the most frequently occurring value in the data set, and *the median*,
the middle value in an ordered list (the middle value if n is odd, or the
average of the two middle values if n is even) (see Table 1.2.2). The order
of observing the values is usually not important, and they may be
arranged in any way which is convenient. The median of the failure data
is 9900 lb; the mode is not unique, since several values appear twice and
none appears more times. These two terms occur commonly in the
literature of other fields and consequently should be understood, but
they only infrequently prove meaningful in engineering problems.

† The reader would do well to plot a number of different histograms of this failure-
load data, both to judge the value of this means for data interpretation in this example
and to illustrate the remarks about histogram intervals made in Sec. 1.1.

Table 1.2.2 Ordered first-crack and ordered failure-load data

First-crack load	Failure load
5,100	9,300
5,800	9,300
6,000	9,400
6,000	9,500
6,000	9,550
6,000	9,550
6,500	9,600
6,500	9,900
6,500	10,100
7,200	10,200
8,450	10,200
9,300	10,300
9,500	10,350
10,350	10,500
10,600	10,600

Measures of dispersion   Given a set of observed data, it is also desirable to be able to summarize in a single number something of the variability of the observations.   In the past the measure most frequently occurring in engineering reports was the *range* of the data.   This number, which is simply the difference between the maximum and minimum values observed, has the virtue of being easy to calculate, but certain obvious weaknesses have led to its being replaced or supplemented.   The range places too much emphasis on the extremes, which are often suspected of experimental error, and neglects the bulk of the data and experimental effort which lies between these extremes.   The range is also sensitive to the size of the sample observed, as will be demonstrated in Sec. 3.3.3.

A far more satisfactory measure of dispersion is found in the *sample variance*.   It is analogous to the moment of inertia in that it deals with squares of distances from a center of gravity, which is simply the sample mean.   The sample variance s^2 is defined here to be

$$s^2 = \frac{1}{n} \sum_{i=1}^{n} (x_i - \bar{x})^2 \tag{1.2.2a}$$

To eliminate the dependence on sample size, the squared distances are divided by n to yield an average squared deviation.   There are sound reasons favoring division by $n - 1$, as will be shown in Chap. 4, but the intuitively more satisfactory form will not be abandoned until the reader can appreciate the reasons. Similarly, small computational changes in other definitions given in this chapter may be found desirable in the light of later discussions on the estimation of *moments of random variables* (Secs. 2.4 and 4.1).

Expansion of Eq. (1.2.2a) yields an expression which will be found far more convenient for computation of s^2:

$$\frac{1}{n} \sum_{i=1}^{n} (x_i - \bar{x})^2 = \frac{1}{n} \left(\sum_{i=1}^{n} x_i^2 - 2\bar{x} \sum_{i=1}^{n} x_i + n\bar{x}^2 \right)$$

But, by Eq. (1.2.1),

$$\sum_{i=1}^{n} x_i = n\bar{x}$$

Therefore,

$$s^2 = \frac{1}{n} \left(\sum_{i=1}^{n} x_i^2 - n\bar{x}^2 \right) \tag{1.2.2b}$$

The positive square root s of the sample variance of the data is termed the *sample standard deviation*. It is analogous to the radius of gyration of a structural cross section; they are both shape- rather than size-dependent parameters. The addition of a constant to all observed values, for example, would alter the sample mean but leave the sample standard deviation unchanged. This number has the same units as the original data, and, next to the mean, it conveys more useful information to the engineer than any other single number which can be computed from the set of data. Roughly speaking, the smaller the standard deviation of the sample, the more clustered about the sample mean is the data and the less frequent are large variations from the average value.

For the beam-failure data the sample variance and sample standard deviation are computed as follows:

$$s^2 = \frac{1}{n} \left(\sum_{i=1}^{n} x_i^2 - n\bar{x}^2 \right)$$

$$= \frac{1}{15}(1,470,064,450 - 1,467,181,500)$$

$$= 192,200 \text{ lb}^2$$

$$s = \sqrt{192,200} = 440 \text{ lb}$$

(Notice that, owing to the required subtraction of two nearly equal numbers, many significant digits must be carried in the sums to maintain accuracy.) In this example, the sample standard deviation might be used to compare the variability of the strength of lab beams with that of field-constructed beams.

When comparing the relative dispersion of more than one kind of data, it is convenient to have a dimensionless description such as the commonly quoted *sample coefficient of variation*. This quantity v is defined as the ratio of the sample standard deviation to the sample mean.

$$v = \frac{s}{\bar{x}} \tag{1.2.3}$$

The sample coefficient of variation of the beam-failure data is

$$v = \frac{s}{\bar{x}} = \frac{440}{9890} = 0.0446$$

while that of the first observed crack is much larger, being

$$v = \frac{s}{\bar{x}} = \frac{1690}{7310} = 0.233$$

The engineer might interpret the difference in magnitudes of these coefficients as an indication that first-crack loads are "more variable" or more difficult to predict closely than failure loads. Such information is important when appearance as well as strength is a design criterion.

Measure of asymmetry   One other numerical summary† of observed data is simply a logical extension of the reasoning leading to the formula for the sample variance.   Where the variance was an average second moment about the mean, so the *sample coefficient of skewness* is related to the third moment about the mean.   To make the coefficient nondimensional, the moment is divided by the cube of the sample standard deviation.

The coefficient of skewness g_1 provides a measure of the degree of

† A fourth numerical summary, the *coefficient of kurtosis*, may also be employed. Without large sample sizes, however, its use is seldom recommended.

The sample coefficient of kurtosis g_2 is related to the "peakedness" of the histogram:

$$g_2 = \frac{(1/n) \sum_{i=1}^{n} (x_i - \bar{x})^4}{s^4}$$

Traditionally the value of this coefficient is compared to a value of $g_2 = 3.0$, the kurtosis coefficient of a commonly encountered bell-shaped continuous curve which we shall learn to call the *normal curve* (Sec. 3.3.3).   For the data here,

$$g_2 = \frac{\frac{1}{15}(836 \times 10^9)}{(440)^4} = 1.64 < 3.0$$

asymmetry about the mean of the data:

$$g_1 = \frac{(1/n) \sum_{i=1}^{n} (x_i - \bar{x})^3}{s^3} \tag{1.2.4}$$

The coefficient is positive for histograms skewed to the right (i.e., with longer tails to the right) and negative for those skewed to the left. Zero skewness results from symmetry but does not necessarily imply it.

For the beam-failure data of Table 1.2.1,

$$g_1 = \frac{\frac{1}{15}(120{,}000{,}000)}{(440)^3} = 0.103 > 0$$

indicating mild skewness to the right. The implication is that there were fewer but larger deviations to the high side than to the low side of the average value. (The sample coefficient of skewness should be calculated using $x_i - \bar{x}$, since an expansion similar to Eq. (1.2.2b) for the sample variance does not prove useful.)

In this case, if the students reported their experimental results in the form of only three numbers, $\bar{x}$, s, and g_1, it would already be sufficient to gain appreciation for the shape of the histogram. The economy in the use of such sample averages to transmit information about data becomes even more obvious as the amount of data increases.

1.3 DATA OBSERVED IN PAIRS

If paired samples of two items of related interest, such as the first-crack load and the failure load of a beam (Table 1.2.1), are available, it is often of interest to investigate the correlation between them. A graphical picture is available in the so-called *scattergram*, which is simply a plot of the observed pairs of values. The scattergram of the reinforced-concrete–beam data is shown in Fig. 1.3.1, where the x_i are values of first-crack loads and the y_i are values of failure loads. There is no clearly defined functional relationship between these observations, even though an engineer might expect larger-than-average values of one load generally to pair with larger-than-average values of the other, and similarly with low values.

A numerical summary of the tendency towards high-high, low-low pairings is provided by the *sample covariance* s_{XY}, defined by

$$s_{X,Y} = \frac{1}{n} \sum_{i=1}^{n} (x_i - \bar{x})(y_i - \bar{y}) \tag{1.3.1}$$

Clearly, if larger (smaller) than average values of x are frequently paired with larger (smaller) than average values of y, most of the terms will be

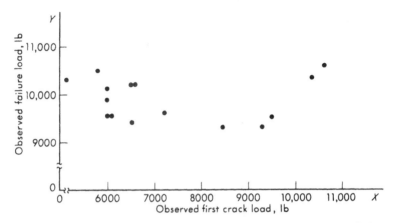

Fig. 1.3.1 Beam-data scattergram; plot shows lack of linear correlation.

positive, while small-large pairings will tend to yield negative values of $s_{X,Y}$.

It is common to normalize the sample covariance by the sample standard deviations, denoted now with subscripts, s_X and s_Y. The result is called the *sample correlation coefficient* $r_{X,Y}$:

$$r_{X,Y} = \frac{s_{X,Y}}{s_X s_Y} = \frac{1}{n} \sum_{i=1}^{n} \left(\frac{x_i - \bar{x}}{s_X}\right)\left(\frac{y_i - \bar{y}}{s_Y}\right) \tag{1.3.2}$$

It can be shown that $r_{X,Y}$ is limited in value to $-1 \le r_{X,Y} \le 1$ and that the extreme values are obtained if and only if the points in the scattergram lie on a perfectly straight line, that is, only if

$$y_i = a + bx_i$$

the sign of $r_{X,Y}$ depending only on the sign of b. In this case the factors are said to be *perfectly correlated*. For other than perfectly linear relationships $|r_{X,Y}|$ is less than 1, the specific value $r_{X,Y} = 0$ being said to indicate that the x's and y's are *uncorrelated*. The x's and y's may, in fact, lie on a very well-defined nonlinear curve and hence be closely, perhaps functionally, related (for example, $y_i = bx_i^2$); in this case, the absolute value of the sample correlation coefficient will be less than one. The coefficient is actually a measure of only the *linear* correlation between the factors sampled.

For the beam data of Table 1.2.1, the sample covariance is

$$s_{X,Y} = \frac{1}{n} \sum_{i=1}^{n} (x_i - 7310)(y_i - 9890) = 94{,}200 \text{ lb}^2$$

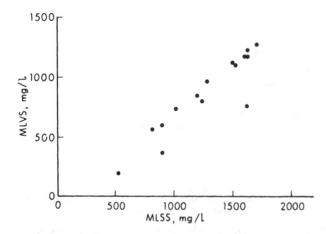

Fig. 1.3.2 Scattergram of MLVS against MLSS.

while the sample correlation coefficient is

$$r_{X,Y} = \frac{s_{X,Y}}{s_X s_Y} = \frac{94{,}200}{(1690)(440)} = +0.12$$

The small value of this coefficient summarizes the qualitative conclusion reached by observing the scattergram, that is, that the first-crack loads and failure loads are not closely related. To the engineer who must judge the ultimate strength of a (moment) cracked beam, the implications are that the crack does not necessarily mean that (moment) failure is imminent, and also that he cannot successfully use the first-crack load to help predict the ultimate load better (Sec. 4.3).

Illustration: Large correlation As a second example consider a problem in mixed-liquor analysis. Suspended solids (MLSS) can be readily measured, but volatile solids (MLVS) prove more difficult. It would be convenient if a measure of MLSS could be interpreted in terms of MLVS. A scattergram between MLVS and MLSS is shown in Fig. 1.3.2. The sample correlation coefficient for the data shown is $r_{X,Y} = 0.92$. This relatively large value of the correlation coefficient follows from the nearly linear relationship between the two factors in the scattergram. It also suggests that the value of the MLVS could be closely estimated given a measurement of MLSS. Methods for carrying out and evaluating this prediction will be investigated in Sec. 4.3.

Discussion In addition to providing efficient summaries of groups of numbers, the data-reduction techniques discussed in this chapter play an important role in the formulation and implementation of probability-based engineering studies. The data in the illustrations and problems in this chapter might be interpreted by the engineer as repeated observations of a random or *probabilistic* mathematical model that he formulated to aid him in describing and predicting some natural physical phenomenon of concern.

The concepts introduced in Chap. 2 are fundamental to the construction and manipulation of these probabilistic models. In Chap. 4 we shall be concerned with the use made of physical-world observed data, such as that seen in this chapter, in fitting the mathematical models, and in verifying their appropriateness.

The importance of the histogram in choosing and in verifying the mathematical model will become evident in Chaps. 2 to 4. Parameters of the mathematical models which are analogous to the sample mean, sample variance, and sample correlation coefficient will be discussed in Sec. 2.4. The relationship between sample averages and these model parameters will be treated in detail in Chap. 4.

1.4 SUMMARY FOR CHAPTER 1

In Chap. 1 we introduced graphical and numerical ways to reduce sets of observed data. The former include various forms of relative frequency and cumulative frequency diagrams. The latter include measures of central tendency, dispersion, skew, and correlation. The most commonly encountered are the sample mean $\bar{x}$, the sample standard deviation s, the sample coefficient of variation v, and the sample correlation coefficient r.

REFERENCES

General

The student will find it desirable to refer to other texts to amplify this chapter. In particular, Hald [1952] will prove most useful.

Bowker, A. H. and G. J. Lieberman [1959]: "Engineering Statistics," Prentice-Hall, Inc., Englewood Cliffs, N.J.

Bryant, E. C. [1960]: "Statistical Analysis," McGraw-Hill Book Company, New York.

Hahn, G. J. and S. S. Shapiro [1967]: "Statistical Models in Engineering," John Wiley & Sons, Inc., New York.

Hald, A. [1952]: "Statistical Theory with Engineering Applications," John Wiley & Sons, Inc., New York.

Neville, A. M. and J. B. Kennedy [1964]: "Basic Statistical Methods for Engineers and Scientists," International Textbook Company, Scranton, Pa.

Wine, R. L. [1964]: "Statistics for Scientists and Engineers," Prentice-Hall, Englewood Cliffs, N.J.

Specific text references

Sturges, H. A. [1926]: The Choice of a Class Interval, *J. Am. Statist. Assoc.*, vol. 21, pp. 65–66.

PROBLEMS

1.1. Ten timber beams were tested on a span of 4 ft by a single concentrated load at midspan. The Douglas fir beams were 2 by 4 in. in nominal section (actual section: 1.63 by 3.50 in.). The purpose of the study was to compare ultimate load and load allowable by building code (394 lb), to compare actual and calculated deflection ($E = 1,600,000$ psi), and to determine if a relationship exists between rigidity and ultimate strength.

Specimen	Deflection at working load, in.	Ultimate load, lb
1	0.160	1750
2	0.130	2350
3	0.155	2050
4	0.134	2100
5	0.135	1525
6	0.123	2000
7	0.168	1450
8	0.130	2100
9	0.150	1475
10	0.132	1675

Compute the sample mean and variance for each set of data. Construct histograms and relative frequency distributions for each set. Plot the scatter diagram and compute the correlation coefficient. What are your conclusions?

1.2. Ten bolted timber joints were tested using a $\frac{1}{2}$-in. bolt placed in double shear through three pieces of 2 by 4 in. (nominal size) Douglas fir. The results of the tests were:

Specimen	Load at yield of $\frac{1}{16}$ in., lb	Load at yield of $\frac{1}{8}$ in., lb	Ultimate load, lb
1	3500	3900	4000
2	3400	3500	3500
3	3100	3600	3800
4	2700	3500	4150
5	2900	3680	4050
6	3400	4250	4500
7	2700	3500	4400
8	3100	3700	4250
9	3100	4100	4550
10	2300	3200	3650

Calculate the sample standard deviations and coefficients of variation. Plot a scatter diagram and compute the sample correlation coefficient between load at yield of $\frac{1}{8}$ in. and ultimate load. Might this simpler, nondestructive test be considered as a prediction of ultimate joint capacity?

1.3. Jack and anchor forces were measured in a prestressed-concrete lift-slab job for both straight and curved cables. The objective of the study was to obtain estimates of the influence of friction and curvature. The data are given in terms of the ratio of jack to anchor force.

Straight cables	Curved cables
1.6609	2.0464
1.5497	1.7496
1.7190	2.0464
1.6534	1.8487
1.7446	2.1328
1.4954	1.9250
1.6836	2.5022
1.4882	2.0830
2.0811	2.4280
1.9819	2.1375
2.1898	2.6610
2.0061	2.3679
	2.6045
	2.3011
	2.8134
	2.4206

Compute the sample mean and variance of each set of data. Construct histograms, frequency distributions, and cumulative frequency distributions. Assuming that the jack forces are constant, what can be said about the influence of nominal curvature on friction loss?

1.4. The following values of shear strength (in tons per square foot) were determined by unconfined compression tests of soil from Waukell Creek, California. Plot histograms of this data using four, six, and eight intervals. In the second case consider three different locations of the interval division points. With a few extreme points, as here, it may be advantageous to consider unequal intervals, longer in the right-hand tail.

0.12	0.21	0.36	0.37
0.39	0.46	0.47	0.50
0.50	0.51	0.53	0.58
0.61	0.62	0.77	0.81
0.93	1.05	1.59	1.73

1.5. Compute the individual sample means, standard deviations, and coefficients of variation for the basement, fourth floor, and ninth floor of the floor-load data in

Table 1.2.1.   How do they compare with each other?   Why might such differences arise?

1.6. The following data for the Ogden Valley artesian aquifer have been collected over a period of years.   Find the sample means, variances, standard deviations, and correlation coefficient.

Year	Measured discharge, acre-ft	Estimated recharge, acre-ft
1935	11,300	11,400
1936	12,800	14,600
1937	12,700	13,600
1938	10,400	10,100
1939	10,800	9,900
1940	11,500	12,200
1941	9,900	9,700
1942	11,900	11,800
1943	13,000	12,700
1944	13,700	13,600
1945	14,100	14,600
1946	15,200	14,900
1947	15,100	14,300
1948	15,400	14,200
1949	16,000	17,400
1950	16,500	16,400
1951	16,700	14,900

1.7. The maximum annual flood flows for the Feather River at Oroville, California, for the period 1902 to 1960 are as follows.   The data have been ordered, but the years of occurrence are also given.

Year	Flood, cfs	Year	Flood, cfs
1907	230,000	1958	102,000
1956	203,000	1903	102,000
1928	185,000	1927	94,000
1938	185,000	1951	92,100
1940	152,000	1936	85,400
1909	140,000	1941	84,200
1960	135,000	1957	83,100
1906	128,000	1915	81,400
1914	122,000	1905	81,000
1904	118,000	1917	80,400
1953	113,000	1930	80,100
1942	110,000	1911	75,400
1943	108,000	1919	65,900
		1925	64,300
		1921	62,300
		1945	60,100
		1952	59,200

Year	Flood, cfs	Year	Flood, cfs
1935	58,600	1918	28,200
1926	55,700	1944	24,900
1954	54,800	1920	23,400
1946	54,400	1932	22,600
1950	46,400	1923	22,400
1947	45,600	1934	20,300
1916	42,400	1937	19,200
1924	42,400	1913	16,800
1902	41,000	1949	16,800
1948	36,700	1912	16,400
1922	36,400	1908	16,300
1959	34,500	1929	14,000
1910	31,000	1952	13,000
		1931	11,600
		1933	8,860
		1939	8,080

Compute sample mean and variance. Plot histogram and frequency distribution. If a 1-year construction project is being planned and a flow of 20,000 cfs or greater will halt construction, what, in the past, has been the relative frequency of such flows?

1.8. The piezometer gauge pressures at a critical location on a penstock model were measured under maximum flow conditions as follows (ordered data in inches of mercury).

12.01	12.76
12.08	12.83
12.18	12.84
12.23	12.88
12.27	12.88
12.37	12.90
12.49	12.92
12.53	13.00
12.58	13.08
12.69	13.35

Compute sample mean, variance, and skewness coefficient. Use a histogram with the mean at an interval boundary to answer the question: did values above and below the mean occur with equal frequency? Were very low readings as common as very high readings?

1.9. Embankment material for zone 1 of the Santa Rosita Dam in Southern Chihuahua, Mexico, will come from a borrow pit downstream from the dam site at a location that is frequently flooded. A cofferdam 800 m long is needed and the contractor needs to know the optimum construction height. Normal flow (200 m³/sec) requires a height of 3 m. Flooding will involve a 3-week delay in construction. Maximum flow rates from 1921 to 1965 were:

Year	Flow rate, m³/sec	Year	Flow rate, m³/sec
1921	1340	1929	1060
1922	1380	1930	412
1923	1450	1931	184
1924	618	1932	1480
1925	523	1933	876
1926	508	1934	113
1927	1220	1935	516
1936	1780	1951	1000
1937	1090	1952	1890
1938	944	1953	611
1939	397	1954	409
1940	282	1955	780
1941	353	1956	674
1942	597	1957	969
1943	995	1958	870
1944	611	1959	329
1945	985	1960	458
1946	1430	1961	1556
1947	778	1962	1217
1948	1280	1963	819
1949	1020	1964	576
1950	1300	1965	1324

The contractor's options are:

Cofferdam height, m	Capacity, m³/sec	Cost
3	200	$15,600
4.5	550	18,600

The cost of a 3-week delay from flooding of the borrow pit is estimated as $30,000.

Compute the sample mean and variance. Will a histogram be useful in the analysis of the contractor's decision? Why? How would you structure the decision situation? How does time enter the problem?

1.10. In heavy construction operations the variation in individual vehicle cycle times can cause delays in the total process. Find the sample means, standard deviations, and coefficients of variation for each of the following specific steps in a cycle and for the total cycle time. Which steps are "most variable?" Which contribute most seriously to the variations in the total cycle time? Is there an indication that some vehicle/driver combinations are faster than others? (Check by calculated haul and return time correlation.) How does the standard deviation of the sum compare to those of the parts? How does the coefficient of variation of the sum compare to those of the parts?

Wait and exchange	Load	Haul	Dump and turn	Return	Total
0.00	1.58	2.08	0.46	2.55	6.67
0.00	1.72	1.95	0.53	2.45	6.65
0.00	2.52	2.18	0.40	3.13	8.23
0.51	1.84	2.08	0.53	2.53	7.49
1.02	1.74	2.15	0.61	2.54	8.06
0.00	1.71	2.20	0.63	2.52	7.06
1.71	1.80	2.17	0.49	2.51	8.68
2.35	1.67	2.25	0.64	2.46	9.37
1.12	2.38	1.99	0.69	2.52	8.70
0.50	2.06	2.07	0.41	2.44	7.48
1.29	1.92	2.18	0.54	2.36	8.29
0.00	2.03	2.17	0.59	2.69	7.48
0.83	1.85	1.96	0.60	2.52	7.76
0.46	2.23	2.16	0.55	2.47	7.87
1.05	1.94	1.99	0.57	2.49	8.04
0.77	2.00	2.13	0.60	2.68	8.18
0.00	1.88	2.03	0.80	2.35	7.06
1.32	1.93	1.99	0.50	2.36	8.10

1.11. Total cycle times of trucks hauling asphaltic concrete on a highway project were observed and found to be (in minutes):

30	18	17	24	20
20	16	24	25	19
24	28	23	23	23
17	18	11	18	

Find the sample mean, standard deviation, skewness coefficient, and coefficient of kurtosis of this set of data. Plot its histogram.

1.12. Fifteen lots of 100 sections each of 108-in. concrete pipe were tested for porosity. The number of sections in each lot failing to meet the standards were:

1	5	6	3	0
7	4	9	4	1
3	2	1	8	6

Compute the sample mean, variance, and coefficient of variation.

If the plant continues to manufacture pipe of this quality, can you suggest a possible technique for quality control of the product? What cost factors enter the problem?

1.13. The times (in seconds) for loading, swinging, dumping, and returning for a shovel moving sandstone on a dam project were measured as shown in table at top of p. 24.

Compute sample mean, variance, and coefficient of variation of each set of data. If the variability in total time is causing hauling problems, which operation should be studied as the primary source of variability in the total? Which of the summary statistics would be most useful in such a study?

Load	Swing	Dump	Return	Total
25	10	2	8	45
17	9	2	9	37
14	8	2	9	33
19	10	2	9	40
18	8	2	10	38
16	10	2	15	43
19	7	2	8	36
22	11	2	8	43
17	9	2	8	36
15	10	2	9	36
20	8	2	11	41
15	25	2	10	52
18	7	2	8	35
15	8	2	10	35
25	10	2	10	47
14	8	2	10	34
14	8	2	9	33
21	7	2	8	38
17	10	2	9	38
15	9	2	11	37
16	12	3	12	43
21	8	2	10	41
13	9	2	9	43
15	10	2	9	36
26	10	2	13	51

1.14. The San Francisco–Oakland Bay Bridge provides an interesting opportunity for the study of highway capacity as a random phenomenon. A device located at the toll booth summarizes traffic flow by direction in 6-min increments. Maximum flow each day has been recorded as follows for a survey duration of 2 days.

	Vehicles in 6 min	
Date	1st day	2d day
October, 1963	804	800
January, 1964	877	874
April, 1964	904	865
July, 1964	810	854
October, 1964	903	915
April, 1965	866	812
October, 1965	866	887

    (a) Check for growth trend.

    (b) Plot the histogram and compute the numerical summaries using all 14 data points.

    (c) What trends will become apparent if the bridge is approaching its maximum daily capacity? Consider the influence of variation in demand associated with users' knowledge of congestion periods.

1.15. The traffic on a two-lane road has been studied in preparation for the start of a construction project during which one lane will be closed. The problem is whether to use an automatic signal with a fixed cycle, a flagman, or go to the expense of constructing a new temporary lane. The data are:

Southbound lane		Northbound lane	
Vehicles/30 sec	*Frequency*	*Vehicles/30 sec*	*Frequency*
0	18	0	9
1	32	1	16
2	28	2	30
3	20	3	22
4	13	4	19
5	7	5	10
6	0	6	3
7	1	7	7
8	1	8	3
≥ 9	0	≥ 9	1

Compute sample mean and variance of each set. What can you say about the traffic pattern that tends to discriminate between an automatic signal with a fixed cycle and a flagman who can follow the traffic demand? How would you make a study of the possible demand for a new temporary lane? *Hint:* How does the detour travel time influence the problem? What data are needed?

1.16. Soil resistivity is used in studies of corrosion of pipes buried in the soil. For example, a resistivity of 0 to 400 ohms/cm represents extremely severe corrosion conditions; 400 to 900, very severe; 900 to 1500, severe; 1500 to 3500, moderate; 3500 to 8000, mild; and 8000 to 20,000, slight risk. There were 32 measurements of soil resistivity made at a prospective construction site.

Station	Resistivity, ohms/cm	Station	Resistivity, ohms/cm
1	1750	17	1490
2	960	18	1610
3	740	19	1110
4	1030	20	1340
5	530	21	2180
6	1170	22	1340
7	5770	23	1680
8	2300	24	1550
9	1240	25	2500
10	510	26	2300
11	910	27	1240
12	840	28	3060
13	1340	29	1880
14	1240	30	6550
15	1370	31	1180
16	1260	32	2760

Compute sample mean and variance. Construct a histogram using the classification bands above as intervals. Illustrate the same relations on the cumulative frequency distribution and indicate the observed relative frequency of occurrence. Note that the mean represents a spatial mean and that the variance is a measure of lack of homogeneity of the resistivity property over the site.

1.17. The water-treatment plant at an air station in California was constructed for a design capacity of 4,500,000 gal/day (domestic use). It is nearly always necessary to suspend lawn irrigation when demand exceeds supply. There are, of course, attendant losses. Measured demands during July and August 1965 (weekdays only) were (in thousands of gallons per day, ordered data):

2298	4536	4905
3205	4565	4908
3325	4591	4923
3609	4657	4941
3918	4666	4993
3992	4670	4998
4057	4724	5035
4188	4737	5041
4289	4763	5058
4363	4784	5142
4377	4816	5152
4448	4817	5152
4450	4852	5330
4524	4887	5535

Compute sample mean and variance. Construct a cumulative histogram in which 4,500,000 gal/day is one of the interval boundaries. On a relative frequency basis, how often did demand exceed capacity?

1.18. The Stanford pilot plant attempts to treat the effluent of an activated sludge sewage-treatment plant alternating between a diatomite filter and a sand filter.

x, hr	y, hr
2.0	1.0
3.0	0.5
3.0	3.0
3.0	1.0
3.5	4.0
4.0	2.0
4.5	0.5
5.0	5.0
5.5	0.5
6.0	2.5
11.5	4.0
12.0	3.0
14.0	2.0
18.0	1.5

The diatomaceous earth filter fouls quickly if the raw water contains too many suspended solids. The sand filter is then used during periods in which the diatomite filter is inoperative. If x is the time of operation of the diatomite filter and y the operating time of the sand filter during cleaning of the diatomite filter, the data are a typical sample of paired observations.

Compute the sample mean and variance of each set and the correlation coefficient. Plot a scatter diagram. Does the plot verify the calculated correlation coefficient?

1.19. Show that the following alternate, more easily computed form of the sample covariance is equivalent to Eq. (1.3.1).

$$s_{X,Y} = \frac{1}{n} \left(\sum_{i=1}^{n} x_i y_i \right) - \bar{x}\,\bar{y} \qquad (1.3.1b)$$

1.20. The percentage frequencies of wind direction and speed on an annual basis for a 10-year study at the Logan Airport, Boston, Massachusetts, are:

Hourly observations of wind speed, mph*

Direction	0–3	4–7	8–12	13–18	19–24	25–31	32–38	39–46	47+	Total	Average speed
N	+	1	2	2	+	+	+	+		5	12.0
NNE	+	1	1	1	+	+	+	+	+	3	13.3
NE	+	1	1	1	1	+	+	+	+	4	14.2
ENE	+	1	1	1	1	+	+	+	+	4	14.1
E	+	1	2	2	+	+	+	+	+	5	12.9
ESE	+	1	2	2	+	+	+			5	12.0
SE	+	1	2	1	+	+	+		+	4	10.8
SSE	+	1	1	1	+	+	+	+	+	3	10.4
S	+	1	2	1	+	+	+	+	+	4	11.0
SSW	+	1	3	2	1	+	+	+	+	7	13.0
SW	+	1	4	5	1	+	+	+		11	13.0
WSW	+	1	3	3	1	+	+	+		8	17.0
W	+	1	2	3	1	+	+	+		7	4.2
WNW	+	1	3	3	2	1	+	+		10	14.6
NW	+	1	3	4	2	1	+	+	+	11	15.2
NNW	+	1	2	3	1	+	+	+		7	13.8
Calm	2									2	
Total	2	16	34	35	11	2	+	+	+	100	13.3

* + indicates frequency less than 1 percent.

(a) Make a scatter diagram for direction versus average speed, one point for each 1 percent. Make a similar study for direction versus 8 to 12 and versus 13 to 18 mph wind speeds. Why is the first diagram different from the other two?

(b) Compute the sample mean and sample variance of the average speed.

(c) Compute the mean and variance of direction (in degrees from N). What should you do with the two percent of calm days?

1.21. One procedure for setting the allowable speed limit on a highway is to assume that the proper limit has a 15 percent chance of being exceeded based on a spot speed study. If this decision rule is used, what speed limit do you recommend for U.S. Route 50 and Main Street? No limit existed when these data were observed:

U.S. Route 50

Speed, mph	Local passenger cars	Other passenger cars	Trucks
34	0	0	1
36	1	0	1
38	0	0	1
40	1	0	3
42	1	0	2
44	1	0	2
46	2	1	2
48	6	0	2
50	7	0	3
52	9	2	2
54	12	1	3
56	8	7	2
58	7	8	2
60	3	9	1
62	4	3	0
64	5	4	1
66	2	3	0
68	3	0	0
70	2	2	0
72	3	1	0
74	0	0	0
76	1	0	0
78	1	0	0
80	0	0	0
82	0	0	0
84	0	1	0

Main street

Speed, mph	Northbound	Southbound
24	0	1
26	5	0
28	4	0
30	12	4
32	13	5
34	23	17
36	14	11
38	13	9
40	8	7
42	1	4
44	2	3
46	3	0
48	0	0
50	1	0

(a) Plot the histograms for speed of passenger cars, trucks, and all vehicles on U.S. Route 50 and on Main Street. On U.S. Route 50, is a dual speed limit for passenger cars and trucks reasonable?

(b) Compute the mean, variance, and coefficient of variation of speed for all vehicles on U.S. Route 50 and for those on Main Street.

(c) If you were presenting your study to the City Council, what would you recommend as the speed limits? Why? If the council is very sensitive to local political pressure, would your recommendation change, assuming that the quoted decision rule is only a rough guide? The Main Street traffic is primarily local traffic.

1.22. Show that in situations such as that in Prob. 1.10, where interest centers on both the parts and their sum, the sample variance s_T^2 of the sum of *two* variables equals:

$$s_T^2 = s_1^2 + s_2^2 + 2r_{12}s_1s_2$$

where s_1^2, s_2^2, and r_{12} are, respectively, the sample variances and sample correlation coefficient of variables 1 and 2. Show, too, that:

$$(s_1 - s_2)^2 \leq s_T^2 \leq (s_1 + s_2)^2$$

In general, for a number n of variables and their sum,

$$s_T{}^2 = \sum_{i=1}^{n} s_i{}^2 + 2 \sum_{j=1}^{n} \sum_{i>1}^{n} r_{ij}s_i s_j$$

Notice that if the r_{ij}'s are small in absolute value,

$$s_T{}^2 \approx \sum_{i=1}^{n} s_i{}^2$$

But in general $s_T{}^2$ may be larger or smaller than this sum.   See Prob. 1.10.

1.23. One of the factors considered when determining the susceptibility to liquefaction of sands below a foundation during earthquake motions is their penetration resistance as measured by blow counts (number of blows per foot of penetration).   These measurements show marked dispersion, especially in soils whose liquefaction susceptibility might be in question.   Therefore histograms of blow-count tests made at different depths and locations in the soil layer are compared in shape with histograms of blow counts at sites where liquefaction has or has not occurred during earthquakes.   At a site under investigation the following data are made available:

Range, blows/ft	No. of tests
1–10	14
11–20	32
21–30	17
31–40	4
41–50	6
51–60	3
	76

(a) Prepare a histogram of this data.   Use three different vertical axes with scales: "Number of tests per 10-blow interval," "frequency of tests per 10-blow interval," and "frequency of tests per 1-blow interval."

(b) The following histograms are presented as shown in a report on the large 1964 Niigata Earthquake in Japan.   As is all too frequently the case in engineering reports, it is not clear how the histograms were constructed.   Assume that the vertical axis is "frequency per five-blow interval" and that the dots represent the midpoints of the intervals -2.4 to $+2.5$, 2.6 to 7.5, 7.6 to 12.5, etc.   In order to aid in judging the susceptibility to liquefaction of the present site, replot these two histograms on the histogram prepared in part (a) above.   Which of the three vertical scales should be used to avoid most easily the difficulties caused by the different numbers of tests and different interval lengths?

(c) Instead of comparing entire histograms, what sample averages might one compute for the various sets of data to aid in this judgment?   Do you think the average is sufficient?   Is the skewness coefficient helpful?

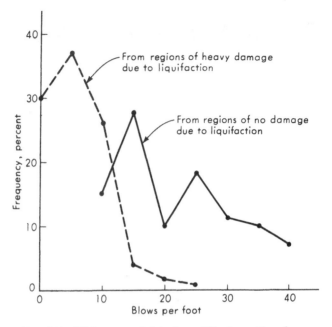

Fig. P1.23   Histograms of data from Niigata earthquake.

1.24. In a study of rainstorms, the following data were observed at a number of recording stations.

No. of storms/(station)(year)	Observed no. of occurrences
0	102
1	144
2	74
3	28
4	10
5	2
>5	0

Source: E. L. Grant [1938], Rainfall Intensities and Frequencies, *ASCE Trans.*, vol. 103, pp. 384–388.

(a) Plot a histogram and a cumulative frequency polygon for this data.   Compute the sample mean and sample standard deviation.

(b) How do these data differ in character from stream flow data?

2
Elements of Probability Theory

Every engineering problem involves phenomena which exhibit scatter of the type illustrated in the previous chapter. To deal with such situations in a manner which incorporates this variability in his analyses, the engineer makes use of the *theory of probability*, a branch of mathematics dealing with uncertainty.

A fundamental step in any engineering investigation is the formulation of a set of mathematical models—that is, descriptions of real situations in a simplified, idealized form suitable for computation. In civil engineering, one frequently ignores friction, assumes rigid bodies, or adopts an ideal fluid to arrive at relatively simple mathematical models, which are amenable to analysis by arithmetic or calculus. Frequently these models are deterministic: a single number describes each independent variable, and a formula (a model) predicts a specific value for the dependent variable. When the element of uncertainty, owing to *natural variation* or *incomplete professional knowledge*, is to be considered explicitly, the models derived are probabilistic and subject to analysis by the rules of probability theory. Here the values of the independent variables are not known with certainty, and thus the variable related to them through the physical model cannot be precisely predicted. In addition, the physical model may itself contain elements of uncertainty. Many examples of both situations will follow.

This chapter will first formalize some intuitively satisfactory ideas about events and relative likelihoods, introducing and defining a number of words and several very useful notions. The latter part of the chapter is concerned with the definition, the description, and the manipulation of the central character in probability—*the random variable*.

2.1 RANDOM EVENTS

Uncertainty is introduced into engineering problems through the variation inherent in nature, through man's lack of understanding of all the causes and effects in physical systems, and through lack of sufficient data. For example, even with a long history of data, one cannot predict the maximum flood that will occur in the next 10 years in a given area. This uncertainty is a product of natural variation. Lacking a full-depth hole, the depth of soil to rock at a building site can only be estimated. This uncertainty is the result of incomplete information. Thus both the depth to rock and the maximum flood are uncertain, and both can be dealt with using the same theory.

As a result of uncertainties like those mentioned above, the future can never be entirely predicted by the engineer. He must, rather, consider the *possibility* of the occurrence of particular events and then determine the likelihood of their occurrence. This section deals with the logical treatment of uncertain events through probability theory and the application to civil engineering problems.

2.1.1 Sample Space and Events

Experiments, sample spaces, and events The theory of probability is concerned formally with *experiments* and their outcomes, where the term experiment is used in a most general sense. The collection of all possible outcomes of an experiment is called its *sample space*. This space consists of a set S of points called *sample points*, each of which is associated with one and only one distinguishable outcome. The fineness to which one makes these distinctions is a matter of judgment and depends in practice upon the use to which the model will be put.

As an example, suppose that a traffic engineer goes to a particular street intersection exactly at noon each weekday and waits until the traffic signal there has gone through one cycle. The engineer records the number of southbound vehicles which had to come to a complete stop before their light turned green. If a minimum vehicle length is 15 ft

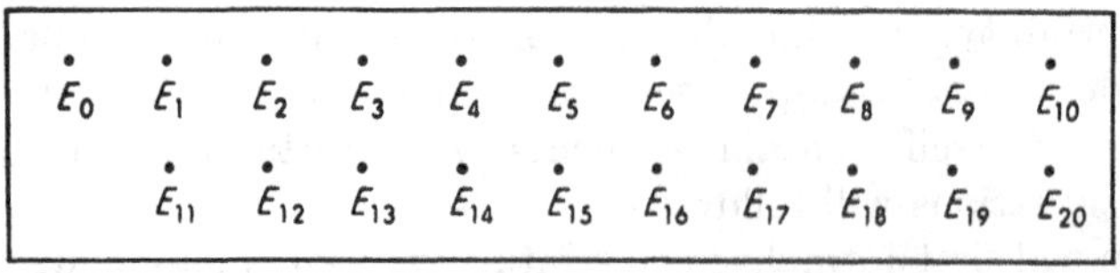

$$
\begin{array}{ccccccccccc}
\bullet & \bullet & \bullet & \bullet & \bullet & \bullet & \bullet & \bullet & \bullet & \bullet & \bullet \\
E_0 & E_1 & E_2 & E_3 & E_4 & E_5 & E_6 & E_7 & E_8 & E_9 & E_{10}
\end{array}
$$

$$
\begin{array}{cccccccccc}
\bullet & \bullet & \bullet & \bullet & \bullet & \bullet & \bullet & \bullet & \bullet & \bullet \\
E_{11} & E_{12} & E_{13} & E_{14} & E_{15} & E_{16} & E_{17} & E_{18} & E_{19} & E_{20}
\end{array}
$$

Fig. 2.1.1 An elementary sample space—E_j implies j vehicles observed.

and the block is 300 ft long, the maximum possible number of cars in the queue is 20. If only the total number of vehicles is of interest, the sample space for this experiment is a set of 21 points labeled, say, E_0, E_1, . . . , E_{20}, each associated with a particular number of observed vehicles. These might be represented as in Fig. 2.1.1. If the engineer needed other information, he might make a finer distinction, differentiating between trucks and automobiles and recording the number of each stopped. The sample space for the experiment would then be larger, containing an individual sample point $E_{i,j}$ for each possible combination of i cars and j trucks such that the maximum value of $i + j = 20$, as in Fig. 2.1.2.

An *event* A is a collection of sample points in the sample space S of an experiment. Traditionally, events are labeled by letters. If the distinction should be necessary, a *simple event* is an event consisting of a single sample point, and a *compound event* is made up of two or more sample points or elementary outcomes of the experiment. The *complement A^c* of an event A consists of all sample points in the sample space

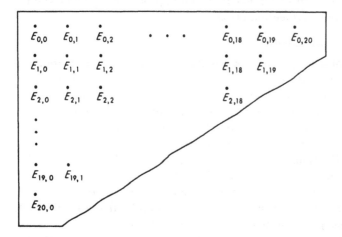

Fig. 2.1.2 Elementary sample space for cars and trucks. E_{ij} implies i cars and j trucks with a maximum of 20 including both types of vehicles.

of the experiment *not* included in the event.   Therefore, the complement of an event is also an event.†

In the experiment which involved counting all vehicles without regard for type, the observation of "no stopped vehicles" is a simple event A, and the finding of "more than 10 stopped vehicles" is a compound event B.   The complement of the latter is the event B^c, "10 or fewer stopped vehicles were observed."   Events defined on a sample space need not be exclusive; notice that events A and B^c both contain the sample point E_0.

In testing the ultimate strength of reinforced-concrete beams described in Sec. 1.2, the load values were read to the nearest 50 lb. The sample space for the experiment consists of a set of points, each associated with an outcome, 0, 50, 100, . . . , or M lb, where M is some indefinitely large number, say, infinity.   A set of events of interest might be A_0, A_1, A_2, . . . , defined such that A_0 contains the sample points associated with loads 0 to 950 lb, A_1 contains those associated with 1000 to 1950 lb, and so forth.

In many physical situations, such as this beam-strength experiment, it is more natural and convenient to define a *continuous sample space*. Thus, as the measuring instrument used becomes more and more precise, it is reasonable to assume that *any* number greater than zero, not just certain discrete points, will be included among the possible outcomes and hence defined as a sample point.   The sample space becomes the real line, 0 to ∞.   In other situations, a finite interval is defined as the sample space.   For example, when wind directions at an airport site are being observed, the interval 0 to 360° becomes the sample space.   In still other situations, when measurement errors are of interest, for example, the line from $-\infty$ to $+\infty$ is a logical choice for the sample space.   To include an event such as ∞ in the sample space does not necessarily mean the engineer thinks it is a possible outcome of the experiment.   The choice of ∞ as an upper limit is simply a convenience; it avoids choosing a specific, arbitrarily large number to limit the sample space.

Events of interest in the beam experiment might be described as follows‡ (when the sample space is defined as continuous):

$D_1 = $ [the observation was 10,101]

$D_2 = $ [the outcome was greater than 10,000, that is, $>10,000$]

$D_3 = $ [the result was greater than 9000 but less than or equal to 10,000, that is, >9000 and $\leq 10,000$]

---

† If A is the *certain event*, i.e., if A is the collection of all sample points in the sample space S, the complement A^c of event A will be the *null event;* i.e., it will contain no sample points.

‡ The square brackets [   ] should be read "the event that."

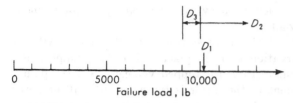

Fig. 2.1.3 Continuous sample space for beam-failure load.

D_1 is a simple event; D_2 and D_3 are compound events. They are shown graphically in the sample space 0 to ∞ in Fig. 2.1.3.

Relationships among events   Events in a sample space may be related in a number of ways.   Most important, if two events contain no sample points in common, the events are said to be *mutually exclusive* or *disjoint*. Two mutually exclusive events—A, "fewer than 6 stopped vehicles were observed" and B, "more than 10 stopped vehicles were observed"—are shown shaded in the sample space of the first vehicle-counting experiment (Fig. 2.1.4).   The events D_1 and D_3 defined above are mutually exclusive. D_2 and D_3 are also mutually exclusive, owing to the care with which the *inequality* ($\leq$) and *strict inequality* ($>$) have been written at 10,000.

The notion of mutually exclusive events extends in an obvious way to more than two events.   By their definition simple events are mutually exclusive.

If a pair of events A and B are not mutually exclusive, the set of points which they have in common is called their *intersection*, denoted $A \cap B$.   The intersection of the event A defined in the last paragraph and the event C, "from four to eight stopped vehicles were observed," is illustrated in Fig. 2.1.5.   The intersection of the events D_1 and D_2 in Fig. 2.1.3 is simply the event D_1 itself.   If the intersection of two

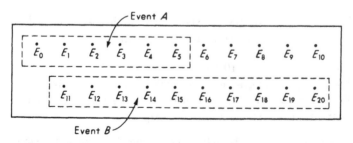

Fig. 2.1.4 Mutually exclusive events.   Compound events A and B are mutually exclusive; they have no sample points in common.

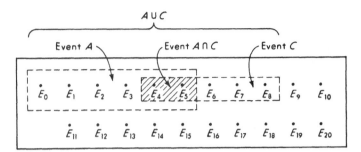

Fig. 2.1.5 Intersection $A \cap C$, comprising events E_4 and E_5, is the intersection of compound events A and C.

events is equivalent to one of the events, that event is said to be contained in the other. This is written $D_1 \subset D_2$.

The *union* of two events A and C is the event which is the collection of all sample points which occur at least once in either A or C. In Fig. 2.1.5 the union of the events A and C, written $A \cup C$, is the event "less than nine stopped vehicles were observed." The union of the events D_2 and D_3 in Fig. 2.1.3 is the event that the failure load is greater than 9000 lb. The union of D_1 and D_2 is simply D_2 itself.

Two-dimensional and conditional sample spaces For purposes of visualization of certain later developments, two other types of sample spaces deserve mention. The first is the *two-* (or higher) *dimensional sample space*. The sample space in Fig. 2.1.2 is one such, where the experiment involves observing two numbers, the number of cars and the number of trucks. It might be replotted as shown in Fig. 2.1.6; there each point on the grid represents a possible outcome, a sample point. In the determination of the best orientation of an airport runway, an experiment might involve measuring both wind speed and direction. The continuous sample space would appear as in Fig. 2.1.7, limited in one dimension and unlimited in the other. Any point in the area is a sample point. An experiment involving the measurement of number and average speed of vehicles on a bridge would lead to a discrete-continuous two-dimensional sample space.

The second additional kind of sample space of interest here is a *conditional sample* space. If the engineer is interested in the possible outcomes of an experiment *given* that some event A has occurred, the set of events associated with event A can be considered a new, reduced sample space. For, conditional on the occurrence of event A, only the simple events associated with the sample points in that reduced space are possible outcomes of the experiment. For example, given that

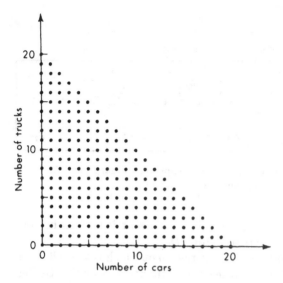

Fig. 2.1.6　Discrete two-dimensional sample space. Each sample point represents an observable combination of cars and trucks such that their sum is not greater than 20.

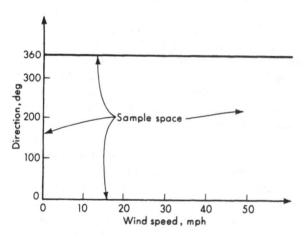

Fig. 2.1.7　Continuous two-dimensional sample space. All possible observable wind speeds and directions at an airport are described.

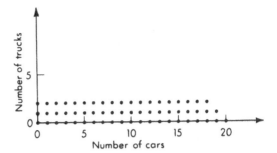

Fig. 2.1.8 Discrete conditional sample space. Given that two or fewer trucks were observed (see Fig. 2.1.6).

exactly one truck was observed, the conditional sample space in the traffic-light experiment becomes the set of events $E_{0,1}, E_{1,1}, \ldots, E_{19,1}$. Given that two or fewer trucks were observed, the conditional sample space is that illustrated in Fig. 2.1.8. Similarly, the airport engineer might be interested only in higher-velocity winds and hence restrict his attention to the conditional sample space associated with winds greater than 20 mph, leading to the space shown in Fig. 2.1.9. Whether a sample space is the primary or the conditional one is clearly often a matter of the engineer's definition and convenience, but the notion of the conditional sample space will prove helpful.

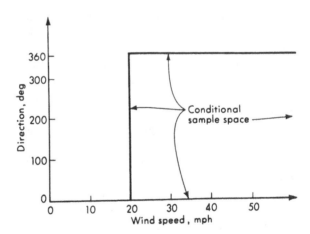

Fig. 2.1.9 Continuous conditional sample space. Describes all possible wind directions and velocities given that the velocity is greater than 20 mph.

For the remainder of Sec. 2.1 we shall restrict our attention to one-dimensional discrete sample spaces, and shall return to the other important cases only after the introduction in Sec. 2.2 of the *random variable*.

2.1.2 Probability Measure

Interpretation of probabilities   To each sample point in the sample space of an experiment we are going to assign a number called a *probability measure*. The mathematical theory of probability is *not* concerned with where these numbers came from or what they mean; it only tells us how to use them in a consistent manner. The engineer who puts probability to work on his models of real situations must be absolutely sure what the set of numbers he assigns means, for the results of a probabilistic analysis of an engineering problem can be helpful only if this input is meaningful.

An intuitively satisfying explanation of the probability measure assigned to a sample point is that of relative frequencies. If the engineer assigns a probability measure of p to a sample point in a sample space, he is usually willing to say that if he could make repeated trials of the same experiment over and over again, say M times, and count the number of times N that the simple event associated with this sample point was observed, the ratio of N to M would be very nearly p. One frequently hears, for example, that the probability of a tossed coin coming up "heads" is one-half. Experience has shown that this is very nearly the fraction of any series of a large number of tosses of a well-balanced coin that will show a head rather than a tail. This interpretation of the probability measure is commonly adopted in the physical sciences. It is considered quite objectively as a property of certain repetitious phenomena. When formalized through limit theorems, the notion of relative frequency can serve as a basis as well as an interpretation of probability (Von Mises [1957]). Relative frequency, when it applies, is without question a meaningful and useful interpretation of the probability measure.

But what is the engineer to do in a situation where such repetition of the experiment is impossible and meaningless? How does one interpret, for instance, the statement that the probability is 0.25 that the soil 30 ft below a proposed bridge footing is not sand but clay? The soil is not going to be clay on 1 out of every 4 days that it is observed; it is either clay or it is not. The experiment here is related to an unknown condition of nature which later may be directly observed and determined once and for all.

The proved usefulness in bringing probability theory to bear in such situations has necessitated a more liberal interpretation of the expression "the probability of event A is p." The probabilities assigned by an engineer to the possible outcomes of an experiment can also be thought of as a set of weights which expresses that individual's measure of the relative likelihoods of the outcomes. That is, the probability of an event might be simply a subjective measure of the degree of belief an engineer has in a judgment or prediction. Colloquially this notion is often expressed as "the odds are 1 to 3 that the soil is clay." Notice that if repetitions are involved, the notions of relative frequencies and degree of belief should be compatible to a reasonable man. Much more will be said of this "subjective" probability in Chap. 5, which includes methods for aiding the engineer in assessing the numerical values of the probabilities associated with his judgements. This interpretation of probability, as an intellectual concept rather than as a physical property, also can serve as a basis for probability theory. Engineering students will find Tribus' recent presentation of this position very appealing (Tribus [1969]).

Like its interpretations, the sources of the probability measure to be assigned to the sample points are also varied. The values may actually be the results of frequent observations. After observing the vehicles at the intersection every weekday for a year, the traffic engineer in the example in the previous section might assign the observed relative frequencies of the simple events, "no cars," "one stopped car," etc., to the sample points $E_0, E_1, \ldots, E_{20}$.

Reflecting the second interpretation of probability, the probability measure may be assigned by the engineer in a wholly subjective manner. Calling upon past experience in similar situations, a knowledge of local geology, and the taste of a handful of the material on the surface, a soils engineer might state the odds that each of several types of soil might be found below a particular footing.

Finally, we shall see that through the theory of probability one can derive the probability measure for many experiments of prime interest, starting with assumptions about the physical mechanism generating the observed events. For example, by making certain plausible assumptions about the behavior of vehicles and knowing something about the average flow rate, that is, by modeling the underlying mechanism, the engineer may be able to calculate a probability measure for each sample point in the intersection experiment without ever making an actual observation of the particular intersection. Such observation is impossible, for example, if the intersection is not yet existent, but only under design. As in deterministic problem formulations, subsequent observations may or may not agree with the predictions of the hypothesized mathematical model. In this manner, models (or theories) are confirmed or rejected.

Axioms of probability   No matter how the engineer chooses to interpret the meaning of the probability measure and no matter what its source, as long as the assignment of these weights is consistent with three simple axioms, the mathematical validity of any results derived through the correct application of the axiomatic theory of probability is assured. We use the notation† $P[A]$ to denote the probability of an event A, which in the context of probability is frequently called a *random event*. The following conditions must hold on the probabilities assigned to the events in the sample space:

Axiom I   The probability of an event is a number greater than or equal to zero but less than or equal to unity:

$$0 \leq P[A] \leq 1$$

Axiom II   The probability of the certain event S is unity:

$$P[S] = 1$$

where S is the event associated with all the sample points in the sample space.

Axiom III   The probability of an event which is the union of two *mutually exclusive* events is the sum of the probabilities of these two events:

$$P[A \cup B] = P[A] + P[B]$$

Since S is the union of all simple events, the third axiom implies that Axiom II could be written:

$$\sum_{\text{all } i} P[E_i] = 1$$

in which the E_i are simple events associated with individual sample points.‡

The first two axioms are simply convenient conventions. All probabilities will be positive numbers and their sum over the simple events (or *any* mutually exclusive, collectively exhaustive set of events) will be normalized to 1. These are natural restrictions on probabilities which arise from observed relative frequencies. If there are k possible

---

† $P[A]$ is read "the probability of the event A."
‡ Upon returning in Sec. 2.2 to continuous sample spaces, where the number of sample points is infinite, we shall not assign probabilities to specific points but to small lengths or areas. Then the integral (sum) over all these regions in the sample space must be unity.

outcomes to an experiment and the experiment is performed M times, the observed relative frequencies F_i are

$$F_1 = \frac{n_1}{M}$$

$$F_2 = \frac{n_2}{M}$$

$$\cdot \ \cdot \ \cdot$$

$$F_k = \frac{n_k}{M}$$

where $n_1, n_2, \ldots, n_k$ are the numbers of times each particular outcome was observed. Since $n_1 + n_2 + \cdots + n_k = M$, each frequency satisfies Axiom I and their sum satisfies Axiom II:

$$\frac{n_1}{M} + \frac{n_2}{M} + \cdots + \frac{n_k}{M} = \frac{n_1 + n_2 + \cdots + n_k}{M} = 1$$

If, alternatively, one prefers to think of probabilities as weights on events indicative of their relative likelihood, then Axioms I and II simply demand that, after assigning to the set of all simple events a set of relative weights, one normalizes these weights by dividing each by the total.

Axiom III is equally acceptable. It requires only that the probabilities are assigned to events in such a way that the probability of any event made up of two mutually exclusive events is equal to the sum of the probabilities of the individual events. If the original assignment of probabilities is made on a set of collectively exhaustive, mutually exclusive events, such as the set of all simple events, there can be no possibility of violating this axiom. If, for example, the source of these assignments is a set of observed relative frequencies, as long as the original set of k outcomes has been properly defined (in particular, not overlapping in any way), the relative frequency of the outcome i or j is

$$\frac{n_i + n_j}{M} = \frac{n_i}{M} + \frac{n_j}{M}$$

Similarly, if an engineer assigns relative weights to a set of possible distinct outcomes, he would surely be inconsistent if he felt that the relative likelihood of either of a pair of disjoint outcomes was anything but the sum of their individual likelihoods.

Suppose that the soils engineer in the above example decides that the odds on there being clay soil at a depth of 30 ft are 1 to 3 (or 1 in 4), and sand is just as likely; and if neither of these is present, the material will surely be sound rock. The implication is that he gives clay a relative weight of 1 and the other outcomes a total weight of 3. These possible outcomes include only sand, with a weight 1, and rock, with weight $3 - 1$, or 2. To be used as probabilities these weights need normalizing by their sum, 4, to satisfy the first and second axioms of probability. Let the event C be "there is clay 30 ft below the footing" and let the events S and R be associated with the presence of sand and rock, respectively. Then

$$P[C] = \tfrac{1}{4} = 0.25$$
$$P[S] = \tfrac{1}{4} = 0.25$$
$$P[R] = \tfrac{1}{2} = 0.50$$

Notice that the three axioms are satisfied and that

$$P[R \cup S] = 0.50 + 0.25 = 0.75$$
$$P[C \cup S] = 0.25 + 0.25 = 0.50$$
$$P[C \cup R] = 0.25 + 0.50 = 0.75$$

Once measures have been assigned in accord with these three axioms to the points in the sample space, these probabilities may be operated on in the manner to be demonstrated throughout this book. The engineer may be fully confident that his results will be mathematically valid, but it cannot be emphasized too strongly that the physical or practical significance of these results is no better than the "data," the assigned probabilities, upon which they are based.

2.1.3 Simple Probabilities of Events

Certain relationships among the probabilities of events follow from the relationships among events and from the axioms of probability. Many simple conclusions are self-evident; others require some derivation. Further, some additional relationships between events are defined in terms of relationships between their probabilities. The remainder of this section treats and illustrates these various relationships.

Probability of an event   Since, in general, an event is associated with one or more sample points or simple events, and since these simple events are mutually exclusive by the construction of the sample space, the *probability of any event is the sum of the probabilities assigned to the sample points with which it is associated.* If an event contains all the sample points with nonzero probabilities, its probability is 1 and it is sure to occur. If an event is impossible, that is, if it cannot happen as the

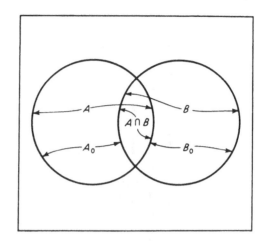

Fig. 2.1.10 Venn diagram. Illustrates decomposition of two events A and B into mutually exclusive events A_0, B_0 and $A \cap B$.

result of the experiment, then the probabilities of all the sample points associated with the event are zero.

Probability of union   The probability of an event which is the union of two events A and B, disjoint or not, can be derived from the material in the previous sections.   The event A can be considered as the union of the intersection $A \cap B$ and a nonoverlapping set of sample points, say, A_0.   Similarly, event B is the union of two mutually exclusive events $A \cap B$ and B_0.   These events are illustrated in Fig. 2.1.10.†   By Axiom III

$$P[A] = P[A \cap B] + P[A_0] \tag{2.1.1a}$$
$$P[B] = P[A \cap B] + P[B_0] \tag{2.1.1b}$$

Now $A \cup B$ can be divided into three mutually exclusive events, A_0, B_0, and $A \cap B$.
   Therefore, by Axiom III,

$$P[A \cup B] = P[A_0] + P[A \cap B] + P[B_0] \tag{2.1.2}$$

Solving Eqs. (2.1.1a) and (2.1.1b) for $P[A_0]$ and $P[B_0]$ and substituting into Eq. (2.1.2),

$$P[A \cup B] = P[A] + P[B] - P[A \cap B] \tag{2.1.3}$$

† Such figures, which exploit the analogy between the algebra of events and the areas in a plane, are called Venn diagrams.   They can be most helpful in visualizing event relationships, but it is important to realize that these diagrams cannot illustrate relationships among numbers, that is, the probabilities of events.

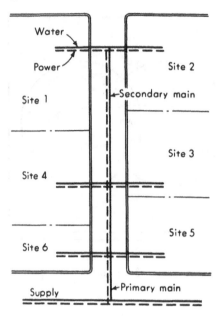

Fig. 2.1.11 Illustration of industrial-park utilities.

In words, the probability of the occurrence of either one event or another or both is the sum of their individual probabilities minus the probability of their joint occurrence. This extremely important result is easily verified intuitively. In summing the probabilities of the events A and B to determine the probability of a compound event $A \cup B$, one has added the probability measure of the sample points in the event $A \cap B$ twice. In the case of mutually exclusive events, when the intersection $A \cap B$ contains no sample points, $P[A \cap B] = 0$ and the equation reduces to Axiom III.

To illustrate this and following concepts, let us consider the design of an underground utilities system for an industrial park containing six similar building sites (Fig. 2.1.11). The sites have not yet been leased, and so the nature of occupancy of each is not known. If the engineer provides water and power capacities in excess of the demand actually encountered, he will have wasted his client's capital; if, on the other hand, the facilities prove inadequate, expensive changes will be required. For simplicity, consider any particular site and assume that the electric power required by the occupant will be either 5 or 10 units, while the water capacity demanded will be either 1 or 2 units. Then the sample space describing an experiment associated with a single occupant consists of four points, labeled (5,1), (10,1), (5,2), or (10,2), according to the combination of levels of power and water demanded. The space can be illustrated in either of two ways, as shown in Fig. 2.1.12. The client,

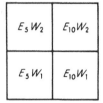

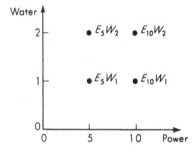

Fig. 2.1.12 Alternate graphical representations of the utilities-demand sample space.

in an interview with the engineer, makes a series of statements about odds and relative weights from which the engineer calculates the following set of probabilities.

Sample point	Simple event label	Owner's estimate of probability
$(5,1)$	$E_5 W_1$	0.1
$(10,1)$	$E_{10} W_1$	0.1
$(5,2)$	$E_5 W_2$	0.2
$(10,2)$	$E_{10} W_2$	0.6

The probability of an event W_2, "the water demand is 2 units," is the sum of the probabilities of the corresponding, *mutually exclusive*, *simple events*.

$$P[W_2] = P[E_5 W_2] + P[E_{10} W_2] = 0.2 + 0.6 = 0.8$$

Also, the probability of power demand being 10 units at a particular site is

$$P[E_{10}] = P[E_{10} W_1] + P[E_{10} W_2] = 0.1 + 0.6 = 0.7$$

The probability that either the water demand is 2 units or the power demand is 10 units may be calculated by Eq. (2.1.3).

$$P[W_2 \cup E_{10}] = P[W_2] + P[E_{10}] - P[W_2 \cap E_{10}]$$

or, since the intersection of events E_{10} and W_2 is the simple event $E_{10}W_2$,

$$P[W_2 \cup E_{10}] = 0.8 + 0.7 - 0.6 = 0.9$$

Notice that the same result is obtained by summing the probabilities of the simple events in which one observes either a water demand of 2 units or a power demand of 10 units or both.

$$P[W_2 \cup E_{10}] = P[E_{10}W_2] + P[E_5W_2] + P[E_{10}W_1]$$
$$= 0.1 + 0.2 + 0.6 = 0.9$$

Conditional probability A concept of great practical importance is introduced into the axiomatic theory of probability through the following definition. The *conditional probability* of the event A given that the event B has occurred, denoted $P[A \mid B]$, is defined as the ratio of the probability of the intersection of A and B to the probability of the event B.

$$P[A \mid B] = \frac{P[A \cap B]}{P[B]} \tag{2.1.4a}$$

(If $P[B]$ is zero, the conditional probability $P[A \mid B]$ is undefined.)

The conditional probability can be interpreted as the probability that A has occurred given the knowledge that B has occurred. The condition that B has occurred restricts the outcome to the set of sample points in B, or the conditional sample space, but should not change the relative likelihoods of the simple events in B. If the probability measure of those points in B that are also in A, $P[A \cap B]$, is renormalized by the factor $1/P[B]$ to account for operation within this reduced sample space, the result is the ratio $P[A \cap B]/P[B]$ for the probability of A given B.

In the preceding illustration the engineer might have need for the probability that a site with a power demand of E_{10} will also require a water demand W_2. In this case

$$P[W_2 \mid E_{10}] = \frac{P[E_{10}W_2]}{P[E_{10}]} = \frac{0.6}{0.7} = 0.86$$

In applications, $P[B]$ and $P[A \mid B]$ often come from a study of the problem, whereas actually the joint probability $P[A \cap B]$ is desired; this is obtained as follows:

$$P[A \cap B] = P[A \mid B]P[B] = P[B \mid A]P[A] \tag{2.1.4b}$$

Where many events are involved, the following expansion is often helpful:

$$P[A \cap B \cap C \cap \cdots N]$$
$$= P[A \mid B \cap C \cap \cdots N]P[B \cap C \cap \cdots N]$$
$$= P[A \mid B \cap C \cap \cdots N]P[B \mid C \cap \cdots N]$$
$$P[C \mid D \cap \cdots N] \cdots P[N] \tag{2.1.5}$$

For example, for three events

$$P[A \cap B \cap C] = P[A \mid B \cap C]P[B \mid C]P[C]$$

Illustrations are given later in the section and in problems.

Independence If two physical events are not related in any way, we would not alter our measure of the probability of one even if we knew that the other had occurred. This intuitive notion leads to the definition of *probabilistic* (or *stochastic*) *independence*. Two events A and B are said to be independent if and only if

$$P[A \mid B] = P[A] \tag{2.1.6}$$

From this definition and Eq. (2.1.4a), the independence of events A and B implies that

$$\frac{P[A \cap B]}{P[B]} = P[A] \tag{2.1.7}$$

$$P[A \cap B] = P[A]P[B] \tag{2.1.8}$$

$$P[B \mid A] = P[B] \tag{2.1.9}$$

Any of these equations can, in fact, be used as a definition of independence.

Within the mathematical theory, one can only prove independence of events by obtaining $P[A]$, $P[B]$, and $P[A \cap B]$ and demonstrating that one of these equations holds. In engineering practice, on the other hand, one normally relies on knowledge of the physical situation to declare that in his model two particular events shall (or shall not) be assumed independent. From the assumption of independence, the engineer can calculate one of the three quantities, say, $P[A \cap B]$, given the other two.

In general, events A, B, C, . . . , N are mutually independent if and only if

$$P[A \cap B \cap C \cap \cdots N] = P[A]P[B]P[C] \cdots P[N] \tag{2.1.10}$$

This is the theorem known as the *multiplication rule*. In words, *if events are independent*, the probability of their joint occurrence is simply the product of their individual probabilities of occurrence.

Returning to the industrial-park illustration, let us assume that the engineer, not content with the loose manner in which his client assigned probabilities to demands, has sampled a number of firms in similar industrial parks. He has concluded that there is no apparent relationship between their power and water demands. A high power demand, for example, does not seem to be correlated with a high water demand.

Based on information easily obtained from the respective utility companies, the engineer assigns the following probabilities:

Event	Engineer's estimate of probability
Power demand	
E_5	0.2
E_{10}	0.8
	1.0
Water demand	
W_1	0.3
W_2	0.7
	1.0

Adopting the assumption of stochastic independence of the water and power demands,† the engineer can calculate the following probabilities for the joint occurrences or simple events.

$$P[E_5 W_1] = P[E_5]P[W_1] = (0.2)(0.3) = 0.06$$
$$P[E_5 W_2] = P[E_5]P[W_2] = (0.2)(0.7) = 0.14$$
$$P[E_{10} W_1] = P[E_{10}]P[W_1] = (0.8)(0.3) = 0.24$$
$$P[E_{10} W_2] = P[E_{10}]P[W_2] = (0.8)(0.7) = 0.56$$
$$1.00$$

Decisions under uncertainty   With no more than the elementary operations introduced up to this point, we can demonstrate the usefulness of probabilistic analysis when the engineer must make economic decisions in the face of uncertainty. We shall present only the rudiments of decision analysis at this point. We chose to continue to use the preceding example to introduce these ideas. For simplicity of numbers, let us concentrate on the water demand only and investigate the design capacity of a secondary main serving a *pair* of similar sites in the industrial park. The occupancies of the two sites represent two repeated *trials* of the experiment described above. Denote the event that the demand of

† Notice that the calculation of $P[W_2 \mid E_{10}]$ on page 48, which is based on the client's four probability estimates, contradicts this assumption, for $P[W_2 \mid E_{10}] = 0.86$ does *not* equal $P[W_2] = 0.8$. Hence, using the client's probabilities, the events are *not* independent. The engineer, on the other hand, has *assumed* independence, and has estimated only $P[E_5]$ and $P[W_1]$, and *calculates* the remaining probabilities of interest. The assumption of independence reduces the number of numerical estimates necessary to describe the phenomenon completely.

each firm is one unit by W_1W_1 and the event that the demand of the first is one unit and the second is two units by W_1W_2, and so forth. Assuming stochastic independence of the demands from the two sites, one can easily calculate the probabilities of various combinations of outcomes.

$$P[W_1W_1] = P[W_1]P[W_1] = (0.3)(0.3) = 0.09$$
$$P[W_1W_2] = P[W_1]P[W_2] = (0.3)(0.7) = 0.21$$
$$P[W_2W_1] = P[W_2]P[W_1] = (0.7)(0.3) = 0.21$$
$$P[W_2W_2] = P[W_2]P[W_2] = (0.7)(0.7) = \underline{0.49}$$
$$1.00$$

Notice that events W_1W_2 and W_2W_1 lead to the same total demand; on the sample space of this two-site experiment, we could define new events D_2, D_3, and D_4 which correspond to total demands of two, three, and four units, respectively (see Fig. 2.1.13). It is most often the case that engineers are interested in outcomes which have associated numerical values. This observation is expanded upon in Sec. 2.2.

The assumption of independence between sites implies that the engineer feels that there would be no reason to alter the probabilities of the demand of the second firm if he knew the demand of the first. That is, knowledge of the demand of the first gives the engineer no new information about the demand of the second. Such might be the case, for example, if the management of the second firm chose its site without regard for the nature of its neighbor. If the demands of all six sites are mutually independent, the probability that all the sites will demand two units of water is:

$$P[W_2W_2W_2W_2W_2W_2] = P[W_2]P[W_2]P[W_2]P[W_2]P[W_2]P[W_2]$$
$$= (0.7)^6 = 0.117$$

How can such information be put to use to determine the best choice for the capacity of the *secondary* pipeline? These estimates of the relative likelihoods must in some way be related to the relative costs

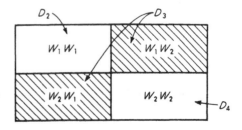

Fig. 2.1.13 Two-site water-demand illustration sample space.

of the designs and the losses associated with possible inadequacies. Suppose the engineer has compiled the following cost table:

Initial costs:

> Capacity three units: \$500 + cost of two units
> Capacity four units: \$1500 + cost of two units

Cost associated with possible later enlargement:

> Two units to three units: \$1200
> Three units to four units: \$1500
> Two units to four units: \$2000

A common method used to account for uncertain losses is to weight the possible losses by the probabilities of their occurrences. Thus to the initial cost of a moderate design of capacity three units the engineer might add the weighted loss $(0.49)(\$1500) = \$735$, which is associated with a possible need for later enlargement if both firms should demand two units (the event W_2W_2). If a two-unit capacity is chosen, either of two future events, D_3 or D_4, will lead to additional cost. The weighted cost of each contributes to the total expected cost of this design alternative. These costs are called "expected costs" for reasons which will be discussed in Sec. 2.4. The validity of their use in making decisions, to be discussed more fully in Chap. 5,† will be accepted here as at least intuitively satisfactory. The following table of expected costs (over the basic initial cost of the two-unit capacity) can be computed for each alternative design.

Initial capacity choice	Incremental initial cost (over cost of two units)	Expected loss due to possible demand of three units (event W_1W_2 or W_2W_1)	Expected loss due to possible demand of four units (event W_2W_2)	Total expected cost
two units	0	$(0.21 + 0.21)(1200)$ $= 500$	$(0.49)(2000) = 980$	1480 + cost of two units
three units	500	0	$(0.49)(1500) = 735$	1235 + cost of two units
four units	1500	0	0	1500 + cost of two units

† In that chapter we shall find that, in general, decisions should be based on *utilities*, which may or may not coincide with dollars.

A design providing an initial capacity of three units appears to provide the best compromise between initial cost and possible later expenses. Notice that the common cost of two units does not enter the decision of choosing among the available alternatives.

As modifications lead to future rather than initial expense, the effect of the introduction of interest rates, and hence the time value of money, into the economic analysis would be to increase the relative merit of this strategy (three units) with respect to the large-capacity (four-unit) design while decreasing its relative advantage over the small-capacity (two-unit) system. A high interest rate could make the latter design more economical.

To design the primary water lines feeding all sites in the industrial park, a similar but more complex analysis is required. In dealing with repeated trials of such two-outcome experiments, one is led to the binomial model to be introduced in Sec. 3.1.2. Where more than two demand levels or more than one service type are needed, the multinomial distribution (Sec. 3.6.1) will be found to apply.

In Chaps. 5 and 6 we shall discuss decision making in more detail. The following two illustrations demonstrate further the computations of probabilities of events using the relationships we have discussed up to this point, and the use of expected costs in decisions under uncertainty.

Illustration: Construction scheduling A contractor must select a strategy for a construction job. Two independent operations, I and II, must be performed in succession. Each operation may require 4, 5, or 6 days to complete. A sample space is illustrated in Fig. 2.1.14. M_4 is the event that operation I

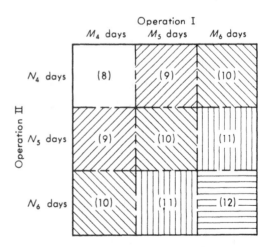

Fig. 2.1.14 Sample space for construction-strategy illustration.

Table 2.1.1

Operation	Rate	Cost per day at rate	Probability of time to completion			
			4 days	5 days	6 days	Sum
I	A	$200	0.2	0.5	0.3	1.0
I	B	240	0.3	0.6	0.1	1.0
I	C	280	0.6	0.4	0	1.0
II	D	200	0.1	0.4	0.5	1.0
II	E	300	0.3	0.4	0.3	1.0
II	F	400	0.6	0.3	0.1	1.0

requires 4 days, N_4 that II requires 4 days, etc. Each operation can be performed at three different rates, each at a different cost, and each leading to different time requirement likelihoods. In addition, if the job is not completed in 10 days, the contractor must pay a penalty of $2000 per day. The total time required for each combination of time requirements is shown in parentheses in Fig. 2.1.14. For example, $M_4 \cap N_6$ requires a total of 10 days.

The contractor judges from experience that by working at rate A his probability of completing phase I in 4 days, event M_4, is only 0.2, in 5 days (or M_5) is 0.5, and in 6 days (M_6) is 0.3. Proceeding in this manner, he assigns a complete set of probabilities (Table 2.1.1) to all possibilities, reflecting that he can probably accelerate the job by working at a more costly rate. He assumes that the events M_4, M_5, and M_6 are independent of N_4, N_5, and N_6.

The expected costs $E[\text{cost}]$ of construction can now be calculated. For level I at rate A,

$$E[\text{cost } A] = P[M_4](\text{cost of 4 days}) + P[M_5](\text{cost of 5}) + P[M_6](\text{cost of 6})$$

or

$$E[\text{cost } A] = (0.2)(4)(200) + (0.5)(5)(200) + (0.3)(6)(200) = \$1020$$

Similarly,

$$E[\text{cost } B] = (0.3)(4)(240) + (0.6)(5)(240) + (0.1)(6)(240) = \$1152$$

Similarly, all expected costs of construction are:

Operation	Rate	$E[cost]$
I	A	$1020
I	B	1152
I	C	1232
II	D	1080
II	E	1500
II	F	1800

The optimum strategy is rate A for operation I and rate D for II to obtain a minimum expected cost of construction.

The possibility of an overtime penalty must also be included in the total cost. The probability of strategy AD requiring 8 days of time is the probability of $(M_4 \cap N_4)$ and, owing to independence, is simply the product of the individual probabilities of each operation requiring 4 days. Assuming independence of the events, under strategy AD,

$$P[8 \text{ days}] = P[M_4 \cap N_4] = (0.2)(0.1) = 0.02$$

A 9-day construction time can occur in two mutually exclusive ways: 4 days required for I and 5 days for II, or 5 days for I and 4 days for II. This event is crosshatched in Fig. 2.1.14.

$$P[9 \text{ days}] = P[(M_4 \cap N_5) \cup (M_5 \cap N_4)] = P[M_4 \cap N_5] + P[M_5 \cap N_4]$$

Under strategy AD, $P[9 \text{ days}] = (0.2)(0.4) + (0.5)(0.1) = 0.13$. Similarly, a 10-day time occurs as shown in Fig. 2.1.14 and, with the probabilities of strategy AD, has probability

$$P[10 \text{ days}] = (0.2)(0.5) + (0.5)(0.4) + (0.3)(0.1) = 0.33$$

Losses are associated with construction times of 11 or 12 days. Using rates A and D,

$$P[11 \text{ days}] = (0.5)(0.5) + (0.3)(0.4) = 0.37$$
$$P[12 \text{ days}] = (0.3)(0.5) = 0.15$$

The expected penalty, if strategy AD is adopted, is, then,

$$E[\text{penalty}] = (0.37)(2000) + (0.15)(2)(2000) = 740 + 600 = 1340$$

The complete penalty results are shown in Table 2.1.2, and the total expected costs, that is, those due to construction plus those due to possible penalties, are collected in Table 2.1.3.

Table 2.1.2

Strategy	Probability for days		Expected penalty for days		Total expected penalty
	11	12	11	12	
AD	0.37	0.15	740	600	$1340
AE	0.27	0.09	540	360	900
AF	0.14	0.03	280	120	400
BD	0.34	0.05	680	200	880
BE	0.22	0.03	440	120	560
BF	0.09	0.01	180	40	220
CD	0.20	0	400	0	400
CE	0.12	0	240	0	240
CF	0.04	0	80	0	80

Table 2.1.3

Strategy	E[construction cost]	E[penalty]	E[total]
A D	1020 + 1080 = 2100	1340	$3440
A E	1020 + 1500 = 2520	900	3420
A F	1020 + 1800 = 2820	400	3220
B D	1152 + 1080 = 2232	880	3112
B E	1152 + 1500 = 2652	560	3212
B F	1152 + 1800 = 2952	220	3172
C D	1232 + 1080 = 2312	400	2712
C E	1232 + 1500 = 2732	240	2972
C F	1232 + 1800 = 3032	80	3112

The optimum strategy has a minimum expected total cost and is in fact strategy CD.

Expected costs simply provide a technique for choosing among alternative strategies.   The contractor's cost will *not* be $2712 if he chooses strategy CD, but rather the cost of either 4, 5, or 6 days with I and with II plus $2000 per actual day of overtime.

Illustration: Analysis of bridge lifetimes†   As an example of the construction of a more complex probability model from simple basic assumptions, consider this problem in bridge design.   Assume that a bridge or culvert is usually replaced either because a flood exceeding the capacity of the structure has occurred or because it becomes obsolete owing to a widening or rerouting of the highway.   The designer is interested in the likelihood that the life of the structure will come to an end in each of the years after construction.‡   Assume that there is a constant probability p that in any year a flow exceeding the capacity of the culvert will take place.   Let r_i be the probability that the structure will become obsolete in year i *given* that it has not become obsolete prior to year i. For most situations this probability grows with time.   It may well be a reasonable engineering assumption that the effects of floods and obsolescence are unrelated and that the occurrences of critical flood magnitudes from year to year are independent events.   Our problem is to determine the probability that the life of the structure comes to an end in year j for the first time.

Each year is a simple experiment with events defined as

A_i = [a critical flood destroys the structure in year i]

B_i = [the structure becomes obsolete in year i]

The elementary events in these simple experiments are

$$\{A_i \cap B_i,\ A_i \cap B_i{}^c,\ A_i{}^c \cap B_i,\ A_i{}^c \cap B_i{}^c\}$$

† This and subsequent sections and illustrations marked with a dagger may be omitted at first reading since they are of a more advanced nature.
‡ This example is based on the treatment developed in M. A. Benson and N. C. Matalas [1965], The Effects of Floods and Obsolescence on Bridge Life, *Proc. ASCE, J. Highway Div.*, HW1, vol. 91, January.

The probability that the structure's life does not end in the first year is, owing to the assumed independence,[†]

$$P[A_i{}^c \cap B_i{}^c] = P[A_i{}^c]P[B_i{}^c] = (1 - p)(1 - r_i)$$

Successive years represent repetitions of such experiments. The probability that the life does not end in either of the first two years is, by Eq. (2.1.4a),

$$P[(A_1{}^c \cap B_1{}^c) \cap (A_2{}^c \cap B_2{}^c)] = P[A_1{}^c \cap B_1{}^c]P[A_2{}^c \cap B_2{}^c \mid A_1{}^c \cap B_1{}^c]$$

Similarly, the last term on the right-hand side can be written

$$P[A_2{}^c \cap B_2{}^c \mid A_1{}^c \cap B_1{}^c] = P[A_2{}^c \mid B_2{}^c \cap (A_1{}^c \cap B_1{}^c)]P[B_2{}^c \mid A_1{}^c \cap B_1{}^c]$$

This is easily verified by writing out the definitions of the conditional probabilities involved, although a simple reading of the statement suggests why it must be so.

Because of the various assumptions of independence made above,

$$P[A_2{}^c \mid B_2{}^c \cap A_1{}^c \cap B_1{}^c] = P[A_2{}^c] = 1 - p$$
$$P[B_2{}^c \mid A_1{}^c \cap B_1{}^c] = P[B_2{}^c \mid B_1{}^c] = 1 - r_2$$

Putting these results together,

$$P[(A_1{}^c \cap B_1{}^c) \cap (A_2{}^c \cap B_2{}^c)] = P[A_1{}^c \cap B_1{}^c](1 - p)(1 - r_2)$$
$$= (1 - p)^2(1 - r_1)(1 - r_2)$$

Clearly, the probability that the structure survives floods and obsolescence through j years is

$$P[\text{survival through } j] = P[A_1{}^c \cap B_1{}^c \cap A_2{}^c \cap B_2{}^c \cap \cdots \cap A_j{}^c \cap B_j{}^c]$$

which by simple extension of the argument above is

$$P[\text{survival through } j] = (1 - p)^j \prod_{i=1}^{j} (1 - r_i)$$

For the structure's life to first come to end in year j, on the other hand, it must have survived $j - 1$ years, which will have happened with probability

$$(1 - p)^{j-1} \prod_{i=1}^{j-1} (1 - r_i)$$

and must then either have become obsolete or met a critical flood in year j. The latter event, $A_j \cup B_j$, has probability, *given previous survival*, of

$$P[A_j \cup B_j \mid \text{previous survival}]$$
$$= P[A_j \cup B_j \mid A_1{}^c \cap B_1{}^c \cap A_2{}^c \cap B_2{}^c \cap \cdots \cap A_{j-1}^c \cap B_{j-1}^c]$$

[†] The engineering reader should take special care to read these equations in terms of their event definitions. The following equation, for example, states that the probability of neither a critical flood nor obsolescence in the first year is the probability that no flood occurs multiplied by the probability that no obsolescence occurs. In the more involved equations to follow, understanding will usually come more quickly in this way than through abstract symbols.

Equation (2.1.3) applies, subject to the conditioning event, previous survival, and

$P[A_j \cup B_j \mid$ previous survival$] = P[A_j \mid$ previous survival$]$
$$+ P[B_j \mid \text{previous survival}] - P[A_j \cap B_j \mid \text{previous survival}]$$

Owing to the various independences,

$P[A_j \cup B_j \mid$ previous survival$] = P[A_j] + P[B_j \mid B_1^c \cap B_2^c \cap \cdots \cap B_{j-1}^c]$
$$- P[A_j]P[B_j \mid B_1^c \cap B_2^c \cap \cdots \cap B_{j-1}^c]$$
$$= p + r_j - pr_j$$

Finally, then,

$P[$life ends in year j for first time$]$

$$= P[\text{no survival in } j \mid \text{previous survival}]P[\text{previous survival}]$$

$$= [p + r_j - pr_j][(1 - p)^{j-1} \prod_{i=1}^{j-1} (1 - r_i)]$$

For example, if the structure is designed for the so-called "50-year flood,"† it implies that $p = \frac{1}{50} = 0.02$, and, if $r_i = 1 - e^{-0.025i}$, $i = 1, 2, 3, \ldots$, then

$P[$life ends in year j for the first time$]$

$$= [0.02 + (1 - e^{-0.025i}) - 0.02(1 - e^{-0.025i})](0.98)^{j-1} \prod_{i=1}^{j-1} e^{-0.025i}$$

$$= (1 - 0.98e^{-.025i})(0.98)^{j-1} \prod_{i=1}^{j-1} e^{-0.025i}$$

A plot of these probabilities for the years $j = 1$ to 22 is given in Fig. 2.1.15.

Combined with economic data these probabilities would permit the engineer to calculate an expected present worth of this design, to be compared with those of other alternate designs of different capacities for flow and perhaps with different provisions to reduce the likelihood of obsolescence.

Total probability theorem   The equation defining conditional probabilities, Eq. (2.1.4a), can be manipulated to yield another important result in the probability of events.   Given a set of mutually exclusive, collectively exhaustive events, $B_1, B_2, \ldots, B_n$, one can always expand the probability $P[A]$ of another event A in the following manner:

$$P[A] = P[A \cap B_1] + P[A \cap B_2] + \cdots P[A \cap B_n]$$

$$= \sum_{i=1}^{n} P[A \cap B_i] \tag{2.1.11}$$

† This concept is discussed in detail in Sec. 3.1.

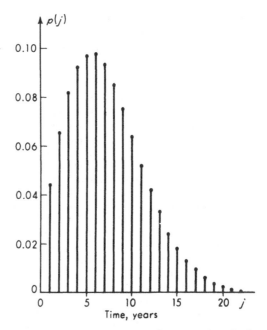

Fig. 2.1.15 Solution to culvert-life example. Probabilities of finding various lengths of life are plotted against length of life.

Figure 2.1.16 illustrates this fact. Each term in the sum can be expanded using Eq. (2.1.4a):

$$P[A] = \sum_{i=1}^{n} P[A \mid B_i]P[B_i] \qquad (2.1.12)$$

This result is called "the theorem of total probabilities." It represents the expansion of the probability of an event in terms of its conditional probabilities, conditioned on a set of mutually exclusive, collectively

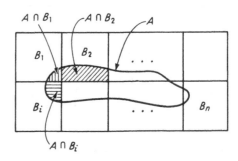

Fig. 2.1.16 Venn diagram for total probability theorem. Event A intersects mutually exclusive and collectively exhaustive events B_i.

exhaustive events.   It is often a useful expansion to consider in problems
when it is desired to compute the probability of an event A, since the
terms in the sum may be more readily obtainable than the probability
A itself.

Illustration: Additive random demands on engineering systems   Consider the generalized civil-engineering design problem of providing "capacity" for a probabilistic "demand."   Depending on the situation, demand may be a loading,
a flood, a peak number of users, etc., while the corresponding capacity may be
that of a building, a dam, or a highway.   In many examples to follow, the
general terminology, that is, demand and capacity, will be used with the
express purpose of encouraging the reader to supply his preferred specific application.   In this example two possible types (primary and secondary) of capacity
at different unit costs are available, and loss is incurred if demand exceeds the
total capacity provided or if it requires the use of some secondary capacity.
This situation faces every designer, for the peak demand is often uncertain and
the design usually cannot economically be made adequate for the maximum
possible demand.   The engineer seeks a balancing of initial cost and potential
future losses.

   For example, a building frame should be able to sustain a moderate seismic
load without visible damage to the structure.   During a rare, major earthquake, however, a properly designed structure will develop secondary resistance
involving large plastic deformations and some acceptable level of damage to
windows and partitions.   Design for zero damage under all possible conditions
is impossible or uneconomical.   Similar problems arise in design of systems
in which provision for future expansion is included.   The future demand is
unknown, so that the estimation of the optimum funds to be spent now to
provide for expansion in the future is a similar type of capacity-demand
situation.

   Assume, in this example, that demand arises from two additive sources A
and B and that the engineer assigns probabilities as shown in Table 2.1.4 to the
various levels from source B and conditional probabilities to various levels
from source A for each level of source B.   The two sources are not
independent.

**Table 2.1.4   Conditional probabilities of $[A_i \mid B_j]$ and the
probabilities of B_j**

| | $P[A \text{ level} \mid B \text{ level}]$ | | | | | |
	A_{100}	A_{200}	A_{300}	A_{400}	Sum	$P[B \text{ level}]$
B_{100}	0	0.6	0.4	0	1.0	0.7
B_{200}	0	0.3	0.5	0.2	1.0	0.2
B_{300}	0	0	0.2	0.8	1.0	0.1
B_{400}	...	...	...	...	...	0
						1.00

The probabilities of the A levels can be found by applying Eq. (2.1.12) for each level from source A. For example,

$$P[A_{200}] = \sum_{\text{all } j} P[A_{200} \mid B_j]P[B_j]$$

$$= (0.6)(0.7) + (0.3)(0.2) + (0)(0.1) + 0 = 0.48$$

Similarly,

$$P[A_{100}] = 0.0$$
$$P[A_{300}] = 0.40$$
$$P[A_{400}] = 0.12$$

Before proceeding with the analysis, note that the engineer has assigned some zero probabilities.† Two interpretations exist of a zero probability. First, the event may simply be impossible. Second, it may be possible, but its likelihood negligible, and the engineer is willing to effectively exclude this level from the current study (while retaining the option to assign later a nonzero probability).

A two-dimensional sample space for the demand levels appears in Fig. 2.1.17. Any simple event is an intersection $A_i \cap B_j$ of an A level and B level. The various simple events which lead to the same total demand D_k are also indicated in this figure, the total values being given in parentheses. To determine the probabilities of an event such as $D_{700} = $ [total demand is 700], we find the probability of this event as the union of mutually exclusive, simple events (and hence as simply the sum of their probabilities). Example:

$$D_{700} = (A_{300} \cap B_{400}) \cup (A_{400} \cap B_{300})$$
$$P[D_{700}] = P[A_{300} \cap B_{400}] + P[A_{400} \cap B_{300}]$$
$$= P[A_{300} \mid B_{400}]P[B_{400}] + P[A_{400} \mid B_{300}]P[B_{300}]$$
$$= 0 + (0.8)(0.1) = 0.08$$

† Note that $P[A_i \mid B_{400}]$ is undefined, since $P[B_{400}] = 0$ [see Eq (2.1.4a)].

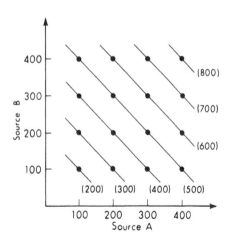

Fig. 2.1.17 Demand-level sample space showing how various total demands may arise.

A similar calculation for D_{500} involves four terms:

$$P[D_{500}] = \sum_{\text{all } j} P[A_j \mid B_{500-j}]P[B_{500-j}]$$

$$= (0)(0) + (0)(0) + (0.5)(0.2) + (0)(0.7) = 0.10$$

Similarly, we find that the probabilities of various demand levels are:

Demand level event D_k	Probability $P[D_k]$
$k = 200$	0
300	0.42
400	0.34
500	0.10
600	0.06
700	0.08
800	$\underline{0.0}$
	1.00

Although questions about the events which lead to particular demand levels are not asked in this decision problem, they might well be of interest in such problems. We ask one such question here to illustrate conditional sample spaces and conditional probabilities. What is the conditional sample space given that the total demand is 600? It is the set of those events which produce such a total demand, namely $A_{200} \cap B_{400}$, $A_{300} \cap B_{300}$, and $A_{400} \cap B_{200}$. This space is illustrated in Fig. 2.1.18. The conditional probabilities of these events, given total demand is 600, are

$$P[A_{400} \cap B_{200} \mid D_{600}] = \frac{P[A_{400} \cap B_{200} \cap D_{600}]}{P[D_{600}]}$$

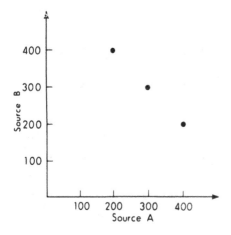

Fig. 2.1.18 Conditional sample space for a given total demand of 600.

Table 2.1.5 Example: expected cost calculations for several possible designs

Design		C_p cost, dollars	C_s cost, dollars	Total initial cost, dollars	Expected damage loss from using C_s: (cost of use) P[use], $	Expected failure loss from demand $> C_p + C_s$: (cost of failure) P[fail], $	Total initial plus expected costs, dollars
Primary capacity C_p	Secondary capacity C_s						
700	0	700,000	0	700,000	0	0	700,000
600	100	600,000	10,000	610,000	$(700-600)(1000)(0.08) =$ 8000	0	618,000
600	0	600,000	0	600,000	0	$(700-600)(2000)(0.08) =$ 16,000	616,000
500	100	500,000	10,000	510,000	$(100)(1000)(0.06) =$ 6,000	$(100)(2000)(0.08) =$ 16,000	532,000
400	200	400,000	20,000	420,000	$(200)(1000)(0.06) +$ $(100)(1000)(0.10) =$ 22,000	$(100)(2000)(0.08) =$ 16,000	458,000
0	0	0	0	0	0	$(700)(2000)(0.08) +$ $(600)(2000)(0.06) +$ $(500)(2000)(0.10) +$ $(400)(2000)(0.34) +$ $(300)(2000)(0.42) =$	808,000
0	700	0	70,000	70,000	$(700)(1000)(0.08) +$ $(600)(1000)(0.06) +$ $(500)(1000)(0.10) +$ $(400)(1000)(0.34) +$ $(300)(1000)(0.42) =$ 404,000	0	474,000

Because D_{600} includes the event $A_{400} \cap B_{200}$, we get

$$P[A_{400} \cap B_{200} \mid D_{600}] = \frac{P[A_{400} \cap B_{200}]}{P[D_{600}]}$$

$$= \frac{P[A_{400} \mid B_{200}]P[B_{200}]}{P[D_{600}]}$$

$$= \frac{(0.2)(0.2)}{0.06} = \frac{0.04}{0.06} = 0.67$$

and

$$P[A_{300} \cap B_{300} \mid D_{600}] = \frac{0.02}{0.06} = 0.33$$

$$P[A_{200} \cap B_{400} \mid D_{600}] = \frac{0}{0.06} = 0$$

Notice that the unconditional probabilities of these events are 0.04, 0.02, and 0. These conditional probabilities are just these same relative values normalized to sum to unity.

Assume that primary capacity can be provided at \$1000 per unit and that the secondary capacity cost is \$100 per unit. If the demand is such that the secondary capacity must be used, the associated ("damage") loss is \$1000 per unit used, and if the demand exceeds total (primary plus secondary) capacity, the ("failure") loss is \$2000 per unit of excess over total capacity. The design alternatives include any combination of 0 to 700 primary capacity and 0 to 700 secondary capacity.

Associated with each design alternative is a primary capacity cost C_p, a secondary capacity cost C_s, and an expected cost associated with potential loss due to excessive demands. The latter cost includes a component due to demand possibly exceeding primary capacity, but not secondary capacity, and a component arising due to demand possibly exceeding the total capacity. {In the latter event, "failure," no "damage" loss is involved; e.g., if primary capacity is 500, secondary capacity is 100, and a demand of 700 occurs, the loss is $2000[700 - (500 + 100)]$, not this plus $1000(700 - 500)$.}

A number of expected cost computations are illustrated in Table 2.1.5. The trend, as capacities are decreased from the most (initially) costly design of 700 units of primary capacity, is toward lowering total cost by accepting higher risks. After a point, the risks become too great relative to the consequences and to the initial costs, and total expected costs rise again.

Bayes' theorem   Continuing the study of the event A and the set of events B_i considered in Eq. 2.1.11 (Fig. 2.1.16), examine the conditional probability of B_j given the event A. By Eq. (2.1.4a), and since, clearly, $P[B_j \cap A] = P[A \cap B_j]$,

$$P[B_j \mid A] = \frac{P[B_j \cap A]}{P[A]} = \frac{P[A \cap B_j]}{P[A]}$$

The numerator represents one term in Eq. (2.1.11) and can be replaced as in Eq. (2.1.12) by the product $P[A \mid B_j]P[B_j]$, and the denominator can be represented by the sum of such terms, Eq. (2.1.12). Substituting,

$$P[B_j \mid A] = \frac{P[A \mid B_j]P[B_j]}{\sum_{i=1}^{n} P[A \mid B_i]P[B_i]} \qquad (2.1.13)$$

This result is known as *Bayes' theorem* or Bayes' rule. Its simple derivation belies its fundamental importance in engineering applications. As will best be seen in the illustration to follow, it provides the method for incorporating new information with previous or, so-called *prior*, probability assessments to yield new values for the engineer's relative likelihoods of events of interest. These new (conditional) probabilities are called *posterior probabilities*. Bayes' theorem will be more fully explained and applied in Chaps. 5 and 6.

Illustration: Imperfect testing Bayes' theorem can be generalized in application by calling the unknown classification the *state*, and by considering that some generalized *sample* has been observed. Symbolically, Eq. (2.1.13) becomes

$$P[\text{state} \mid \text{sample}] = \frac{P[\text{sample} \mid \text{state}]P[\text{state}]}{\sum_{\text{all states}} P[\text{sample} \mid \text{state}]P[\text{state}]}$$

To illustrate the generalization, assume that an existing reinforced concrete building is being surveyed to determine its adequacy for a new future use. The engineer has studied the appearance and past performance of the concrete and, based on professional judgment, decides that the concrete quality can be classified as either 2000, 3000, or 4000 psi (based on the usual 28-day cylinder strength). He also assigns relative likelihoods or probabilities to these states:

State, psi	Prior probabilities
2000	0.3
3000	0.6
4000	0.1
	1.0

Concrete cores are to be cut and tested to help ascertain the true state. The engineer believes that a core gives a reasonably reliable prediction, but that it is not conclusive. He consequently assigns numbers reflecting the reliability of the technique in the form of conditional probability measure on the possible core-strength values z_1, z_2, or z_3 (in this case, core strength† of, say, 2500, 3500, or 4500 psi) as predictors of the unknown state:

P [core strength | state]

	State		
Core strengths	2000 psi	3000 psi	4000 psi
z_1 (favors 2000 psi)	0.7	0.2	0
z_2 (favors 3000 psi)	0.3	0.6	0.3
z_3 (favors 4000 psi)	0	0.2	0.7
	1.0	1.0	1.0

In words, if the true 28-day concrete quality classification is 3000 psi, the technique of taking a core will indicate this only 60 percent of the time. The total error probability is 40 percent, divided between z_1 and z_3. That is, the technique will significantly overestimate or underestimate the true quality 4 times in 10, on the average. Controlled experiments using the technique on concrete of known strength are used to produce such reliability information.

A core is taken and found to have strength 2500 psi favoring a 28-day strength of 2000 psi; that is, z_1 is observed. The conditional probabilities of the true strength are then [Eq. (2.1.13)]

$$P[2000 \mid z_1] = \frac{(0.7)(0.3)}{(0.7)(0.3) + (0.2)(0.6) + (0)(0.1)} = \frac{0.21}{0.33} = 0.635$$

$$P[3000 \mid z_1] = \frac{(0.2)(0.6)}{0.33} = 0.365$$

$$P[4000 \mid z_1] = \frac{(0)(0.1)}{0.33} = 0$$

The sample outcome causes the exclusion of 4000 as a possible state and shifts the relative weights more towards the indicated state.

In the light of the test's limitations, the engineer chooses to take a sample of two independent cores. In this case, it makes no difference if the calculation of posterior probabilities is made for each core in succession or for both cores simultaneously. Consider the latter approach first. Assume that the first core indicated z_1 and the second core indicated z_2. The probability of finding the sample outcome $\{z_1, z_2\}$ if the state is really 2000 (or 3000 or 4000) psi is the product of two conditional probabilities (since the core results are assumed independent). Thus,

$$P[\text{sample} \mid 2000] = P[z_1 \mid 2000]P[z_2 \mid 2000] = (0.7)(0.3)$$
$$P[\text{sample} \mid 3000] = P[z_1 \mid 3000]P[z_2 \mid 3000] = (0.2)(0.6)$$
$$P[\text{sample} \mid 4000] = P[z_1 \mid 4000]P[z_2 \mid 4000] = (0)(0.3)$$

† Since concrete strength increases with time, a strength of, say, 3500 psi now suggests that the 28-day strength (which is used as a design standard) was probably about 3000 psi.

Recall that the probabilities of state prior to this sample of two cores were 0.3, 0.6, and 0.1. The posterior probabilities then become

$$P[2000 \mid \text{sample}] = \frac{(0.7)(0.3)(0.3)}{(0.7)(0.3)(0.3) + (0.2)(0.6)(0.6) + (0)(0.3)(0.1)}$$

$$= \frac{0.063}{0.063 + 0.072 + 0} = \frac{0.063}{0.135} = 0.47$$

$$P[3000 \mid \text{sample}] = \frac{(0.2)(0.6)(0.6)}{0.135} = \frac{0.072}{0.135} = 0.53$$

$$P[4000 \mid \text{sample}] = \frac{(0)(0.3)(0.1)}{0.135} = \frac{0}{0.135} = 0$$

The role of Bayes' theorem as an "information processor" is revealed when it is recognized that the engineer might have taken the first core only, found z_1 (favoring 2000 psi), computed the posterior probabilities of state as (0.635, 0.365, 0), and only then decided that another core was desirable. At this point his prior probabilities (prior, now, only to the second core) are 0.635, 0.365, 0. The posterior probabilities, given that the second core favored 3000, become

$$P[2000 \mid z_2] = \frac{(0.3)(0.635)}{(0.3)(0.635) + (0.6)(0.365) + (0.3)(0)}$$

$$= \frac{0.19}{0.19 + 0.22 + 0} = 0.47$$

$$P[3000 \mid z_2] = \frac{(0.6)(0.365)}{0.41} = \frac{0.22}{0.41} = .53$$

$$P[4000 \mid z_2] = \frac{(0.3)(0)}{0.41} = 0$$

As they must, these probabilities are the same as those computed by considering the two cores as a single sample. If a third core (or several more cores) were taken next, these probabilities 0.47, 0.53, 0 would become the new prior probabilities. Bayes' theorem will permit the continuous up-dating of the probabilities of state as new information becomes available. The information might next be of some other kind, for example, the uncovering of the lab data obtained from test cylinders cast at the time of construction and tested at 28 days. Such information, of course, would have a different set of conditional probabilities of sample given state (quite possibly, in this case, with somewhat smaller probabilities of "errors," that is, with smaller probabilities of producing samples favoring a state other than the true one).

This simple example illustrates how all engineering sampling and experimentation can better be viewed in a probabilistic formulation. For reasons of expediency and economy, most testing situations measure a quantity which is only indirectly related to the quantity that is of fundamental engineering interest. In this example, the engineer has measured an extracted core's ultimate compressive strength in order to estimate the concrete's 28-day strength, which in turn is known to be correlated with its compressive strength in bending strength, shear (or diagonal tension) strength, corrosion resistance, durability, etc. The soils engineer may measure the density of a compacted fill not because he wishes to estimate its density, but because he wishes to estimate its strength and ultimately the embankment's stability. Various hardness-testing apparatuses exist because testing for hardness is a simple, nondestructive way to estimate the factor of direct interest, that is, a material's strength.

Both in the actual making of the measurement and in the assumed relationship between the measurable quantity (sample) and factor of direct interest (state) there may be experimental error and variation, and hence uncertainty. For example, accompanying a hardness-testing procedure is usually a graph, giving a strength corresponding to an observed value of a hardness measurement. But repeated measurements on a specimen may not give the same hardness value, and, when the graph was prepared for a given hardness there may have been some variation in the strength values about the average value through which the graph was drawn. It is this uncertainty which is reflected in the conditional probabilities P[sample | state]. A very special case is that where there is negligible measurement error and where the relationship between the sampled quantity and the factor of interest is exact (for example, if the state can be measured directly). In this case, the engineer will logically assign P[sample | state] $= 1$ if the sample value indicates (i.e., "favors") the state, and 0 if it does not. These assignments are implicit in any nonprobabilistic model of testing. With these special conditional probability assignments, inspection of Bayes' theorem will reveal that the only nonzero term in the denominator is the prior probability of the state favored by the observed sample, while the numerator is the same for this state and zero for all states not favored by the sample. Hence, if the entire sampling procedure is, indeed, perfect, the posterior probabilities of the states are zero except for the state favored or indicated by the sample. This state has probability 1. If, however, there is any uncertainty in the procedure, whether due to experimental error or inexact relationships between predicted quantity and predicting quantity, at least some of the other states will retain nonzero probabilities.† The better the procedure, the more sharply centered will be its conditional probabilities on the sample-favored state and the higher will be the posterior probability of this state.

This fundamental engineering problem of predicting one quantity given an observation of another related or correlated one will occur

† Unless, of course, the engineer excluded their possibility at the outset by setting their prior probabilities to zero.

throughout this text. In Secs. 2.2.2 and 2.4.3, and in Sec. 3.6.2, the probabilistic aspects of the problem are discussed. In Sec. 4.3 the problem is treated statistically; that is, questions of determining from observed data the "best" way to predict one quantity given another are discussed. In Chaps. 5 and 6, we shall return to the decision aspects of the problem.

2.1.4 Summary

In Sec. 2.1 we have presented the basic ideas of the probability of events. After defining events and the among-events relationships of union, intersection, collectively exhaustive, and mutually exclusive, we discussed the assignment of probabilities to events. We found the following:

1. The probability of the union of mutually exclusive events was the sum of their probabilities.
2. The probability of the union of two nonmutually exclusive events was the sum of their probabilities minus the probability of their intersection.
3. The probability of the intersection of two events is the product of their probabilities *only if* the two events are *stochastically independent*. In general, a *conditional probability* must appear in the product.

These basic definitions were manipulated to obtain a formula for the probability of the intersection of several events, the total probability theorem, and Bayes' theorem.

The use of simple probability notions in engineering decisions was illustrated, using the weighted or expected-cost criterion.

2.2 RANDOM VARIABLES AND DISTRIBUTIONS

Most civil-engineering problems deal with quantitative measures. Thus in the familiar deterministic formulations of engineering problems, the concepts of mathematical variables and functions of variables have proved to be useful substitutes for less precise qualitative characterizations. Such is also the case in probabilistic models, where the variable is referred to as a *random variable*. It is a numerical variable whose specific value cannot be predicted with certainty before an experiment. In this section we will first discuss its description for a single variable, and then for two or more variables jointly.

2.2.1 Random Variables

The value assumed by a random variable associated with an experiment depends on the outcome of the experiment.   There is a numerical value of the random variable associated with every simple event defined on the sample space, but different simple events may have the same associated value of the random variable.† Every compound event corresponds to one or more or a range of values of the random variable.

In most engineering problems there is seldom any question about how to define the random variable; there is usually some "most natural" way.   The traffic engineer in the car-counting illustration (Sec. 2.1.1) would say, "Let X equal the number of cars observed."   In other situations the random variable in question might be Y, the daily discharge of a channel, or Z, stress at yield of a steel tensile specimen.   In fact, a random variable is usually the easiest way to describe most engineering experiments.   The cumbersome subscripting of events found in previous illustrations could have been avoided by dealing directly with random variables such as demand level, concrete strength, etc., rather than with the events themselves.

The behavior of a random variable is described by its *probability law*, which in turn may be characterized in a number of ways.   The most common way is through the *probability distribution* of the random variable. In the simplest case this may be no more than a list of the values the variable can take on (i.e., the possible outcomes of an experiment) and their respective probabilities.

Discrete probability mass function (PMF)   When the number of values a random variable can take on is restricted to a countable number, the values 1, 2, 3, and 4, say, or perhaps all the positive integers, 0, 1, 2, . . . , the random variable is called *discrete*, and its probability law is usually presented in the form of a *probability mass function*, or PMF.   This function $p_X(x)$ of the random variable X‡ is simply the mathematical form of the list mentioned above:

$$p_X(x) = P[X = x] \tag{2.2.1}$$

For example, having defined the random variable X to be the number of vehicles observed stopped at the traffic light, the engineer may have

---

† More precisely, a random variable is a function defined on the sample space of the experiment.   It assigns a numerical value to every possible outcome.   (See, for example, Parzen [1960].)

‡ In general a capital letter will be used for a random variable, and the same letter, in lowercase, will represent the values which it may take on.

assigned probabilities to the events (Fig. 2.1.1) and corresponding values
of X such that

$$p_X(x) = \begin{cases} 0.1 \\ 0.2 \\ 0.3 \\ 0.2 \\ 0.1 \\ 0.1 \\ 0 \end{cases} \text{for} \begin{cases} x = 0 \\ x = 1 \\ x = 2 \\ x = 3 \\ x = 4 \\ x = 5 \\ x = 6, 7, \ldots \end{cases}$$

$$\text{Sum} = \overline{1.0}$$

The probability mass function is usually plotted as shown in Fig. 2.2.1a,
with each bar or spike being proportional in height to the probability
that the random variable takes on that value.

To satisfy the three axioms of probability theory the probability
mass function clearly must fulfill three conditions:

$$0 \le p_X(x) \le 1 \qquad \text{for all } x \qquad\qquad\qquad (2.2.2a)$$

$$\sum_{\text{all } x_i} p_X(x_i) = 1 \qquad\qquad\qquad (2.2.2b)$$

$$P[a \le X \le b] = \sum_{x_i \ge a}^{x_i \le b} p_X(x_i) \qquad\qquad\qquad (2.2.2c)$$

The sums in Eqs. (2.2.2b) and (2.2.2c) are, of course, only over those
values of x where the probability mass function is defined.

Cumulative distribution function (CDF)   An equivalent means† by which
to describe the probability distribution of a random variable is through
the use of a *cumulative distribution function*, or CDF.   The value of this
function $F_X(x)$ is simply the probability of the event that the random
variable takes on value equal to or less than the argument:

$$F_X(x) = P[X \le x] \qquad\qquad\qquad (2.2.3)$$

For discrete random variables, i.e., those possessing probability
mass functions, this function is simply the sum of the values of the
probability mass function over those values less than or equal to x that
the random variable X can take on.

$$F_X(x) = \sum_{\text{all } x_i \le x} p_X(x_i) \qquad\qquad\qquad (2.2.4)$$

† Two alternate ways of specifying a probability law—moment-generating and
characteristic functions—are discussed in Prob. 2.62, but will not be used in this text.

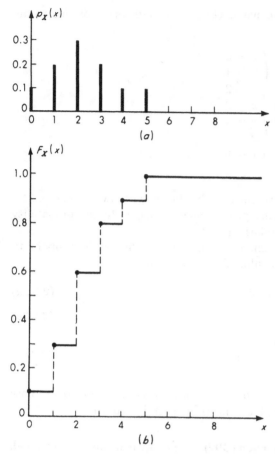

Fig. 2.2.1 Probability law for traffic illustration. (*a*) Probability mass function PMF; (*b*) cumulative distribution function CDF.

The CDF of the random variable X, the number of stopped cars, with the PMF described on page 73, is a step function:

$$F_X(x) = \begin{cases} 0 \\ 0.1 \\ 0.3 \\ 0.6 \\ 0.8 \\ 0.9 \\ 1.0 \end{cases} \quad \text{for} \quad \begin{cases} x < 0 \\ 0 \leq x < 1 \\ 1 \leq x < 2 \\ 2 \leq x < 3 \\ 3 \leq x < 4 \\ 4 \leq x < 5 \\ 5 \leq x \end{cases}$$

Although clumsy to specify analytically, such functions are easy to visualize. This discontinuous function is graphed in Fig. 2.2.1*b*. One would read from it, for example, that the probability of finding a line containing two or fewer vehicles is $F_X(2) = 0.6$ [which equals $p_X(0) + p_X(1) + p_X(2) = 0.1 + 0.2 + 0.3$] or that $F_X(3) = 0.8$ [or $F_X(2) + p_X(3) = 0.6 + 0.2$].

The PMF can always be recovered if the CDF is given, since the former simply describes the magnitudes of the individual steps in the CDF. Formally,

$$p_X(x_i) = F_X(x_i) - F_X(x_i - \epsilon)$$

where ϵ is a small positive number.

Continuous random variable and the PDF　Although the discrete random variable is appropriate in many situations (particularly where items such as vehicles are being counted), the *continuous random variable* is more frequently adopted as the mathematical model for physical phenomena of interest to civil engineering. Unlike the discrete variable, the continuous random variable is free to take on any value on the real axis.†　Strictly speaking, one must be extremely careful in extending the ideas of sample spaces to the continuous case, but conceptually the engineer should find the continuous random variable more natural than the discrete. All the engineer's physical variables—length, mass, and time—are usually dealt with as continuous quantities. A flow rate might be 1000, 1001, 1001.1, or 1001.12 cfs. Only the inadequacies of particular measuring devices can lead to the rounding off that causes the measured values of such quantities to be limited to a set of discrete values.

The problem of specifying the probability distribution (and hence the probability law) of a continuous random variable X is easily managed. If the x axis is separated into a large enough number of short intervals each of infinitesimal length dx, it seems plausible that we can define a function $f_X(x)$ such that the probability that X is in interval x to $x + dx$ is $f_X(x)\,dx$. Such a function is called the *probability density function*, or PDF, of a continuous random variable.

Since occurrences in different intervals are mutually exclusive events, it follows that the probability that a random variable takes on a value in an interval of finite length is the "sum" of probabilities or the integral of $f_X(x)\,dx$ over the interval. Thus the area under the PDF in

---

† As will be seen, this does not imply that the random variable *must* take on values over the entire axis. Intervals, such as the negative range, for example, can be excluded (i.e., assigned zero probability).

an interval represents the probability that the random variable will take on a value in that interval

$$P[x_1 \leq X \leq x_2] = \int_{x_1}^{x_2} f_X(x)\, dx \tag{2.2.5}$$

The probability that a continuous random variable X takes on a specific value x is zero, since the length of the interval has vanished. The value of $f_X(x)$ is *not* itself a probability; it is only the measure of the density or intensity of probability at the point. It follows that $f_X(x)$ need not be restricted to values less than 1, but two conditions must hold:

$$f_X(x) \geq 0 \tag{2.2.6}$$

$$\int_{-\infty}^{\infty} f_X(x)\, dx = 1 \tag{2.2.7}$$

These properties can be verified by inspection for the following example.

For illustration of the PDF see Fig. 2.2.2a. Here the engineer has called the yield stress of a standard tensile specimen of A36 steel a random variable Y. He has assigned this variable the triangular probability density function. The bases of the assumption of this particular form are observed experimental data and the simplicity of its shape. Its *range* (35 to 55 ksi) and *mode* (41 ksi) define a triangular PDF inasmuch as the area must be unity. Although simple in shape, this function is somewhat awkward mathematically

$$f_Y(y) = \begin{cases} \dfrac{2}{55-35}\dfrac{y-35}{41-35} & 35 \leq y \leq 41 \\[2ex] \dfrac{2}{55-35}\left(1 - \dfrac{y-41}{55-41}\right) & 41 \leq y \leq 55 \\[2ex] 0 & \text{elsewhere}\dagger \end{cases}$$

The shaded area between y_2 and y_3 represents the probability that the yield strength will lie in this range, and the shaded region from $y = 35$ to y_1 is equal in area to the probability that the yield strength is less than y_1. For the values of y_1, y_2, and y_3 in the ranges shown in Fig. 2.2.2a,

$$P[Y \leq y_1] = \tfrac{2}{120} \int_{35}^{y_1} (y-35)\, dy = \frac{(y_1-35)^2}{120} \qquad 35 \leq y_1 \leq 41$$

$$P[y_2 \leq Y \leq y_3] = \tfrac{2}{280} \int_{y_2}^{y_3} (55-y)\, dy$$

$$= \frac{55(y_3-y_2) - \tfrac{1}{2}(y_3{}^2 - y_2{}^2)}{140} \qquad 41 \leq y_2 \leq y_3 \leq 55$$

† This line will usually be omitted, the PDF (or PMF or CDF) being tacitly defined as equal to zero in regions where it is not specifically defined (except that the CDF is 1 for values of the argument *larger* than the indicated range).

Continuous random variable and the CDF Again the cumulative distribution function, or CDF, is an alternate form by which to describe the probability distribution of a random variable. Its definition is unchanged for a continuous random variable:

$$F_X(x) = P[X \leq x] \tag{2.2.8}$$

The right-hand side of this equation may be written $P[-\infty \leq X \leq x]$ and thus, for continuous random variables [by Eq. (2.2.5)],

$$F_X(x) = P[-\infty \leq X \leq x] = \int_{-\infty}^{x} f_X(u)\, du \tag{2.2.9}$$

where u has been used as the dummy variable of integration to avoid confusion with the limit of integration x [the argument of the function $F_X(x)$]. The CDF of the steel yield stress random variable is shown in Fig. 2.2.2b.

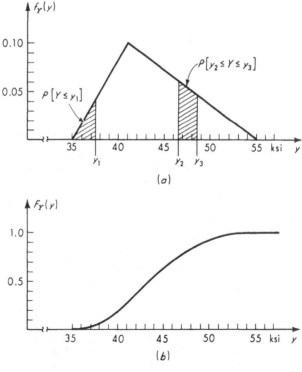

Fig. 2.2.2 Steel-yield-stress illustration. (a) Probability density function; (b) cumulative distribution function.

In addition, the PDF can be determined if the CDF is known, since $f_X(x)$ is simply the slope or derivative of $F_X(x)$:

$$\frac{dF_X(x)}{dx} = \frac{d}{dx}\left[\int_{-\infty}^{x} f_X(u)\ du\right] = f_X(x) \qquad (2.2.10)$$

It is sometimes desirable to use as models *mixed random variables,* which are a combination of the continuous and discrete variety. In this case one can always define a meaningful (discontinuous) CDF, but its derivative, a PDF, cannot be found without resort to such artifices as Dirac delta functions.† The mixed random variable will seldom be considered explicitly in this work, since an understanding of the discrete and continuous variables is sufficient to permit the reader to deal with this hybrid form. One use is pictured in Fig. 2.2.5.

The cumulative distribution function of any type of random variable—discrete, continuous, or mixed—has certain easily verified properties which follow from its definition and from the properties of probabilities:‡

$$0 \le F_X(x) \le 1 \qquad (2.2.11)$$

$$F_X(-\infty) = 0 \qquad (2.2.12)$$

$$F_X(\infty) = 1 \qquad (2.2.13)$$

$$F_X(x + \epsilon) \ge F_X(x) \text{ for any positive } \epsilon \qquad (2.2.14)$$

$$F_X(x_2) - F_X(x_1) = P[x_1 < X \le x_2] \qquad (2.2.15)$$

Equation 2.2.14 implies that the CDF is a function which is monotonic and nondecreasing (it may be flat in some regions). Cumulative distribution functions for discrete, continuous, and mixed random variables are illustrated in Figs. 2.2.3 to 2.2.5.

Histograms and probability distribution models   Although they may be similar in appearance, the distinction between histograms (Chap. 1) and density functions (Chap. 2) and the distinction between cumulative frequency polygons and cumulative distribution functions must be well understood. The figures presented in Chap. 1 are representations of

---

† These functions are zero everywhere except at a single point where they are infinite, although the integral under the "curve" is finite. They are analogous to the concentrated load, pure impulse, or infinite source employed elsewhere in civil engineering.
‡ Notice that in general some care must be taken with the inequalities. If a jump does not occur in $F_X(x)$ at x, then the inequalities ($\le$ and $\ge$) and strict inequalities ($<$ and $>$) may be interchanged freely without altering the value of the right-hand side of the equation. This is always the case if X is a continuous random variable.

observed empirical data; the functions defined here are descriptions of the probability laws of mathematical variables.

The histogram in Fig. 2.2.6, for example, might represent the observed annual runoff data from the watershed of a particular stream. [The first bar includes six observations (years) in which the runoff was so small as not to be measurable.] In constructing a mathematical model of a river basin, the engineer would have use for the stream information represented here. Letting the random variable X represent the annual runoff of this area, the engineer can construct any number of plausible mathematical models of this phenomenon. In particular, any one of the probability laws pictured in Figs. 2.2.3 to 2.2.5 might be adopted.

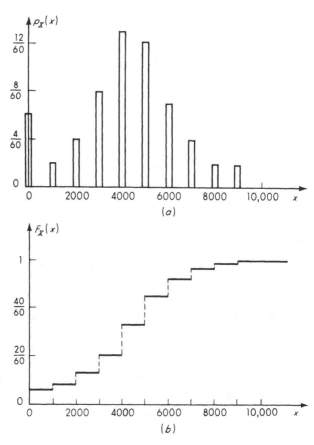

Fig. 2.2.3 Discrete random variable model reproducing histogram of Fig. 2.2.6. (*a*) Probability mass function; (*b*) cumulative distribution function.

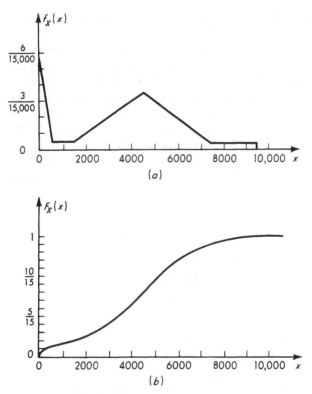

Fig. 2.2.4 Continuous random variable approximately modeling histogram of Fig. 2.2.6. (*a*) Probability density function; (*b*) cumulative distribution function.

The first model, Fig. 2.2.3, within the restriction of a discrete random variable, reproduces exactly the frequencies reported in the histogram. The engineer may have no reason to alter his assigned probabilities from the observed relative frequencies, even though another sequence of observations would change these frequencies to some degree. The second model, Fig. 2.2.4, enjoys the computational convenience frequently associated with continuous random variables. In assigning probabilities to the mathematical model, the observed frequencies have been smoothed to a series of straight lines to facilitate their description and use. A third possible model, Fig. 2.2.5, employs, like the second model, a continuous random variable to describe the continuous spectrum of physically possible values of runoff, but also accounts explicitly for the important possibility that the runoff is *exactly* zero. Over the continuous range the density function has been given a smooth, easily described curve whose general mathematical form may be determined by arguments† about the physical process leading to the runoff.

Unfortunately it is not possible to state in general which is the "best" mathematical model of the physical phenomenon. The questions of constructing models and the relationship of observed data to such models will be discussed throughout this work. Commonly used models will be discussed in Chap. 3. The use of data to estimate the parameters

---

† Additional comment is made in Secs. 2.3.3, 3.3.1, and 3.5.3 on this particular runoff problem.

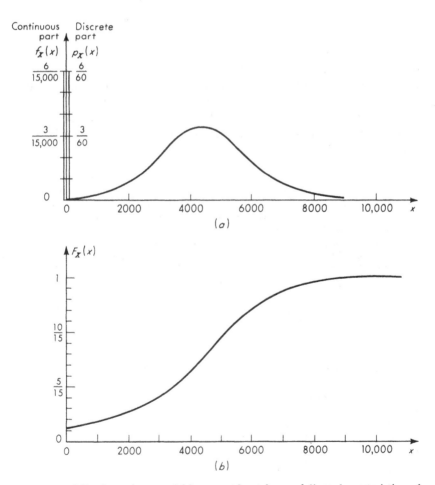

Fig. 2.2.5 Mixed random variable approximately modeling characteristics of histogram of Fig. 2.2.6. (a) Graphical description of a mixed probability law; (b) cumulative distribution function.

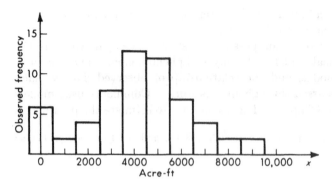

Fig. 2.2.6 Histogram of observed annual runoff (60-year record).

of the model will be considered in Chap. 4, where we also discuss techniques for choosing and evaluating models when sufficient data is available.

Often distributions of interest are developed from assumptions about underlying components or "mechanisms" of the phenomenon. Two examples follow. A third illustration draws the PDF from a comparison with data.

Illustration: Load location An engineer concerned with the forces caused by arbitrarily located concentrated loads on floor systems might be interested in the distribution of the distance X from the load to the nearest edge support. He assumes that the load will be located "at random," implying here that the probability that the load lies in any region of the floor is proportional only to the area of that region. He is considering a square bay $2a$ by $2a$ in size.

From this assumption about the location of the load, we can conclude (see Fig. 2.2.7) that

$$F_X(x) = P[X \le x] = \frac{\text{shaded area}}{\text{total area}} = \frac{4a^2 - (2a - 2x)^2}{4a^2}$$

$$= 1 - \frac{(a - x)^2}{a^2} \qquad 0 \le x \le a \tag{2.2.16}$$

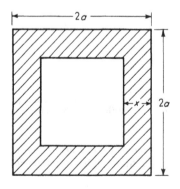

Fig. 2.2.7 Floor-system illustration. Shaded region is area in which distance to nearest edge is equal to or less than x.

The density function is thus

$$f_X(x) = \frac{dF_X(x)}{dx} = \frac{2(a-x)}{a^2} \qquad 0 \leq x \leq a \tag{2.2.17}$$

which is simply a triangle. That this is a proper probability distribution is verified by noting that at $x = 0$, $F_X(0) = 0$, and at $x = a$, $F_X(a) = 1$.

Illustration: Quality control Specification limits on materials (e.g., concrete, asphalt, soil, etc.) are often written recognizing that there is a small, "acceptable" probability p that an individual specimen will fail to meet the limit even though the batch is satisfactory. As a result more than one specimen may be called for when controlling the quality of the material. What is the probability mass function of N, the number of specimens which will fail to meet the specifications in a sample of size three when the material is satisfactory? The probability that any specimen is unsatisfactory is p.

   Assuming independence of the specimens,

$$p_N(0) = P[N = 0] = P[\text{all specimens are satisfactory}] = (1 - p)^3$$

$$p_N(3) = P[\text{all specimens are unsatisfactory}] = p^3$$

$$p_N(1) = P[\text{one specimen is unsatisfactory}] = 3p(1 - p)^2$$

This last expression follows from the fact that any one of the *three* sequences $\{s,s,u\}$, $\{s,u,s\}$, $\{u,s,s\}$ (where s indicates a satisfactory specimen and u an unsatisfactory one) will lead to a value of N equal to 1, and each sequence has probability of occurrence $p(1 - p)^2$. Similarly,

$$p_N(2) = 3p^2(1 - p)$$

These four terms can be expressed as a function:

$$p_N(n) = \frac{3!}{n!(3-n)!}p^n(1-p)^{3-n} \qquad n = 0, 1, 2, 3 \tag{2.2.18}$$

This function is plotted in Fig. 2.2.8 for several values of the *parameter p*. (We will study distributions of this general form in Sec. 3.1.2.)

   That the PMF is proper for any value of p can be verified by expanding the individual terms and adding them together by like powers of p. The sum is unity.

   In quality-control practice, of course, the engineer must make a decision based on an observation of, say, two bad specimens in a sample size of three about whether the material meets specifications or not. If, under the assumption that the material is satisfactory, the likelihood of such an event is, in fact, calculated to be very small, the engineer will usually decide (i.e., act as if) the material is *not* satisfactory.

Illustration: Annual maximum wind velocity A structural engineer is interested in the design of a tall tower for wind loads. He obtains data for a number of years of the maximum annual wind velocity near the site and finds that when a histogram of the data is plotted, it is satisfactorily modeled from a probability viewpoint by a continuous probability distribution of the negative

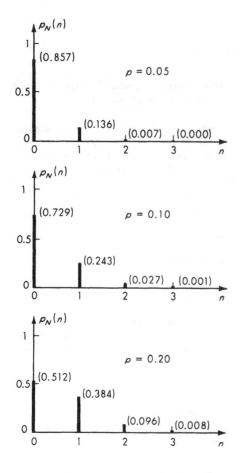

Fig. 2.2.8 Quality-control illustration. Plots of

$$p_N(n) = \frac{3!}{n!(3-n)!}\, p^n(1-p)^{3-n}.$$

exponential form.† If X is maximum annual wind velocity, the PDF of X is of the form:

$$f_X(x) = ke^{-\lambda x} \qquad x \geq 0$$

where k is a constant which can be found by recognizing that the integral of $f_X(x)$ over 0 to ∞ must equal unity. Hence

$$\int_0^\infty ke^{-\lambda x}\, dx = \frac{-k}{\lambda}\, e^{-\lambda x}\, \Big|_0^\infty = \frac{k}{\lambda} = 1$$

or

$$k = \lambda$$

yielding

$$f_X(x) = \lambda e^{-\lambda x} \qquad x \geq 0 \tag{2.2.19}$$

† A more commonly adopted distribution for maximum wind velocities will be discussed in Sec. 3.3.3.

The CDF is found by integration:

$$F_X(x) = \int_0^x f_X(u) \, du = \int_0^x \lambda e^{-\lambda u} \, du$$

$$= -e^{-\lambda u} \Big|_0^x = 1 - e^{-\lambda x} \qquad x \geq 0 \qquad\qquad (2.2.20)$$

The record shows that the probability of maximum annual wind velocities less than 70 mph is approximately 0.9. This estimate affords an estimate of the parameter λ. (Other methods for parameter estimation are discussed in Chap. 4.)

$$P[0 \leq X \leq 70] = 0.9$$
$$0.9 = 1 - e^{-\lambda 70}$$
$$e^{-70\lambda} = 0.1$$
$$-70\lambda = -2.3$$
$$\lambda = 0.033$$

Then,

$$f_X(x) = 0.033 \, e^{-0.033x} \qquad\qquad (2.2.21)$$
$$F_X(x) = 1 - e^{-0.033x} \qquad\qquad (2.2.22)$$

The PDF and CDF of X are shown in Fig. 2.2.9.

One minus $F_X(x)$ is important because design decisions are based on the probability of large wind velocities. Define† the *complementary distribution function* as

$$G_X(x) = 1 - F_X(x)$$
$$G_X(x) = e^{-0.033x} \qquad\qquad (2.2.23)$$

$G_X(x)$ is the probability of finding the maximum wind velocity in any year greater than x. The probability of a maximum annual wind velocity between 35 and 70 mph is indicated on the PDF of Fig. 2.2.9 along with the probability of a maximum annual wind velocity equal to or greater than 140 mph. Equation (2.2.23) can be used to determine numerical values of these probabilities. Their use in engineering design will be discussed in Secs. 3.1 and 3.3.3.

2.2.2 Jointly Distributed Random Variables

In Sec. 2.1 we discussed examples, such as counting cars and counting trucks, in which two-dimensional sample spaces are involved. When two or more random variables are being considered simultaneously, their joint behavior is determined by a *joint probability* law, which can in turn be described by a *joint cumulative distribution function*. Also, if both random variables are discrete, a *joint probability mass function* can be used to describe their governing law, and if both variables are continuous, a *joint probability density function* is applicable. Mixed joint distributions are also encountered in practice, but they require no new techniques.

† Note that for continuous random variables, $G_X(x) = P[X \geq x]$ and for discrete random variables, $G_X(x) = P[X > x]$.

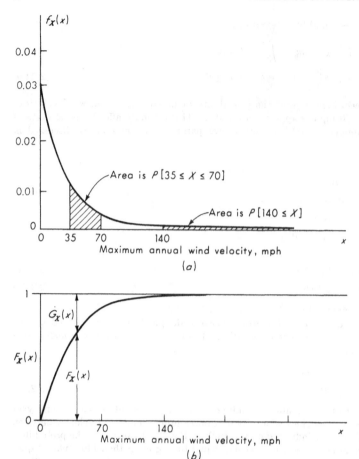

Fig. 2.2.9 Wind-velocity illustration. (a) PDF of X and (b) CDF of X (X is maximum annual wind velocity).

Joint PMF   The joint probability mass function $p_{X,Y}(x,y)$ of two discrete random variables X and Y is defined as

$$p_{X,Y}(x,y) \equiv P[(X = x) \cap (Y = y)] \qquad (2.2.24)$$

It can be plotted in a three-dimensional form analogous to the two-dimensional PMF of a single random variable. The joint cumulative distribution function is defined as

$$F_{X,Y}(x,y) \equiv P[(X \leq x) \cap (Y \leq y)] = \sum_{x_i \leq x} \sum_{y_i \leq y} p_{X,Y}(x_i, y_j) \quad (2.2.25)$$

Consider an example of two discrete variables whose joint behavior must be dealt with. One, X, is the random number of vehicles passing a point in a 30-sec time interval. Variability in traffic flow is the cause of delays and congestion. The other random variable is Y, the number of vehicles in the *same* 30-sec interval actually recorded by a particular, imperfect traffic counter. This device responds to pressure on a cable placed across one or more traffic lanes, and it records the total number of such pressure applications during successive 30-sec intervals. It is used by the traffic engineer to estimate the number of vehicles which have used the road during given time intervals. Owing, however, to dynamic effects (causing wheels to be off the ground) and to mechanical inability of the counter to respond to all pulses, the actual number of vehicles X and the recorded number of vehicles Y are not always in agreement. Data were gathered in order to determine the nature and magnitude of this lack of reliability in the counter. Simultaneous observations of X and Y in many different 30-sec intervals led to a scattergram (Sec. 1.3). The engineer adopted directly the observed relative frequencies as probability assignments in his mathematical model,† yielding the joint probability mass function graphed in Fig. 2.2.10. Notice by the strong

† As mentioned in Sec. 2.2.1, the reader should remember that the engineer's mathematical model is conceptually different in kind from observed data, histograms, scattergrams, observed relative frequencies, and the like. In situations such as this, however, it may be quite reasonable simply to adopt a model with probability assignments equal in numerical value to previously observed relative frequencies.

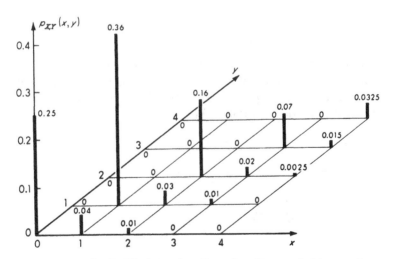

Fig. 2.2.10 Joint PMF of actual traffic and traffic recorded by counter.

probability masses on the diagonal (0, 0; 1, 1; etc.) that the counter is usually correct. Note too that it is not possible for Y to take on a value greater than X in this example. The joint CDF could also be plotted, appearing always something like an exterior corner of an irregular staircase, but it seldom proves useful to do so.

The probability of any event of interest is found by determining the pairs of values of X and Y which lead to this event and then summing over all such pairs. For example, the probability of C, the event that an arbitrary count is *not* in error, is

$$P[C] = P[Y = X] = \sum_{\text{all } x_i} P[(X = x_i) \cap (Y = x_i)]$$

$$= \sum_{\text{all } x_i} p_{X,Y}(x_i, x_i)$$

$$= p_{X,Y}(0,0) + p_{X,Y}(1,1) + p_{X,Y}(2,2) + p_{X,Y}(3,3) + p_{X,Y}(4,4)$$

$$= 0.25 + 0.36 + 0.16 + 0.07 + 0.0325$$

$$= 0.8725$$

The probability of an error by the counter is

$$P[C^c] = P[Y \neq X] = \sum_{\text{all } x_i} \sum_{\text{all } y_j \neq x_i} p_{X,Y}(x_i, y_j)$$

This probability is most easily calculated as

$$P[C^c] = 1 - P[C] = 1 - 0.8725 = 0.1275$$

since, to be properly defined,

$$\sum_{\text{all } x_i} \sum_{\text{all } y_j} p_{X,Y}(x_i, y_j) = 1 \qquad (2.2.26)$$

Marginal PMF   A number of functions related to the joint PMF are of value. The behavior of a particular variable irrespective of the other is described by the *marginal* PMF. It is found by summing over all values of the disregarded variable. Formally,

$$p_X(x) \equiv P[X = x] = \sum_{\text{all } y_i} p_{X,Y}(x, y_i) \qquad (2.2.27)$$

$$F_X(x) \equiv P[X \leq x] = \sum_{x_i \leq x} p_X(x_i)$$

$$= \sum_{x_i \leq x} \sum_{\text{all } y_j} p_{X,Y}(x_i, y_j)$$

$$= F_{X,Y}(x, \infty) \qquad (2.2.28)$$

Similar expressions hold for the marginal distribution of Y.

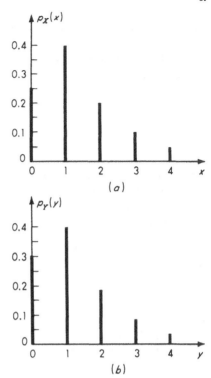

Fig. 2.2.11 (a) Marginal PMF of actual traffic X and (b) marginal PMF counter response Y.

In the example here the distribution of X, the actual number of cars, is found for each value x by summing all spikes in the y direction. For example,

$$p_X(3) = \sum_{y=0}^{4} p_{X,Y}(3,y) = 0.01 + 0.02 + 0.07 = 0.1$$

The marginal distributions of X and Y are plotted on Fig. 2.2.11. It should be pointed out that generally the marginal distributions are *not* sufficient to specify the joint behavior of the random variables. In this example, the joint PMF requires specification of $(5)(5) = 25$ numbers, while the two marginal distributions contain only $5 + 5 = 10$ pieces of information. The conditions under which the marginals *are* sufficient to define the joint will be discussed shortly.

Conditional PMF A second type of distribution which can be obtained from the joint distribution is also of interest. If the value of one of the variables is known, say $Y = y_0$, the relative likelihoods of the various values of the other variable are given by $p_{X,Y}(x,y_0)$. If these values are renormalized so that their sum is unity, they will form a proper distribution function. This distribution is called the *conditional probability mass function* of X given Y, $p_{X|Y}(x,y)$. The normalization is performed by dividing each of the values by their sum. For Y given equal to any particular value y,

$$p_{X|Y}(x,y) = P[X = x \mid Y = y] = \frac{P[(X = x) \cap (Y = y)]}{P[Y = y]}$$

$$= \frac{p_{X,Y}(x,y)}{\sum_{\text{all } x_i} p_{X,Y}(x_i,y)} = \frac{p_{X,Y}(x,y)}{p_Y(y)} \qquad (2.2.29)$$

(The function is undefined if $p_Y(y)$ equals zero.) Notice that the denominator is simply the marginal distribution of Y evaluated at the given value of Y. The conditional PMF of X is a proper distribution function in x, that is,

$$0 \leq p_{X|Y}(x,y) \leq 1 \qquad (2.2.30)$$

and

$$\sum_{\text{all } x_i} p_{X|Y}(x_i,y) = 1 \qquad (2.2.31)$$

The conditional distribution of Y given X is, of course, defined in a symmetrical way.

It is the relationship between the conditional distribution and the marginal distribution that determines how much an observation of one variable helps in the prediction of the other. In the traffic-counter example, our interest is in X, the actual number of cars which have passed in a particular interval. The marginal distribution of X is initially our best statement regarding the relative likelihoods of the various possible values of X. An observation of Y (the mechanically recorded number) should, however, alter these likelihoods. Suppose the counter reads $Y = 1$ in a particular interval. The actual number of cars is not known with certainty, but the relative likelihoods of different values are now given by $p_{X,Y}(x,1)$ or 0, 0.36, 0.03, 0.01, and 0 for x equal, to 0, 1, 2, 3, and 4, respectively. Normalized by their sum, these likelihoods become the conditional distribution $p_{X|Y}(x,1)$, which is plotted in Fig. 2.2.12. As expected, the probability is now high that X equals 1, but the imperfect nature of the counter does not permit this statement to be made with absolute certainty. The better the counter, the more peaked it will make this conditional distribution relative to the marginal distribution. In the words of Sec. 2.1, the more closely the measured quantity (here Y) is related or correlated with the quantity of interest (here X), the more "sharply" can we predict X given a measured value Y.

Other conditional distributions It is of course true that any number of potentially useful conditional distributions can be constructed from a

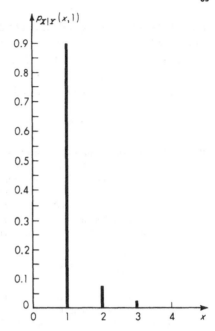

Fig. 2.2.12 Conditional PMF of traffic given that the counter reads unity.

joint probability law depending on the event conditioned on. Thus one might seek the distribution of X given that Y is greater than y:

$$p_{X|[Y>y]}(x,y) = P[X = x \mid Y > y]$$

$$= \frac{\sum_{y_i > y} p_{X,Y}(x,y_j)}{\sum_{y_i > y} p_Y(y_j)} \qquad (2.2.32)$$

or the conditional distribution of Y given that X was in a particular interval:

$$p_{Y|[a<X\leq b]}(y) = P[Y = y \mid a < X \leq b]$$

$$= \frac{\sum_{a<x_j\leq b} p_{X,Y}(x_j,y)}{\sum_{a<x_j\leq b} p_X(x_j)} \qquad (2.2.33)$$

We will encounter examples of such conditional distributions at various points in this text. Their treatment and meaning, however, is obvious once the notation is defined. In this notational form the fundamental conditional PMF [Eq. (2.2.29)] becomes

$$p_{X|Y}(x,y) = p_{X|[Y=y]}(x,y)$$

Joint PMF from marginal and conditional probabilities As mentioned before with regard to events, conditional probabilities are often more readily determined in practice than are the joint probabilities. Together with one of the marginal PMF's a conditional PMF can be used to compute the joint PMF, since, from Eq. (2.2.29),

$$p_{X,Y}(x,y) = p_{X|Y}(x,y)p_Y(y) = p_{Y|X}(y,x)p_X(x) \qquad (2.2.34)$$

Illustration: Analytical approach to traffic-counter model We shall illustrate the use of the previous equation by showing that the joint distribution in the traffic-counter example might have been determined without data by constructing a mathematical model of the probabilistic mechanism generating the randomness. A commonly used model of traffic flow (to be discussed in Sec. 3.2.1) suggests that the number of cars passing a fixed point in a given interval of time has a discrete distribution of the mathematical form

$$p_X(x) = \frac{\nu^x e^{-\nu}}{x!} \qquad x = 0, 1, 2, \ldots, \infty \qquad (2.2.35)$$

where ν is the average number of cars in all such intervals. If it is assumed, too, that each vehicle is recorded only with probability p, then in Sec. 3.1.2 we shall learn that the *conditional* distribution of Y given $X = x$ must be

$$p_{Y|X}(x,y) = \frac{x!}{y!(x - y)!} [p^y(1 - p)^{x-y}] \qquad y = 0, 1, \ldots, x \qquad (2.2.36)$$

[The argument is a generalization of that which led to Eq. (2.2.18).] Hence we need data or information sufficient only to estimate ν, which is related to the average flow, and p, the unreliability of the counter, rather than all the values in the joint PMF. This is true because the joint PMF follows from the marginal and conditional above:

$$p_{X,Y}(x,y) = p_{Y|X}(y,x)p_X(x) = \frac{x!}{y!(x - y)!} [p^y(1 - p)^{x-y}] \frac{\nu^x e^{-\nu}}{x!}$$

$$= \frac{\nu^x e^{-\nu}}{y!(x - y)!} [p^y(1 - p)^{x-y}] \qquad x = 0, 1, 2 \ldots, \infty;$$

$$y = 0, 1, 2, \ldots, x \qquad (2.2.37)$$

Without being able to argue through the conditional PMF, it would have been extremely difficult for the engineer to derive this complicated joint PMF directly. For $\nu = 1.3$ cars and $p = 0.9$, its shape is very nearly that given in Fig. 2.2.11, although the sample space now extends beyond 4 to infinity. It is not expected or important at this stage that the reader absorb the details of this model. It is brought up here to illustrate the utility of the conditional distribution and to display the "more mathematical" form of some commonly encountered discrete distributions. It points out that, if the physical process is well understood, the engineer may attempt to construct a reasonable model of the generating mechanism involving only a few parameters (here ν and p), rather than rely on observed data to provide estimates of all the probability values of the distribution function.

Joint PDF and CDF   The functions associated with jointly distributed continuous random variables are totally analogous with those of discrete variables, but with density functions replacing mass functions.   Using the same type of argument employed in Sec. 2.2.1, the probability that X lies in the interval $\{x, x + dx\}$ and Y lies in the interval $\{y, y + dy\}$ is $f_{X,Y}(x,y)\, dx\, dy$.   This function $f_{X,Y}(x,y)$ is called the *joint probability density function*.

The probability of the joint occurrence of X and Y in some region in the sample space is determined by integration of the joint PDF over that region.   For example,

$$P[(x_1 \leq X \leq x_2) \text{ and } (y_1 \leq Y \leq y_2)]$$
$$= \int_{x_1}^{x_2} \int_{y_1}^{y_2} f_{X,Y}(x,y)\, dy\, dx \quad (2.2.38)$$

This is simply the volume under the function $f_{X,Y}(x,y)$ over the region. If the region is not a simple rectangle, the integral may become more difficult to evaluate, but the problem is no longer one of probability but calculus (and all its techniques to simplify integration, such as change of variables, are appropriate).

Clearly the joint PDF must satisfy the conditions

$$f_{X,Y}(x,y) \geq 0 \quad\quad\quad (2.2.39)$$

$$\int\!\!\int_{-\infty}^{+\infty} f_{X,Y}(x,y)\, dx\, dy = 1 \quad\quad\quad (2.2.40)$$

The joint cumulative distribution function is defined as before and can be computed from the density function by applying Eq. (2.2.38). Thus

$$F_{X,Y}(x,y) \equiv P[(X \leq x) \text{ and } (Y \leq y)]$$
$$= P[(-\infty \leq X \leq x) \text{ and } (-\infty \leq Y \leq y)]$$
$$= \int_{-\infty}^{x} \int_{-\infty}^{y} f_{X,Y}(x_0,y_0)\, dy_0\, dx_0 \quad (2.2.41)$$

in which dummy variables of integration, x_0 and y_0, have been used to emphasize that the arguments of the CDF, x and y, appear in the limits of the integral.   The properties of the function are analogous to those given in Eqs. (2.2.11) to (2.2.15) for the CDF of a single variable.

It should not be unexpected that, as in Sec. 2.2.1, the density function is a derivative of the cumulative function, now a partial derivative,

$$f_{X,Y}(x,y) = \frac{\partial^2}{\partial x\, \partial y} F_{X,Y}(x,y) \quad\quad\quad (2.2.42)$$

Marginal PDF As with the discrete random variables, one frequently has need for marginal distributions. To eliminate consideration of Y in studying the behavior of X, one need only integrate the joint density function over all values of Y and determine the *marginal PDF* of X, $f_X(x)$:

$$f_X(x) \equiv \int_{-\infty}^{+\infty} f_{X,Y}(x,y)\, dy \qquad\qquad (2.2.43)$$

The marginal cumulative distribution function of X, $F_X(x)$, is, consequently,

$$F_X(x) = P[X \le x] = \int_{-\infty}^{x} f_X(x_0)\, dx_0$$

or

$$F_X(x) = F_{X,Y}(x, \infty) \qquad\qquad (2.2.44)$$

which implies

$$f_X(x) = \frac{\partial}{\partial x} F_{X,Y}(x, \infty)$$

Symmetrical results hold for the marginal distribution of Y.

Conditional PDF If one is given the value of one variable, $Y = y_0$, say, the relative likelihood of X taking a value in the interval x, $x + dx$ is $f_{X,Y}(x,y_0)\, dx$. To yield a proper density function (i.e., one whose integral over all values of x is unity) these values must be renormalized by dividing them by their sum, which is

$$\int_{-\infty}^{+\infty} f_{X,Y}(x,y_0)\, dx = f_Y(y_0)$$

In this manner we are led, plausibly if not rigorously,† to the definition of the conditional PDF of X given Y as

$$f_{X|Y}(x,y) = \frac{f_{X,Y}(x,y)}{f_Y(y)} \qquad\qquad (2.2.45a)$$

The conditional cumulative distribution is

$$F_{X|Y}(x,y) = P[X \le x \mid Y = y] = \int_{-\infty}^{x} f_{X|Y}(u,y)\, du \qquad (2.2.45b)$$

As with discrete variables, distributions based on other conditioning events, for example, $f_{X|[Y>y]}(x,y)$, may also prove useful. Their definitions and interpretations should be obvious.

Sketches of some joint PDF's appear in Fig. 2.2.13 in the form of contours of equal values. Graphical interpretations of marginal and conditional density functions also appear there.

---

† The theoretical difficulties here revolve around the fact that the probability that Y takes on any specific value is zero. See, for example, Parzen [1960].

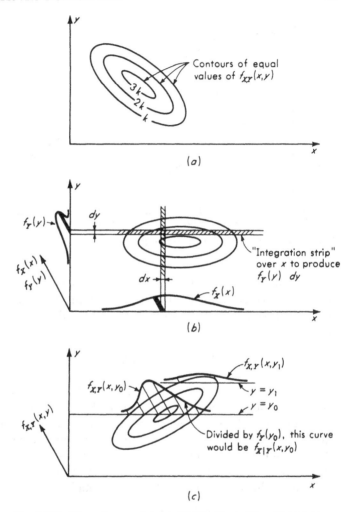

Fig. 2.2.13 Three types of joint PDF's illustrating the relationships between the two random variables. (*a*) Contours (higher
values of X usually occur with lower values of Y); (*b*) marginal
PDF's (higher values of X occur about equally with higher and
lower values of Y); (*c*) cross sections at $y = y_0$ and $y = y_1$ (higher
values of X usually occur with higher values of Y).

As an illustration of joint continuous random variables, consider
the flows X and Y in two different streams in the same day. Interest in
their joint probabilistic behavior might arise because the streams feed
the same reservoir. If, as in the case in Fig. 2.2.13*c*, high flows in one
stream are likely to occur simultaneously with high flows in the other,

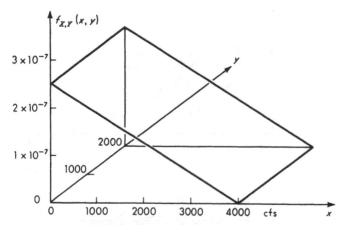

Fig. 2.2.14 Joint PDF of two stream flows.

their joint influence on the reservoir will not be the same as it would if
this were not the case. Assume that the joint distribution in this simplified illustration is that shown in Fig. 2.2.14. By inspection, its equation
is

$$f_{X,Y}(x,y) = C\,\frac{4000 - x}{4000} \qquad \begin{array}{l} 0 \le x \le 4000 \\ 0 \le y \le 2000 \end{array}$$

The constant, $C = 2.5 \times 10^{-7}$, was evaluated after the shape was fixed
so that the integral over the entire sample space would be unity.

The probability of such events as "the flow X is more than twice as
great as the flow Y" can be found by integration over the proper region.
The portion of the sample space where this is true is shown shaded in
Fig. 2.2.15a. The probability of this event is the volume under the
PDF over this region, Fig. 2.2.15b. Formally,

$$P[X \ge 2Y] = \int_0^{4000} \int_0^{0.5x} \frac{C}{4000}\,(4000 - x)\,dy\,dx$$

By carrying out the integrations or by inspection of the volume, knowing
it to be ($\frac{1}{3}$) (area of base) (height):

$$P[X \ge 2Y] = \frac{1}{3}\,\frac{(2000)(4000)}{2}\,(C) = \frac{1}{3}$$

The marginal distributions of X and Y are formally

$$f_X(x) = \int_0^{2000} C\,\frac{4000 - x}{4000}\,dy$$

$$= 2000C\,\frac{4000 - x}{4000} \qquad 0 \le x \le 4000$$

and

$$f_Y(y) = \int_0^{4000} C\,\frac{4000 - x}{4000}\,dx = 2000C \qquad 0 \le y \le 2000$$

These functions are plotted in Fig. 2.2.16. Their shapes could have been anticipated by inspection of the joint PDF.

Without further information, prediction of the flow X must be based on $f_X(x)$. It might be, however, that the engineer has in mind using an observation of Y (the flow in the "smaller" stream) to predict X more "accurately," just as the traffic engineer used an inexpensive, but unreliable, mechanical counter to provide a sharper knowledge of the number of vehicles which passed. As we have seen, the relationship between the conditional distribution function and the marginal distribution provides a measure of the information about X added by the

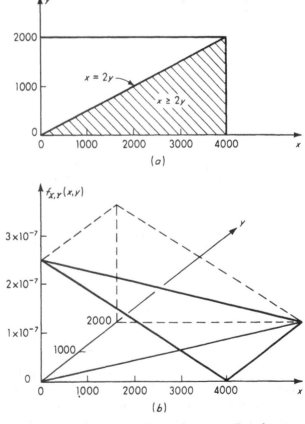

Fig. 2.2.15 Stream-flow illustration. (a) Sample space showing the event $X \ge 2Y$; (b) volume equal to $P[X \ge 2Y]$

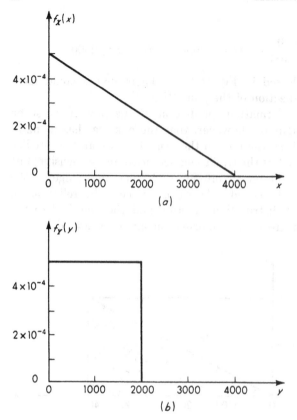

Fig. 2.2.16 Marginal distributions of stream flow from joint PDF of Fig. 2.2.14. (a) Marginal PDF of X; (b) Marginal PDF of Y.

knowledge of Y. The conditional PDF of X given Y is

$$f_{X|Y}(x,y) = \frac{f_{X,Y}(x,y)}{f_Y(y)}$$

$$= \frac{C(4000 - x)/4000}{2000C}$$

$$= \frac{1}{2000}\frac{4000 - x}{4000} \qquad 0 \leq x \leq 4000$$

Notice that this is unchanged from the marginal distribution $f_X(x)$. The knowledge of Y has *not* altered the distribution of X and hence has not provided the engineer any new information in his quest to predict or describe the behavior of X. This case is an example of a very important notion.

Independent random variables   In general, if the conditional distribution $f_{X|Y}(x,y)$ is identical to the marginal distribution $f_X(x)$, X and Y are said to be (*stochastically*) *independent random variables*. Similarly, for discrete variables, if, for all values of y

$$p_{X|Y}(x,y) = p_X(x) \tag{2.2.46}$$

then X and Y are independent.   The notion of independence of random variables is directly parallel to that of independence of events, for recall that Eq. (2.2.46) can be written in terms of probabilities of events as

$$P[X = x \mid Y = y] = P[X = x] \tag{2.2.47}$$

If the random variables X and Y are independent, events related to X are independent of those related to Y.   As a result of the definition above, the following statements hold if X and Y are independent and continuous (analogous relationships for discrete or mixed random variables can be derived):

$$f_{X|Y}(x,y) = f_X(x) \tag{2.2.48}$$

$$f_{Y|X}(y,x) = f_Y(y) \tag{2.2.49}$$

$$f_{X,Y}(x,y) = f_X(x)f_Y(y) \tag{2.2.50}$$

$$F_{X,Y}(x,y) = F_X(x)F_Y(y) \tag{2.2.51}$$

$$F_{X|Y}(x,y) = F_X(x) \tag{2.2.52}$$

As with events, in engineering practice independence of two or more random variables is usually a property *attributed* to them *by the engineer* because he thinks they are unrelated.   This assumption permits him to determine joint distribution functions from only the marginals [Eq. (2.2.50)].   In general this is not possible, and both a marginal and a conditional are required.

The concept of probabilistic independence is central to the successful application of probability theory.   From a purely practical point of view, the analysis of many probabilistic models would become hopelessly complex if the engineer were unwilling to adopt the assumption of independence of certain random variables in a number of key situations. Many examples of assumed independence of random variables will be found in this text.

Three or more random variables   Attention has been focused in this section on cases involving only two random variables but, at least in theory, the extensions of these notions to any number of jointly distributed random variables should not be difficult.   In fact, however, the calculus—the partial differentiation and multiple integrations over bounded regions—may become unwieldy. Many new combinations of functions become possible when more variables are considered. A few should be mentioned for illustration. For example, with only three variables X, Y, Z with joint CDF $F_{X,Y,Z}(x,y,z)$ there might be interest in joint marginal CDF's, such as

$$F_{X,Y}(x,y) = F_{X,Y,Z}(x,y,\infty) \tag{2.2.53}$$

as well as in simple marginal CDF's, such as

$$F_X(x) = F_{X,Y,Z}(x,\infty,\infty) \tag{2.2.54}$$

Joint probability density functions are, as above, partial derivatives:†

$$f_{X,Y,Z}(x,y,z) = \frac{\partial^3}{\partial x\,\partial y\,\partial z} F_{X,Y,Z}(x,y,z) \tag{2.2.55}$$

$$f_{X,Y}(x,y) = \frac{\partial^2}{\partial x\,\partial y} F_{X,Y,Z}(x,y,\infty) \tag{2.2.56}$$

$$f_X(x) = \frac{\partial}{\partial x} F_{X,Y,Z}(x,\infty,\infty) \tag{2.2.57}$$

Joint conditional PDF's are also now possible, and they are defined as might be expected:

$$f_{X,Y|Z}(x,y,z) = \frac{f_{X,Y,Z}(x,y,z)}{f_Z(z)} \tag{2.2.58}$$

The simple conditional PDF follows this pattern:

$$f_{X|Y,Z}(x,y,z) = \frac{f_{X,Y,Z}(x,y,z)}{f_{Y,Z}(y,z)} \tag{2.2.59}$$

If the random variables X, Y, and Z are *mutually independent*,‡

$$f_{X,Y,Z}(x,y,z) = f_X(x)f_Y(y)f_Z(z) \tag{2.2.60}$$

and conditional distributions reduce to marginal or joint PDF's. For example,

$$f_{X,Y|Z}(x,y,z) = f_{X,Y}(x,y) = f_X(x)f_Y(y) \tag{2.2.61}$$

Illustration: Reliability of a system subjected to n random demands In a capacity-demand situation such as those discussed in Sec. 2.1, it is often the case that the system is subjected to a succession of n demands (annual maximum flows,

---

† Only continuous random variables are illustrated here, but in fact continuous, discrete, and mixed random variables might well appear simultaneously as jointly distributed random variables.

‡ Although rare in practice, the reader should be aware that *pairwise independence* of random variables (for example, X and Y, X and Z, and Y and Z) does *not* necessarily imply *mutual independence* (X, Y, and Z). Problem 2.18 demonstrates the parallel fact for events.

or extreme winds, for example). Assuming that these random variables $D_1, D_2, \ldots, D_n$ are independent and identically distributed, and that the random capacity C is independent of all these demands, we can find the probability of the "failure" event, $A = $ [at least one of the n demands exceeds the capacity], as follows. Assume the random variables are all discrete.

$$P[A] = 1 - P[\text{all } D_i \leq C] \tag{2.2.62}$$

Expanding, using Eq. (2.1.12),

$$P[A] = 1 - \sum_{\text{all } c} P[\text{all } D_i \leq c \mid C = c]P[C = c] \tag{2.2.63}$$

Since capacity C is assumed independent of the D_i,

$$P[A] = 1 - \sum_{\text{all } c} P[\text{all } D_i \leq c]P[C = c] \tag{2.2.64}$$

Since the D_i are assumed mutually independent,

$$P[\text{all } D_i \leq c] = F_{D_1, D_2, \ldots, D_n}(c, c, \ldots, c)$$
$$= F_{D_1}(c)F_{D_2}(c) \cdots F_{D_n}(c)$$

and since the D_i are assumed identically distributed (say, all distributed like a random variable denoted simply D),

$$P[\text{all } D_i \leq c] = [F_D(c)]^n \tag{2.2.65}$$

Thus

$$P[A] = 1 - \sum_{\text{all } c} [F_D(c)]^n p_C(c) \tag{2.2.66}$$

For example, if the PMF's of D and C have values at the relative positions shown in Fig. 2.2.17, then

$$P[A] = 1 - (0)(p_2') - (1 - p_1 - p_1')^n(p_2) - (1 - p_1')^n(1 - p_2 - p_2') \tag{2.2.67}$$

Assuming that np_1 and np_1' are small compared with unity,

$$P[A] \approx 1 - (1 - np_1 - np_1')(p_2) - (1 - np_1')(1 - p_2 - p_2')$$
$$= np_1' + p_2' + np_1p_2 - np_1'p_2' \tag{2.2.68}$$

Assuming that p_2' is small compared with 1, $np_1'p_2'$ is small compared to np_1',

$$P[A] \approx np_1' + p_2' + np_1p_2 \tag{2.2.69}$$

Depending on the magnitudes of p_1p_2, p_1', and p_2', the last term may or may not be negligible.

The relative magnitudes and locations of the probability spikes in Fig. 2.2.17 are intended to be suggestive of the common design situation where c_3 is the "intended" capacity expected by the designer and c_1' is the anticipated or typical "design" demand. Failure of the system will occur during the design lifetime of n demands if any of the following happens:

1. A demand c_3', rare in magnitude or unanticipated in kind, occurs at some time during the lifetime.

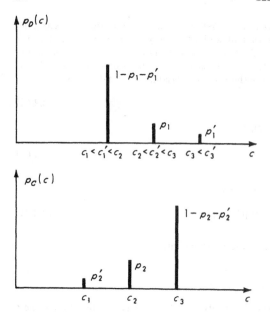

Fig. 2.2.17 Distributions of demand and capacity.
Positions of spikes on c axis are relative only.

2. An unusually low capacity c_1 is obtained either through construction inadequacy or through a gross misjudgement by the engineer as to the design capacity which would result from his specified design.
3. There exists a combination or joint occurrence of a moderately higher demand c_2' and a capacity of the only moderately lower value c_2.

After simplification and approximation these three events make their appearance as the three terms in Eq. (2.2.69).

Notice that in general the probability of a system failure increases with design life n; a system intended to last longer is exposed to a greater risk that one of the higher load values occurs in its lifetime. One term, p_2', does not grow with n, however, since, if this very low capacity is obtained, the system will fail with certainty under the first demand. If the capacity is not this low, further demands do not increase the likelihood of failure due to such a source.

2.3 DERIVED DISTRIBUTIONS

Much of science and engineering is based on functional relationships which predict the value of one (dependent) variable given any value of another (independent) variable. Static pressure is a function of fluid density; yield force, a function of cross-sectional area; and so forth. If, in a probabilistic formulation, the independent variable is considered to be a random variable, this randomness is imparted to those variables which are functionally dependent upon it. The purpose of this section is to develop methods of determining the probability law of functionally dependent random variables when the probability law of the independent† variable is known.

2.3.1 One-variable Transformations: $Y = g(X)$

The nature of such problems is best brought out through a simple, discrete example.

Solution by enumeration For several new high-speed ground-transportation systems it has been proposed to utilize small, separate vehicles that can be dispatched from one station to another, not on a fixed schedule, but whenever full. If the number of persons arriving in a specified time period at station A who want to go to station B is a random variable X, then Y, the number of vehicles dispatched from A to B in that period, is also a random variable, *functionally related* through the vehicles' capacities to X. For example, if each vehicle carries two persons, then $Y = 0$ if $X = 0$ or 1; $Y = 1$ if $X = 2$ or 3; $Y = 2$ if $X = 4$ or 5, etc. (assuming that no customer is waiting at the beginning of the time period). A graph of this functional relationship is shown in Fig. 2.3.1a.

The probability mass function of Y, $p_Y(y)$, can be determined from that of X by ennumeration of the values of X which lead to a particular value of Y. For example,

$$p_Y(0) = P[Y = 0] = P[(X = 0) \cup (X = 1)]$$
$$= p_X(0) + p_X(1)$$
$$p_Y(1) = P[(X = 2) \cup (X = 3)]$$
$$= p_X(2) + p_X(3)$$

Or, in general,

$$p_Y(y) = p_X(2y) + p_X(2y + 1) \qquad y = 0, 1, 2, 3, \ldots \qquad (2.3.1)$$

A numerical example is shown in Fig. 2.3.1b and c.

In this and in all derived distribution problems, it is even easier

---

† It is important not to confuse the notions of functional dependence (the more familiar) and stochastic or probabilistic dependence (Secs. 2.1.3 and 2.2.2). The first implies the second, but the converse is not true. In fact, functional dependence can be thought of as "perfect" stochastic dependence. If Y is *functionally* related to X by the equation $Y = g(X)$, the conditional distribution of Y given that $X = x$ is a unit mass at $y = g(x)$ and zero elsewhere (assuming $g(x)$ is single valued). That is, if there is functional dependence, the joint probability density contours (Fig. 2.2.13) "squeeze" together and merge into the line $y = g(x)$.

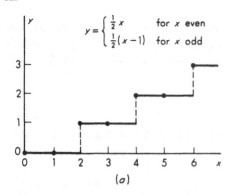

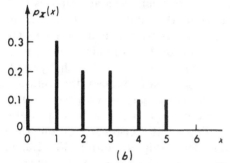

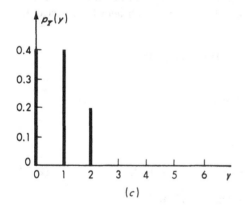

Fig. 2.3.1 High-speed transportation-system example. (*a*) Functional relationship between the number of arriving customers X and the number of dispatched vehicles Y; (*b*) PMF of number of arriving customers; (*c*) PMF of number of dispatched vehicles.

to obtain the cumulative distribution function $F_Y(y)$ from that of X. Thus

$$F_Y(0) = P[Y \le 0] = P[X \le 1] = F_X(1)$$
$$F_Y(1) = P[X \le 3] = F_X(3)$$

or

$$F_Y(y) = F_X(2y + 1) \qquad y = 0, 1, 2, 3, \ldots \qquad (2.3.2)$$

(It should be stated that this dispatching policy may not be a totally satisfactory one; it should perhaps be supplemented by a rule that dispatches half-full vehicles if the customer on board has been waiting more than a certain time.)

In this section we shall be concerned, then, with deriving the probability law of a random variable which is directly or functionally related to another random variable whose probability law is known. Owing to their importance in practice and in subsequent chapters, we shall concentrate on analytical solutions to problems involving continuous random variables. Problems involving discrete random variables can always be treated by enumeration as demonstrated above. In practical situations it may also prove advantageous to compute solutions to continuous problems in this same way, after having first approximated continuous random variables by discrete ones (Sec. 2.2.1).

One-to-one transformations For a commonly occurring class of problems it is possible to develop simple, explicit formulas as follows. If the function $y = g(x)$ relating the two random variables is such that it always increases as x increases† and is such that there is only a single value of x for each value of y and vice versa,‡ then Y *is less than or equal to some value y_0 if and only if X is less than or equal to some value x_0,* namely, that value for which $y_0 = g(x_0)$. (See Fig. 2.3.2.)

† Such a function is said to be a "monotonically increasing" function.
‡ Such a function is said to express a "one-to-one" transformation between the variables x and y. Note, for example, that the relationship between X and Y in the preceding dispatching example is *not* one-to-one. $Y = 2$ could result from X equal to either 4 or 5.

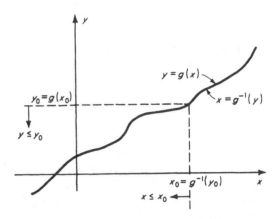

Fig. 2.3.2 A monotonically increasing one-to-one function relating Y to X.

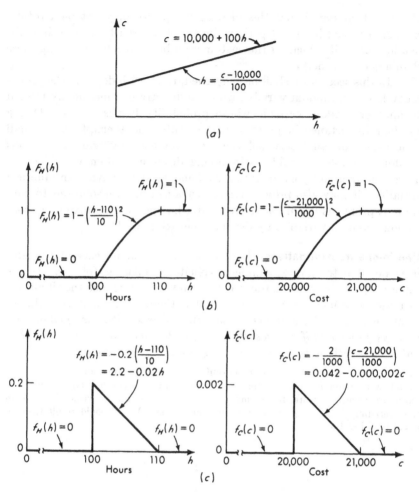

Fig. 2.3.3 Derived distributions, construction-cost illustration, $C = 10{,}000 + 100H$. (a) Functional relationship between cost and time; (b) cumulative distribution function of H, given, and C, derived; (c) probability density function of H, given, and C, derived.

Suppose, for example, that for purposes of developing a bid, the distribution of the total cost C of constructing a retaining wall is desired. Assume that this cost is the cost of materials, which can be accurately predicted to be \$10,000, plus the cost of a labor crew at \$100/hr. The number of hours H to complete such a job is uncertain. The total cost is functionally related to H:

$$C = \$10{,}000 + \$100H$$

The relationship is monotonically increasing and one-to-one (Fig. 2.3.3). The cost is less than any particular value c only if the number of hours is less than a particular value h. This is the key to solving such problems.

Generally we can solve $y = g(x)$ for x to find the *inverse function*† $x = g^{-1}(y)$, which gives the value of x corresponding to any particular value of y. (See Figs. 2.3.2 and 2.3.3.) For example, if the value of c corresponding to any given h is $c = g(h) = 10{,}000 + 100h$, the value of h corresponding to any given c is $h = g^{-1}(c) = (c - 10{,}000)/100$. If, as another example, $y = g(x) = ae^{bx}$, then $x = g^{-1}(y) = (1/b) \ln (y/a)$, in which $\ln (u)$ denotes the natural logarithm of u.

Under these conditions we can solve directly for the CDF of the dependent variable Y, since the probability that Y is less than any value of y is simply the probability that X is less than the corresponding value of x, that is, $x = g^{-1}(y)$. This probability can be obtained from the known CDF of X:

$$F_Y(y) = P[Y \le y]$$
$$= P[X \le g^{-1}(y)]$$

or

$$F_Y(y) = F_X(g^{-1}(y)) \tag{2.3.3}$$

Thus, in the construction-cost example,

$$F_C(c) = F_H \left(\frac{c - 10{,}000}{100} \right)$$

Suppose, for example, that the CDF of H is the parabolic function shown in Fig. 2.3.3b. Then, in the range of interest, we find $F_C(c)$ by simply substituting $(c - 10{,}000)/100$ for h in $F_H(h)$:

$$F_C(c) = F_H \left(\frac{c - 10{,}000}{100} \right)$$
$$= 1 - \left\{ \frac{[(c - 10{,}000)/100] - 110}{10} \right\}^2$$
$$= 1 - \left(\frac{c - 21{,}000}{1000} \right)^2 \qquad 20{,}000 \le c \le 21{,}000$$

In general, when finding distributions of functionally related random variables, we must work with their CDF's. To obtain a PDF of Y, the CDF must be found and then differentiated. A major advantage of the class of functional relationships‡ we are considering, however, is

---

† This is a confusing, but standard, notation consistent with $\sin^{-1} (u)$ for the arcsine or inverse sine function. In general, $g^{-1}(u)$ does *not* equal $1/g(u)$.

‡ An additional restriction, continuity of $y = g(x)$, is also necessary for this next step.

that we may pass directly from the PDF of one to the PDF of the other. Analytically we simply need to take the derivative of the CDF, Eq. (2.3.3):

$$f_Y(y) = \frac{d}{dy} F_Y(y)$$

$$= \frac{d}{dy} [F_X(g^{-1}(y))] = \frac{d}{dy} \left[\int_{-\infty}^{g^{-1}(y)} f_X(x) \, dx \right]$$

which can be shown to be

$$f_Y(y) = \frac{dg^{-1}(y)}{dy} f_X(g^{-1}(y)) \tag{2.3.4}$$

or, replacing $g^{-1}(y)$ by x, we obtain the more suggestive forms

$$f_Y(y) = \frac{dx}{dy} f_X(x) \tag{2.3.5}$$

or

$$f_Y(y) \, dy = f_X(x) \, dx \tag{2.3.6}$$

In words, the likelihood that Y takes on a value in an interval of width dy centered on the value y is equal to the likelihood that X takes on a value in an interval centered on the corresponding value $x = g^{-1}(y)$, but of width $dx = dg^{-1}(y)$. As shown graphically in Fig. 2.3.4, these interval widths are generally not equal, owing to the slope of the function $g(x)$ or $g^{-1}(y)$ at the value of y of interest. This slope and hence the ratio of dx to dy may or may not be the same for all values of y.

In our construction-cost example, the ratio dh/dc is constant because the relationship between C and H is linear (Fig. 2.3.3a):

$$\frac{dh}{dc} = \frac{dg^{-1}(c)}{dc} = \frac{d(c - 10{,}000)/100}{dc} = \frac{1}{100}$$

Therefore, given the PDF of H, the PDF of C follows directly from Eq. (2.3.4) or Eq. (2.3.5):

$$f_C(c) = \frac{dh}{dc} f_H(h) = \frac{dg^{-1}(c)}{dc} f_H(g^{-1}(c))$$

$$= \frac{1}{100} f_H \left(\frac{c - 10{,}000}{100} \right)$$

For example, the CDF of H given in Fig. 2.3.3b implies that the PDF of H is triangular (Fig. 2.3.3c):

$$f_H(h) = -0.2 \frac{h - 110}{10}$$

$$= 2.2 - 0.02h \qquad 100 \le h \le 110$$

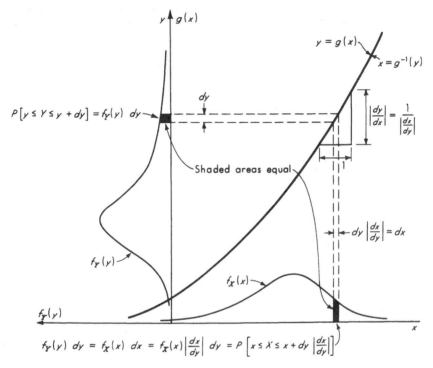

Fig. 2.3.4 Graphical interpretation of Eqs. (2.3.5) and (2.3.12): $f_Y(y) = |dx/dy| f_X(x)$.

Therefore, substituting $g^{-1}(c) = (c - 10{,}000)/100$ for h and multiplying by $dh/dc = \frac{1}{100}$ give

$$f_C(c) = \frac{1}{100}\left(2.2 - 0.02\,\frac{c - 10{,}000}{100}\right)$$

$$= 0.042 - 0.000{,}002c \qquad 20{,}000 \le c \le 21{,}000$$

Note that care must be taken to obtain the region on the c axis in which the density function holds.   In this case it is found simply by calculating the values of c corresponding to $h = 100$ and 110, the ends of the region on the h axis.   The PDF of C is also shown in Fig. 2.3.3c.   In this simple linear case, the shape of the density function has been left unchanged. As is demonstrated in the following illustration this is not generally the case.

Illustration: Bacteria growth for random time   We illustrate here separately the two procedures, passing from CDF to CDF and passing from PDF to PDF.   Under constant environmental conditions the number of bacteria Q in a tank is known to increase proportionately to $e^{\lambda T}$, where T is the time, λ is the growth rate, and the proportionality constant k is the population at time $T = 0$:†

$$Q = ke^{\lambda T} \tag{2.3.7}$$

If the time permitted for growth (as determined, say, by the time required for the charge of water in which the bacteria are living to pass through a filter) is a random variable with distribution function $F_T(t)$, $t \geq 0$, then the final population has distribution function

$$
\begin{aligned}
F_Q(q) &= P[Q \leq q] \\
&= P[ke^{\lambda T} \leq q] \\
&= P\left[e^{\lambda T} \leq \frac{q}{k}\right] \\
&= P\left[\lambda T \leq \ln\frac{q}{k}\right] \\
&= F_T\left(\frac{1}{\lambda}\ln\frac{q}{k}\right) \quad q \geq k
\end{aligned}
\tag{2.3.8}
$$

Here, $g^{-1}(q) = (1/\lambda) \ln (q/k)$.

Now, considering passing from PDF to PDF,

$$\frac{dg^{-1}(q)}{dq} = \frac{1}{\lambda} q^{-1}$$

Note that the derivative of the inverse function is a function of q, not a constant. Then,

$$f_Q(q) = \frac{1}{\lambda q} f_T\left(\frac{1}{\lambda}\ln\frac{q}{k}\right) \quad q \geq k \tag{2.3.9}$$

Clearly the population Q will be no smaller than the initial population k.

Suppose, as a specific example, that $f_T(t)$ is of a decaying type, say,

$$f_T(t) = 2e^{-t^2/2} \quad t \geq 0 \tag{2.3.10}$$

Then

$$
\begin{aligned}
f_Q(q) &= \frac{1}{\lambda q} f_T\left(\frac{1}{\lambda}\ln\frac{q}{k}\right) \\
&= \frac{2}{\lambda q} \exp\left(-\frac{1}{2\lambda^2}\ln^2\frac{q}{k}\right) \\
&= \frac{2}{\lambda q}\frac{q^{-[(\lambda^2/2)\ln(q/k)]}}{k} \quad q \geq k
\end{aligned}
\tag{2.3.11}
$$

These two distributions are sketched in Fig. 2.3.5. Note the change in shape in passing from one PDF to the other.

† This result is true on the average. In fact, for any given time $T = t$, the population Q may better be considered a random variable. Such a model is an example of a random function of t or a "stochastic process." Note that the population size, actually an integer number, is being treated for convenience as a continuous variable, since it is a large number.

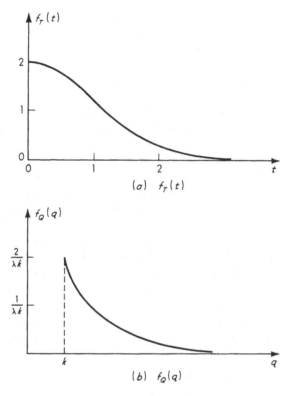

Fig. 2.3.5 Bacteria-growth illustration.

If the relationship between Y and X is one-to-one, but monotonically *decreasing*, Y will take on values less than any particular value y_0 only if X takes on values *greater than* the corresponding value x_0. The implications for the previous derivations can be shown easily by the reader. The final result for the relationship between PDF's is simply that an absolute value sign needs to be placed about the dx/dy or $dg^{-1}(y)/dy$ in previous equations, yielding the general result:

$$f_Y(y) = \left| \frac{dx}{dy} \right| f_X(x) \tag{2.3.12}$$

$$f_Y(y) = \left| \frac{dg^{-1}(y)}{dy} \right| f_X(g^{-1}(y)) \tag{2.3.13}$$

The simplest and single most important application of this formula is for the case when the relationship between X and Y is linear:

$$Y = a + bX \tag{2.3.14}$$

Simple scale changes, such as feet to inches or Fahrenheit to Centigrade, are commonly arising examples. In this case, $X = (Y - a)/b$ and $|dx/dy| = |1/b|$. Hence,

$$f_Y(y) = \left| \frac{1}{b} \right| f_X \left(\frac{y - a}{b} \right) \tag{2.3.15}$$

The effect of such a linear transformation is to retain the basic shape of the function, but to shift it and to expand or contract it.

General transformations   Relationships $Y = g(X)$, for which Eq. (2.3.13) holds, are very common in engineering, where an increase in one variable usually means either an increase or decrease in some dependent variable. Nevertheless in many situations a more complicated nonmonotonic or non-one-to-one relationship may be involved. The simple discrete illustration at the beginning of this section is an example. Although the relationship was nondecreasing, it was not one-to-one; values of X of either 0 or 1, for example, led to a single value of Y, namely, 0. In all such problems the reader is urged to deal directly with the *cumulative distribution function* of Y, as follows.

Mathematically, the problem is to find $F_Y(y)$ given that $Y = g(X)$ and given that X has CDF $F_X(x)$. Conceptually the problem is simple. By definition

$$F_Y(y) = P[Y \leq y]$$

but

$$P[Y \leq y] = P[X \text{ takes on any value } x \text{ such that } g(x) \leq y]$$

$$= \int_{R_y} f_X(x) \, dx \tag{2.3.16}$$

where R_y is that region where $g(x)$ is less than or equal to y. In Fig. 2.3.6a a representation of a general function $y = g(x)$ is sketched. Any particular value of y, say y_0, is represented by a straight horizontal line. To determine $F_Y(y_0)$ is to determine the probability that the random variable X falls in any of those intervals where the curve $g(x)$ falls below the horizontal line, $y = y_0$. If, for example, X were described by a continuous PDF as shown in Fig. 2.3.6b, $F_Y(y_0)$ would be equal to the crosshatched area under the density function in these intervals. Solution of the problem requires relating these intervals to each and every value y_0.

Consider the following example. The kinetic energy K of a moving mass, say a simple structural frame vibrating under a dynamically imposed wind or vehicle load, is proportional to the square of its velocity V:

$$K = \frac{mV^2}{2} \tag{2.3.17}$$

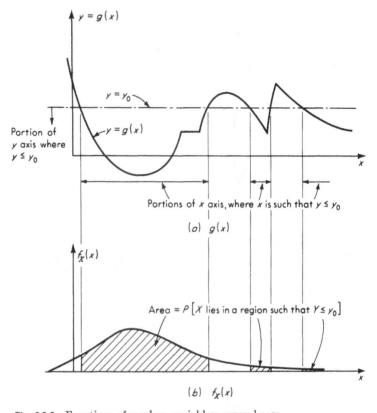

Fig. 2.3.6 Functions of random variables: general case.

in which m is the known mass of the object. Velocity may, of course, be positive or negative (e.g., right or left, up or down). The relationship is neither monotonic nor one-to-one, as shown in Fig. 2.3.7a. The probability that the kinetic energy will be less than a certain value k_0 is evidently

$$P[K \le k_0] = P\left[-\sqrt{\frac{2k_0}{m}} \le V \le +\sqrt{\frac{2k_0}{m}}\right] \qquad (2.3.18)$$

The probability of the complementary event $K > k_0$ is more easily calculated. This is $1 - F_K(k)$:

$$1 - F_K(k) = P\left[\left(V \le -\sqrt{\frac{2k}{m}}\right) \cup \left(V \ge +\sqrt{\frac{2k}{m}}\right)\right]$$

Since these events are mutually exclusive,

$$1 - F_K(k) = P\left[V \le -\sqrt{\frac{2k}{m}}\right] + P\left[V \ge +\sqrt{\frac{2k}{m}}\right]$$

This result is true no matter what the distribution of V. As a specific case, suppose the velocity has been repeatedly measured and the engineer claims that the triangular-shaped distribution shown in Fig. 2.3.7b represents a good model of the random behavior of the velocity at any instant. By the symmetry of the distribution of V,

$$1 - F_K(k) = 2P\left[V \le -\sqrt{\frac{2k}{m}}\right] = 2F_V\left(-\sqrt{\frac{2k}{m}}\right)$$

$$F_K(k) = 1 - 2\left(\frac{1}{3}\right)\frac{\left(3 - \sqrt{\frac{2k}{m}}\right)^2}{6}$$

$$= 1 - \frac{1}{9}\left(3 - \sqrt{\frac{2k}{m}}\right)^2 \qquad 0 \le k \le \frac{9}{2}m \qquad (2.3.19)$$

The limits of the values that K can take on, namely, 0 and $\frac{9}{2}m$, follow from the minimum (0) and maximum (3) values of the absolute magnitude of the velocity. The density function of K can be found by differentiation of the CDF:

$$f_K(k) = \frac{d}{dk}[F_K(k)]$$

$$= \frac{1}{9}\sqrt{\frac{2}{m}}\left(3k^{-\frac{1}{2}} - \sqrt{\frac{2}{m}}\right) \qquad 0 \le k \le \frac{9}{2}m \qquad (2.3.20)$$

This density function is shown in Fig. 2.3.7c.

In more complicated cases involving non-one-to-one transformations, required regions of integration may become difficult to determine and sketches of the problem are strongly recommended. If discrete or mixed random variables are involved (see, for example, the pump-selection illustration to follow in Sec. 2.4), sums may have to replace or supplement integration, but the formulation in terms of the cumulative distribution of Y remains unchanged. The determination of the PDF or PMF from the CDF follows by the techniques discussed in Sec. 2.2.

2.3.2 Functions of Two Random Variables

Frequently a quantity of interest to the engineer is a function of two or more variables which have been modeled as random variables. The implication is that the dependent variable is also random. For example, the spatial average velocity V in a channel is the flow rate Q divided by the area A. If Q and A have a given joint probability distribution, the problem becomes to find the probability law of V. The total number of vehicles Z on a particular transportation link may be the sum of those vehicles on two feeder links, X and Y.

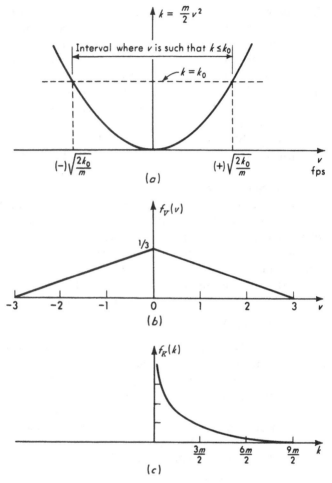

Fig. 2.3.7 Energy-velocity illustration, Nonmonotonic Case. (a) Functional relationship between energy and velocity; (b) given PDF of velocity V; (c) derived PDF of kinetic energy K.

Clearly in a discrete case like the latter example we can use enumeration to calculate the PMF or CDF of Z. The probability that Z equals any particular value z, say 4, is the sum of the probabilities of all the pairs of values of X and Y which sum to 4, that is, (0,4), (1,3), (2,2), (3,1), and (4,0). These individual probabilities are given in the known joint PMF of X and Y. Similarly the CDF of Z at value z could be found by summing the probabilities of all pairs (x,y) for which $x + y \le z$. In difficult cases this approach may also prove computationally advantageous when dealing in an approximate way with continuous random variables. In this section we deal, however, with analytical solutions of problems involving continuous random variables.

Two approaches are considered and illustrated. One begins, as is generally appropriate, with the determination of the CDF of the dependent variable; the other makes use of the notions of the conditional distribution (Sec. 2.2.2) to go directly after the PDF of the dependent random variable.

Our problem, in general, is to find $F_Z(z)$ when we know $Z = g(X,Y)$ and when the joint probability law of X and Y is also known.

Direct method   In Sec. 2.3.1 we found the CDF of the dependent variable Z, say, at any particular value z_0 by finding the probability that the independent variable X, say, would take on a value in those intervals of the x axis where $g(x) \leq z_0$. Now, by extension, since Z is a function of both X and a second variable Y, we must calculate $F_Z(z_0)$ by finding the probability that X and Y lie in a region where $g(x,y) \leq z_0$.

An easily visualized example is the following. We wish to determine the CDF of Z, where Z is the larger of X and Y, or

$$Z = \max [X,Y] \tag{2.3.21}$$

If, for example, X and Y are the magnitudes of two successive floods or of the queue lengths of a two-lane toll station, then primary interest may lie not in X or Y in particular, but in the Z, the greater of the two, whichever that might be. To determine $F_Z(z)$ for any value z, it is necessary to integrate the joint PDF of the (assumed continuous) random variables X and Y over the region where the maximum of x and y is less than z. This is the same as the region where *both* x and y are less than z, that is, the region shown in Fig. 2.3.8.

$$F_Z(z) = P[Z \leq z] = P[\max [X,Y] \leq z]$$

$$= \int\int_{-\infty}^{z} f_{X,Y}(x,y) \ dx \ dy \tag{2.3.22}$$

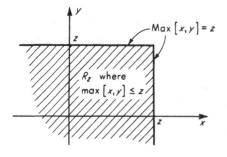

Fig. 2.3.8 $Z = \max [X,Y]$ illustration.

The integral of $f_{X,Y}(x,y)$ over this region is, of course, simply $F_{X,Y}(z,z)$ [Eq. (2.2.41)]. If, as a special case, X and Y are independent and identically distributed† with common CDF $F_X(x)$,

$$F_Z(z) = F_{X,Y}(z,z) = F_X(z)F_X(z) = F_X^2(z) \qquad (2.3.23)$$

In this case it is possible to determine the PDF of Z explicitly by differentiation

$$f_Z(z) = \frac{d}{dz} F_Z(z) = \frac{d}{dz} [F_X^2(z)] = 2F_X(z) \frac{d}{dz} F_X(z)$$

$$= 2F_X(z)f_X(z) \qquad (2.3.24)$$

In words, this last result can be interpreted as the statement that "the probability that the maximum of two variables is in a small region about z is proportional to the probability that one variable X or Y is less than or equal to z while the other is in a small region about z." Since this can happen in two mutually exclusive ways, the (equal) probabilities are added. The treatment of the maximum of a set of several random variables will be discussed more fully in Sec. 3.3.3.

If, for example, X and Y are independent and identically distributed annual stream flows, with common distribution,

$$f_X(x) = \lambda e^{-\lambda x} \qquad x \geq 0 \qquad (2.3.25)$$

implying

$$F_X(x) = 1 - e^{-\lambda x} \qquad x \geq 0 \qquad (2.3.26)$$

then the distribution of Z, the larger of the two flows, is

$$F_Z(z) = F_X^2(z) = (1 - e^{-\lambda z})^2 \qquad z \geq 0 \qquad (2.3.27)$$

and

$$f_Z(z) = 2\lambda e^{-\lambda z}(1 - e^{-\lambda z}) \qquad z \geq 0 \qquad (2.3.28)$$

These density functions are sketched in Fig. 2.3.9 for $\lambda = \frac{1}{1000}$ acre-ft.

Illustration: Distribution of the quotient As a second example, consider the determination of the probability law of Z when

$$Z = \frac{X}{Y} \qquad (2.3.29)$$

In specific problems Z might be a cost-benefit ratio or a "safety factor," the ratio of capacity to load. Proceeding as before to find the CDF of Z, assuming

† For example, this is a common engineering assumption in the case where X and Y represent the largest river flows in each of 2 successive years.

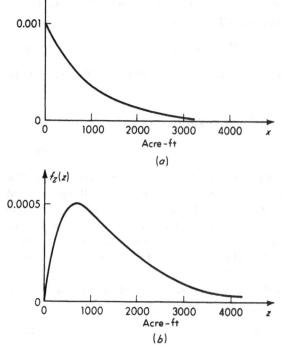

Fig. 2.3.9 PDF of stream flow of each of two rivers and PDF of largest stream flow. (a) Density function of stream flow, $f_X(x) = 0.001e^{-0.001x}$; (b) density function of largest stream flow.

that X and Y are jointly distributed,

$$F_Z(z) = P[Z \leq z] = P\left[\frac{X}{Y} \leq z\right]$$

$$= \iint\limits_{R_z} f_{X,Y}(x,y)\, dx\, dy \tag{2.3.30}$$

where R_z is that region of the xy plane, the sample space of X and Y, where x/y is less than z. Such a region is shown shaded in Fig. 2.3.10 for a particular value of z. Other values of z would lead to other lines $x/y = z$ through the origin. The limits of integration are of the same form for any value of z and follow from inspection of this figure:

$$F_Z(z) = \int_{-\infty}^{0} dy \int_{zy}^{\infty} dx\, f_{X,Y}(x,y) + \int_{0}^{\infty} dy \int_{-\infty}^{zy} dx\, f_{X,Y}(x,y)$$

$$= -\int_{-\infty}^{0} dy \int_{\infty}^{zy} dx\, f_{X,Y}(x,y) + \int_{0}^{\infty} dy \int_{-\infty}^{zy} dx\, f_{X,Y}(x,y) \tag{2.3.31}$$

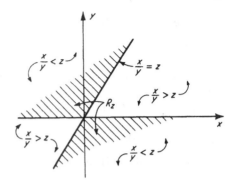

Fig. 2.3.10 Derived-distribution illustration, $Z = X/Y$.

To carry out this integration may in fact be a troublesome task, particularly as the joint PDF of X and Y may be defined by different functional forms over different regions of the x and y plane. We shall give complete examples later in this section, but we prefer to avoid further details of calculus at this point as they serve only to confuse the probability theoretic aspects of the problem, which are complete in this statement, Eq. (2.3.31).

In this case we can again find an expression for the PDF of Z by differentiating with respect to z *before* carrying out the integration. Consider the first term

$$\frac{d}{dz}\left[-\int_{-\infty}^{0} dy \int_{\infty}^{zy} dx\, f_{X,Y}(x,y)\right] = -\int_{-\infty}^{0}\left\{\frac{d}{dz}\left[\int_{\infty}^{zy} f_{X,Y}(x,y)\,dx\right]\right\} dy$$

$$= \int_{-\infty}^{0} -yf_{X,Y}(zy,y)\,dy \quad (2.3.32)$$

The second term is similar, lacking the minus sign. The terms may be combined to form

$$f_Z(z) = \frac{d}{dz} F_Z(z) = \int_{-\infty}^{\infty} |y| f_{X,Y}(zy,y)\,dy \qquad (2.3.33)$$

Again for specific functions $f_{X,Y}(x,y)$ the subsequent evaluation of the integral may be tedious, but it is not fundamentally a probability problem.

Distribution of sum X and Y by conditional distributions It is useful to have available the important results for the case when the dependent variable is the *sum of two random variables*. Sums of variables occur frequently in practice, e.g., sums of loads, of settlements, of lengths, of waiting times, or of travel times. Furthermore, their study is an important part of the theories of probability and mathematical statistics, as we shall see in Chaps. 3 and 4. In determining these results, an alternate, but less general, approach to finding directly the formula for the PDF of a continuous function of continuous random variables will be demonstrated. This method employs conditional probability density functions and a line of argument that often proves enlightening as well as productive.

We wish to determine the density function of

$$Z = X + Y \qquad (2.3.34)$$

when the joint PDF of X and Y is known. Let us consider *first* the conditional density function of Z given that Y equals some particular value y. Given that $Y = y$, Z is

$$Z = y + X \qquad (2.3.35)$$

and the (conditional) distribution of X is

$$f_{X|Y}(x,y) = \frac{f_{X,Y}(x,y)}{f_Y(y)} \qquad (2.3.36)$$

The density function of such a linear function of a *single* random variable (y being treated as a constant for the moment) is given in Eq. (2.3.15). Using the results found there, conditional on $Y = y$, the *conditional* PDF of Z is

$$f_{Z|Y}(z,y) = f_{X|Y}(z - y, y) \qquad (2.3.37)$$

The *joint* density function of any two random variables Z and Y can always be found by multiplying the *conditional* PDF of Z given Y by the *marginal* PDF of Y [Eq. (2.2.34)]:

$$f_{Z,Y}(z,y) = f_{Z|Y}(z,y) f_Y(y) \qquad (2.3.38)$$

Substituting Eq. (2.3.37),

$$f_{Z,Y}(z,y) = f_{X|Y}(z - y, y) f_Y(y) \qquad (2.3.39)$$

But the right-hand side, being the product of a *conditional* and a *marginal*, is only the *joint* PDF of X and Y evaluated at $z - y$ and y. Thus

$$f_{Z,Y}(z,y) = f_{X,Y}(z - y, y) \qquad (2.3.40)$$

In words, this result says, roughly, that the likelihood that $Y = y$ and $Z = X + Y = z$ equals the likelihood that $Y = y$ and $X = z - y$.

The marginal of Z follows upon integration over all values of y [Eq. (2.2.43)]:

$$f_Z(z) = \int_{-\infty}^{\infty} f_{Z,Y}(z,y)\, dy \qquad (2.3.41)$$

or

$$f_Z(z) = \int_{-\infty}^{\infty} f_{X,Y}(z - y, y)\, dy \qquad (2.3.42)$$

For the important special case when X and Y are independent, their joint PDF factors into

$$f_Z(z) = \int_{-\infty}^{\infty} f_X(z - y) f_Y(y) \, dy \qquad (2.3.43)$$

By the symmetry of the argument, it follows that it is also true that

$$f_Z(z) = \int_{-\infty}^{\infty} f_{X,Y}(x, z - x) \, dx \qquad (2.3.44)$$

or, if X and Y are independent,

$$f_Z(z) = \int_{-\infty}^{\infty} f_X(x) f_Y(z - x) \, dx \qquad (2.3.45)$$

Roughly speaking, this last equation states that the probability that Z lies in a small interval around Z is proportional to the probability that X lies in an interval x to $x + dx$ times a factor proportional to the probability that Y lies in a small interval around $z - x$, the value of Y necessary to make $X + Y$ equal z. This product is then summed over all values that X can take on. [Equation (2.3.45) is the type of integral known as the *convolution integral*. This form occurs time and time again in engineering, and it will be familiar already to the student who has studied the dynamics of linear structural, hydraulic, mechanical, or electrical systems.]

As mentioned above, equations such as Eqs. (2.3.24), (2.3.33), and (2.3.45) are in fact complete answers to the probability theory part of the problem of finding the distribution of the function of other random variables, but the completion of the problem, that is, carrying out the indicated integrations, may prove challenging, particularly since many practical PDF's are defined by different functions over different regions. Illustrations follow.

Illustration: Distribution of total waiting time   When a transportation system user must travel by two modes, say bus and subway, to reach his destination, a portion of this trip is spent simply waiting at stops or terminals for the arrivals of the two vehicles. Under certain conditions to be discussed in Sec. 3.2, it is reasonable to assume that the individual waiting times X and Y have distributions of this form

$$f_X(x) = \alpha e^{-\alpha x} \qquad x \geq 0$$
$$f_Y(y) = \beta e^{-\beta y} \qquad y \geq 0$$

To determine the properties of the *total* time spent waiting for the vehicles to arrive we must find the PDF of $Z = X + Y$. Assuming independence of X and Y, we can apply Eq. (2.3.45):

$$f_Z(z) = \int_{-\infty}^{\infty} f_X(x) f_Y(z - x) \, dx$$

Substitution of the functions takes some care, since they are, in fact, zero for negative values of their arguments. Since $f_X(x)$ is zero for negative values of x:

$$f_Z(z) = \int_0^\infty \alpha e^{-\alpha x} f_Y(z - x)\, dx$$

Since $f_Y(y)$ is zero for y negative, $f_Y(z - x)$ is zero for $z - x$ negative or for x greater than z. Therefore,

$$f_Z(z) = \int_0^z \alpha e^{-\alpha x} \beta e^{-\beta(z-x)}\, dx$$

$$= \alpha \beta e^{-\beta z} \int_0^z e^{(\beta - \alpha)x}\, dx$$

$$= \frac{\alpha \beta}{\beta - \alpha} (e^{-\alpha z} - e^{-\beta z}) \qquad z \geq 0 \tag{2.3.46}$$

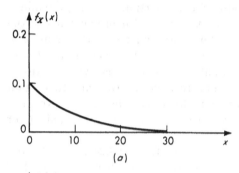

(a)

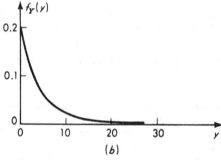

(b)

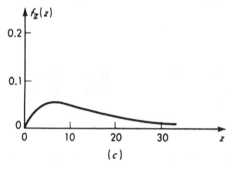

(c)

Fig. 2.3.11 Waiting-time illustration with $Z = X + Y$. (a) Bus waiting time, minutes; (b) subway waiting time, minutes; (c) total waiting time, minutes.

The density functions of X, Y, and Z are sketched in Fig. 2.3.11 for $\alpha = 0.1$ min^{-1} and $\beta = 0.2$ min^{-1}.

Illustration: Earthquake intensity at a site We are interested as the designers of a dam which is in an earthquake-prone region in studying the intensity of ground motion at the site given that an earthquake occurs somewhere in a circular region of radius r_0 surrounding the site. The information about the geology of the region is such that it is reasonable to assume that the location of the source of the disturbance (the epicenter) is equally likely to be anywhere in the region; that is, the engineer assigns an equal probability to equal areas in the region and hence a constant density function over the circle. The constant value is $1/(\pi r_0{}^2)$, yielding a unit volume under the function. The implication is that the density function of R, the radial distance from the site to the epicenter, has a triangular distribution

$$\frac{1}{\pi r_0{}^2} 2\pi r = \frac{2r}{r_0{}^2}$$

$$f_R(r) = \begin{cases} \dfrac{2}{r_0{}^2} r & 0 \le r \le r_0 \\ 0 & \text{elsewhere} \end{cases} \qquad (2.3.47)$$

a fact the reader can verify formally by the techniques of this section.

We shall assume that historical data for the region suggest a density function of the form†

$$f_Y(y) = \begin{cases} \frac{3}{64}(9 - y)^2 & 5 \le y \le 9 \\ 0 & \text{elsewhere} \end{cases} \qquad (2.3.48)$$

for Y, the Richter magnitudes of earthquakes of significant size to be of concern to the engineer. Magnitudes greater than 9 have not been observed. The equation relating an empirical measure of the intensity X of the ground motion *at the site* to the magnitude and distance of the earthquake we shall assume to be

$$X = g(Y,R) = c_1 + c_2 Y - c_3 \ln R \qquad (2.3.49)$$

reflecting the attenuation of intensity with distance from the epicenter.‡ The distribution of X is the desired result.

We seek first the CDF of X,

$$F_X(x) = P[X \le x] = \int_{R_x} f_{Y,R}(y,r)\, dy\, dr \qquad (2.3.50)$$

in which R_x is the region in the ry plane where $g(y,r) = c_1 + c_2 y - c_3 \ln r$ is less than x. For a given value of x this region is as shown in Fig. 2.3.12a. For such fixed values of x, it is possible to solve for the equation of the line bounding

---

† For a discussion of this question see, for example, Rosenblueth [1964].
‡ This problem is considered in Esteva and Rosenblueth [1964]. For Southern California it has been found empirically that $c_1 = \ln 280/\ln 2$, $c_2 = 1/\ln 2$, and $c_3 = 1.7/\ln 2$ with R in kilometers. This empirical relationship is not valid for small values of r (less than about 40 km), but it will be retained here for simplicity.

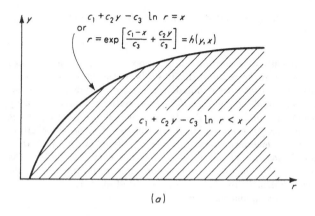

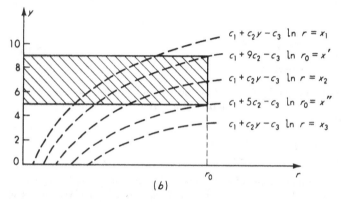

Fig. 2.3.12 Earthquake-intensity illustration. (a) R_x: the region where $g(y,r) = c_1 + c_2 y - c_3 \ln r \leq x$ for a particular value of x; (b) region where $f_{Y,R}(y,r)$ is nonzero.

the region as a function of r:

$$r = \exp\left(\frac{c_1 - x}{c_3} + \frac{c_2 y}{c_3}\right) = h(y,x)$$

Hence, in general, the CDF of X is (for nonnegative random variables Y and R) found by integrating over this region:

$$F_X(x) = \int_0^\infty dy \int_{h(y,x)}^\infty dr \, f_{Y,R}(y,r)$$

This completes the formal probability aspects of the problem. The calculus of the integral's evaluation proves more awkward.

In this particular case the joint density function of Y and R is positive only over the region shown in Fig. 2.3.12b. This causes the limits of the integrals to become far more complicated. As shown in Fig. 2.3.12b three cases, x_1, x_2, or x_3, exist depending on whether the particular value of x creates a bounding curve passing out through the top of the rectangle, through its side, or not at all. In the first situation, where x is larger than $x' = c_1 + 9c_2 - c_3 \ln r_0$,

$$F_X(x) = \int_5^9 dy \int_{h(y,x)}^{r_0} dr \left[\tfrac{3}{64}(9 - y)^2 \frac{2r}{r_0^2} \right]$$

$$= 1 - ke^{-2x/c_3} \qquad x > x'$$

$$f_X(x) = \frac{d}{dx} F_X(x)$$

$$= \frac{2}{c_3} ke^{-2x/c_3} \qquad x > x' \tag{2.3.51}$$

For $x'' \le x \le x'$ complicated forms of $F_X(x)$ and $f_X(x)$ result. For $x \le x''$, both functions are zero. For the values of the constants appropriate for Southern California (with $r_0 = 300$ km) the CDF and PDF of X, the intensity of the ground motion at a site given that an earthquake of magnitude 5 to 9 occurs within this radius, are plotted in Fig. 2.3.13. Notice that as long as one is interested only in intensities in excess of 7.1 (i.e., in intensities with probability of occurrence, given an earthquake, of less than 2.5 percent) it is possible to deal with the distribution in the region in which its form is simple and tractable. (Other problems will be found throughout this text dealing with this question of seismic risk.)

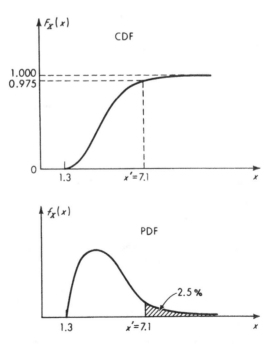

Fig. 2.3.13 Probability distribution of intensity at site.

A variety of methods and problems in deriving distributions of random variables will go unillustrated here. The awkard analytical aspects far outweigh any new insights they provide into probability theory.†

2.3.3 Elementary Simulation‡

The probabilistic portion of the derivation of the distributions of a function of one or more random variables may be a rather direct problem. The calculus needed to evaluate the resulting integrals, however, is frequently not tractable. In such circumstances one can often resort to numerical integration or other techniques beyond the interest of this work. One approximate method of solution of derived distribution problems is of immediate interest, however, because it makes direct use of their probabilistic nature§ to obtain artificially the results of many repeated experiments. The histogram of these results will approximate the desired probability distribution. This, the *Monte Carlo method*, or *simulation*, is best presented by example.

Simulating a discrete random variable First let us find simply the distribution of the ratio of the number of wet weeks to the number of dry weeks in a year on a given watershed, when N, the total number of rainy weeks with at least a trace of rain, is a random variable with a probability mass function given by ¶

$$p_N(n) = \frac{20^n e^{-20}}{n!} \qquad n = 0, 1, 2, \ldots, 52 \qquad (2.3.52)$$

We seek the distribution, then, of the ratio of wet to dry, or

$$Y = \frac{N}{52 - N} \qquad (2.3.53)$$

We could use, following the methods of Sec. 2.3.1, simple enumeration of all possible values of N and the corresponding values of Y to find

---

† The reader is referred, however, to any of several texts (e.g., Freeman [1963], page 82) for a particularly efficient technique for obtaining the joint PDF of n random variables which are (simultaneous) functions of n other jointly distributed random variables whose PDF is known; the method applies to the case where the functional relationship is one-to-one and continuous. It is sometimes called the method of Jacobeans. Equation (2.3.12) is a special case when $n = 1$.

‡ This section can be omitted at first reading.

§ It should be noted that probabilistic Monte Carlo methods are also used to evaluate multidimensional integrals which arise from problems which are not probabilistic at all. (See Hammersly and Hanscomb [1960].)

¶ There is a negligible error here; as described, $\Sigma p_N(n)$ does not equal 1 unless $n = 1, 2, \ldots, \infty$. See Sec. 3.2, where the distribution will be introduced as the "Poisson distribution." The total probability mass of the neglected terms is small in this case.

the PMF of Y. For illustrative purposes we are going to determine an approximation of the distribution of Y by carrying out a series of experiments, each of which simulates a year or an observation of N. In each experiment we shall artificially *sample* the distribution of the random variable N to determine a number n, the number of rainy weeks which occurred in that "year." Then we can calculate $y = n/(52 - n)$ to find the ratio of the number of wet to dry weeks in that "year," that is, an observation of the random variable Y. A sufficient number of such experiments will yield enough data to draw a histogram (Sec. 1.1), which we can expect to approximate the shape of the mass function $p_Y(y)$. It remains to be shown how the sampling is actually carried out.

The experimental sampling of mathematically defined random variables is accomplished by selecting a series of *random numbers*, each of which is associated with a particular value of the random variable of interest. Random numbers are generated in such a manner that each in the set is equally likely to be selected. The numbers 1 to 6 on a perfect die are, for example, a set of random numbers. Divide the circumference of a dial into 10 equal sectors labeled 0 to 9 (or 100 equal sectors labeled 0 to 99, etc.), attach a well-balanced spinner, and you have constructed a mechanical *random number generator*. Various ingenious electrical and mechanical devices to generate random numbers have been developed, and several tables of such numbers are available (see Table A.8), but most random number generation in practice is accomplished on computers through numerical schemes.†

To relate the value of the random variable of interest N to the value obtained from the random number generator, one must assign a table of relationships (a mapping) which matches probabilities of occurrence. Assume that a random number generator is available which produces any of the numbers 0 to 9999 with equal probability 1/10,000. On the other hand, the probability that N takes any of the various values 0, 1, 2, . . . , 52 can be determined by evaluating its probability mass function. For example,

$$p_N(0) \approx 0.0000$$
$$p_N(1) \approx 0.0000$$
$$. \quad . \quad . \quad . \quad . \quad . \quad . \quad .$$
$$p_N(5) \approx 0.0001$$
$$p_N(6) \approx 0.0002$$
$$p_N(7) \approx 0.0005$$
$$. \quad . \quad . \quad . \quad . \quad . \quad . \quad .$$

† One of many schemes is described in Prob. 2.53.

$$p_N(19) \approx 0.0888$$
$$p_N(20) \approx 0.0888$$
$$p_N(21) \approx 0.0846$$

. . . .

Then values of the random numbers and the random variable N might be assigned as follows:

Random numbers	Probability of occurrence	Corresponding values n of the random variable N
None	0	0 to 4
0000	0.0001	5
0001–0002	0.0002	6
0003–0007	0.0005	7
. . . .	. . . . . .	. . . . .
3814–4701	0.0888	19
4702–5589	0.0888	20
5590–6435	0.0846	21
. . . .	. . . . . .	. . . . .

If the first random number generated happened to be 4751, the number n of rainy weeks in that simulated year would be taken as 20. Then the corresponding value of Y would be $y = 20/(52 - 20) = \frac{5}{8}$. Repetition of these "experiments," a process ideally suited for digital computers, will yield y_1, y_2, y_3, . . . , a sample of observed values of Y. A histogram (Sec. 1.1) of these values will approximate the true PMF

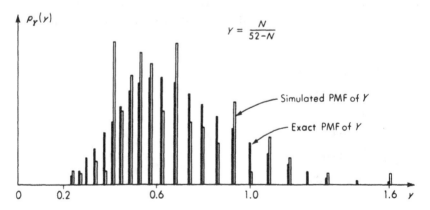

Fig. 2.3.14 Probability mass functions of Y, exact and by simulation.

of Y. For example, the 100 random numbers in Table 2.3.1 lead to the indicated values of N, Y, and the histogram in Fig. 2.3.14. The exact PMF is plotted in the same figure.

Note that if another set of 100 values were generated, a histogram similar in shape but different in details is to be expected. In short, the histograms are themselves observations of random phenomena. Practical questions which the simulator must face are, "How much variability is there in such histograms?" "What can I safely conclude about the true PMF from one observed histogram?" "How does the shape of an observed histogram behave probabilistically as the sample size increases?" These and similar questions will be considered in Secs. 4.4 and 4.5.

Illustration: Total annual rainfall Monte Carlo techniques become particularly useful when the relationships among variables are complicated functions, involve many random variables, or, as in the next example, themselves depend upon a random element. In this illustration it is desired to determine the distribution of total annual rainfall T when the rainfall R_i in the ith rainy week is given by

$$f_{R_i}(r) = 2e^{-2r} \qquad r \geq 0 \tag{2.3.54}$$

and when the total number of rainy weeks in a year is N, as before, a random variable with the distribution given in Eq. (2.3.52). This problem might be part of a large study of the performance of a proposed irrigation system.

The relationship between T, N, and the R_i's is:

$$T = \sum_{i=1}^{N} R_i \tag{2.3.55}$$

That is, T is the sum of a *random* number of random variables. (It is assumed, for convenience, that rainfalls in rainy weeks are identically distributed, mutually independent, and independent, too, of the number of rainy weeks in the year.)

In Sec. 2.3.2, we saw that the determination of the sum of even two random variables can be difficult enough. Analytical solutions to the type of problem here, are, in fact, feasible (Sec. 3.5.3), but we propose to determine the distribution of T experimentally by simulating a number of years and producing a sample of values of T. In each year we shall sample the random variable N to determine a particular number of rainy weeks n. Then the random variables R_i will be sampled that number of times and the n values summed to get one observation of T. This total experiment will be repeated as accuracy requires.

The sampling of the discrete variable N has been discussed. Several schemes for relating values of random numbers to values of the continuous random variable R_i can be considered. One approach simply approximates the continuous distribution of R_i by a discrete one. Equal intervals of r, say 0.0 to 0.9, 0.10 to 0.19, etc., might be considered, the probability of R_i taking on a value in each interval evaluated, and a corresponding proportion of the random numbers assigned to be associated with the interval. The middle value† of

---

† Or better, some kind of average value, say the centroid of the area under the probability density curve.

Table 2.3.1

Random number	Value of N	Value of Y	Random number	Value of N	Value of Y
5134	21	0.678	4627	19	0.577
3347	18	0.530	8457	25	0.927
1319	15	0.405	6113	21	0.678
1408	15	0.405	1372	15	0.405
7340	23	0.792	8310	24	0.858
1997	16	0.445	6017	21	0.678
1741	16	0.445	4438	19	0.577
3244	18	0.530	7522	23	0.792
3710	18	0.530	4031	19	0.577
9761	30	1.332	7678	23	0.792
1072	15	0.405	1208	15	0.405
3953	19	0.577	3672	18	0.530
8859	25	0.927	7094	22	0.734
3004	18	0.530	1409	15	0.405
2489	17	0.486	3096	18	0.530
1248	15	0.405	0749	14	0.368
4672	19	0.577	3263	18	0.530
7406	23	0.792	7601	23	0.792
8678	25	0.927	2669	17	0.486
9933	32	1.600	8215	24	0.858
5941	21	0.678	6135	21	0.678
5862	21	0.678	9284	27	1.080
5201	20	0.625	4015	19	0.577
5845	21	0.678	6429	21	0.678
3990	19	0.577	2532	17	0.486
7983	24	0.858	1177	15	0.405
1713	16	0.445	6699	22	0.734
9458	27	1.080	6042	21	0.678
2804	17	0.486	0539	13	0.333
0549	13	0.333	4910	20	0.625
1252	15	0.405	6851	22	0.734
6781	22	0.735	2875	17	0.486
3404	18	0.530	4621	19	0.577
5981	21	0.678	3707	18	0.530
1347	15	0.405	8686	25	0.927
6349	21	0.678	5375	20	0.625
5936	20	0.625	4555	19	0.577
0052	10	0.238	2351	17	0.486
2377	17	0.486	3170	18	0.530
9362	27	1.080	6443	25	0.927
0174	11	0.268	3249	18	0.530
5936	21	0.678	9082	26	1.000
1438	15	0.405	9647	28	1.168
8869	25	0.927	5234	20	0.625
8083	22	0.734	9528	27	1.080
1962	16	0.445	5581	20	0.625
8613	25	0.927	1375	15	0.405
2006	16	0.445	2292	17	0.486
2135	16	0.445	4113	19	0.577
9616	28	1.168	2346	17	0.486

each interval can then be used as the single representative value of the entire interval. If a random number associated with the interval were drawn, this representative value would be accumulated with others to determine an observed value of T.

Alternatively, a preferable scheme calls for dividing the range of r into intervals of *equal probability*. The fineness of the intervals chosen depends only on the accuracy desired.† Assume here that 10 intervals are considered sufficient. Then the dividing lines between intervals are 0, r_1, r_2, . . . , r_{10} such that

$$F(r_1) = 0.1 = 1 - e^{-2r_1}$$

$$F(r_2) = 0.2 = 1 - e^{-2r_2}$$

. . . .

$$F(r_{10}) = 1.0 = 1 - e^{-2r_{10}}$$

From tables of e^{-x} we obtain intervals:

Interval no. j	Interval	Random number	Representative value of R $\bar{r}_j$
1	0 to 0.052	0	0.026
2	0.052 to 0.111	1	0.07
3	0.111 to 0.178	2	0.13
4	0.178 to 0.255	3	0.21
5	0.255 to 0.346	4	0.30
. . .	. . . .	. . .	. . . . .
9	0.802 to 1.15	8	0.95
10	1.15 to ∞	9	1.65

Corresponding values of the random numbers are assigned as shown.

In this case, where the number of intervals is small, the determination of the representative value $\bar{r}_j$ for each interval is most important. What, for example, should be used as the representative value of the last interval? The center of gravity is a logical choice. The centroid of the segment of the density function over the last interval is

$$\bar{r}_{10} = \frac{\int_{r_9}^{\infty} r f_R(r) \, dr}{\int_{r_9}^{\infty} f_R(r) \, dr} = r_9 + \tfrac{1}{2} = 1.65$$

† There can be as many intervals as there are members in the set of random numbers from which the generator is selecting, and this number is indefinitely large, since successive one-digit numbers can always be strung together to make higher-order numbers. If decimal fractions between 0 and 1 (obtained by dividing the random number by the total number members in the set) are employed, in the limit one obtains a continuous, uniform distribution of the random numbers on the interval 0 to 1.

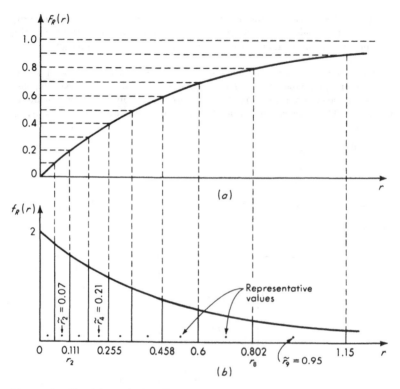

Fig. 2.3.15 Partition of the distribution of rainfall R for simulation. (a) CDF of R; (b) PDF of R.

The use here of these centroids will be justified in the following section, where it will be known as the "conditional mean" of R_i given that $1.15 \le R_i \le \infty$. This and other representative values appear in the previous list, and the dissection of the PDF and CDF of R is illustrated in Fig. 2.3.15a and b.

In computer-aided simulation a large number of intervals are normally used (e.g., 1 million, if six-digit random numbers are generated). In this case it is common to use simply the division values r_j as the representative values $\tilde{r}_j$. A random number is generated on the interval 0 to 1, say 0.540012. The corresponding value r is found by solving $F(r) = 0.540012$ for r, that is, by finding the inverse of $F(r)$. In this example we have

$$F(r) = 1 - e^{-2r} = 0.540012$$

implying

$$r = -\tfrac{1}{2} \ln (1 - 0.540012)$$

$$r = 0.388$$

This same procedure is valid no matter what the form of the CDF of the random variable.† It is not always feasible, however, to obtain a closed-form analytical

† This method implicitly makes use of the so-called "probability integral transform," Prob. 2.34.

expression for the inverse of the CDF of a random variable. In this case one can find the value of x corresponding to the randomly selected value of $F_X(x)$ by hand through graphical means or by computer through "table lookup" schemes. Various other techniques are also used to generate sample values of random variables with particular distributions. For example, a sample of a "normally distributed" random variable (Sec. 3.3.1) is usually obtained in computer simulation by adding 12 or more random numbers. The justification for this procedure will become clear in Sec. 3.3.1.

In the first experiment in this example a random number 4751 was generated, and a sample value of N, $n = 20$, determined. Consequently, to complete this first experiment, 20 random numbers must be generated and the corresponding values of the R_i listed and accumulated to determine a sample value of T. Such a sequence is shown in Table 2.3.2 using the divisions shown in Fig. 2.3.15. The table represents "1 year" of real-time observation.

Again, repetition of such experiments will yield a sample of values of T whose histogram is an approximation of $f_T(t)$ and whose frequency polygon† is an approximation to $F_T(t)$. Figure 2.3.16 shows the results of such a series of experiments for several sample sizes and two different numbers of histogram intervals of r. Analytical treatment of aspects of this same problem will follow in Sec. 2.4 and Chap. 3.

† As we shall see in Chap. 4, the sample mean and sample variance (Sec. 1.2) also have significance with respect to the random variable T. They are estimates of the first two moments (Sec. 2.4.1) of T.

Table 2.3.2 Simulated values of the rainfall in the 20 rainy weeks of 1 year

Week no.	Random number generated	Corresponding value of R
1	0	0.026
2	7	0.685
3	2	0.130
4	0	0.026
5	6	0.520
6	9	1.650
7	8	0.950
8	3	0.210
9	8	0.950
10	3	0.210
11	0	0.026
12	8	0.950
13	4	0.300
14	0	0.026
15	9	1.650
16	0	0.026
17	4	0.030
18	0	0.026
19	5	0.400
20	1	0.070
		$t = 8.871$ in.

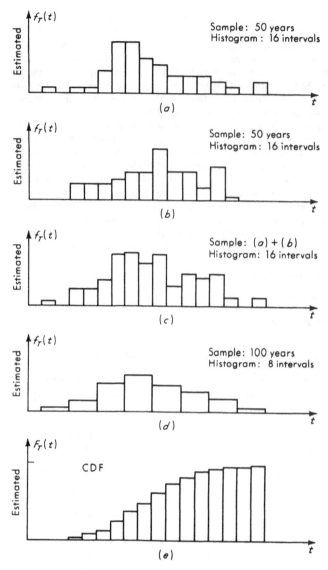

Fig. 2.3.16 Histograms of several simulation runs reduced to estimated density functions and an estimated cumulative distribution function.

It should be evident even from these simple illustrations that many complex engineering problems can be analyzed by Monte Carlo simulation techniques. The models need not be restricted to functional relationships, but can involve complicated situations in which the distributions of variables depend upon the (random) state of the physical system or in which the whole sequence of steps to be followed may depend upon which particular value of a random variable is observed. Elaborate probabilistic models of many vehicles on complicated networks of streets and intersections have proved most useful to traffic engineers trying to predict the performance of a proposed design. The uncertainty and variation in drivers' reaction time, driving habits, and origin-destination demands can all be treated on a probabilistic basis. Long histories of the operation of whole river-basin systems of water-resource controls have been simulated in minutes (Hufschmidt and Fiering [1966]). Rainfall and runoff values form probabilistic inputs to chains of dams and channels (real or designed) whose proposed operating policies (e.g., the degree to which reservoirs should be lowered in anticipation of flood runoffs) are being evaluated for long-term consequences, likelihoods of poor performance, and economics. Simulation, combined with numerical integration of the equations of motion, has been used to obtain approximate distributions of the maximum dynamic response of complex, non-linear structures to the chaotic, random ground motions during earthquakes (Goldberg et al. [1964]).

The successful application of simulation depends upon the appropriateness of the model and the interpretation of the results as much as on the sophistication of the simulation techniques used. The former problems are the ones the engineer usually faces in using mathematical models of natural phenomena; for the latter problems, those of technique, the engineer can find help and documented experience in a number of references. (See, for example, Tocher [1963] or Hammersley and Handscomb [1964].) Many methods are available, for example, to generate random numbers, to account for dependence among variables, or to reduce the effort needed to get the desired accuracy.

2.3.4 Summary

This section presents methods for deriving the distributions of random variables which are functionally dependent upon random variables whose distributions are known. In general, one should seek the CDF of the dependent random variable Z. For any particular value z, $F_Z(z)$ is found by calculating in the sample space of X (or X and Y) the probability of all those events where $g(x)$ [or $g(x,y)$] is less than or equal to z:

$$F_Z(z) = P[Z \leq z] = P[g(X) \leq z]$$

This can be done by enumeration if the distribution of X (or X and Y) is discrete. If the distribution of X is continuous, integration is required. In certain circumstances (monotonic, increasing, one-to-one relationships), this procedure simply reduces to

$$F_Z(z) = F_X(g^{-1}(z))$$

The density function of Z can be found by differentiation of $F_Z(z)$ (assuming that Z is a continuous random variable). In certain cases the differentiation can be carried out explicitly *before* particular probability laws $F_X(x)$ are considered, in which case one can obtain a formula for the PDF of Z directly. The two most important examples are

1. Under monotonic, one-to-one conditions:

$$f_Z(z) = \left| \frac{d}{dz} g^{-1}(z) \right| f_X(g^{-1}(z))$$

2. If $Z = X + Y$:

$$f_Z(z) = \int_{-\infty}^{\infty} f_{X,Y}(x, z - x) \, dx$$

Certain computational methods are available for cases where analytical solutions are difficult or impossible. They include

1. Approximating continuous distributions by discrete ones, and solving by enumeration
2. Applying simulation techniques to obtain, by repeated experimentation, observed histograms which approximate desired results.

2.4 MOMENTS AND EXPECTATION

Because of the very nature of a random variable it is not possible to predict the exact value that it will assume in any particular experiment, but a complete description of its behavior is contained in its probability law, as presented in the CDF (or PMF or PDF, if applicable) of the variable. This complete information can be communicated only by stipulating an entire function, e.g., the PDF. In many situations this much information may not be necessary or available. More concise descriptors, summarizing only the dominant features of the behavior of a random variable, are often sufficient for the engineering purpose at hand. One or more simple numbers are used in place of a whole probability density function. These numbers usually take the form of weighted averages of certain functions of the random variable. The weights used are the PMF or PDF of the variable, and the average is called the *expectation* of the function.

We will find that, compared with entire probability laws, these expectations are much easier to work with in the analysis of uncertainty, as well as much easier to obtain estimates of from available data. Therefore, in engineering applications, where expedience often dictates that approximate but fast answers are better than none at all, averages and expectations prove invaluable.

2.4.1 Moments of a Random Variable

Mean Every engineer is familiar with averages of observed numerical data. The sample mean and sample variance (Sec. 1.2) are the most common examples. Although they do not communicate all the available information, they are concise descriptors of the two most significant properties of the batch of observed data, namely, its central value and its scatter or dispersion.

On the other hand, given a solid body, e.g., a rod of nonuniform shape or density, the engineer is accustomed to determining certain numerical descriptions of the body such as the location of its center of mass and a moment of inertia about that point. Not complete in their description, these quantities are nonetheless sufficient to enable the engineer to predict a great deal about the gross static and dynamic behavior of the body.

Both these examples deserve being kept in mind when we define the *mean* m_X or the *expected value* $E[X]$ of a discrete random variable X as

$$m_X = E[X] = \sum_{\text{all } x_i} x_i p_X(x_i) \tag{2.4.1}$$

or, for a continuous random variable, as

$$m_X = E[X] = \int_{-\infty}^{\infty} x f_X(x) \, dx \tag{2.4.2}$$

In the mean (or *mean value*) we are condensing the information in the probability distribution *function* into a single *number* by summing over all possible values of X the product of the value x and its likelihood $p_X(x)$ or $f_X(x) \, dx$.

Recall from Chap. 1 that the sample mean of n numbers was defined as

$$\bar{x} = \frac{1}{n} \sum_{i=1}^{n} x_i \tag{2.4.3}$$

If several observations of each value x_i are found, this definition can be written

$$\bar{x} = \sum_{i=1}^{r} \frac{n_i}{n} x_i = \sum_{i=1}^{r} x_i f_i \tag{2.4.4}$$

in which r is the total number of distinct values observed, n_i is the total number of observations at value x_i, and $f_i = n_i/n$ is the observed frequency of the value x_i in the sample.

Notice the close proximity in appearance between the definition of the mean of a (discrete) random variable, Eq. (2.4.1), and the sample mean of a batch of observed numbers, Eq. (2.4.4). This similarity helps make clear the notion of the mean of a random variable (especially when repeated observations of the random variable are anticipated), but the student should be most careful to avoid confusing the two means. The sample mean is computed from given observations, and the mean (or expectation) is computed from the mathematical probability law (e.g., from the PDF, CDF, or PMF) of a random variable. The latter is sometimes called the *population mean* to distinguish it from the sample mean. Geometrically, it is clear from its definition, Eqs. (2.4.1) or (2.4.2), that the mean defines the center of gravity of the shape defined by the PDF or PMF of a random variable.

In a physical problem, where some phenomenon has been modeled as a random variable, the mean value of that variable is usually the most significant single number the engineer can obtain. It is a measure of the central tendency of the variable, and, often, of the value about which scatter can be expected if repeated observations of the phenomenon are to be made. The sample mean of many such observations will, with high probability,[†] be very close to the (population) mean of the underlying random variable. For these and other reasons to be seen, when probabilistic model and a deterministic (nonprobabilistic) model of a physical phenomenon are compared, it is usually the mean of the probabilistic model which one compares with the single value of a deterministic model.

The mean of the discrete random variable adopted in Sec. 2.2.1 to model the annual runoff (see Fig. 2.2.3) is computed as follows, using Eq. (2.4.1):

x_i	$p_X(x_i)$	$x_i p_X(x_i)$
0	$6/60$	0
1,000	$2/60$	2,000/60
2,000	$4/60$	8,000/60
3,000	$8/60$	24,000/60
4,000	$13/60$	52,000/60
5,000	$12/60$	60,000/60
6,000	$7/60$	42,000/60
7,000	$4/60$	28,000/60
8,000	$2/60$	16,000/60
9,000	$2/60$	18,000/60
		250,000/60 = 4167 acre-ft

[†] For a more formal statement of this, the *law of large numbers*, see Prob. 2.43.

The mean runoff is 4167 acre-ft. This is a central value, a value about which observed values of X will tend in the long run to be scattered. It is probably the number the engineer would use if he were restricted to using only a single number to describe the runoff, that is, if he had to treat runoff deterministically rather than probabilistically in his analysis.

The mean value of the rainfall in a rainy week, the random variable R in the illustration in Sec. 2.3.3, is calculated by integration, after Eq. (2.4.2). Substituting for the PDF the function $2e^{-2r}$, $r \geq 0$,

$$m_R = \int_{-\infty}^{\infty} r f_R(r) \, dr \qquad (2.4.5)$$

$$= \int_0^{\infty} r 2e^{-2r} \, dr$$

$$= \tfrac{1}{2} \text{ in.} \qquad (2.4.6)$$

If it rains in a week, $\tfrac{1}{2}$ in. is the mean value or "expected" value of the rainfall. Comparing this value with the density function of R (Fig. 2.3.15), it is apparent that in this case, the mean of R is not central, in that it does not correspond to a peak in the distribution. Nor is the mean the value which will be exceeded half of the time.† (That is, $1 - F_R(\tfrac{1}{2} \text{ in.}) = e^{-1} = 0.368 \neq 0.5$. Solving $1 - F_R(u) = 0.5$ yields the median $u = 0.346$ in.) Nonetheless, even here the mean value yields "order of magnitude" information as to weekly rainfall in rainy weeks, and, if the observed records of many such weeks are averaged, this sample average would almost certainly be very close‡ to $\tfrac{1}{2}$ inch (assuming always that the mathematical model is a good representation of the physical phenomenon).

Variance The mean describes the central tendency of a random variable, but it says nothing of that behavior which leads engineers to study probability theory at all, namely, uncertainty or randomness. Here we seek a descriptor, a single number, which will give an indication of the scatter or dispersion, or, loosely, of the "randomness" in the random variable's behavior.

Several such measures are possible. The *range* of the random variable is one example, although it is frequently a rather uninformative

---

† These two descriptors of a random variable, namely, the "most probable" value, that at the peak of its PDF, and the "midpoint," the value of x at which the CDF equals $\tfrac{1}{2}$, are known, respectively, as the *mode* and the *median*. They are only seldom used (see Sec. 3.3.2 for one use) and will not be considered further here. Clearly the mean, mode, and median coincide if a distribution has a single peak and is symmetrical. Note, however, that neither mode nor median may have a unique value for some distributions.

‡ This notion can be made more explicit through the law of large numbers, Prob. 2.43.

pair of numbers such as $-\infty$ to ∞ or 0 to ∞. Even if the range is two finite numbers, say, a to b, it gives no indication of the relative frequency of extreme values as compared with central values. It is desirable therefore to measure the dispersion from a central value, the mean, and to weight all deviations from the mean by their relative likelihoods.

The most common and most useful such measure of the dispersion of a random variable is the *variance* σ_X^2, or Var $[X]$. It is defined as the weighted average of the squared deviations from the mean:

$$\sigma_X^2 = \text{Var } [X] = \sum_{\text{all } x_i} (x_i - m_X)^2 p_X(x_i) \qquad (2.4.7)$$

or

$$\sigma_X^2 = \int_{-\infty}^{\infty} (x - m_X)^2 f_X(x) \, dx \qquad (2.4.8)$$

The variance of a random variable bears the same relationship to the sample variance of a set of numbers (Sec. 1.2) as the mean does to the sample mean, and a comparison analogous to that made above between Eqs. (2.4.1) and (2.4.4) could be made with ease. A more meaningful analogy to draw is that between the variance of a random variable and the central moment of inertia of a bar of variable density (and unit mass). The variance σ_X^2 is the second central moment of the area of the PDF or PMF with respect to its center of gravity m_X.

In Fig. 2.4.1 are shown probability density functions of the same basic shape. The curves in Figs. 2.4.1a and b differ only in their mean, while the curves in Figs. 2.4.1b and c differ only in their variances. Smaller variances generally imply "tighter" distributions, less widely spread about the mean.

Standard deviation The positive square root of the variance is given the name *standard deviation:*

$$\sigma_X = \sqrt{\sigma_X^2} = (+) \left[\int_{-\infty}^{\infty} (x - m_X)^2 f_X(x) \, dx \right]^{1/2} \qquad (2.4.9)$$

The conventional form of the standard notation, σ and σ^2, seems to indicate that in practice the standard deviation is given more importance than the variance. That is, in fact, the case.† The standard deviation has the same units as the variable X itself and can be compared easily and quickly with the mean of the variable to gain some feeling for the degree and gravity of the uncertainty associated with the random variable.

---

† In the analysis and development of models, on the contrary, the variance is a more fundamental notion and also is easier to work with, as will be seen in subsequent sections.

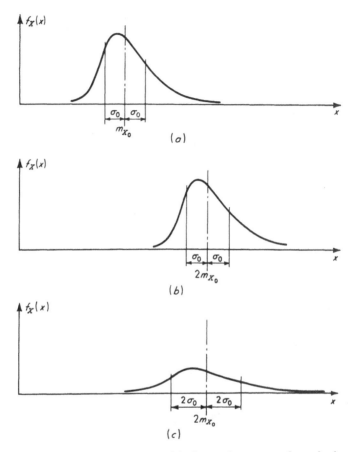

Fig. 2.4.1 Changes in PDF's with changes in means and standard deviations.

Coefficient of variation A unitless characteristic that formalizes this comparison and that also facilitates comparisons among a number of random variables of different units is the *coefficient of variation* V_X:

$$V_X = \frac{\sigma_X}{m_X} \tag{2.4.10}$$

The coefficient of variation of the strength of the concrete produced by a given contractor is often assumed to be a constant for all mean strengths and to be a measure of the quality control practiced in his work (see ACI [1965] and Prob. 2.45). Thus a contractor practicing "good" control might produce concrete with a coefficient of variation of

0.10 or 10 percent, implying that concrete of mean strength 4000 psi would have a standard deviation of 400 psi, whereas "5000-psi concrete" would have a standard deviation of 500 psi.

The variance of the discrete runoff random variable can be computed from Eq. (2.4.7) as follows:

$x_i - m_X$	$p_X(x_i)$	$(x_i - m_X)^2 p_X(x_i)$
$-4,167$	$6/60$	$1,740,000$
$-3,167$	$2/60$	$333,000$
$-2,167$	$4/60$	$313,000$
$-1,167$	$8/60$	$181,000$
-167	$13/60$	$6,000$
833	$12/60$	$139,000$
$1,833$	$7/60$	$390,000$
$2,833$	$4/60$	$531,000$
$3,833$	$2/60$	$487,000$
$4,833$	$2/60$	$687,000$
		Total: $4,798,000$ acre-ft$^2 = \sigma_X{}^2$

The standard deviation of this variable is

$$\sigma_X = \sqrt{\sigma_X{}^2} = 2200 \text{ acre-ft}$$

and the coefficient of variation is

$$V_X = \frac{\sigma_X}{m_X} = \frac{2200}{4167} = 0.528$$

The variance of the rainfall variable R is found by applying Eq. (2.4.8):

$$\sigma_R{}^2 = \int_{-\infty}^{\infty} (r - m_R)^2 f_R(r) \, dr$$
$$= \int_{0}^{\infty} (r - \frac{1}{2})^2 2e^{-2r} \, dr = \frac{1}{4} \text{ in.}^2$$

while

$$\sigma_R = \sqrt{\sigma_R{}^2} = \frac{1}{2} \text{ in.}$$

and

$$V_R = \frac{\sigma_R}{m_R} = \frac{\frac{1}{2}}{\frac{1}{2}} = 1.0$$

Illustration: Sigma bounds and the Chebyshev inequality It is common in engineering applications of probability theory to speak of the "one-, two-, and three-sigma" bounds of a random variable. The range between the two-sigma bounds, for

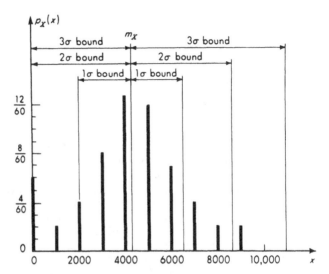

Fig. 2.4.2 Sigma bounds: runoff model.

example, is the range between $m_X - 2\sigma_X$ and $m_X + 2\sigma_X$. The two-sigma bounds on the runoff variable X are $4167 - 2(2200)$ and $4167 + 2(2200)$ or 0 and 8567.† The three sets of bounds for this variable are shown in Fig. 2.4.2.

In absence of the knowledge of the complete PDF, but knowing mean and variance, it is frequently stated in engineering applications (for reasons that will be clear in Sec. 3.3.1) that the probability that a variable lies within the one-sigma bounds of its mean is approximately 65 percent; within the two-sigma bounds, about 95 percent; and within the three-sigma bounds, about 99.5 percent. That even rough probability figures can be given with so little information is indicative of the value of the standard deviation and the variance as measures of dispersion. In fact, these approximate statements should only be used when it is known that the distribution is roughly bell-shaped.

More formally we can show that the mean and standard deviation alone are sufficient to make certain exact statements on the probability of a random variable lying within given bounds. The Chebyshev inequality‡ states that

$$P[(m_X - h\sigma_X) \le X \le (m_X + h\sigma_X)] \ge 1 - \frac{1}{h^2} \qquad (2.4.11)$$

Note that, corresponding to the one-, two-, and three-sigma bounds, h equals 1, 2, and 3.

For example, the runoff variable has mean 4167 and standard deviation 2200. The Chebyshev inequality states

$$P[(4167 - 2200h) \le X \le (4167 + 2200h)] \ge 1 - \frac{1}{h^2}$$

† Since negative values are not meaningful in this problem, the lower two-sigma bound on this variable is given as 0 rather than -233.

‡ The proof of this inequality is elementary and is outlined in a hint to Prob. 2.42.

If $h = 2$

$P[-233 \leq X \leq 8567] \geq 0.75$

Or, recognizing in this particular case that X, the runoff, is nonnegative,

$P[X \leq 8567] \geq 0.75$

or, in another form,

$P[X \geq 8567] \leq 0.25$

The rule of thumb for the two-sigma bounds suggests that the former probability, that is, $P[X \leq 8567]$, should be about 95 percent, while the exact value, Fig. 2.4.2, is $1 - \frac{2}{60} = 0.967$.

The Chebyshev inequality does not yield very sharp bounds. [Consider, for example, any value of h less than 1, when the right-hand side of Eq. (2.4.11) is negative.] It has the advantage, however, that it requires no assumption on the part of the engineer regarding the shape of the distribution. Hence the probability statements are totally conservative, within the qualification, of course, that m_X and σ_X are known with certainty. When these parameters' estimates are based on only a small amount of observed data, they, in fact, are not known with high confidence; this uncertainty is a subject of Chaps. 4 and 6.

A host of less general, but more precise inequalities are available. (See, for example, Parzen [1960], Freeman [1963].) As the engineer makes more and more assumptions regarding the shape of the distribution, such as its being unimodal (single-peaked), having "high-order contact" with the x axis in the extreme tails,† being symmetrical, having known values of higher moments (see Sec. 2.4.2), etc., the sharper his probability statements can be. Finally, of course, if he is willing to stipulate $F_X(x)$ itself, he can state exactly the proportion of the probability mass lying inside or outside any interval. This progression is common in applied probability theory; the more the engineer is willing to assume to be known, given, or hypothesized information, the more precise and penetrating can he be in his subsequent probabilistic analysis.

Discussion   It should be emphasized that from the point of view of applications, two quite opposite problems have been discussed in this section. The first is that of calculating the mean and variance of a random variable knowing its distribution, and the second is that of making some kind of statement about the behavior of the variable when only the mean and variance are known.

The latter case is frequent. It arises in many situations. Data is often available only as summarized by its sample mean and sample variance, which are the natural estimates of the corresponding two model parameters. Also, in the analysis of complex models, the mean and variance of the dependent random variable are often easily obtainable when the complete story, i.e., the whole distribution, is lost in a maze

† If only these first two conditions are assumed, for example, h^2 can be replaced by $2.25h^2$ in Eq. (2.4.11) (Freeman [1963]).

of intractable integrals (Sec. 2.3). As will be discussed later in this section, we can often determine these two parameters as simple functions of the same two parameters of the independent random variables involved, implying that the engineer may never need to commit himself to unnecessarily detailed models of these variables. As shown in the next section, the mean and variance alone contain a substantial amount of information on which to base engineering decisions. Finally, they represent the first and most important step beyond deterministic engineering models in that they characterize not only the typical value, as the traditional model does, but also the dispersion. Recognition of the existence of the variance indicates that the probabilistic aspects of the engineering problem won't be ignored. Study of the mean and variance, and their later counterparts, is referred to as a *second-order-moment analysis;* there will be an emphasis on this notion throughout our study of applied probability and statistics.

Summary In Sec. 2.4.1 we have defined the mean, variance, standard deviation, and coefficient of variation of a random variable. They are defined as sums of the possible values of the random variable or as sums of the squared deviations from the mean, weighted by their probabilities of occurrence. The mean and variance are the center of gravity of the probability mass and its moment of inertia. They can be calculated from given probability distributions, or used alone, without knowledge of the entire distribution, as summaries of the predominant characteristics of central value and dispersion.

2.4.2 Expectation of a Function of a Random Variable

In Sec. 2.3 we emphasized the importance in engineering applications of being able to determine the behavior of a (dependent) random variable Y, which is a function $g(X)$ of another (independent) random variable X, whose behavior is known. The computational difficulties involved in actually finding, say, the PDF of Y from the PDF of X have also been encountered. Fortunately, as mentioned above in the justification for a concentrated study of the mean and variance, no such computational complications arise if one seeks only these two moments of Y, given the probability law of the independent variable X. Often even less is needed; for example, often only the mean and variance of X can be used to find corresponding moments of Y.

Expectation of a function If we know the probability law of X and our interest is in Y, where

$$Y = g(X)$$

then the expected value or *expectation* of Y is, by definition [Eq. (2.4.2)],

$$E[Y] = \int_{-\infty}^{\infty} y f_Y(y) \, dy \tag{2.4.12}$$

The method for computing $f_Y(y)$ and its attendant difficulties were the subject of Sec. 2.3. It is a fundamental, but difficult to prove,† result of probability theory that this expectation can be evaluated by the much easier computation‡

$$E[Y] = E[g(X)] \tag{2.4.13}$$

where the notation $E[g(X)]$ is defined as:

$$E[g(X)] = \int_{-\infty}^{\infty} g(x) f_X(x) \, dx \tag{2.4.14}$$

For a discrete variable,§ the expectation of $g(X)$ is defined as

$$E[g(X)] = \sum_{\text{all } x_i} g(x_i) p_X(x_i) \tag{2.4.15}$$

It is helpful to see a simple discrete illustration. Suppose a random variable X has PMF as shown in Fig. 2.4.3a. Then the PMF of $Y = g(X) = X^2$ is as shown in Fig. 2.4.3b. This result can be verified by straightforward enumeration in this simple example. The expected value of Y is, by definition, $E[Y] = (\frac{2}{3})(1) + (\frac{1}{3})(4) = 2$. Equation (2.4.14) states, however, that this expected value can also be found without first determining $p_Y(y)$. Using Eq. (2.4.14),

$$E[Y] = E[X^2] = (\frac{1}{3})(-1)^2 + (\frac{1}{3})(1)^2 + (\frac{1}{3})(2)^2 = 2$$

as before.

Notice that the two quantities defined in Sec. 2.4.1, the mean and variance, can in fact be interpreted as merely special cases of Eq. (2.4.14) with $g(X) = X$ for the mean and $g(X) = (X - m_X)^2$ for the variance. This equivalence suggests that two quite different interpretations might be given to Eq. (2.4.14). In some situations one may think of $E[g(X)]$ as representing the mean of a random variable $Y = g(X)$ conveniently calculated by this equation rather than by Eq. (2.4.12). This interpretation is common when interest centers on $Y = g(X)$ as

---

† No proof will be attempted here. It is hoped that the student can mentally generalize upon the discrete example which is to follow well enough to convince himself of the plausibility of the result.

‡ The expectation is said to "exist" if and only if the integral in Eq. (2.4.14) is absolutely convergent, that is, if $\int_{-\infty}^{\infty} |g(x)| f_X(x) \, dx < \infty$.

§ In this section, subsequent definitions will be stated only in terms of continuous random variables. The extension to the discrete case will be obvious.

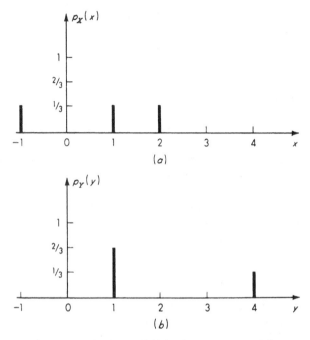

Fig. 2.4.3 PMF's of X and derived $Y = X^2$. (a) Discrete PMF of X; (b) discrete PMF of $Y = X^2$.

a dependent random variable functionally related to another random variable X (with known distribution function), as, for example, when X is velocity and $Y = aX^2$ is kinetic energy. On the other hand, if interest centers on X itself, then the expectation of $g(X)$ as given by Eq. (2.4.14) is usually interpreted as weighted average of $g(X)$, over X, that is, as the sum of the values of the function of X evaluated at the various possible values of X and weighted by the likelihoods of those values of X. The variance of X, for example, is usually thought of in this light, rather than as the mean of a random variable $Y = (X - m_X)^2$. The distinction between these two interpretations is not of fundamental importance, but it may prove helpful when considering the uses to which the material to follow will be put.

Moments   In keeping with the latter interpretation of $E[g(X)]$ as a weighting of $g(x)$ by $f_X(x)$, we introduce a family of averages of X, called *moments*, that prove useful as numerical descriptors of the behavior of X. We call

$$m_X{}^{(n)} = E[X^n] = \int_{-\infty}^{\infty} x^n f_X(x) \, dx \tag{2.4.16}$$

the *nth moment of X*. Notice that this moment corresponds to the nth moment of the area of the PDF with respect to the origin. If $n = 1$, when the superscript is usually omitted, we have the mean of the variable.

It is possible, of course, to consider moments of areas about any point. In particular, moments with respect to the mean are called *central moments:*

$$\mu_X^{(n)} = E[(X - m_X)^n] = \int_{-\infty}^{\infty} (x - m_X)^n f_X(x)\, dx \qquad (2.4.17)$$

Such moments correspond to the familiar moments of areas with respect to their centroids. The most important particular case is for $n = 2$, when $\mu_X^{(2)} \equiv \sigma_X^2$, the variance. The first central moment is, of course, always zero. If the asymmetry of a distribution is of interest, this property is often quantified or characterized by the third central moment $\mu_X^{(3)}$, or by the corresponding dimensionless *coefficient of skewness* γ_1:

$$\gamma_1 = \frac{\mu_X^{(3)}}{\sigma_X^3} \qquad (2.4.18)$$

If a distribution is symmetrical, this coefficient is zero (although the converse is not necessarily true). Positive values of γ_1 usually correspond to PDF's with dominant tails on the right; negative values to long tails on the left (see Fig. 2.4.4).

A less common coefficient γ_2, the *coefficient of kurtosis* (flatness), is defined similarly:

$$\gamma_2 = \frac{\mu_X^{(4)}}{\sigma_X^4} \qquad (2.4.19)$$

It is often compared to a "standard value" of 3.†

† $(\gamma_2 - 3)$ is called the *coefficient of excess*. The value of 3 is chosen only because it is the value of the kurtosis coefficient of a particular, commonly used distribution (see Sec. 3.3.1).

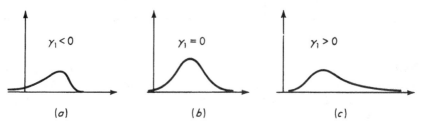

Fig. 2.4.4 Variation of shape of PDF with coefficient of skewness γ_1. (*a*) Negative skewness; (*b*) zero skewness; (*c*) positive skewness.

For example, the central moments of the rainfall variable R, defined in Sec. 2.3.3, are, in general,

$$\mu_R{}^{(n)} = \int_{-\infty}^{\infty} (r - m_R)^n f_R(r)\, dr$$

$$= \int_0^{\infty} (r - \tfrac{1}{2})^n e^{-2r}\, dr$$

In the previous section, it was found that

$$\mu_R{}^{(2)} = \sigma_R{}^2 = \tfrac{1}{4} \text{ in.}^2$$

Also

$$\mu_R{}^{(3)} = \int_0^{\infty} 2(r - \tfrac{1}{2})^3 e^{-2r}\, dr = \tfrac{1}{8} \text{ in.}^3$$

and

$$\mu_R{}^{(4)} = \int_0^{\infty} 2(r - \tfrac{1}{2})^4 e^{-2r}\, dr = \tfrac{9}{16} \text{ in.}^4$$

Hence the skewness and kurtosis coefficients are

$$\gamma_1 = \frac{\tfrac{1}{8}}{(\tfrac{1}{2})^3} = 1$$

indicating a positive skewness or long right-hand tail (see Fig. 2.3.15), and

$$\gamma_2 = \frac{\tfrac{9}{16}}{(\tfrac{1}{2})^4} = 9$$

Properties of expectation   No matter which interpretation of the expectation $E[g(X)]$ is involved, several general properties of the operation can be pointed out. For example, the expectation of a constant c is just the constant itself. This fact is easily shown. Simply by writing out the definition, we have

$$E[c] = \int_{-\infty}^{\infty} c f_X(x)\, dx = c \int_{-\infty}^{\infty} f_X(x)\, dx = c \tag{2.4.20}$$

Similarly the following properties can be verified with ease for constants a, b, and c:

$$E[cX] = cE[X] \tag{2.4.21}$$

$$E[a + bX] = a + bE[X] \tag{2.4.22}$$

$$E[g_1(X) + g_2(X)] = E[g_1(X)] + E[g_2(X)] \tag{2.4.23}$$

The implication of this last equation is that expectation, like differentiation or integration, is a *linear operation*. This linearity property is very useful computationally. It can be used, for example, to find the following formula for the variance of a random variable in terms of more easily calculated quantities.

$$\text{Var}\,[X] = E[(X - m_X)^2]$$

Expanding the square,

$$\text{Var}\,[X] = E[X^2 - 2m_X X + m_X{}^2]$$

Each term in the sum can be treated separately as a function of X and, according to Eq. (2.4.23), *the expectation of their sum is the sum of their expectations:*

$$\text{Var}\,[X] = E[X^2] - E[2m_X X] + E[m_X{}^2]$$

Using other properties of expectation [Eqs. (2.4.20) and (2.4.21)],

$$\text{Var}\,[X] = E[X^2] - 2m_X E[X] + m_X{}^2$$

But $E[X] = m_X$; therefore

$$\text{Var}\,[X] = E[X^2] - 2m_X{}^2 + m_X{}^2 = E[X^2] - m_X{}^2 \qquad (2.4.24)$$

In an alternate form, the variance is said to be the "mean square" minus the "squared mean":

$$\text{Var}\,[X] = E[X^2] - E^2[X] \qquad (2.4.25)$$

Given the PDF of X, the simplest way to evaluate the variance is usually to calculate $\int_{-\infty}^{\infty} x^2 f_X(x)\,dx$, that is, $E[X^2]$, and then to subtract the squared mean.

Note that this last derivation took place *without any reference to a particular form* of the PDF and even without indication as to whether a discrete or continuous variable was involved. It is this ability to work with expectation without specifying the PDF that often permits us to determine relationships among the moments of two functionally related variables, X and $Y = g(X)$, before specifying or even without knowledge of the PDF of X. For example, in concrete the quadratic relationship

$$Y = bX - cX^2 \qquad (2.4.26)$$

holds between compressive stress Y and unit strain X well beyond the linear elastic range (Hognestad [1951]). If the unit strain applied to a specimen by a testing machine is a random variable, owing, say, to uncertainties in recording, then so is the stress in the concrete. How might the expected or mean stress be determined? If the PDF of X is known, the mean stress m_Y *could* be found by using the methods of Sec. 2.3 to find $f_Y(y)$, and then applying Eq. (2.4.2). Alternatively,

$E[Y]$ could be calculated using Eq. 2.4.14, since

$$m_Y = E[Y] = E[bX - cX^2] = \int_{-\infty}^{\infty} (bx - cx^2)f_X(x)\, dx \qquad (2.4.27)$$

But, if the mean and variance of the strain are available, even if the PDF is not, the mean of Y can be found much more directly as simply

$$m_Y = E[Y] = E[bX - cX^2] = bE[X] - cE[X^2]$$

Using Eq. (2.4.24),

$$E[X^2] = \sigma_X^2 + m_X^2$$

Therefore

$$m_Y = bm_X - c(m_X^2 + \sigma_X^2) \qquad (2.4.28)$$

Thus the mean and variance of X are sufficient to determine the mean of Y. An important practical distinction between deterministic and probabilistic analysis is also illustrated by this example. In a deterministic formulation of this problem, one would assume only a single value of strain was possible, say the typical value m_X. The predicted stress would then be $bm_X - cm_X^2$, which does *not* coincide with the mean value of Y, unless, of course, the dispersion in X is truly zero. The greater the uncertainty in the strain, the greater the *systematic* error in the predicted value of the stress.

Formalizing this observation: in general, we cannot find the expectation of a function of X by substituting in the function $E[X]$ for X, or

$$E[g(X)] \neq g(E[X]) \qquad (2.4.29)$$

For example, the mean of $1/X$ is not $1/m_X$.

The linearity property of expectation, which made the previous derivations so direct, does not carry over to variances. The variance of $Y = 2X$ is *not* two times the variance of X. This fact is easily demonstrated:

$$\text{Var}\,[Y] = \text{Var}\,[2X] = E[(2X - E[2X])^2]$$
$$= E[4X^2 - 4XE[2X] + E^2[2X]] = 4E[X^2] - 4E[X]2E[X]$$
$$+ 4E^2[X] = 4(E[X^2] - E^2[X]) = 4\,\text{Var}\,[X]$$

Several useful general properties of variances can be stated (and easily verified) however:

$$\text{Var}\,[c] = 0$$
$$\text{Var}\,[cX] = c^2\,\text{Var}\,[X]$$
$$\text{Var}\,[a + bX] = b^2\,\text{Var}\,[X]$$

That is, a constant has no variance, and the standard deviation of a linear function of X, $a + bX$, is just $|b|\sigma_X$. A simple example of such a linear function is a problem in which there is a change of units.

Conditional expectation   In Sec. 2.2 we discussed briefly the construction of conditional distributions which are formed conditionally on the occurrence of some event. For example, in a design situation we might be interested in the distribution of the demand or load X, given that it is larger than some threshold value x_0, say, the nominal design demand. Then, by definition of conditional probabilities,

$$F_{X|[X>x_0]}(x) = \frac{P[(X < x) \cap (X > x_0)]}{P[X > x_0]}$$

$$= \begin{cases} \dfrac{F_X(x) - F_X(x_0)}{1 - F_X(x_0)} & x \geq x_0 \\ 0 & x < x_0 \end{cases} \quad (2.4.30)$$

The conditional PDF is found by differentiation with respect to x:

$$f_{X|[X>x_0]}(x) = \frac{f_X(x)}{1 - F_X(x_0)} \qquad x \geq x_0$$

The conditional distributions satisfy all the necessary conditions to be proper probability distributions, and may be used as such.

It is therefore also meaningful to consider conditional means, conditional variances, and in general any expectation that is *conditional* on the prescribed event. So the conditional mean of the demand X, *given* that it is larger than the nominal demand x_0, is

$$E[X \mid X > x_0] = \int_{x_0}^{\infty} x f_{X|[X>x_0]}(x)\, dx \qquad (2.4.31)$$

and its conditional variance is defined as

$$\text{Var}\,[X \mid X > x_0]$$
$$= \int_{x_0}^{\infty} (x - E[X \mid X > x_0])^2 f_{X|[X>x_0]}(x)\, dx \quad (2.4.32)$$

In general, for some event A,

$$E[g(X) \mid A] = \int_{-\infty}^{\infty} g(x) f_{X|[A]}(x)\, dx \qquad (2.4.33)$$

Note that there is a trivial case:

$$E[g(X) \mid X = x_0] = g(x_0) \qquad (2.4.34)$$

A numerical example of the use of conditional distributions and expectations in engineering design will be found in an illustration to follow.

Expected costs and benefits A common use of expectation of a function of a random variable arises from the practice of basing engineering decisions, in situations involving risk, on *expected costs*. Frequently a portion of the total cost of a proposed design depends upon the more or less uncertain future magnitudes of such phenomena as local rainfall, traffic volume, or unit bid prices. If the engineer describes the uncertainty associated with these variables by treating them as random variables, then, when comparing alternate designs, the question arises of how to combine those portions of the costs or benefits that depend upon these random variables with the other, nonrandom, components of cost. How do you compare a more expensive spillway design with one of smaller initial cost, but of smaller capacity, and hence of greater risk of inadequacy during peak flows?

As demonstrated in previous illustrations (e.g., the industrial park example in Sec. 2.1), the *expected cost* (or benefit) related to the variable is usually used to obtain a single number; this cost reflects the sum of all possible values of the random cost weighted by the likelihood of their occurrence. The use of expected cost in making decisions is the subject of much experimental as well as theoretical investigation (Fishburn [1965]), and it will be discussed more fully in Chap. 5. For the time being it can be accepted as an intuitively rational description of the economic consequences associated with the random variable.

In general, every value of the random variable leads to a corresponding cost or benefit. In other words, we can define cost as a function of the variable:

$$C = C(X) \qquad\qquad (2.4.35)$$

Examples of the shapes of such cost functions are sketched in Fig. 2.4.5.

In many cases the cost function is at least approximately linear. The cost of evacuating 100,000 yd^3 of earth, for instance, depends linearly on X, the price bid per cubic yard, Fig. 2.4.5c. Then $C(X)$ is of the form (in Fig. 2.4.5c the constant a is zero)

$$C(X) = a + bX \qquad\qquad (2.4.36)$$

and the expected cost

$$E[C] = E[a + bX] = a + bm_X \qquad\qquad (2.4.37)$$

That is, in this case the expected cost depends only on the mean of the random variable.

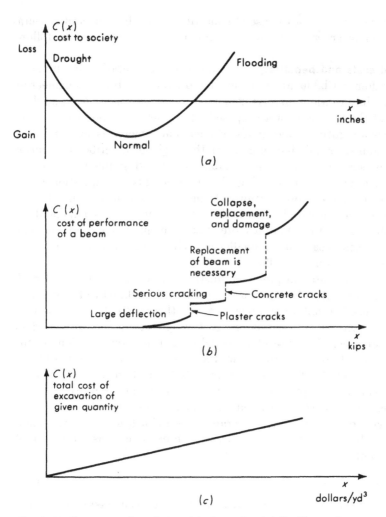

Fig. 2.4.5 Some cost functions. (*a*) Annual rainfall; (*b*) maximum load on a concrete beam; (*c*) unit bid price.

In other situations the cost can be approximated by a quadratic relationship. Figure 2.4.5*a* is a possible example. In this case the expected cost depends only on the mean and variance of the random variable:

$$C(X) = a + bX + cX^2 \qquad (2.4.38)$$

$$E[C] = E[a + bX + CX^2] = a + bm_X + cE[X^2]$$
$$= a + bm_X + c(\sigma_X^2 + m_X^2) \qquad (2.4.39)$$

The importance of the first- and second-order moments, i.e., means and variances, of random variables is magnified when it is recognized that frequently the ultimate engineering use of the probabilistic model will be in a decision-making context where a linear or quadratic cost function is a valid approximation. In such cases, the mean and variance are sufficient information on which to base the decision if an expected cost criterion is used.

Illustration: Pump-capacity decision This extended illustration is designed to demonstrate a number of the concepts discussed to this point in the text and to serve as a discussion of others. In particular, a mixed random variable will be encountered, a result here of a function $g(X)$ which takes on the same value for many different values of X.

The demand X during the peak summer hour at a water-pumping station has a triangular distribution given by

$$f_X(x) = \begin{cases} \dfrac{1}{(100)^2}(x - 50) & 50 \leq x \leq 150 \text{ cfs} \\ \dfrac{2}{100} - \dfrac{1}{(100)^2}(x - 50) & 150 \leq x \leq 250 \text{ cfs} \\ 0 & \text{elsewhere} \end{cases} \qquad (2.4.40)$$

and sketched in Fig. 2.4.6a.

The existing pump is adequate for any demand from 0 to 150 cfs. A new pump is to be added; it will be operated only if demand exceeds the capacity of the existing pump. Hence a demand of up to 150 cfs involves no load on the new pump, while a demand of 250 cfs would impose a load of 100 cfs on the new pump. Let Y be a random variable, "the load on the new pump" (i.e., the excess of demand over existing capacity):

$$Y = g_1(X) = \begin{cases} 0 & X \leq 150 \\ X - 150 & X > 150 \end{cases} \qquad (2.4.41)$$

This function is shown in Fig. 2.4.6b. It is *not* a one-to-one function.

Let us review how the distribution of Y can be determined. After Eq. (2.3.16), we have

$$F_Y(y) = \int_{R_y} f_X(x)\, dx$$

where R_y is the region in which $Y = g(X) \leq y$. Y is clearly never negative, but it is zero when X is less than 150; thus

$$F_Y(0) = F_X(150) = \int_{-\infty}^{150} f_X(x)\, dx$$

By the symmetry of $f_X(x)$ about 150,

$$F_Y(0) = \tfrac{1}{2}$$

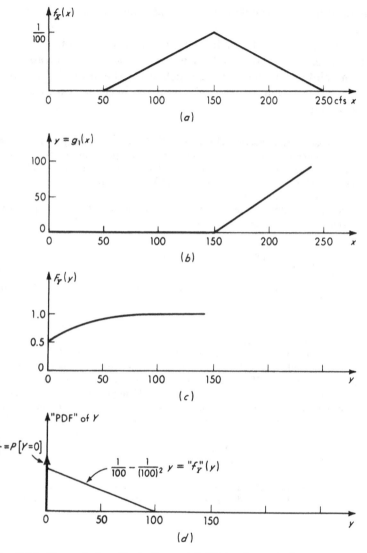

Fig. 2.4.6 Pump-design illustration. (*a*) Water-demand distribution; (*b*) load on new pump versus demand; (*c*) CDF of load on new pump; (*d*) PDF of load on new pump.

Y is never greater than $250 - 150$, or 100, but the probability that Y is less than any value in the range 0 to 100, say, 30, is the probability that X is less than $30 + 150 = 180$, or, in general, for $0 \le y \le 100$,

$$F_Y(y) = F_X(y + 150) = \int_{-\infty}^{y+150} f_X(x)\, dx \qquad 0 \le y \le 100$$

Substituting for $f_X(x)$ in the range $x = 150$ to 250, since $F_X(150) = \frac{1}{2}$,

$$F_Y(y) = \frac{1}{2} + \int_{150}^{y+150} \left[\frac{2}{100} - \frac{1}{(100)^2} (x - 50) \right] dx$$

$$= 1 - \frac{1}{2} \frac{(100 - y)^2}{(100)^2} \qquad 0 \le y \le 100 \qquad (2.4.42)$$

This function is, of course, unity for $y > 100$ and zero for $y < 0$, as shown in Fig. 2.4.6c.

The distribution of Y is of the mixed type; that is, there is a finite probability, $\frac{1}{2}$, that Y will be exactly equal to zero, but elsewhere it behaves as a continuous random variable. Strictly, it has neither a PMF nor a PDF,† but a graphical representation of the latter is possible, as shown in Fig. 2.4.6d. The spike at zero represents the finite probability of that value occurring. The area under the "continuous part," which we will denote as "f_Y"(y), is, of course, only $\frac{1}{2}$, not unity. This continuous part is given by

$$"f_Y"(y) = \frac{d}{dy} F_Y(y) = \frac{1}{100} - \frac{y}{(100)^2} \qquad 0 \le y \le 100 \qquad (2.4.43)$$

It is helpful to recognize that any mixed distribution of practical interest can be represented as the weighted sum of two proper distributions, one discrete and one continuous:

$$F_Y(y) = p F_{Y_d}(y) + (1 - p) F_{Y_c}(y) \qquad (2.4.44)$$

in which $0 \le p \le 1$ and in which F_{Y_d} is associated with the discrete values which Y can take on and F_{Y_c} is associated with the continuous range. In the example here, $p = \frac{1}{2}$,

$$F_{Y_d}(y) = \begin{cases} 0 & y < 0 \\ 1 & y \ge 0 \end{cases}$$

and

$$F_{Y_c}(y) = \begin{cases} 0 & y < 0 \\ 1 - \dfrac{(100 - y)^2}{(100)^2} & 100 \ge y \ge 0 \end{cases}$$

The function denoted "$f_Y(y)$" in Eq. (2.4.43) is just $(d/dy) [(1 - p)F_{Y_c}(y)]$. The treatment of mixed distributions does not deserve special discussion in this applied text, not because they are not useful, but because their treatment follows in an obvious manner from that of discrete and continuous variables.

Consider now the task of choosing a capacity for the new pump. A "conservative" design would be 100 cfs, the maximum value the pump could possibly be called on to provide. But the occurrence of large values of demand are rare, and so alternate design rules might prove more economical in balance. One such rule might be to match the new pump's capacity to the expected demand upon the pump, that is, $E[Y]$.

This value can be obtained using the definition of expectation. Here an obvious extension of both Eqs. (2.4.1) and (2.4.2) is needed to provide for the mixed

† This is true unless one employs Dirac delta functions, as suggested in Sec. 2.2.1.

distribution of Y:

$$E[Y] = 0P[Y = 0] + \int_0^{100} y \left[\frac{1}{100} - \frac{y}{(100)^2} \right] dy$$

$$= 0(\tfrac{1}{2}) + (16.7) = 16.7 \text{ cfs}$$

Alternatively, if this expected value were the only objective, we could have obtained it directly using Eq. (2.4.14) without the necessity of finding the probability law of Y. In this case

$$E[Y] = E[g_1(X)] = \int_{-\infty}^{\infty} g_1(x) f_X(x) \, dx$$

Substituting first for $g_1(x)$ [Eq. (2.4.41)],

$$E[g_1(X)] = \int_{-\infty}^{150} 0 f_X(x) \, dx + \int_{150}^{\infty} (x - 150) f_X(x) \, dx$$

and then for $f_X(x)$,

$$E[g_1(X)] = \int_{150}^{250} (x - 150) \left[\frac{2}{100} - \frac{(x - 50)}{(100)^2} \right] dx = 16.7 \text{ cfs} \qquad (2.4.45)$$

This rule suggests a second pump of capacity 16.7 cfs.

An alternate design rule might call for a pump capacity equal to the average of the nonzero demands on the pump, i.e., the expected demand given that the pump must be turned on at all. Formally, this value is the expected value of Y given that Y is greater than zero. The conditional distribution of Y given that Y is strictly greater than zero is, in this case, just the continuous part of its probability law renormalized to unit area or

$$f_{Y|[Y>0]}(y) = 2^{\prime\prime} f_Y^{\prime\prime}(y) = \frac{d}{dy} F_{Y_c}(y) \qquad 0 < y \le 100 \qquad (2.4.46)$$

Hence, the expected value of Y given $Y > 0$, written $E[Y \mid Y > 0]$, is

$$E[Y \mid Y > 0] = \int_{-\infty}^{\infty} y f_{Y|[Y>0]}(y) = \int_0^{\infty} 2y \left[\frac{1}{100} - \frac{y}{(100)^2} \right] dy$$

$$= 33.3 \text{ cfs} \qquad (2.4.47)$$

This design rule suggests a pump twice as large as the first rule.

As a less arbitrary and more explicitly cost-oriented approach, the engineer might seek that pump capacity z which promises the lowest expected total cost. Costs (in arbitrary units) are thought to be described by a cost of $100 + 10z$ (representing a fixed installation cost and the cost of the pump itself) plus a cost associated with failing to meet the peak demand. Determination of this latter factor is complicated by the possibility of large, but off-peak, demands at other times in the same year and by the fact that the system will be in place over a number of years. The engineer might summarize these effects, however, by saying that they are represented in a cost of failing to meet the peak demand in any *one* arbitrary year given by $10 + (Y - z)^2$, if $Y - z$, the amount by which the pump of capacity z fails to meet a random demand of Y, is positive, and zero otherwise. Formally, for a given $z > 0$, the cost as a function of the random variable Y is

$$C = g_2(Y) = \begin{cases} 100 + 10z & Y \le z \\ 100 + 10z + 10 + (Y - z)^2 & Y > z \end{cases} \qquad (2.4.48)$$

When $z = 0$, there is no initial cost—only the cost of failing to meet the demand.

For a pump of capacity $0 < z < 100$, then, the expected value of the total cost $C(z)$ can be found by using either the probability law of Y with Eq. (2.4.14) [and (2.4.15)] or the probability law of X with Eq. (2.4.31). The former path is taken here:†

$$E[C(z)] = (100 + 10z) + \int_z^{100} [10 + (y - z)^2] \left(\frac{1}{100} - \frac{1}{(100)^2} y \right) dy$$

$$= 100 + 10z + 10P[Y > z] + \frac{1}{100} \int_z^{100} (y - z)^2 \, dy$$

$$- \frac{1}{(100)^2} \int_z^{100} y(y - z)^2 \, dy = 100 + 10z + \frac{10(100 - z)^2}{2(100)^2} + \frac{1}{300} (100 - z)^3$$

$$- \frac{1}{3(100)^2} (100^3 - z^3) + \frac{z}{2(100)^2} (100^2 - z^2) \qquad 0 < z \le 100$$

The minimum cost must occur either (1) when no pump is installed, $z = 0$,

$$E[C(0)] = (10) \frac{1}{2} + \frac{(100)^3}{300} - \frac{(100)^3}{3(100)^2} = 3305$$

or (2) when the largest peak is matched, $z = 100$,

$$E[C(100)] = 100 + (10)(100) = 1100$$

or (3) at some intermediate value found by differentiating the expression above, setting it equal to zero, and solving for the optimal value z_0,

$$\frac{dE[C(z_0)]}{dz} = 10 - \frac{10(100 - z_0)}{(100)^2} - \frac{1}{100} (100 - z_0)^2 + \frac{z_0^2}{(100)^2}$$

$$+ \frac{1}{2} - \frac{3}{2} \frac{z_0^2}{(100)^2} = 0$$

Solving,

$$z_0 = 100 \pm 32.5$$

The root of interest is

$$z_0 = 67.5 \text{ cfs}$$

The expected cost at $z = 67.5$ cfs is about 883 units. (Therefore the extremum found is, in fact, a minimum and not a maximum.) Because this cost is less than the expected cost (3305) associated with the "do-nothing" ($z = 0$) policy, and less than the expected cost (1100) of the most conservative ($z = 100$) policy, the best design is to provide a pump with a capacity of 67.5 cfs.

† Despite the superficial appearance of linear and quadratic relationships, $g_1(X)$ and $g_2(Y)$, and the implication that the mean and variance of X would suffice to determine $E[C]$, it should be noted that in fact these functions are not linear or quadratic owing to their varying definitions over various ranges (see Fig. 2.4.6b).

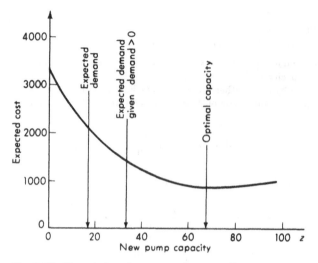

Fig. 2.4.7 Expected cost versus pump capacity.

The other two design rules, suggesting capacities of 16.7 and 33.3 cfs, have, of course, expected costs larger than the optimum value. (In particular, the expected cost of a $z = 33.3$ cfs capacity decision is 1402 units.) The function $E[C(z)]$ is plotted in Fig. 2.4.7. In words, such small-capacity designs, although cheaper initially, run a large risk of a very high penalty due to insufficient capacity. The high initial cost of a very large pump (say, 100 cfs) is apparently not justified in the light of the small likelihood of its full capacity being needed. The indicated optimum design achieves a balance between these two factors.

2.4.3 Expectation and Jointly Distributed Random Variables

When two or more random variables must be dealt with simultaneously, their behavior is described by their joint probability distribution (Sec. 2.2.2). The notions of expectation are also extended easily to this situation. Owing to the difficulty in practice of dealing with entire joint distributions, analysis of the joint behavior of random variables is often restricted to their moments. Therefore this subject will be discussed in some detail.

Expectation of a function Generalizing Eqs. (2.4.13) and (2.4.14), we obtain the expectation of a function $Z = g(X,Y)$ of two jointly distributed random variables as

$$E[Z] = E[g(X,Y)] \tag{2.4.49}$$

$$E[g(X,Y)] = \iint\limits_{-\infty}^{\infty} g(x,y) f_{X,Y}(x,y) \, dx \, dy \tag{2.4.50}$$

if X and Y are continuous random variables, and

$$E[Z] = \sum_{\text{all } x_i} \sum_{\text{all } y_j} g(x_i, y_j) p_{X,Y}(x_i, y_j) \tag{2.4.51}$$

if they are discrete. As in the previous section it is possible but difficult to show that $E[Z]$ calculated from the appropriate equation above is identical to that which would be obtained if, by the technique of Sec. 2.3.2, the distribution $f_Z(z)$ were first derived from that of X and Y and then the mean of Z were obtained by Eq. (2.4.2).

As before, we can distinguish between two interpretations of expectation. One looks upon $E[g(X,Y)]$ as the average of a function of X and Y over all values x and y, weighted everywhere by the probability that $X = x$ and $Y = y$. Moments fall into this category. The second interpretation sees $E[g(X,Y)]$ as the expected value or mean of a random variable $Z = g(X,Y)$.

Moments   We have already seen that moments, examples of the "averaged function of X and Y" interpretation of $E[g(X,Y)]$, facilitate the study of the marginal behavior of individual random variables. We shall find them of even greater value in studying the joint behavior of two random variables. Consider the function of X and Y of this form:

$$g(X,Y) = X^l Y^n$$

Then its expectation

$$E[X^l Y^n] = \iint_{-\infty}^{\infty} x^l y^n f_{X,Y}(x,y) \, dx \, dy \tag{2.4.52}$$

is called a *moment of order* $l + n$ of the random variables X and Y. The most important are the two first-order moments: ($l = 1$, $n = 0$) and ($l = 0$, $n = 1$). In the former case, $g(X,Y) = X$, and

$$E[g(X,Y)] = E[X] = \iint_{-\infty}^{\infty} x f_{X,Y}(x,y) \, dx \, dy \tag{2.4.53}$$

Here we have simply the mean or expected value of X. Notice that this may be written as

$$E[X] = \int_{-\infty}^{\infty} x \, dx \left[\int_{-\infty}^{\infty} f_{X,Y}(x,y) \, dy \right] \tag{2.4.54}$$

The second integral is simply the marginal distribution $f_X(x)$ [Eq. (2.2.43)]; thus

$$E[X] = \int_{-\infty}^{\infty} x f_X(x) \, dx = m_X \tag{2.4.55}$$

which is the same definition given in Eq. (2.4.2). Thus the expected value of X, Eq. (2.4.53), is the average value of X "without regard for the value of Y." The definition and meaning of $E[Y]$ are, of course, similar.

It may be helpful for the civil engineer to observe that m_X and m_Y locate the center of mass of a two-dimensional plate of variable density or the horizontal coordinates of the center of mass of the "hill" or terrain whose surface elevations are given by $f_{X,Y}(x,y)$. This follows from the definition of these terms and the fact that the volume under $f_{X,Y}(x,y)$ is unity.

By the same analogy, the second-order moments ($l = 2$, $n = 0$), ($l = 0, n = 2$), and ($l = 1, n = 1$) correspond respectively to the moment of inertia about the x axis, the moment of inertia about the y axis, and the product moment of inertia with respect to these axes.

Central moments As for a single variable, and as in mechanics, the more useful second (and higher) moments are those with respect to axes passing through the center of mass. Taking $g(X,Y)$ equal to

$$(X - m_X)^l(Y - m_Y)^n$$

we find

$$E[g(X,Y)] = E[(X - m_X)^l(Y - m_Y)^n]$$

$$= \iint_{-\infty}^{\infty} (x - m_X)^l(y - m_Y)^n f_{X,Y}(x,y) \, dx \, dy \qquad (2.4.56)$$

which is called a *central moment of order* $l + n$ of the random variables X and Y. The first central moments are, of course, both zero.

The most valuable central moments are the set of second-order moments: ($l = 2$, $n = 0$), ($l = 0$, $n = 2$), and ($l = 1$, $n = 1$). The first two cases, as above with the means, reduce to the marginal variances. With ($l = 2, n = 0$), for example,

$$E[(X - m_X)^2] = \iint_{-\infty}^{\infty} (x - m_X)^2 f_{X,Y}(x,y) \, dx \, dy$$

$$= \int_{-\infty}^{\infty} (x - m_X)^2 \, dx \left[\int_{-\infty}^{\infty} f_{X,Y}(x,y) \, dy \right]$$

$$= \int_{-\infty}^{\infty} (x - m_X)^2 f_X(x) \, dx$$

$$= \text{Var}\,[X] \qquad (2.4.57)$$

The result for Y, ($l = 0, n = 2$), is similar.

Covariance   The new type of second central moment that is found when joint random variables are considered is that involving their product, that is, $(l = 1, n = 1)$.   This moment also has a name, *the covariance of X and Y:* Cov $[X,Y]$ or† $\sigma_{X,Y}$.

$$\sigma_{X,Y} = \text{Cov}\,[X,Y] = E[(X - m_X)(Y - m_Y)]$$

$$= \iint_{-\infty}^{\infty} (x - m_X)(y - m_Y)f_{X,Y}(x,y)\,dx\,dy \qquad (2.4.58)$$

If the variances correspond to the moments of inertia about axes in the x and y direction passing through the centroid of a thin plate of variable density, then the covariance corresponds to its product moment of inertia with respect to these axes.

Correlation coefficient   A normalized version of the covariance, called the correlation coefficient $\rho_{X,Y}$, is found by dividing the covariance of X and Y by the product of their standard deviations:

$$\rho_{X,Y} = \frac{\text{Cov}\,[X,Y]}{\sigma_X \sigma_Y} \qquad (2.4.59)$$

It can be shown that this coefficient has the interesting property that

$$-1 \le \rho_{X,Y} \le 1 \qquad (2.4.60)$$

Before discussing the interpretations of the covariance and the correlation coefficient (and its bounds, ± 1), let us illustrate the computation of these numbers in a simple discrete case.   Consider the discrete joint distribution of X and Y sketched in Fig. 2.4.8, where the distribution might represent a discrete model of the maximum annual flows at gauge points in two different, but neighboring, streams.   The engi-

---

† If X and Y have the same dimensions, the dimensions of $\sigma_{X,Y}$ are the same as those of σ_X^2 or σ_Y^2; nonetheless convention dictates the notation $\sigma_{X,Y}$ rather than $\sigma_{X,Y}^2$.

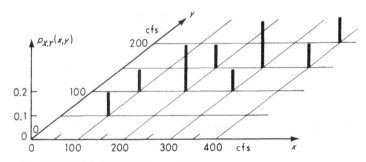

Fig. 2.4.8   Joint PMF of stream flows.

neer's interest in their joint behavior might arise from his concern over flooding of the river which the streams feed or from his desire to estimate flow in one stream by measuring only the flow in the other.

The means and variances are found from the marginal PMF's, or as follows:

$$E[X] = 100(0.1 + 0.1) + 200(0.2 + 0.1)$$
$$+ 300(0.1 + 0.2) + 400(0.1 + 0.1) = 250 \text{ cfs} \quad (2.4.61)$$

$$\text{Var }[X] = (100 - 250)^2(0.2) + (200 - 250)^2(0.3)$$
$$+ (300 - 250)^2(0.3) + (400 - 250)^2(0.2)$$
$$= 10{,}500 \text{ (cfs)}^2 \quad (2.4.62)$$

$$\sigma_X = \sqrt{10{,}500} = 102 \text{ cfs} \quad (2.4.63)$$

Similarly

$$E[Y] = 125 \text{ cfs} \quad (2.4.64)$$
$$\text{Var }[Y] = 1625 \text{ (cfs)}^2 \quad (2.4.65)$$
$$\sigma_Y = \sqrt{1625} = 40.3 \text{ cfs} \quad (2.4.66)$$

The covariance is found using the discrete analog of Eq. (2.4.58) or

$$\text{Cov }[X,Y] = \sum_x \sum_y (x - m_X)(y - m_Y)p_{X,Y}(x,y) \quad (2.4.67)$$

which here becomes

$$\text{Cov }[X,Y] = (100 - 250)(50 - 125)(0.1)$$
$$+ (100 - 250)(100 - 125)(0.1)$$
$$+ (200 - 250)(100 - 125)(0.2)$$
$$+ (200 - 250)(150 - 125)(0.1)$$
$$+ (300 - 250)(100 - 125)(0.1)$$
$$+ (300 - 250)(150 - 125)(0.2)$$
$$+ (400 - 250)(150 - 125)(0.1)$$
$$+ (400 - 250)(200 - 125)(0.1)$$
$$= 1125 + 375 + 250 - 125$$
$$- 125 + 250 + 375 + 1125$$
$$= 3250 \text{ (cfs)}^2 \quad (2.4.68)$$

The correlation coefficient ρ_{XY} is

$$\rho_{X,Y} = \frac{\text{Cov }[X,Y]}{\sigma_X \sigma_Y} = \frac{3250}{(102)(40.3)} = 0.79 \quad (2.4.69)$$

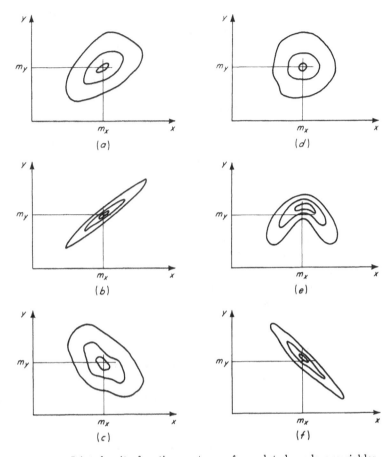

Fig. 2.4.9 Joint density-function contours of correlated random variables. (a) Positive correlation $\rho > 0$; (b) high positive correlation $\rho \approx 1$; (c) negative correlation $\rho < 0$; (d) (e) low correlation $\rho \approx 0$; (f) large negative correlation $\rho \approx -1$.

Discussion: Correlation In Fig. 2.4.9 some joint density functions are sketched and the corresponding values of the correlation coefficients are indicated. Note that the marginal distributions of X and Y remain the same from case to case. Comparing these figures with Fig. 2.2.13, it is clear that large values of ρ imply peaked conditional distributions, such as $f_{Y|X}$; the reverse is not necessarily true (e.g., case e). The implication is that if ρ is large, an observation of X will be very useful in refining and in reducing the uncertainty in predicting Y. Prediction will be discussed more thoroughly later (see page 175).

Examples of phenomena with large positive correlation coefficients (*positively correlated random variables*) include the length and weight of vehicles, and the yield strength and hardness of steels. *Negatively correlated random variables* include the speed and weight of vehicles, the ultimate compressive stress and ultimate compressive strain of concrete, and the capacity remaining in a dam and rainfall on its watershed for the past month. Random phenomena displaying small or zero correlation coefficients (*uncorrelated random variables*) are successive annual maximum floods at a site or the dead and live load on a structure.

Like the mean and variance, the correlation coefficient is often available when the complete probability law is not. This situation may result from the intractability of complete distribution calculations (Sec. 2.3), or from the lack of complete information about the variables. Rather than a well-defined joint probability law, we may have reliable estimates of only the first- and second-order moments. In this case, for example, pairs of observations from two variables may have yielded a scattering of points and an estimate of their correlation coefficient. For the proper interpretation of this number, it is important to know what it says and does *not* say about the joint behavior of the two variables.

First, from inspection of its definition [Eqs. (2.4.58) or (2.4.67)], the covariance (and hence the correlation coefficient) will be positive if larger than mean values of X are likely to be paired with larger than mean values of Y (and smaller with smaller). In this case the product $(x - m_X)(y - m_Y)$ will be positive where $f_{X,Y}(x,y)$ or $p_{X,Y}(x,y)$ is significantly large. If larger than average values of X usually appear with smaller than average values of Y (and vice versa), the covariance and correlation coefficient will be negative. In either case it can be said that at least some linkage or (stochastic) dependence exists between X and Y. This dependence may not be causal; in metals, greater hardness does not cause greater yield stresses, but the high correlation between the two characteristics is useful. It means that nondestructive hardness tests can be used to predict the strength of a piece of metal with a certain, but imperfect, reliability.

Secondly, if the two variables are independent, their covariance and correlation coefficient are zero. To verify this statement consider Eq. (2.4.58). Independence implies that $f_{X,Y}(x,y)$ factors into $f_X(x)f_Y(y)$; the integral becomes simply the product $E[X - m_X]E[Y - m_Y]$, both factors of which are zero. In words, if the variables are unrelated, their correlation coefficient is zero.

Unfortunately, in practice the error is often made of adopting the converse to this last statement; the third and most neglected fact about the correlation coefficient is that a small value does *not* imply that the X and Y are independent. Stochastic dependence (as reflected, say, by the sharpening of conditional distributions relative to marginals, Sec. 2.2.2) may be high even if ρ is zero or near zero.

More specifically, the correlation coefficient is a measure of the *linear* dependence between two random variables, in the following sense. An extreme value of ρ (± 1) obtains if and only if there is a linear *functional* relationship† between X and Y, that is, if and only if $Y = a + bX$. Hence, if ρ is one in absolute value, an observation of X, for example, will permit a perfect prediction of Y; that is, the stochastic dependence is "perfect"—the conditional distribution of Y given $X = x$ is simply a unit spike at $y = a + bx$. However, as will be demonstrated in an illustration to come, such perfect (or functional) dependence may exist, but, being nonlinear (for example, $Y = aX^2$) may yield a small correlation coefficient. Less than functional dependence but a close (nonlinear) relationship such as indicated in Fig. 2.4.9e, can also yield a small or zero correlation coefficient. Stochastic dependence in such a case is clearly large; given a value of $X = x$, the conditional distribution of Y will greatly differ from its marginal distribution. But if one mistakenly concludes a lack of strong stochastic dependence from a small value of the correlation coefficient, he will not reach the proper conclusion in such cases.

In summary, given only the value of the correlation coefficient, one can conclude from a high value that the stochastic dependence is high and furthermore that X and Y have a joint linear tendency, whereas a small value implies only the weakness of a linear trend and not necessarily a weakness in stochastic dependence.

The definitions of moments may be generalized in a straightforward manner to more than two jointly distributed random variables, but the use of higher than second-order moments is seldom encountered in practice. Further discussion of moments will be postponed until they are needed.

Properties of expectation   As suggested, Eq. (2.4.50) provides a more efficient way of computing the expected value of a random variable, $Z = g(X,Y)$, than that of first finding its probability law and then finding its expectation. That is, if, in the discrete example of stream flows above, the mean of the total at the maximum flows‡ $Z = X + Y$, is desired, it could be found by finding $p_Z(z)$ and then averaging. Alter-

---

† One side of this contention is easily verified by assuming that $Y = a + bX$ and showing that $\rho = 1$ for $b > 0$ and -1 for $b < 0$. The other side, that $\rho = +1$ implies $Y = a + bX$, is demonstrated in Freeman [1963].

‡ To be conservative, the engineer might assume that the maximum flows always occur simultaneously, when $Z = X + Y$ would be the maximum flow in the confluence of the two streams.

natively and more simply, using Eq. (2.4.50),

$$E[Z] = \sum_x \sum_y (x + y)p_{X,Y}(x,y)$$

$$= (100 + 50)(0.1) + (100 + 100)(0.1)$$
$$+ (200 + 100)(.2) + (200 + 150)(0.1)$$
$$+ (300 + 100)(0.1) + (300 + 150)(0.2)$$
$$+ (400 + 200)(0.1) + (400 + 150)(0.1)$$

$$= 375 \text{ cfs} \tag{2.4.70}$$

Even more simply we can make use of the easily verified linearity property of expectation

$$E[g_1(X,Y) + g_2(X,Y)] = E[g_1(X,Y)] + E[g_2(X,Y)] \tag{2.4.71}$$

For example, in this special case,

$$E[X + Y] = E[X] + E[Y] = 250 + 125 = 375 \text{ cfs}$$

The linearity property can be used, as in the preceeding section, to find relationships among expectations which simplify computations, or which, because they are independent of the underlying distributions, are valuable when only these expectations are known.

For example, the following relationship often simplifies covariance computation

$$\text{Cov } [X,Y] = E[(X - m_X)(Y - m_Y)]$$
$$= E[XY - Xm_Y - Xm_X + m_X m_Y]$$

which, using the linearity property, reduces to

$$\text{Cov } [X,Y] = E[XY] - m_Y E[X] - m_X E[Y] + m_X m_Y$$
$$= E[XY] - E[X]E[Y] \tag{2.4.72}$$

Illustration: Correlated demand and capacity   We illustrate the use of the linearity property in situations where moments are known, but distributions are not, by the following example.   If the capacity C of an engineering system and the demand upon it D are random variables, the margin $M = C - D$ is a measure of the performance of the system, whether it be inadequate, $M < 0$, or "overdesigned," M large.   Even if the joint distribution of C and D is unknown, the mean and variance of M can be found from first and second moments of X and Y:

$$E[M] = E[C - D]$$

which, using the linearity property, is

$$E[M] = E[C] - E[D] \tag{2.4.73}$$

The variance, on the other hand, is

$$\text{Var } [M] = E[M^2] - E^2[M]$$
$$= E[(C - D)^2] - (E[C] - E[D])^2$$
$$= E[(C^2 - 2CD + D^2)] - E^2[C] + 2E[C]E[D] - E^2[D]$$

Applying the linearity property to the first term and then grouping as

$$\text{Var } [M] = (E[C^2] - E^2[C]) + (E[D^2] - E^2[D]) - 2(E[CD] - E[C]E[D])$$

we recognize that

$$\text{Var } [M] = \text{Var } [C] + \text{Var } [D] - 2 \text{ Cov } [C,D]$$
$$= \sigma_C^2 + \sigma_D^2 - 2\rho_{C,D}\sigma_C\sigma_D$$

Demand and capacity may frequently be correlated. Both depend, for example, upon the average travel velocity of a highway system. Notice that if the correlation is positive, i.e., if higher than expected demand tends to occur with higher than expected capacity (as may be the case of a highway system where the better the system performs the more it is used), the variance of M is *reduced* from the sum of the variances. This latter value, the sum, obtains, of course, only if C and D are uncorrelated, or, in particular, if they are independent. If $\rho_{C,D} = 0$:

$$\text{Var } [M] = \text{Var } [C] + \text{Var } [D]$$

Notice that the variance of the difference between (independent) C and D is not the difference of their variances, but the sum. The uncertainty in both variables "contributes" to the dispersion in their difference.

Even though their joint distribution is not known, the second-order moments of C and D are sufficient to obtain a lower bound on the system reliability, the probability that the capacity exceeds the demand. Clearly

$$P[C > D] = P[C - D > 0] = P[M > 0] = 1 - P[M < 0]$$

The probability of "failure" of the system, $P[M < 0]$, has an upper bound given by the Chebyshev inequality [Eq. (2.4.11)] as

$$P[M < 0] \le P[|M - m_M| \ge m_M] \le \left(\frac{\sigma_M}{m_M}\right)^2 \tag{2.4.74}$$

Hence the system reliability is at least $1 - V_M^2$ where V_M, the coefficient of variation of M, is

$$V_M = \frac{\sigma_M}{m_M} = \frac{\sqrt{\sigma_C^2 + \sigma_D^2 - 2\rho_{C,D}\sigma_C\sigma_D}}{m_C - m_D} \tag{2.4.75}$$

For example, a highway engineer, who estimates that the demand on his system 10 years hence will have a mean 1200 vehicles per hour (vph) and a standard deviation of 400 vph, might consider a design with an expected capacity of 2200 vph and standard deviation 300 vph. Assume that the correlation coefficient in such situations has been measured to be $+10$ percent; then

$$m_M = 2200 - 1200 = 1000 \text{ vph} \qquad (2.4.76)$$

$$\sigma_M{}^2 = 300^2 + 400^2 - 2(0.1)(300)(400)$$
$$= 90,000 + 160,000 - 24,000$$
$$= 226,000 \text{ (vph)}^2 \qquad (2.4.77)$$
$$\sigma_M = \sqrt{226,000} = 475 \text{ vph} \qquad (2.4.78)$$

The likelihood that the capacity will exceed the demand is at least

$$P[C > D] = 1 - P[M < 0] \geq 1 - (475/1000)^2 = 1 - 0.226 = 0.774 \quad (2.4.79)$$

It is assumed here that the engineer knows the moments exactly. If they are the result of estimates from small samples, there arises another order of uncertainty, that is, statistical uncertainty (Chap. 4).

Moments of linear functions   The linearity property, Eq. (2.4.71), can also be employed to derive some general results for the very important case when a linear relationship among two or more jointly distributed random variables is involved. Interest lies in the variable Y, where

$$Y = g(X_1, X_2, \ldots, X_n) = \sum_{i=1}^{n} a_i X_i \qquad (2.4.80)$$

The expectation of Y is

$$E[Y] = E\left[\sum_{i=1}^{n} a_i X_i\right] = \sum_{i=1}^{n} a_i E[X_i] \qquad (2.4.81a)$$

The result is valid whether or not the X_i are independent. In words, the equation states that "the mean of the sum is the sum of the means."

An analogous statement does *not* hold for variances unless the random variables are all uncorrelated. In general,† by carrying out the now familiar steps, we can show that

$$\text{Var}[Y] = \sum_{i=1}^{n} a_i{}^2 \text{Var}[X_i] + 2\sum_{i=1}^{n}\sum_{j=i+1}^{n} a_i a_j \text{Cov}[X_i, X_j] \qquad (2.4.81b)$$

which reduces *for uncorrelated random variables* to, simply,

$$\text{Var}[Y] = \sum_{i=1}^{n} a_i{}^2 \text{Var}[X_i] \qquad (2.4.81c)$$

The special case for $n = 2$ is worth displaying if only because it appears so frequently in practice. If $Z = aX + bY$, then, according to

† The variance of Y is the sum of the terms in the $n \times n$ array of covariances, but owing to the symmetry, that is, $\text{Cov}[X_i, X_j] = \text{Cov}[X_j, X_i]$ and to the fact that the diagonal terms are the variances, that is, $\text{Cov}[X_i, X_i] = \text{Var}[X_i]$, the sum reduces to the form shown.

the previous equation,

$$\text{Var}\,[Z] = \text{Var}\,[aX + bY] = a^2\,\text{Var}\,[X]$$
$$+ b^2\,\text{Var}\,[Y] + 2ab\,\text{Cov}\,[X,Y] \quad (2.4.82)$$

This equation can be verified in a manner parallel to the technique used in determining the variance of $M = C - D$. In fact, notice that that illustration was a special case of the result above with $a = 1$ and $b = -1$.

Illustration: Total capacity of a system of components Some of the implications of Eq. (2.4.82) are best understood by example. Consider the relative variability and hence relative reliability of two contending systems: the first consists of two elements, the sum of whose capacities or strengths, X and Y, determines the capacity of the system; the second consists of a single element of capacity Z. A simple example is the situation where two smaller bars of ductile steel or one larger bar will be used to provide the tensile capacity of a reinforced concrete beam. Other examples include simple pipe or transportation networks. Assume that mean system capacities are, by design, equal (that is, $E[Z] = E[X] + E[Y] = m$) and that the mean small-element capacities are equal $(E[X] = E[Y] = \frac{1}{2}m)$. How do the dispersions in the two systems' capacities compare if the coefficients of variation of X, Y, and Z are equal†? In the latter case,

$$\text{Var}\,[Z] = \sigma_Z{}^2 = V^2 m^2 \quad (2.4.83)$$

while in the case of the two-element system,

$$\begin{aligned}
\text{Var}\,[X + Y] &= \sigma_X{}^2 + \sigma_Y{}^2 + 2\rho\sigma_X\sigma_Y \\
&= V^2(\tfrac{1}{2}m)^2 + V^2(\tfrac{1}{2}m)^2 + 2\rho V^2(\tfrac{1}{2}m)^2 \\
&= \tfrac{1}{2}V^2 m^2(1 + \rho) \quad (2.4.84)
\end{aligned}$$

Notice that if X and Y are perfectly positively correlated, $\rho = 1$, the variability of the system capacity is equal to that in the one element case. But under any other conditions, the two- (parallel-) element system has *less* variance and generally higher reliability than the system with one larger element. In particular, if X and Y are independent, the variance of $X + Y$ is $\frac{1}{2}$ of that of Z. (There is evidence that two bars from the same batch of steel may have a ρ as large as 98 percent.)

Moments of a product The general properties of expectation in one final common case deserve mention, for it is one of the few functions outside of the linear ones considered above which lends itself to such simple treatment. If $Z = XY$, then, from Eq. (2.4.72),

$$E[Z] = E[XY] = \text{Cov}\,[X,Y] + E[X]E[Y] \quad (2.4.85)$$

† The coefficients of variation will be equal in the bar-strength case if each yield force is the product of bar area and yield stress, with all stochastic variation being in the latter. The doubling in mean strength from X to Z is obtained by doubling the bar area, which also doubles the standard deviation.

If and only if the variables are uncorrelated, the expectation of the product is the product of the expectations

$$E[XY] = E[X]E[Y] \tag{2.4.86}$$

If X and Y are independent, expansion will show that

$$\text{Var}\,[XY] = m_X{}^2\sigma_Y{}^2 + m_Y{}^2\sigma_X{}^2 + \sigma_X{}^2\sigma_Y{}^2 \tag{2.4.87}$$

and that

$$V_Z{}^2 = V_X{}^2 + V_Y{}^2 + V_X{}^2V_Y{}^2 \tag{2.4.88}$$

Illustration: Correlation versus functional dependence   The correlation coefficient alone is frequently used to make deductions about the joint behavior of random variables. In many fields, particularly the biological and social sciences, where basic laws are difficult to derive otherwise, it is common to measure repeatedly two variables which are suspected or hypothesized to be related and then to estimate† their correlation coefficient. If the coefficient is near unity in absolute value, strong dependence is assumed verified; if the coefficient is low, it is concluded that one variable has little or no effect on the other.

Such techniques of verification or determination of suspected relationships are becoming more and more common in civil engineering, where many of our common materials—e.g., concrete and soil—and many of our systems—e.g., watersheds and traffic—are so complex that relationships among variables can only be determined empirically. To use such an approach correctly, it is necessary to have a sound understanding of its implications and its potential errors. The following extended illustration is designed both as an exercise in using expectation operations and as an aid to understanding the effects on the correlation coefficient caused by such factors as the character of the functional relationship and the presence of other causes of variation. It is important to appreciate the fact that the conclusions deduced from this study do not depend on the shape of the joint probability distribution. This fact demonstrates the power and generality of working with means, variances, and correlations, without detailed probability law specifications.

Suppose that an engineer is investigating the possible relationship between density of a particular highway subbase material and the road's performance. In each of a number of segments of a road of nominally uniform design, he measures both the density of the subbase material (found in a core drilled through the pavement) and the value of some index of the segment's performance, namely, the smoothness of the ride provided to passing vehicles.‡ Let the random variable Y be the "ridability" or performance index and let X be the deviation of the subbase density from the specified (say, the mean) density.§

---

† Estimation of moments from data is a subject of Chap. 4.

‡ Such indices are normally highly subjective, but more objective techniques, including statistical ones, are possible and are under investigation (see Hutchinson [1965]).

§ This definition of X as the *deviation* about the mean of the subbase density is for algebraic convenience only. The density is $Z = m_Z + X$. Clearly $\sigma_Z{}^2 = \sigma_X{}^2$, and it is easily shown that $\rho_{Z,Y} = \rho_{X,Y}$. Later assumptions will be interpreted in terms of Z as well as X.

The engineer could estimate from the data the correlation coefficient, $\rho_{X,Y}$. Generally, a low value of this coefficient would lead him to conclude that performance (ridability) is virtually independent of subbase density. A relatively high value of $\rho_{X,Y}$, on the other hand, would normally cause the conclusion that subbase density strongly affects performance (and hence, perhaps, must be well controlled in future jobs, if high performance is to be obtained).

In order to understand the general validity of such conclusions, we want to presume that, in fact, a physical law or function (unknown to the engineer, of course) governs the relationship between performance Y and subbase density X. The law we shall assume is of the form

$$Y = a + bX + cX^2 + W \tag{2.4.89}$$

in which a, b, and c are constants and in which W is a random variable with mean zero, stochastically independent of X, and W represents the variation in Y attributable to factors other than the subbase density X. These might include such factors as settlement of the soil fill below the subbase or variations in pavement placement. Assuming the average effect of these factors is accounted for in the constant a, W need only represent the effect on Y of the random variations about the mean values of these factors. Thus the mean zero assumption is reasonable.

Calculation of the Correlation Coefficient   First, as an exercise in dealing with expectation we seek the correlation coefficient $\rho_{X,Y}$ given that Eq. (2.4.89) defines the relationship between Y, X, and the other factors. From Eq. (2.4.72),

$$\text{Cov}\,[X,Y] = E[XY] - E[X]E[Y]$$

Evaluating the first term with the use of the linearity property,

$$E[XY] = E[aX + bX^2 + cX^3 + WX] = aE[X] + bE[X^2] + cE[X^3] + E[WX]$$

Owing to the assumed independence of W and X, $E[WX] = E[W]E[X]$. Since $E[W]$ was assumed to be zero, the last term is zero. Assuming that X, the deviation from the specified subbase density, is symmetrically distributed about zero,† $E[X] = 0$, $E[X^3] = 0$ and $\text{Var}\,[X] = E[X^2]$. Thus

$$\text{Cov}\,[X,Y] = E[XY] - E[X]E[Y] = b\sigma_X^2 \tag{2.4.90}$$

The sign of the covariance in this example depends only on the sign of b.

To find the correlation coefficient, we need the standard deviations of X and Y:

$$E[Y] = a + cE[X^2] \tag{2.4.91}$$

$$E[Y^2] = E[a^2 + b^2X^2 + c^2X^4 + W^2 + 2abX + 2acX^2 + 2aW$$
$$+ 2bcX^3 + 2bXW + 2cX^2W]$$

Owing to the assumptions above, many terms drop from this expression. Combining these results and grouping terms,

$$\text{Var}\,[Y] = E[Y^2] - E^2[Y]$$

$$\sigma_Y^2 = b^2E[X^2] + c^2(E[X^4] - E^2[X^2]) + E[W^2]$$

† This is equivalent to assuming that Z, the subbase density itself, is symmetrically distributed about its mean m_Z.

Simply to shorten the expression, we substitute† Var $[X^2]$ for $E[X^4] - E^2[X^2]$:

$$\sigma_Y^2 = b^2\sigma_X^2 + c^2 \text{ Var } [X^2] + \sigma_W^2 \qquad (2.4.92)$$

Finally,

$$\rho_{X,Y} = \frac{\text{Cov } [X,Y]}{\sigma_X \sigma_Y}$$

$$= \frac{b\sigma_X^2}{\sigma_X \sqrt{b^2\sigma_X^2 + c^2 \text{ Var } [X^2] + \sigma_W^2}} \qquad (2.4.93)$$

Let us concentrate on ρ^2, since it is simpler in form than ρ. Of course, $\rho = 0$ implies $\rho^2 = 0$ and $\rho = \pm 1$ implies $\rho^2 = 1$; hence the same implications regarding presence or lack of dependence follow from ρ^2 exactly as from ρ. Simplifying,

$$\rho_{X,Y}^2 = \frac{1}{1 + (c/b)^2 \text{ Var } [X^2]/\sigma_X^2 + (1/b^2)(\sigma_W^2/\sigma_X^2)} \qquad (2.4.94)$$

Note that no use has been made of integration or density functions in obtaining these results.

Values of ρ Implied by the Form of the True Law   We are now prepared to use this equation to demonstrate how various conditions in the true underlying law, Eq. (2.4.89), affect the value of the correlation coefficient and hence the validity of conclusions drawn from its value. The joint density functions implied by each case are sketched in Fig. 2.4.10.

Case 1: $Y = a + bX$ Notice first that the *only* way a value of ρ (or ρ^2) equal to unity will be found is if b is not equal to zero but both c and σ_W^2 are, that is, if a linear functional relationship,‡ $Y = a + bX$, and hence perfect stochastic dependence, exists. Values of ρ^2 less than 1 will be found in all other situations.

Case 2: $Y = a + bX + W$ Let us look next at the case when $c = 0$, that is, when a linear relationship between X and Y exists, but other sources of randomness Y are present:

$$Y = a + bX + W \qquad (2.4.95)$$

Then

$$\rho^2 = \frac{1}{1 + (1/b^2)(\sigma_W^2/\sigma_X^2)} \qquad (2.4.96)$$

In this case, if the variance of the other contributions to the variation in Y is relatively small, the correlation coefficient will be very nearly unity. The

---

† In terms of $Z = m_Z + X$, Var $[X^2] = \mu_Z^{(4)} - \sigma_Z^4$. Note that Eq. (2.4.92) could have been obtained directly from Eq. (2.4.89) using Eq. (2.4.81c), since, interestingly, X and $U = X^2$ are uncorrelated random variables (if $E[X]$ and $E[X^3]$ are zero). The reader can verify this in one line.

‡ A linear functional relationship between Y and X implies that a linear form also relates Y and Z; $Y = a + bX = a + bm_Z + bZ$.

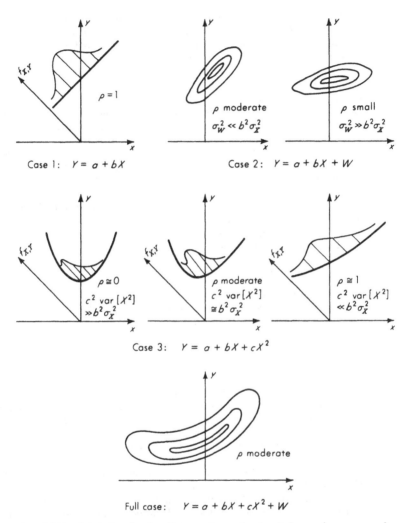

Fig. 2.4.10 Joint density-function contours $f_{X,Y}(x,y)$ for various cases of a physical law governing Y and X. In cases 1 and 3, the PDF's degenerate into "walls" which are only suggested graphically.

engineer's conclusions, based on observing $\rho \approx 1$, that strong stochastic dependence (almost a simple functional dependence) exists between X and Y would be valid. If, however, the contributions to the scatter in Y come predominantly from other factors, and/or if the functional dependence between X and Y is weak (b is small), then $\sigma_W{}^2 \gg b^2\sigma_X{}^2$. As a result the correlation coefficient will be very small. A conclusion (based on observing a small value of ρ) that there is no relationship between X and Y would not be valid, as a linear relationship does exist. But ρ does remain a good indicator of the stochastic dependence between X and Y, for this will be small too; the conditional distribution of Y given $X = x$ will have variance $\sigma_W{}^2$, very nearly as large as that of marginal distribution of Y, which is $b^2\sigma_X{}^2 + \sigma_W{}^2$. In this case the random variation W, which is the result of other factors, masks the linear Y verses X relationship. The implication for field practice and betterment of performance is simply that, even though there may be a functional relationship between performance and subbase density, it is so weak or the variance of subbase density is so small that, compared with the influence of the other factors, the deviations in subbase density contribute little to the variations in performance. Hence the conclusion, based on observation of a small value of ρ^2, that for nominally similar designs (i.e., same mean density) there is little effect of subgrade density on performance may be a valid one operationally.

Case 3: $Y = a + bX + cX^2$ Next we look at another limiting case, namely when $\sigma_W{}^2 = 0$, that is, when the variation in other factors is negligible or at least has negligible effect upon performance. Now a functional relationship between X and Y exists:

$$Y = a + bX + cX^2 \tag{2.4.97}$$

and so does perfect stochastic dependence; i.e., the conditional distribution of Y given $X = x$ is a spike of unit mass at the value $a + bx + cx^2$. But the correlation coefficient may well prove misleading:

$$\rho^2 = \frac{1}{1 + (c/b)^2(\text{Var } [X^2]/\sigma_X{}^2)} \tag{2.4.98}$$

If the squared term dominates over the linear one, that is, if $c \gg b$, the correlation coefficient will be small even though dependence is perfect. In this situation an engineer's conclusion, based on an observed small value of ρ^2, that subgrade density does not significantly affect performance, would be incorrect. Subsequent action based on neglecting the existing dependence could lead to relaxing of quality control standards on subgrade construction and consequent deterioration in performance (assuming b and c are negative). In this case a plot of the data points will usually exhibit the strong functional relationship which the correlation coefficient fails to suggest. Note, on the other hand, that if the linear term predominates, i.e., if $b \gg c$, the correlation will be high (but never unity) and will properly indicate strong dependence. This illustrates why it was stated that the correlation coefficient should properly be called a measure of *linear* dependence.† It must be pointed out that these last conclusions from the illustration depend on the assumption that $E[X^3] = 0$.‡ If this condition does not hold, the correlation coefficient is generally enhanced as a measure of dependence. No matter what the relative values of c and b,

---

† Notice that if c were much greater than b and this strong parabolic relationship were suspected by the engineer, he could estimate the correlation coefficient between Y and $U = X^2$. Then a strong linear dependence between Y and U would exist and a large correlation coefficient $\rho_{Y,U}$ would be found. Note, however, that $c \gg b$, implying that $Y = a + cX^2$ does not imply that $Y = a' + c'Z^2$. The latter relationship contains a linear term in X: $Y = a' + c'm_Z{}^2 + c'2m_Z X + c'X^2$.

‡ That is, $\mu_Z{}^{(3)} = 0$, which holds if Z is symmetrically distributed.

if additional sources of variation W also exist, they will tend to decrease ρ^2 [Eq. (2.4.94)] and mask further the functional relationship between X and Y.

Spurious Correlation   It is important, too, to point out another potential source of misinterpretation associated with the use of correlation studies.   The discussion above revealed that small correlation does not necessarily imply weak dependence, and it is also true that *spurious correlation* may invite unwarranted conclusions of a cause and effect relationship between variables when, in fact, none exists.   In the example here, for instance, some other variable factor, such as moisture content of the soil or volume of heavy trucks, might cause variation in both subbase density and performance.   Hence higher-than-average values of density might, in fact, usually be paired with higher-than-average performance (and low with low), yielding a positive correlation that is not a result of subbase density's beneficial effect on performance but is rather a result of both factors being increased (or decreased) by presence of the high (or low) value of the third factor.   This fact does not, however, reduce the value of using an observation of one variable to help predict the other, as long as the factor causing the correlation is not altered.

Benson [1965] reported on spurious correlation and on its potential and actual presence in many civil-engineering investigations, even those which are deterministic rather than probabilistic in formulation.   He cites as a particularly common source of spurious correlation that introduced by the engineer who seeks to normalize his data by dividing it by a factor which is itself a random variable.   For example, correlation between X, accidents, and Y, out-of-town drivers, is not necessarily implied by correlation (observed in a number of cities) between X/Z, the number of accidents per registered vehicle, and Y/Z, the number of commuters per registered vehicle.   Formally, X and Y may be independent, but X/Z and Y/Z are not, and high correlation between the latter pair does not imply correlation between the former pair.   Similarly, correlation between live load per square foot and cost per square foot in structures need not imply that cost and load are dependent, because both may depend upon the normalizing factor, that is, total floor area.

Summary   The correlation coefficient provides a very useful measure of dependence between variables, but the engineer must be careful in its interpretation.   Large values of ρ^2 imply strong stochastic dependence and a near-linear functional relationship, but not necessarily a cause-and-effect relationship.   Small values may result from the predominance of other sources of variation (and hence low stochastic dependence), or, if a functional relationship exists, from its lack of strong linearity.

Conditional expectation and prediction†   As with single random variables there is often advantage in working with expected values *conditioned* on some event A.   In general we have

$$E[g(X,Y) \mid A] = \iint g(x,y) f_{X,Y|[A]}(x,y) \, dx \, dy \qquad (2.4.95)$$

in which $f_{X,Y|[A]}(x,y)$ is the joint conditional distribution of X and Y given the event A [for example, Eq. (2.2.58)].

† This material may be bypassed on first reading.   However, it is necessary for a thorough understanding of Secs. 3.6.2 and 4.3.

The single most important application of conditional expectation is in *prediction*. If Y is a maximum stream flow or the peak afternoon traffic demand on a bridge, the engineer may be asked to "predict what Y will be." To "predict Y" is to state a *single number*, which, in some sense, is the *best prediction* of what value the random variable Y will take on. Without additional information, the (marginal) mean of Y, or m_Y, is conventionally used to predict Y. Intuitively it is a reasonable choice. More formally it is the predictor of Y which has a minimum expected squared error (or "mean square error"). The *error* in predicting Y by m_Y is simply $Y - m_Y$. Its expected square is

$$E[(Y - m_Y)^2]$$

or the (marginal) variance of Y. No other predictor will give a smaller mean square error.† It should be emphasized that the mean square error criterion for evaluating a predictor is simply a mathematically convenient engineering convention. More generally the engineer should use a predictor which minimizes the expected cost, the cost being some function of the error. If this cost is approximately proportional to the square of the error, whether positive or negative, then the mean square error criterion is also a minimum expected cost criterion and the mean is the best predictor. If, on the other hand, there are larger costs associated with underestimating Y than with overestimating it, the minimum cost predictor will be greater than the mean. Such economic questions are more properly treated in Chaps. 5 and 6; we restrict attention here to the conventional mean square error criterion.

Suppose now that the engineer learns that another random variable X has been observed to be some particular value, say x. The hydraulic engineer may have learned the maximum flow in a neighboring river or at an upstream point, or the traffic engineer may have learned that the morning peak flow was $x = 1000$ cars per hour. Conditional on this new information, what value should the engineer use to predict Y? Applying the same argument used above to the *conditional* distribution of Y given that $X = x$, one concludes that the "best" predictor of Y is the *conditional mean*, or $m_{Y|X}$.

$$m_{Y|X} = E[Y \mid X = x] = \int_{-\infty}^{\infty} y f_{Y|X}(y,x) \, dy \qquad (2.4.96)$$

† Any other predictor, say a, has mean square error:

$$E[(X - a)^2] = E[[(X - m_X) + (m_X - a)]^2] = E[(X - m_X)^2]$$
$$+ 2(m_X - a)E[X - m_X] + (m_X - a)^2 = \text{Var}[X] + (m_X - a)^2$$

Clearly this is greater than Var $[X]$ for any predictor other than $a = m_X$.

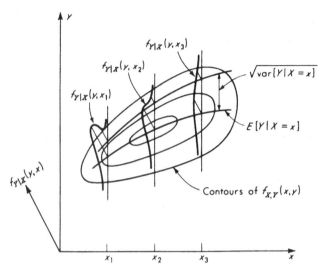

Fig. 2.4.11 Illustration of conditional means and variances of Y given X.

Note that in general this conditional mean† depends on the value x. This predictor has a mean square error equal to the *conditional variance* of Y given $X = x$:

$$\text{Var}\,[Y \mid X = x] = E[(Y - m_{Y|x})^2 \mid X = x]$$

$$= \int_{-\infty}^{\infty} (y - m_{Y|x})^2 f_{Y|X}(y,x)\, dy \qquad (2.4.97)$$

This variance is also a function of x, in general. These relationships are sketched in Fig. 2.4.11, where the square root of the mean square error (or *rms*) is indicated rather than the mean square error itself.

In Sec. 3.6.2 we will discuss in detail an illustration of conditional prediction applied to a particular, commonly adopted, joint distribution of X and Y. In that case we will find, upon carrying out the integrations indicated above, that

$$m_{Y|x} = E[Y \mid X = x] = m_Y + \rho \frac{\sigma_Y}{\sigma_X}(x - m_X) \qquad (2.4.98a)$$

and that

$$\text{Var}\,[Y \mid X = x] = (1 - \rho^2)\sigma_Y{}^2 \qquad (2.4.98b)$$

We note in this special application that the predictor $m_{Y|x}$ is linear in x and that the mean square error is a constant, independent of x. The

† This conditional mean is also called the "regression of Y on X."

importance, once again, of the correlation coefficient ρ is observed. For this particular joint distribution,† given $X = x$, the amount the conditional predictor $m_{Y|X}$ will be altered from the (marginal) predictor m_Y is directly proportional to ρ. Also, the uncertainty in predicting Y given $X = x$ (as measured by the mean square error or conditional variance) is smaller than the marginal uncertainty $\sigma_Y{}^2$ by an amount proportional to the square of ρ.

Illustration: Sum of a random number of random variables Conditional expectation is also valuable when studying probability models. Often moments of important variables may be difficult to determine unless one takes advantage of conditional arguments.

For example, recall that in Sec. 2.3.3 we assumed that T, the total rainfall on a watershed, was the sum of the (independent, identically distributed) random rainfalls, R_1, R_2, . . . , in N "rainy" weeks, where the number of rainy weeks is itself random (and independent of the R_i's):

$$T = \sum_i^N R_i \tag{2.4.99}$$

The estimation of the distribution of T by simulation was found to be a long numerical task. In addition, it required complete knowledge of the distribution of N and R_i. Having knowledge *only* of the means and variances of N and R_i, it would be desirable to be able to calculate explicitly the mean and variance of T. They are

$$E[T] = E\left[\sum_{i=1}^N R_i\right] \tag{2.4.100}$$

and

$$\text{Var}[T] = E[T^2] - E^2[T] = E\left[\left(\sum_{i=1}^N R_i\right)^2\right] - E^2\left[\sum_{i=1}^N R_i\right] \tag{2.4.101}$$

The randomness in N makes a direct attack on this problem troublesome. But, *given* the number of rainy weeks (that is, conditional on $N = n$), we can apply familiar formulas [Eqs. (2.4.81a) and (2.4.81b)] for the sum of a known number of (independent) random variables:

$$E_R[T \mid N = n] = E\left[\sum_{i=1}^n R_i\right] = \sum_{i=1}^n E[R_i] = nm_R$$

---

† We must pass over many interesting topics in prediction, where the role of correlation is central. In particular, if, as is usually done in practice to avoid analytical difficulties, we restrict attention to predictors of Y which are simply linear in the observed value x, it can be shown that the best (linear) predictor of Y is, for *any* joint distribution, the function given above, Eq. (2.4.98a). This equation is called the *linear* regression of Y on X. The mean square error of that best linear predictor, when averaged over all values of x, is also for *any* distribution the same as the function given above, Eq. (2.4.98b). Therefore in *linear* prediction, second moments alone define all the necessary information. (See, for example, Freeman [1963].)

and

$$E_R[T^2 \mid N = n] = \mathrm{Var}\,[T \mid N = n] + E^2[T \mid N = n]$$

$$= \mathrm{Var}\left[\sum_{i=1}^{n} R_i\right] + (nm_R)^2 = \sum_{i=1}^{n} \mathrm{Var}\,[R_i] + n^2 m_R^2$$

$$= n\sigma_R^2 + n^2 m_R^2 \qquad (2.4.102)$$

in which σ_R^2 and m_R are the variance and mean of any R_i. The subscript R on the expectation symbol is simply a reminder that this expectation is with respect to the R's. Now, recognizing that N is in fact random, we have two simple functions of the random variable N:

$$E_R[T \mid N]\dagger = g_1(N) = m_R N$$
$$E_R[T^2 \mid N] = g_2(N) = \sigma_R^2 N + m_R^2 N^2$$

To find $E[T]$ and $E[T^2]$, we need simply take expectation of $g_1(N)$ and $g_2(N)$ with respect to N:

$$E[T] = E_N[g_1(N)] = E_N[m_R N] = m_N m_R \qquad (2.4.103)$$

and

$$E[T^2] = E_N[g_2(N)] = E[\sigma_R^2 N + m_R^2 N^2] = m_N \sigma_R^2 + m_R^2 E[N^2]$$

This implies

$$\mathrm{Var}\,[T] = E[T^2] - E^2[T]$$
$$= m_N \sigma_R^2 + \sigma_N^2 m_R^2 \qquad (2.4.104)$$

In words, the expected value of the sum of a random number of random variables is just the mean number times the mean of each variable. The variance of the sum is, however, *larger* than the mean number times the variance of each variable (as it would be if N were not random but known to equal its mean). The additional term is just the variance of the mean of each variable times N; that is,

$$\mathrm{Var}\left[\sum_{i=1}^{N} m_R\right] = \mathrm{Var}\,[m_R N] = \sigma_N^2 m_R^2$$

In short,

$$\mathrm{Var}\left[\sum_{i=1}^{N} R_i\right] = \mathrm{Var}\left[\sum_{i=1}^{m_N} R_i\right] + \mathrm{Var}\left[\sum_{i=1}^{N} m_R\right] \qquad (2.4.105)$$

This result is intuitively satisfying but hardly easy to anticipate. It shows clearly the potential danger (or error) in oversimplifying probabilistic models. If the engineer had tried, in this case, to simplify his model either by replacing the number of rainy weeks by their average number or by replacing the rainfall

† Note the subtle change in notation from $E[T \mid N = n]$ to $E[T \mid N]$ which accompanies this step.

in a rainy week by its mean value, he would have underestimated the uncertainty (variance) in the total rainfall. If, however, either $\sigma_N{}^2$ or $\sigma_R{}^2$ were small enough, the error induced by the corresponding simplification would not be significant.

It is important to appreciate that again we have been able to analyze a probability model using only means and variances. No assumptions were made in this example about the *shapes* of the distributions of R and N. Expectations alone were used, and this was done symbolically (or operationally) *without* resort to formal integration.

2.4.4 Approximate Moments and Distributions of Functions†

Approximate solutions to the problem of determining the behavior of dependent random variables are usually possible. The approximations have the advantage of always giving moments of dependent variables only in terms of functions of moments of the independent variables. As stressed before, these moments may be all that are available or all that are necessary for the engineer's purposes. Simple approximations to the distribution of dependent random variables are also possible.

Approximate moments: $Y = g(X)$ If the relationship $Y = g(X)$ is sufficiently well behaved, and if the coefficient of variation of X is not large,‡ the following approximations are valid

$$E[Y] = E[g(X)] \approx g(E[X]) \tag{2.4.106}$$

$$\text{Var}\,[Y] = \text{Var}\,[g(X)] \approx \text{Var}\,[X]\left[\frac{dg(x)}{dx}\bigg|_{m_X}\right]^2 \tag{2.4.107}$$

or

$$\sigma_Y \approx \left|\frac{dg(x)}{dx}\bigg|_{m_X}\right|\sigma_X \tag{2.4.108}$$

where $dg(x)/dx\big|_{m_X}$ signifies the derivative of $g(x)$ with respect to x, evaluated at m_X.

If, for example, $Y = a + bX + cX^2$, then the approximations state that

$$E[Y] \approx a + bm_X + cm_X{}^2 \tag{2.4.109}$$

$$\text{Var}\,[Y] \approx (b + 2cm_X)^2\,\text{Var}\,[X] \tag{2.4.110}$$

or

$$\sigma_Y \approx |b + 2cm_X|\sigma_X \tag{2.4.111}$$

---

† Reading of this section is not critical for understanding of subsequent chapters. It may be bypassed on first reading. The material is most important, however, when approximate models and analyses are sufficient for the engineering purpose.
‡ How large the coefficient may be depends upon the degree of nonlinearity of $g(X)$ in the region around m_X and upon the degree of acceptable approximation error.

That the approximations are not exact can be observed by looking at the last term in Eq. (2.4.109) which should be $cE[X^2]$ or $c(m_X{}^2 + \sigma_X{}^2)$ or $cm_X{}^2(1 + V_X{}^2)$. Clearly if the coefficient of variation of X is less than 10 percent, the error involved in this approximation is less than 1 percent.

The justification for these approximations lies in the observation that if V_X is small, X is very likely to lie close to m_X and hence a Taylor-series expansion of $g(X)$ about m_X is suggested:

$$g(X) = g(m_X) + (X - m_X) \frac{dg(x)}{dx}\bigg|_{m_X}$$

$$+ \frac{(X - m_X)^2}{2} \frac{d^2g(x)}{dx^2}\bigg|_{m_X} + \cdots \quad (2.4.112)$$

Keeping only the first two terms in the expansion and taking the expectation of both sides, we obtain the stated approximation for $E[g(X)]$, since $E[X - m_X] = 0$. Similarly, keeping the same terms and finding the variance of both sides yields the approximation in Eq. (2.4.107), since

$$\text{Var}\,[g(m_X)] = 0$$

and

$$\text{Var}\left[\left[\frac{dg(x)}{dx}\bigg|_{m_X}\right](X - m_X)\right] = \left[\frac{dg(x)}{dx}\bigg|_{m_X}\right]^2 \text{Var}\,[X]$$

It is of course possible to increase accuracy by keeping additional terms in the expansion and by involving higher moments in the approximations. The mean is better approximated in general by using three terms of the expansion when taking expectations yields

$$E[g(X)] \approx g(m_X) + \frac{1}{2}\left[\frac{d^2g(x)}{dx^2}\bigg|_{m_X}\right]\sigma_X{}^2 \quad (2.4.113)$$

Notice that in the quadratic case, $Y = a + bX + cX^2$, this "approximation" is exact:

$$E[g(X)] = a + bm_X + cm_X{}^2 + \frac{1}{2}(2c)\sigma_X{}^2$$

$$= a + bm_X + c(m_X{}^2 + \sigma_X{}^2)$$

$$= a + bE[X] + cE[X^2] \quad (2.4.114)$$

This is true, of course, because the higher-order derivatives and hence the higher-order terms of the Taylor-series expansion are all zero. In practice this second-order approximation of the mean [Eq. (2.4.113)] and the first-order approximation of the variance and standard deviation

[Eqs. (2.4.107) and (2.4.108)] are commonly used, since they depend upon only the mean and variance of X.

Approximate distribution: $Y = g(X)$ Similar reasoning can be used to find an approximate distribution for Y (at least in the region of the mean) from the Taylor expansion of $Y = g(X)$. Keeping only the first two or the linear terms in X, we can write

$$Y \approx g(m_X) - m_X \left[\frac{dg(x)}{dx} \bigg|_{m_X} \right] + \left[\frac{dg(x)}{dx} \bigg|_{m_X} \right] X \qquad (2.4.115)$$

which is of the linear form

$$Y \approx a + bX$$

Applying Eq. (2.3.15), for a continuous distribution of X,

$$f_Y(y) \approx \left| \frac{1}{b} \right| f_X \left(\frac{y - a}{b} \right)$$

$$\approx \left| \frac{1}{dg(x)/dx \,|_{m_X}} \right| f_X \left\{ \frac{y - g(m_X) + m_X[dg(x)/dx \,|_{m_X}]}{dg(x)/dx \,|_{m_X}} \right\} \qquad (2.4.116)$$

The implication is that the approximate $f_Y(y)$ has the same shape as $f_X(x)$, only stretched and shifted. In essence the technique replaces the true relationship between X and Y by the linear one associated with the tangent to $g(X)$ at m_X. Consequently one can expect that the approximate distribution will be reasonably close to the exact one at those values of y where the tangent provides a reasonable approximation to $g(X)$. If, owing to the shape of $f_X(x)$, X is confined to lie in this same region (at least with high likelihood), the indicated distribution of Y will be a good approximation to the true $f_Y(y)$ almost everywhere. A quick sketch of $f_X(x)$ and $g(X)$ should determine if these conditions hold.

For illustration assume that $Y = X^2$ and find the approximate distribution of $f_Y(y)$ when X has the distribution

$$f_X(x) = \tfrac{1}{2} \qquad 4 \le x \le 6 \qquad (2.4.117)$$

with $m_X = 5$, by inspection. Then

$$g(m_X) = 25$$

and

$$\frac{dg(x)}{dx} \bigg|_{m_X} = 2x \,|_{m_X} = 10$$

and

$$Y \approx 25 - 5(10) + 10X = -25 + 10X$$

Hence

$$f_Y(y) \approx \tfrac{1}{10} f_X\left(\frac{y-25}{10}\right) = \tfrac{1}{20} \qquad 15 \le y \le 35 \qquad (2.4.118)$$

Notice that the limits on the definition also must come from the approximate relationship, e.g., from $y \ge -25 + 10(4) = 15$, not from $y \ge 4^2 = 16$, to guarantee that the resulting distribution is a proper one. The true distribution, using Eq. (2.3.4),† is

$$f_Y(y) = \frac{1}{2\sqrt{y}} f_X(\sqrt{y}) = \frac{1}{4\sqrt{y}} \qquad 16 \le y \le 36 \qquad (2.4.119)$$

This situation is sketched in Fig. 2.4.12, in a manner analogous to that used for Fig. 2.3.4. There is also shown in the figure the case where $f_X(x)$ is $\tfrac{1}{2}$ in the region $0 \le x \le 2$, a region in which the nonlinearity

† The transformation is one-to-one only because X is restricted to positive values by the form of $f_X(x)$.

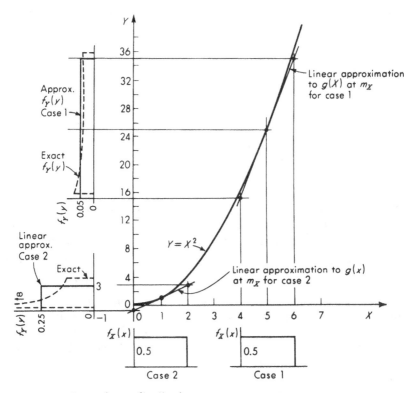

Fig. 2.4.12 Approximate distributions.

of $g(x)$ is more severe and where the approximate distribution is therefore less precise.

Multivariate approximations Similar approximations can be made in multivariate situations. Now a multidimensional Taylor-series expansion is necessary before expectations are taken.

The second-order approximation to the expected value of

is
$$Y = g(X_1, X_2, \ldots, X_n) \tag{2.4.120}$$

$$E[Y] \approx g(m_{X_1}, m_{X_2}, \ldots, m_{X_n})$$

$$+ \frac{1}{2} \sum_{i=1}^{n} \sum_{j=1}^{n} \frac{\partial^2 g}{\partial x_i\, \partial x_j} \bigg|_m \text{Cov } [X_i, X_j] \tag{2.4.121}$$

in which $(\partial^2 g / \partial x_i\, \partial x_j) \big|_m$ is the mixed second partial derivative of $g(x_1, x_2, \ldots, x_n)$ with respect to x_i and x_j evaluated at $m_{X_1}, m_{X_2}, \ldots, m_{X_n}$. The second term (which reduces to one-half the sum of the variances times the second derivatives if the X_i are uncorrelated) is negligible if the coefficients of variations of the X_i and the nonlinearity in the function are not large.

The first-order approximation to the variance of Y is

$$\text{Var } [Y] \approx \sum_{i=1}^{n} \sum_{j=1}^{n} \frac{\partial g}{\partial x_i} \bigg|_m \frac{\partial g}{\partial x_j} \bigg|_m \text{Cov } [X_i, X_j] \tag{2.4.122}$$

which, if the X_i are uncorrelated, is simply

$$\text{Var } [Y] \approx \sum_{i=1}^{n} \left(\frac{\partial g}{\partial x_i} \bigg|_m \right)^2 \text{Var } [X_i] \tag{2.4.123}$$

Notice the form of this equation. It may be interpreted as meaning that each of the n random variables X_i contributes to the dispersion of Y in a manner proportional to its own variance Var $[X_i]$ and proportional to a factor $[(\partial g / \partial x_i) \big|_m]^2$, which is related to the sensitivity of changes in Y to changes in X. We can use this interpretation to evaluate quantitatively the common practice of using engineering judgement to simplify problems. In many cases it is sufficiently accurate to treat some of the independent variables as deterministic rather than as stochastic. This formula indicates that the effect of this action is to neglect a contribution to the variance of Y and that the simplification is a justified approximation if either the variance or the "sensitivity" factor of that variable is small enough that their product is negligible compared to the contributions of other variables in the problem.   In the study of the variance of the ultimate moment capacity of reinforced-concrete beams, for example, it is found that the "most uncertain" independent variable,† namely, the ultimate concrete strength, can in practice be treated as a constant because, owing to a small "sensitivity" factor, it contributes relatively little to the variance of the ultimate moment.   For similar reasons it is reasonable to treat the width of the beam as deterministic, whereas the depth to the steel reinforcement makes a major contribution to variation in the ultimate moment.

To complete the spectrum of second-order moment approximations we need the covariance between two functions, Y_1 and Y_2, of the X's:

$$Y_1 = g_1(X_1, X_2, \ldots, X_n) \tag{2.4.124}$$

$$Y_2 = g_2(X_1, X_2, \ldots, X_n) \tag{2.4.125}$$

In this case,

$$\text{Cov}\,[Y_1, Y_2] \approx \frac{1}{2} \sum_{i=1}^{n} \sum_{j=1}^{n} \frac{\partial g_1}{\partial x_i}\Big|_m \frac{\partial g_2}{\partial x_j}\Big|_m \text{Cov}\,[X_i, X_j] \tag{2.4.126}$$

It should be pointed out that, again, first- and second-order moments are sufficient to provide at least approximations of the same moments of functionally dependent random variables.

For example, if X_1 through X_n are mutually independent, we have, generalizing Eq. (2.4.86), product

$$E[X_1 X_2 \cdots X_n] = E[X_1]E[X_2] \cdots E[X_n] \tag{2.4.127}$$

The answer based on the approximation technique is the same.   Exact expressions for variances are more cumbersome (although possible), but the approximate result is simply

$$\text{Var}\,[X_1 X_2 \cdots X_n] \approx \sum_{i=1}^{n} \left(\prod_{\substack{j=1 \\ j \neq i}}^{n} m_{X_j}^2 \right) \sigma_{X_i}^2 \tag{2.4.128}$$

For $n = 2$,

$$\text{Var}\,[X_1 X_2] \approx m_{X_1}^2 \sigma_{X_2}^2 + m_{X_2}^2 \sigma_{X_1}^2$$

$$\approx m_{X_1}^2 m_{X_2}^2 (V_{X_1}^2 + V_{X_2}^2) \tag{2.4.129}$$

Compare this result with the exact result, Eq. (2.4.87), which has an added term $\sigma_{X_1}^2 \sigma_{X_2}^2$, which is relatively small if the coefficients of varia-

† The "most uncertain" variable is that variable with the largest coefficient of variation (see Prob. 2.46).

tion are small. Notice that, approximately,

$$V^2_{X_1X_2} = \frac{\text{Var}\,[X_1X_2]}{(m_{X_1}m_{X_2})^2} \approx V_{X_1}{}^2 + V_{X_2}{}^2 \tag{2.4.130}$$

This useful result, true in general for two or more variables, says that
the square of the coefficient of variation of the product of uncorrelated
random variables is approximately equal to the sum of the squares of
the coefficients of variation of the variables. Compare this with the
analogous statement for the variance of the sum of uncorrelated random
variables [Eq. (2.4.81c)].

The previous results state, for illustration, that the mean force F
on a body submerged in a moving fluid (a truss bar in the wind or a
grillage in a stream of water, for example) is

$$E[F] = E[R]E[C]E[A]E[S^2] \tag{2.4.131}$$

when $F = RCAS^2$. Here R is the density of the fluid, S is the velocity
of the fluid, C is the body's shape-dependent, empirical drag coefficient,
A is its exposed area, and the variables are assumed mutually independent.
The coefficient of variation of the force is approximately

$$V_F \approx \sqrt{V_R{}^2 + V_C{}^2 + V_A{}^2 + V_{S^2}{}^2} \tag{2.4.132}$$

The coefficient of variation of the square of S is, by definition,

$$V_{S^2} = \frac{\sqrt{\text{Var}\,[S^2]}}{m_{S^2}}$$

which, in the light of Eqs. (2.4.106) and (2.4.108), is, approximately,

$$V_{S^2} \approx \frac{2m_S\sigma_S}{m_S{}^2} = 2V_S \tag{2.4.133}$$

In practice, one or more of these coefficients might be sufficiently small
that the corresponding variable could be treated as a deterministic
constant rather than as a random variable.

Approximate distributions of functions of jointly distributed random
variables can also often be found more easily than the exact distribution
by using the linear part of a Taylor-series expansion and the relatively
simple relationships† for the distributions of linear functions of random
variables. The procedure is straightforward; no further discussion will
be given here.

2.4.5 Summary

This section presents in detail methods of analysis of probabilistic models
based on moments rather than on entire distributions. Defining the

† See, for example, Freeman [1963] or Wadsworth and Bryan [1960].

expectation $E[g(X)]$ of a function $g(X)$ of a random variable X as a weighted average, that is,

$$E[g(X)] = \int_{-\infty}^{\infty} g(x) f_X(x) \, dx$$

we recognize that the two most important cases are first and second moments, the *mean* of a random variable

$$m_X = E[X] = \int_{-\infty}^{\infty} x f_X(x) \, dx$$

and the *variance*

$$\sigma_X{}^2 = \text{Var } [X] = E[(X - m_X)^2] = \int_{-\infty}^{\infty} (x - m_X)^2 f_X(x) \, dx$$

The *standard deviation* σ_X is the positive square root of the variance, and the *coefficient of variation* V_X is the ratio σ_X/m_X.

These formulas show that these expectations can be calculated by integration (or summation) if the probability law $f_X(x)$ is known. More important is the fact that even without knowledge of complete distributions, useful practical analysis of engineering problems can be carried out with expectations alone. Certain rules of thumb and inequalities permit probabilities to be related to these moments. Economic analyses can be based on *expected costs*. Owing primarily to the *linearity property* of the expectations operation,

$$E[g_1(X) + g_2(X)] = E[g_1(X)] + E[g_2(X)]$$

moment analysis of models can often be completed with relative ease and without explicit use of integration.

In particular, this property leads to important formulas for the mean and variance of a sum of random variables. For two variables,

$$E[aX_1 + bX_2] = aE[X_1] + bE[X_2]$$

$$\text{Var } [aX_1 + bX_2] = a^2 \text{ Var } [X_1] + b^2 \text{ Var } [X_2] + 2ab \text{ Cov } [X_1, X_2]$$

The *covariance* is defined as

$$\text{Cov } [X_1, X_2] = E[(X_1 - m_{X_1})(X_2 - m_{X_2})]$$

$$= \int\!\!\int_{-\infty}^{\infty} (x_1 - m_{X_1})(x_2 - m_{X_2}) f_{X_1, X_2}(x_1, x_2) \, dx_1 \, dx_2$$

The *correlation coefficient* ρ is defined as

$$\rho = \frac{\text{Cov } [X_1, X_2]}{\sigma_{X_1} \sigma_{X_2}}$$

The latter coefficient is bounded between -1 and $+1$ and is a useful measure of linear dependence when properly interpreted.

The concept of *conditional expectation* and the availability of approximate expressions for moments (Sec. 2.4.4) extend the range of probabilistic models to which moment analysis can be applied without complete distribution information.

2.5 SUMMARY FOR CHAPTER 2

In Chap. 2 we have presented all the basic theory necessary to accomplish meaningful probabilistic analyses of engineering problems. We have defined, interpreted, and operated on the following important factors: sample spaces, events, probabilities (marginal and conditional), random variables (simple and joint), probability distributions (discrete, continuous, and mixed; marginal, joint, and conditional), random variables which are functions of random variables $[Y = g(X)]$, expectation, and moments (marginal, joint, and conditional).

These represent the fundamental tools for construction and analysis of stochastic models. The next chapter will present a number of the most frequently encountered models. It is emphasized throughout Chap. 3 that we are merely applying there the methods of this chapter in conjunction with various sets of assumptions. These assumptions represent the engineer's model of the physical phenomenon.

In Chap. 4 we discuss the problem of estimating the parameters of these models when presented with a limited quantity of real data. We shall find that again we are only applying the methods of Chap. 2 to a particular class of random variables called *statistics*. Having analyzed the probabilistic behavior (mean, variance, distribution) of a statistic, we attempt to quantify the confidence we can place on inferences about the model that are drawn from an observed value of the statistic.

Chapters 5 and 6 again apply the tools of this chapter to a particular class of problems, namely, decision making in the face of uncertainty.

REFERENCES

General

Feller, W. [1957]: "An Introduction to Probability Theory and Its Applications," vol. I, 2d ed., John Wiley & Sons, Inc., New York.

Freeman, H. [1963]: "Introduction to Statistical Inference," Addison-Wesley Publishing Company, Inc., Reading, Mass.

Hahn, G. J. and S. S. Shapiro [1967]: "Statistical Models in Engineering," John Wiley & Sons, Inc., New York.

Hald, A. [1952]: "Statistical Theory with Engineering Applications," John Wiley & Sons, Inc., New York.

Hammersley, J. M. and D. C. Handscomb [1964]: "Monte Carlo Methods," Methuen & Co., Ltd., London.

Parzen, E. [1960]: "Modern Probability Theory and Its Applications," John Wiley & Sons, Inc., New York.

Tocher, K. D. [1963]: "The Art of Simulation," D. Van Nostrand Company, Inc., Princeton, New Jersey.

Tribus, M. [1969]: "Rational Descriptions, Decisions, and Designs," Pergamon Press, New York.

Von Mises, R. [1957]: "Probability, Statistics, and Truth," 2d ed., The Macmillan Company, New York.

Wadsworth, G. P. and J. G. Bryan [1960]: "Introduction to Probability and Random Variables," McGraw-Hill Book Company, New York.

Specific text references

ACI Standard Recommended Practice for Evaluation of Compression Test Results of Field Concrete (ACI Standard 214-65) [1965], American Concrete Institute, Detroit, Michigan.

Benson, M. A. [1965]: Spurious Correlation in Hydraulics and Hydrology, *ASCE Proc., J. Hydraulics Div.*, vol. 91, no. HY4, July.

Blum, A. M. [1964]: Digital Simulation of Urban Traffic, *IBM Systems J.*, vol. 3, no. 1, p. 41.

Esteva, L. and E. Rosenblueth [1964]: Espectros de Temblores a Distancias Moderadas y Grandes, *Bol. Soc. Mex. Ing. Sismica*, vol. 2, no. 1, March.

Fishburn, P. C. [1964]: "Decision and Value Theory," John Wiley & Sons, Inc., New York.

Goldberg, J. E., J. L. Bogdanoff, and D. R. Sharpe [1964]: Response of Simple Non-linear Systems to a Random Disturbance of the Earthquake Type, *Bull. Seismol. Soc. Am.*, vol. 54, no. 1, pp. 263–276, February.

Hognestad, E. [1951]: A Study of Combined Bending and Axial Load in Reinforced Concrete Members, *Eng. Exptl. Sta. Bull.*, no. 399, University of Illinois.

Hufschmidt, M. and M. B. Fiering [1966]: "Simulation Techniques for Design of Water Resource Systems," Harvard University Press, Cambridge, Miss.

Hutchinson, B. G. [1965]: The Evaluation of Pavement Structural Performance, Ph. D. thesis, Department of Civil Engineering, University of Waterloo, Waterloo, Ontario.

Rosenblueth, E. [1964]: Probabilistic Design to Resist Earthquakes, *J. Eng. Mech. Div., ASCE Proc.*, vol. 90, paper 4090, October.

PROBLEMS

2.1. An engineer is designing a large culvert to carry the runoff from two separate areas. The quantity of water from area A may be 0, 10, 20, 30 cfs and that from B may be 0, 20, 40, 60 cfs. Sketch the sample spaces for A and B jointly and for A and B separately. Define the following events graphically on the sketches.

    (a) $C = A > 10$ cfs

    (b) $D = B \geq 20$ cfs

    (c) $E = C^c$

    (d) $F = C \cap D$

    (e) $G = (C \cup D)^c$

2.2. A warehouse floor system is to be designed to support cartons filled with canned food. The cartons are cubic in shape, 1 ft on a side and weigh 100 lb each. Consider that the cartons may be stacked to a height of 8 ft.

(a) Sketch the sample space for total weight on a square foot of floor area assuming that it is loaded by one stack of boxes. How would this sample space be changed if the area in question can be loaded by half the weight of each of two stacks of boxes?

(b) Sketch the sample space for total load on two adjacent floor areas each 1 by 1 ft, assuming that each such area supports a single stack of boxes.

(c) Define the following events on the sketches.

    (i) C = total load $\geq$ 300 psf

    (ii) D = total load $\leq$ 400 lb

    (iii) E = total load > 500 lb

    (iv) $F = C \cap E$

    (v) $G = D \cup E$

2.3. (a) Sketch a sample space for the following experiment. The number of vehicles on a bridge at a particular instant is going to be counted and weighed; only the total number and total weight of the vehicles are to be recorded (observed). The maximum number of vehicles which can be found is five; the maximum weight of a single vehicle is 5 tons and the minimum weight is 2 tons.

(b) Indicate on the sketch the regions corresponding to each of the following events:

    (i) A = [the number of vehicles was less than four]

    (ii) B = [the total weight was greater than 10 tons]

    (iii) C = [three vehicles each of maximum weight were found]

    (iv) $D = A \cap B$

    (v) $E = A \cup B$

    (vi) $F = A^c \cap B^c$

    (vii) $G = A^c \cap C$ (Possible?)

2.4. (a) Sketch a sample space for the following experiment: A timber pile will be chosen from a supply of assorted lengths L, the longest of which is 60 ft. The pile will be driven into the ground in an area where the solid-rock-bearing stratum is at a variable depth D, the maximum being 60 ft.

(b) On a sequence of such sketches shade the following events:

    (i) $A = [L > D]$ = [pile of adequate length]

    (ii) $B = [D > 50]$ = [it is a relatively long drive to bearing]

    (iii) $C = [L > 50]$ = [the pile is relatively difficult to handle]

    (iv) $D = B \cup C$ = [there is some kind of relative difficulty]

    (v) $E = B \cap C$ = [both kinds of difficulty arise]

    (vi) $F = (B \cup C) \cap A^c$ = [there is some kind of difficulty and the pile was of inadequate length]

2.5. A question of the acceptability of an existing concrete culvert to carry an anticipated flow has arisen. Records are sketchy, and the engineer assigns estimates of annual maximum flow rates and their likelihoods of occurrence (assuming that a maximum of 12 cfs is possible) as follows:

    Event A = [5 to 10 cfs] $P[A] = 0.6$

    Event B = [8 to 12 cfs] $P[B] = 0.6$

    Event $C = A \cup B$ $P[C] = 0.7$

(a) Construct the sample space. Indicate events A, B, C, $A \cap C$, $A \cap B$, and $A^c \cup B^c$ on the sample space.

(b) Find $P[A \cap B]$, $P[A^c]$, $P[B \cup A^c]$.

(c) Find $P[A \mid B]$, $P[B \mid A]$, $P[B \mid A^c]$.

2.6. (a) An engineer has observed that two of the designers in his office work at different rates and have a different frequency of making errors. If designer A will take 6, 7, or 8 hr to do a particular job with the possibility of 0, 1, 2 errors, sketch the sample space associated with this designer. Designer B is faster but more prone to errors. He will require 5, 6, or 7 hr and may make 0, 1, 2, 3 errors. Sketch the sample space for designer B.

If designer A is equally likely to have any one of the nine possible combinations occur, sketch the probability attributes on the sample space for A.

If designer B is twice as likely to make 1 error as either 0, 2, or 3 errors and twice as likely to require 6 hr as either 5 or 7 hr, sketch the probability attributes on the sample space for designer B. Assume independence.

(b) Sketch following events for designer A and determine probabilities.

    (i) 7 hr and 0, 1, or 2 errors

    (ii) 8 hr and 0 or 1 errors

    (iii) 6 or 8 hr and 0 errors

(c) Indicate the following events for designer B on sample space and determine probabilities.

    (i) $[A] = [5 \text{ hr time}]$

    (ii) $[B] = [2 \text{ or } 3 \text{ errors}]$

    (iii) $[C] = [7 \text{ hr and } 2 \text{ errors}]$

    (iv) $[D] = [A \cup B]$

    (v) $[E] = [B \cup C]$

    (vi) $[F] = [B \cap C]$

(d) Compare the probabilities of various job times for the two designers (sketch).

(e) If both designers are working at the same time on the same kind of task, can the probabilities of two 7-hr times be added to determine the probability of a 14-hr total? Why?

(f) Determine the probabilities of various total time requirements in (e) if the designers work independently. Repeat for total number of errors.

2.7. If the occurrences of earthquakes and high winds are unrelated, and if, at a particular location, the probability of a "high" wind occurring throughout any single minute is 10^{-5} and the probability of a "moderate" earthquake during any single minute is 10^{-8}:

(a) Find the probability of the joint occurrence of the two events during any minute. Building codes do not require the engineer to design the building for the combined effects of these loads. Is this reasonable?

(b) Find the probability of the occurrence of one or the other or both during any minute. For rare events, i.e., events with small probabilities of occurrence, the engineer frequently assumes

$$P[A \cup B] \approx P[A] + P[B]$$

Comment.

(c) If the events in succeeding minutes are mutually independent, what is the probability that there will be no moderate earthquakes in a year near this location? In 10 years? Approximate answers are acceptable.

2.8. Revise the water-supply situation (page 52) for a lateral with four tributary sites. Determine probabilities of various water demand levels. What is the optimum design size if capacity costs are linear at $500 per unit provided and enlarging costs are $1000 per unit provided? Formulate a solution for arbitrary costs if enlarging costs are double initial capacity costs and both are linear with quantity.

2.9. Find the probability that a pile will reach bedrock n ft below without hitting a rock if the probability is p that such a rock will be struck in any foot and the occurrences of rocks in different 1-ft levels are mutually independent events.

Evaluate this probability for $p = \frac{1}{20}$ and $n = 10$ and 20 feet, and for $p = \frac{1}{10}$ and $n = 10$ and 20 ft. Notice that even in the last case it is not certain that a rock will be struck.

2.10. Consider the possible failure of a water supply system to meet the demand during any given summer day.

(a) Use the equation of total probability [Eq. (2.1.12)] to determine the probability that the supply will be insufficient if the probabilities shown in the table are known.

	Demand level, gal/day	$P[level]$	$P[inadequate$ $supply \mid level]$
D_1	100,000 gd	0.6	0
D_2	150,000 gd	0.3	0.1
D_3	200,000 gd	0.1	0.5
		1.0	

(b) Find the probability that a demand level of 150,000 gal/day was the "cause" of the system's failure to meet demand if an inadequate supply was observed. (Clearly the word "cause" is not appropriate in such a situation, but this interpretation of Bayes theorem is often adopted. It should be used with caution.)

(c) The likelihood of a pump failing and causing the system to fail is 0.02 regardless of the demand level. What does the equation of total probabilities reduce to in this case, that of independence?

(d) The system may fail in one and only one of three possible modes: M_1, inadequate supply; M_2, a pump failure; or M_3, overload of the purification plant. We have the following information:

Demand level, gal/day	$P[M_3 \mid level]$
100,000	0
150,000	0
200,000	0.1

(The other probabilities are as given above.) Find the probabilities of each of the various possible causes (modes) if a system failure takes place when the demand level is 150,000 gal/day. *Hint:* Modify Bayes theorem [Eq. (2.1.13)] to read

$$P[B_i \mid A \cap C_j] = \frac{P[A \mid B_i \cap C_j]P[B_i \mid C_j]}{P[A \mid C_j]}$$

in which here the B_i's are the modes of failure, the C_j is the demand level, and A is the event a failure took place. Verify this "conditional" Bayes theorem. It can be interpreted as a Bayes-theorem application in the conditional sample space (i.e., given C_j).

(*e*) What are the probabilities of the various causes (modes) if the demand level at failure was 100,000 gal/day? In general, what does Bayes theorem reduce to if A can occur if and only if B_j occurs? Recognize that within the familiar deterministic, one-cause-one-effect view of phenomena, the determination of cause by observation of effect can be interpreted as just such a special case of the application of Bayes theorem.

2.11. An engineer concerned with providing a continuous water supply to a critical operation is considering installing a second "backup" pump to take the place of the primary pump in the event of its failure. Let F_1 be the event that the primary pump fails once during a given period. (The likelihood of two or more such failures is negligible). Let F_2 be the event that the second pump fails to function if it is switched on.

(*a*) What is the relationship between these events and the event F_0 that the system fails to provide continuous service during the period?

(*b*) What is the reliability of the system $P[F_0^c]$ in terms of the reliabilities of the individual pumps $P[F_1^c]$ and $P[F_2^c]$ if the events F_1 and F_2 are independent? How does this answer compare with the reliability of a series-type system in which *both* components must operate simultaneously for the system to function?

(*c*) What is $P[F_0]$ if $P[F_1] = P[F_2] = 10^{-1}$ and the conditions in (*b*) hold?

(*d*) If the failure of one component in a redundant system is caused by an overload, the failure of the stand-by unit will probably not be independent of the failure of the first. Generally, $P[F_2 \mid F_1] > P[F_2]$. Find the reliability of the system above if $P[F_2 \mid F_1] = 2 \times 10^{-1}$.

The system described here is an example of a parallel-type system, in which redundant elements are provided to reduce the likelihood of a system failure.

2.12. In the study of a storage-dam design, it is assumed that quantities can be measured sufficiently accurately in units of $\frac{1}{4}$ of the dam's capacity. It is known from past studies that at the beginning of the first (fiscal) year the dam will be either full, $\frac{3}{4}$ full, $\frac{1}{2}$ full, or $\frac{1}{4}$ full, with probabilities $\frac{1}{3}$, $\frac{1}{3}$, $\frac{1}{6}$, and $\frac{1}{6}$, respectively. During each year water is released. The amount released is $\frac{1}{2}$ the capacity if at least this much is available; it is all that remains if this is less than $\frac{1}{2}$ the capacity. After release, the inflow from the surrounding watershed is obtained. It is either $\frac{1}{2}$ or $\frac{1}{4}$ of the dam's capacity with probabilities $\frac{2}{3}$ and $\frac{1}{3}$, respectively. Inflow causing a total in excess of the capacity is spilled. Assuming independence of annual inflows, what is the probability distribution of the total amount of water at the beginning of the third year?

2.13. A large dam is being planned, and the engineer is interested in the source of fine aggregate for the concrete. A likely source near the site is rather difficult to survey accurately. From surface indications and a single test pit, the engineer believes that the magnitude of the source has the possible descriptions: 50 percent of adequate; adequate; or 150 percent of possible demand. He assigns the following probabilities of these states.

State	Probability
50% of adequate	0.5
Adequate	0.4
150% of demand	0.1
	1.0

Prior to ordering a second test pit, the engineer decides that the various likelihoods of the sample's possible indications (Z_1, Z_2, Z_3) depend upon the (unknown) true state as follows:

P [sample indication | state]

	50%	Adequate	150%
Z_1, favoring 50%	0.50	0.40	0.20
Z_2, favoring adequate	0.40	0.50	0.40
Z_3, favoring 150%	0.10	0.10	0.40
	1.00	1.00	1.00

What are the probabilities of observing the various events Z_1, Z_2, and Z_3?

The second test pit is dug and the source appears adequate from this pit. Compute the posterior probabilities of state. If another test pit gives the same result, calculate the second set of posterior state probabilities. Compare prior and posterior state probabilities.

2.14. The following "urn model" has been proposed to model the occurrence (black ball) or nonoccurrence (red ball) of rainfall on successive days. There are three urns: the "initial urn" containing n_1 black and m_1 red balls, the "dry urn" containing n_2 black and m_2 red balls, and the "rainy urn" containing n_3 black and m_3 red balls. To simulate (sample) a sequence, one draws a ball from the initial urn, its color indicating occurrence or not of rain on the first day. The weather on the second day is found by sampling from the rainy or dry urn depending on the outcome of the first trial. Subsequent draws are made from the rainy or dry urn depending on the weather on the immediate past day. The ball is returned to its appropriate urn after each draw. The model is devised to simulate the persistence of rainy or dry spells. This is an example of what we will come to know as a Markov chain (Sec. 3.7).

Assume that all the balls in any urn are equally likely to be drawn and $n_1 = n_2 = n_3 = 10$ and $m_1 = 70$, $m_2 = 90$, and $m_3 = 40$.

(a) What is the probability that the first four days in a row will be observed to be rainy?

(b) What is the probability that at least three dry days will follow a rainy day?

Source: E. H. Wiser [1965], Modified Markov Probability Models of Sequences of Precipitation Events, *Monthly Weather Review*, vol. 93, pp. 511–516. This reference contains many suggested urn models of more complicated varieties.

2.15. At a traffic signal, the number N of cars that arrive during the red-green cycle on the northbound leg has a PMF of $p_N(n)$, $n = 0, 1, 2, \ldots$. At most three cars can pass through the intersection in a cycle. The engineer is disturbed by his choice of cycle times any time there is a car left at the end of the green phase.

(a) What is the probability that the engineer is disturbed with any particular cycle if at the end of the previous green phase no cars are present?

(b) What is the probability that he is "disturbed" at least twice in succession? (Assume the same zero starting conditions as in (a) at the beginning of the *first* of these two cycles.)

(c) Evaluate (a) and (b) if

$$p_N(n) = \frac{(\nu)^n e^{-\nu}}{n!} \qquad n = 0, 1, 2 \ldots, \infty; \nu = 2$$

(See Table A-2.)

2.16. A quality-control plan for the concrete in a nuclear reactor containment vessel calls for casting 6 cylinders for each batch of 10 yd^3 poured and testing them as follows:

　　1 at 7 days
　　1 at 14 days
　　2 at 28 days
　　2 more at 28 days if any of first four cylinders is "inadequate"
　　The required strength is a function of age.

If the cylinder to be tested is chosen at random from those remaining (i. e., with equal likelihoods):

(a) What is the probability that all six will be tested if in fact one inadequate cylinder exists in the six?

(b) If the batch will be "rejected" if two or more inadequate cylinders are found, what is the likelihood that it will *not* be rejected given that exactly two are in fact inadequate? (Rejection will lead to more expensive coring and testing of concrete in place.)

(c) A "satisfactory" concrete batch gives rise to an inadequate cylinder with probability $p = 0.1$. (This value is consistent with present recommended practice.) What is the probability that there will be one or more inadequate cylinders in the six when the batch is "satisfactory"? (Assume independence of the quality of the individual cylinders.)

(d) Given that the batch is satisfactory ($p = 0.1$), what is the probability that the batch will be rejected? What is the probability that an unsatisfactory batch (in particular, say, $p = 0.3$) will *not* be rejected? Clearly a quality control plan wants to keep both these probabilities low, while also keeping the cost of testing small.

2.17. A new transportation system has three kinds of vehicles with seating capacities 2, 4, and 8. They become available to the dispatcher at a terminal in mixed trains having from one to four cars. If each of the possible train lengths is equally likely and if the three vehicles appear independently and in equal relative frequencies, what is the probability that exactly 10 seats will be available for dispatch in an arbitrary train?

2.18. Pairwise independence does not imply mutual independence. Assume that the particular scales used in dry-batching a concrete mix are such that both aggregate and cement weights are subject to error. The two weights are measured independently. The aggregate is equally likely to be measured exactly correct or 20 pounds too large. The cement is equally likely to be measured exactly correct or 20 pounds too small. The total weight of aggregate and cement is desired. Let event A be no error in aggregate weight measurement, event B be no error in cement weight, and event C be no error in total weight.

(a) Show that the pairs of events A and B, A and C, and B and C are independent.

(b) Use (a) as a counterexample to show that pairwise independence does not imply mutual independence.

(c) Is the inverse statement true?

2.19. A major city transports water from its storage reservoir to the city via three large tunnels. During an arbitrary summer week there is a probability q that the reservoir level will be low. Owing to the occasional call to repair a tunnel or its control valves, etc., there are probabilities $p_i(i = 1, 2, 3)$ that tunnel i will be out of service during any particular week. These calls to repair particular tunnels are independent of each other and of the reservoir level.

The "safety performance" of the system (in terms of its potential ability to meet heavy emergency fire demands) in any week will be satisfactory if the reservoir level is high and if all tunnels are functioning; the performance will be poor if more than one tunnel is out of service or if the reservoir is low *and* any tunnel is out of service; the performance will be marginal otherwise.

(a) Define the events of interest. In particular, what events are associated with marginal performance?

(b) What is the probability that exactly one tunnel fails?

(c) What is the probability of marginal performance?

(d) What is the probability that any particular week of marginal performance will be caused by a low reservoir level rather than by a tunnel being out of service?

2.20. At a certain intersection, of all cars traveling north, the relative frequency of cars continuing in the same direction is p. The relative frequency of those turning east is q; all others turn west.

Assume that drivers behave independently of one another. A small group of n cars enters the intersection. For this group

(a) What is the marginal distribution of Y, the number of cars turning west? Find the conditional distribution of X, the number of cars turning east, given that Y equals y. *Hint:* what is the probability that any car not turning west will turn east?

(b) Find the joint distribution of X and Y. Be careful with the limits of validity.

(X, Y, and Z, the number going straight, have a joint "multinomial" distribution which is studied in Sec. 3.6.1.)

2.21. Two kinds of failure of reinforced-concrete beams concern the engineer: one, "the under-reinforced moment" failure, is preceded by large deflections which give warning of its imminence; the other, the "diagonal-tension or shear" failure, occurs suddenly and without warning, not permitting persons to remove the cause of the overload or to evacuate the structure.

A structural consultant has been retained to observe a suspect beam in a building. From the engineer's experience he estimates that about 5 percent of all beams proportioned according to the building code in use at the time the building was designed will fail owing to a weakness in the shear manner, if tested to failure, while the others will fail in the moment manner. From laboratory experience, however, the engineer knows that at some load prior to failure 8 of 10 beams destined to fail in the shear manner will exhibit small characteristic diagonal cracks near their ends. On the other hand, only 1 of 10 beams which would finally fail in the moment manner shows similar cracks prior to failure.

Suppose that the relative consequences of the sudden shear failure versus warning-giving moment type of failure are such that the expensive replacement of the beam is justified only if a sudden failure is more likely than a moment failure. Then, if upon inspecting the beam, the engineer observes these characteristic diagonal cracks, should he demand the repairs or conclude that the risk is too small to justify the repairs (without further study)?

2.22. A preliminary investigation of a site leads an engineer to state that the relative weights are 3 to 5 to 2 (respectively) that the unconfined compressive strength of the soil below is 1200, 1000, or 800 psf (the only three values considered possible, for simplicity). "Undisturbed" samples of the soil will be obtained by boring and tested to gain further information. Owing to the difficulties in obtaining such samples and owing to testing inaccuracies, the following frequencies of indicated strengths are considered applicable for each specimen:

P [**indicated strength | state**]

Indicated strength	State		
	800	1000	1200
800	0.8	0.4	0.2
1000	0.2	0.5	0.3
1200	0	0.1	0.5
	1.0	1.0	1.0

The engineer calls for a sampling plan of two independent specimens.

(a) Find the conditional probabilities of each of the possible outcomes of this sample of size two given that the true strength is 1200 psf.

(b) If the results of the sampling were one specimen indicating 1000 and one indicating 800 psf, find the engineer's posterior probabilities of the strength.

(c) Suppose that after these two specimens the engineer continued sampling and found an uninterrupted sequence of specimens indicating 1200 psf. After how many could he stop:

(i) Confident that the strength was not actually 800?

(ii) At least "90 percent confident" that the strength was actually 1200?

2.23. Consider the following problem associated with synchronizing traffic lights. A particular traffic light has a cycle as follows:

Red = 1 min

Green = 1.5 min

Yellow = 0.5 min

Some distance before this light—light 1—is another light—light 2. Owing to varying drivers and conditions, the travel time between the two varies from vehicle to vehicle. Data suggest that 40 percent of all cars leaving the location of light 2 at a time when light 1 is red are not delayed by light 1 when they reach it, 80 percent of all cars leaving light 2 during a green cycle of light 1 are not delayed, and 20 percent of all cars leaving during a yellow cycle are not delayed.

Given that a car was delayed by light 1, what is the probability that it left light 2 while light 1 was red? green? yellow?

2.24. A machine to detect improper welds in a fabricating shop detects 80 percent of all improper welds, but it also incorrectly indicates an improper weld on 5 percent of all satisfactory welds. Past experience indicates that 10 percent of all welds are improper. What is the probability that a weld which the machine indicates to be defective is in fact satisfactory?

2.25. The cost of running an engineering office is a function of office size X. Assume that an engineer is trying to make a projection of cost for the next year's operations. He believes the demand will require an X of from 1 to 6.

x	$p(x)$
1	0.05
2	0.20
3	0.20
4	0.25
5	0.20
6	0.10

(a) Cost varies with X according to:

$$Y = 1000X + 100 \sqrt{X}$$

The first term represents space and salary expense while the second term represents overhead. Find the PMF of Y.

(b) The gross income to the owner Z is jointly distributed with X:

$$p_{X,Z}(x,z) = \begin{cases} & x & z \text{ (in thousands)} \\ 0.05 & 1 & 10 \\ & 2 & 10, 12, 14, 16 \\ & 3 & 12, 14, 16, 18 \\ & 4 & 12, 14, 16, 18, 20 \\ & 5 & 14, 16, 18, 20 \\ & 6 & 18, 20 \\ 0 & \text{elsewhere} \end{cases}$$

Find the PMF of the net income:

$$I = Z - Y$$

2.26. Show that the function below is the PDF of R, the distance between the epicenter of an earthquake and the site of a dam, when the epicenter is equally likely to be at any location along a neighboring fault. You may restrict your attention to a length of the fault l that is within a distance r_0 of the site because earthquakes at greater distances will have negligible effect at the site.

$$f_R(r) = \frac{2r}{l} (r^2 - d^2)^{-\frac{1}{2}} \qquad d \leq r \leq r_0$$

Sketch this function.

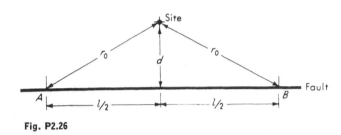

Fig. P2.26

2.27. A system has a certain capacity R and must meet a maximum demand L; both can be treated as nonnegative, continuous random variables, with joint density $f_{R,L}(r,l)$.

Write an integral expression in terms of $f_{R,L}(r,l)$ for the probability of failure, $P_F = P[R < L]$. A sketch of the sample space will help. Evaluate the integral if

$$f_{R,L}(r,l) = \begin{cases} \frac{1}{100} & 0 \leq r \leq 10, \ 0 \leq l \leq 10 \\ 0 & \text{elsewhere} \end{cases}$$

2.28. It has been found that the risk of an accident (in terms of expected number per 100 million vehicle miles) depends on the speed at which a vehicle travels. This risk is a minimum if one travels at 10 mph above the average speed a of all vehicles on the highway. In general, the risk for a vehicle traveling at a "constant" speed v is approximately

$$r = ce^{b[v-(a+10)]^2}$$

Assume that any vehicle in a particular class of vehicles (say certain commercial vehicles) travels at a "constant" speed V, which has the probability distribution (over all the vehicles in this class):

$$f_V(v) = \frac{1}{d} \qquad a - d \leq v \leq a$$

Sketch this relationship and this distribution. Find the PDF of R, the risk experienced by a vehicle from this class.

2.29. Owing to variations in raw materials, preparation, etc., the quality Y of concrete varies from batch to batch. For a batch of concrete of given quality ($Y = y$), a specimen taken from the batch and tested by a standard procedure (itself subject to variation) will indicate "O.K." with probability p_y, some increasing function of the quality. Thus this probability P varies from batch to batch.

Assume that P has been found to have a quadratic distribution for a particular set of production and testing conditions

$$f_P(p) = 3p^2 \qquad 0 \le p \le 1$$

(a) What is the distribution of the number N of O.K. specimens in a sample of three (independent) specimens, if the quality is exactly equal to the desired quality y_0 (with its associated probability p_0)?

(b) Express the joint distribution of N and P.

(c) What is the distribution of the number of O.K. specimens in an arbitrary batch?

(d) What is the probability that the quality of a particular batch is less than the desired quality given that two out of three specimens were found to be O.K.?

(e) A desirable property of a quality-control plan is that it not "indicate" good quality when quality is in fact low [e.g., part (d) above]. State and demonstrate, qualitatively, ways in which this probability can be reduced, using this example. Consider both the reliability of the testing procedure and the number of specimens per sample.

2.30. Earthquakes of many sizes (magnitudes) occur. The density function of magnitudes is known to be

$$f_X(x) = \lambda e^{-\lambda x} \qquad x \ge 0$$

The design for a nuclear power plant near an active source of earthquakes will be based on the assumption that an earthquake of magnitude x_0 will occur, where x_0 is chosen such that $P[X \ge x_0] = p$. If this magnitude (x_0) occurs, the structure will be designed so that uninterrupted operation will continue. To avoid a nuclear incident, however, the designer wants to design to resist collapse of the structure if some larger earthquake occurs. He wants to choose this other value x_1 such that, *given* that an earthquake larger than x_0 occurs, the probability that it will exceed x_1 is again p. Find x_1 in terms of x_0, λ, and p (or fewer parameters, if possible). (This distribution is considered in detail in Sec. 2.2.2.) Find the conditional distribution of X given that $X \ge x_0$. Find the conditional mean and variance of X given that $X \ge x_0$.

2.31. The stress-strain relationship for a certain type of concrete has the form shown in the accompanying diagram, which can be adequately approximated by saying that strain is proportional to the square of the stress. In structural laboratories strains rather than stresses are measured. Many specimens of a structural model were tested to failure, and the strain at failure appeared to have a probability density function of the *form*

$$\left[1 - \cos\left(2\pi \frac{x}{r} \right) \right] \qquad 0 \le x \le r$$

What is the *form* of the probability density function of the stress at failure?

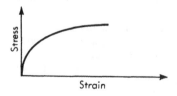

Strain   **Fig. P2.31**

2.32. The manager of a dam is said to be following a "normal operating policy" if he releases in a year an amount of water Y, which depends on the amount of water available Z, through the relationship shown in the accompanying diagram.† In the figure c is the capacity of the dam and d is the "target outflow," that is, the amount planned on or expected by users. Slopes are unity. (Other policies, i.e., other functional relationships between y and z, must, of course, lie between the dashed lines to be feasible.)

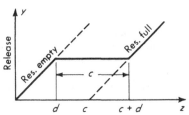

Fig. P2.32

(a) The amount of water available in any year is a random variable Z with a distribution given by

$$
f_Z(z) = \begin{cases} \dfrac{2\lambda}{d(\lambda d + 2)} z & 0 < z < d \\[2ex] \dfrac{2\lambda}{\lambda d + 2} e^{-\lambda(z-d)} & z \geq d \end{cases}
$$

Sketch this function and show that it is a proper probability density function.

(b) Find the cumulative distribution function of Y, the amount of water released in a year, if the "normal operating policy" is followed.

2.33. An engineer states that the error X in a measurement has probability distribution in the shape of a cosine curve:

$$
f_X(x) = \begin{cases} N \cos \dfrac{\pi x}{2\epsilon} & -\epsilon \leq x \leq \epsilon \\[2ex] 0 & \text{elsewhere} \end{cases}
$$

(a) Find the normalizing factor N that will make this a proper probability density function.

(b) Sketch the density function.

(c) Find and sketch $F_X(x)$.

(d) What is the probability that X is greater than $\epsilon/2$?

(e) What is the probability that the error is greater than $\epsilon/2$ in absolute magnitude?

(f) Find $f_X(x)$ from $F_X(x)$.

2.34. Probability integral transformation. Consider a random variable X with cumulative function $F_X(x)$, $-\infty \leq x \leq \infty$. Now define a new random variable U to be a particular function of X, namely,

$$
U = F_X(X)
$$

† M. B. Fiering [1967], "Streamflow Synthesis," Harvard University Press, Cambridge, Mass.

For example, if $F_X(x) = 1 - e^{-\lambda x}$, then $U = 1 - e^{-\lambda X} = g(X)$. Show [at least for reasonably smooth $F_X(x)$] that the random variable U has a constant density function on the interval 0 to 1 and is zero elsewhere. *Hint:* Convince yourself graphically that $g(g^{-1}(u)) = u$ and assume that $F_X(x)$ satisfies the conditions needed to apply Eq. (2.3.3).

To illustrate this notion, sketch the $F_X(x)$ and $F_X^{-1}(u)$ for several CDF's, including the case where X is a mixed random variable and the case where $F_X(x)$ equals a constant, say 0.5, over an interval. [These cases suggest why the definition of the inverse function $F_X^{-1}(u)$ is better stated as "smallest value of x for which $F(x)$ is *greater than or equal to u*. With this definition of the inverse function, Eq. (2.3.3) will hold for any practical CDF.]

The practical importance of this, the "probability integral transformation," lies in the fact that in simulation studies (Sec. 2.3.3) one can sample the variable X by the easier process of first sampling the variable U, finding the value u, and then calculating the corresponding value of X by

$$x = F_X^{-1}(u)$$

If $F_X^{-1}(u)$ cannot be evaluated explicitly, the finding of x given u can always be done graphically on a plot of $F_X(x)$ (or in a computer by "table lookup").

For the example distribution above (with $\lambda = 1$), find the values of the random variable X corresponding to the random numbers 0.51, 0.20, and 0.98 drawn from a table. These are observations of U.

2.35. In tests on a model structure, which has been instrumented to find the maximum strain in each member caused by windstorm loadings, a number of data points have been found for X, the wind load strain in the first member. X is judged to have a *gamma distribution* (Sec. 3.2.3)

$$f_X(x) = \frac{c^{b+1}x^b e^{-cx}}{b!} \qquad 0 \le x \le \infty$$

with parameters $b = 1$, $c = 2$. Assuming a linear model, the total stress Y is the known dead load stress y_0 plus the wind load stress rX, where r is the modulus of elasticity of the material:

$$Y = y_0 + rX$$

(a) Find $f_Y(y)$.
(b) Find $E[Y]$ two ways, using both $f_Y(y)$ and $f_X(x)$.

2.36. (a) *The Bernoulli distribution:*

$$p_X(x) = \begin{cases} p & x = 1 \text{ "success"} \\ 1 - p & x = 0 \text{ "failure"} \\ 0 & \text{elsewhere} \end{cases}$$

for $0 \le p \le 1$. Find the mean and variance of X in terms p. For what value of p is the variance a maximum? Evaluate at $p = 0.5$ and 0.1.

(b) *The Poisson distribution:*

$$p_X(x) = \begin{cases} \dfrac{e^{-\lambda}\lambda^x}{x!} & x = 0, 1, 2, \ldots \\ 0 & \text{elsewhere} \end{cases}$$

for $\lambda > 0$. Find the mean and standard deviation of X. Evaluate at $\lambda = 0.5$ and 2. Sketch the PMF for these cases.

(c) A triangular distribution:

$$f_X(x) = \begin{cases} \dfrac{2}{a(a+b)}\,(x+a) & -a \le x \le 0 \\ \dfrac{2}{b(a+b)}\,(b-x) & 0 \le x \le b \\ 0 & \text{elsewhere} \end{cases}$$

for a and b nonnegative. Find mean and standard deviation of X. It will prove simplest to make use of tables of moments of areas of simple shapes and parallel axes transformation theorems.

2.37. Owing to the gradual accumulation of strain, the likelihood (given that the last earthquake took place in year zero) of a major earthquake on a particular fault in year i grows with i. Specifically, it can perhaps be assumed that

$$P\left[\text{occurrence in year } i \,\middle|\, \begin{matrix} \text{an occurrence in year 0 but no} \\ \text{occurrences in intervening years} \end{matrix}\right]$$
$$= 1 - pa^i \qquad i = 1, 2, \dots$$

in which a is a constant between 0 and 1.

(a) What is the probability that the first occurrence will take place in year k?

(b) What is the cumulative distribution function of X, the year of the first occurrence?

2.38. Population concentration in cities has been found to obey the law

$$C(r) = ae^{-br} \qquad a, b, r \ge 0$$

as a function of radius r from the center. Convert the law to a PDF and find the outermost radius necessary in a public transportation network that will serve 75 percent of the residents.

Find the average distance from the center of the city of

(a) A resident

(b) A resident served by the network above

2.39. In a construction project, units are arriving at an operation which takes time or causes delay. If a unit arrives *within* b sec of the arrival of the previous unit, departure of the later unit will be delayed until a given time c sec after the arrival of the

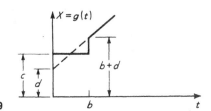

Fig. P2.39

preceding object $(c > b)$. If a unit arrives *more than* b sec after the previous arrival, it will be delayed until d sec $(d < c)$ after its own arrival time. If the distribution of independent interarrival times is uniform

$$f_T(t) = \frac{1}{a} \qquad 0 \le t \le a, a > b$$

what is the cumulative distribution of X, the length of the interval between the time of the previous unit's arrival and the time of the later unit's departure? The stated functional relationship between X and T is sketched in the accompanying diagram.

2.40. Treat the accompanying table of data as the joint probability mass function of wind velocity X, and wind direction Y, at Logan Airport, Boston, Massachusetts. Assume that conditions are similar at a nearby site for a nuclear reactor. For safety studies relating to the area downwind which would be covered by a potential cloud of radioactivity released in the event of an accident and the penetration of the containment vessel, sketch the region on a sample space representation and determine

    (a) The probability that $\frac{7}{8}\pi \leq Y < \frac{5}{4}\pi$ and $X \geq 10$

    (b) The probability that $X \geq 15.5$

    (c) The PMF of X

    (d) The PMF of X given that $Y = 1\frac{3}{4}\pi$

    (e) The PMF of Y given that $X \geq 21.5$

	\multicolumn{6}{c}{x, mph}
y, radians	1.5	5.5	10	15.5	21.5	28
North						
0	1	1	2	2	0	0
$\frac{1}{8}\pi$	0	1	1	1	0	0
$\frac{1}{4}\pi$	0	1	1	1	1	0
$\frac{3}{8}\pi$	0	1	1	1	1	0
East						
$\frac{1}{2}\pi$	0	1	2	2	0	0
$\frac{5}{8}\pi$	0	1	2	2	0	0
$\frac{3}{4}\pi$	0	1	2	1	0	0
$\frac{7}{8}\pi$	0	1	1	1	0	0
South						
1π	0	1	2	1	0	0
$\frac{9}{8}\pi$	0	1	3	2	1	0
$\frac{5}{4}\pi$	0	1	4	5	1	0
$1\frac{1}{8}\pi$	0	1	3	3	1	0
West						
$1\frac{1}{2}\pi$	0	1	2	3	1	0
$1\frac{3}{8}\pi$	0	1	3	4	2	1
$1\frac{3}{4}\pi$	0	1	3	4	2	1
$1\frac{5}{8}\pi$	0	1	2	3	1	0

$$(100)p_{X,Y}(x,y)$$

This data was obtained by observing the wind velocity and direction each hour on the hour for 10 years. Speeds indicated are middle values in recording intervals. Higher speeds than 28 occurred with negligible frequency.

2.41. Trucks of three types are being used for a long haul on a large earth-dam construction project in such proportions that the likelihood that a foreman arriving at the construction site at some arbitrary point in time observes the next arriving truck to be type A is $\frac{1}{2}$, type B, $\frac{1}{3}$, and type C, $\frac{1}{6}$.† If the round-trip times for these types are the minimum trip time, 20 min, *plus* exponentially distributed‡ random additional times with parameters $\frac{1}{3}$ min^{-1}, $\frac{1}{5}$ min^{-1}, and $\frac{1}{8}$ min^{-1} for types A, B, and C, respectively, what is the probability that the foreman will have to wait more than 10 min beyond the minimum trip time after the first arriving truck until he sees it return a second time? Knowing the planned minimum trip time, an observed trip time exceeding this by 10 min might be a signal in an operation-control plan to start a more careful check to see if there is a systematic slowdown in the operation which needs correcting. The probability found is the likelihood that the "signal" will be made even though the operation is running as planned. *Suggestion:* Use the theorem of total probability, Eq. (2.1.12).

2.42. Chebyshev inequality. Show that the Chebyshev inequality, Eq. (2.4.11) holds by:

(a) Splitting the integral defining σ^2 into three intervals, $-\infty$ to $m - h\sigma$, $m - h\sigma$ to $m + h\sigma$, and $m + h\sigma$ to ∞.

(b) Showing that

$$\sigma^2 \geq \int_{-\infty}^{m-h} \sigma^2 h^2 f(x)\,dx + \int_{m+h}^{\infty} \sigma^2 h^2 f(x)\,dx$$
$$\geq \sigma^2 h^2 (P[X \leq m - h] + P[X \geq m + h])$$

and hence that Eq. (2.4.11) holds.

If a random variable has known m equal to 5000 and known σ^2 equal to $(1000)^2$ but with unknown distribution (owing, say, to the intractable mathematics involved in its derivation), find the ranges within which the variable will lie with probabilities at least 0.5, 0.75, 0.90, and 0.99.

2.43. Law of large numbers. Find the mean and standard deviation of the random variable

$$Y = \frac{1}{n}(X_1 + X_2 + X_3 + \cdots + X_n)$$

when the random variables $X_1, X_2, \ldots, X_n$ are mutually independent, have the same means m, and the same variances σ^2. Notice that if the X_i represent n independent samples from the same distribution, then Y is the random variable which is the average of the observed random variables. The variance of the random variable Y decreases as n (the size of the sample) increases. Hence Y can be expected (or proved, by Chebyshev's inequality) to be very near its mean (and the mean of all the X_i) if n is large (at least with high probability). Y appears to be a good choice as an estimate of m if this parameter is not known.

† This does not imply that the *numbers* of trucks are in exactly these proportions; the faster trucks make more trips. For an analysis of this point see F. A. Haight [1963], "Mathematical Theories of Traffic Flow," Academic Press, Inc., New York.
‡ That is,

$$f_X(x) = \lambda e^{-\lambda x} \qquad x \geq 0$$

where λ is the parameter of the distribution.

More precisely, show, using the Chebyshev inequality, that

$$P[|Y - m| \geq \epsilon] \leq \frac{\sigma^2}{n\epsilon^2}$$

Thus no matter how small ϵ is chosen, given large enough n, the probability that the sample average Y lies within ϵ of the mean of the random variable m can be made arbitrarily close to unity. This statement is one form of the law of large numbers.

2.44. The horizontal distance X from a given structure to the epicenter of the next large earthquake within r_0 miles is distributed:

$$f_X(x) = \frac{2}{r_0^2} x \qquad 0 \leq x \leq r_0$$

The magnitude Y of large earthquakes is distributed:

$$f_Y(y) = \frac{3}{64}(9 - y)^2 \qquad 5 \leq y \leq 9$$

Assume that X and Y are independent. What is the probability that the next large earthquake within r_0 will have a magnitude greater than 8 and that its epicenter will lie within $\frac{1}{2}r_0$ of the structure?

2.45. In designing a concrete mix to meet specified "minimum strength" (ultimate compressive stress) requirements, the American Concrete Institute Code requires that the mix be designed such that the mean strength resulting is 1.5 standard deviations above the specified strength. It has been shown experimentally that the coefficient of variation of concrete strength remains almost constant when the mean is varied by adjusting the proportions—e.g., water-to-content ratio—of the mix design.

Find an expression for the required mean strength (the value the mix designer uses) in terms of the specified strength S and the coefficient of variation V.

The value of the coefficient of variation depends on the material quality, the working conditions, and the contractor's mixing and pouring policies. All else being equal, it serves as a measure of the latter, i.e., of the contractor's quality control, and usually remains constant from job to job. V may be as high as 20 percent for some contractors, as low as 10 percent for others. What reduction in (mean) mix design strength—and hence in material costs—can a contractor providing 5000 psi concrete (minimum specified strength) achieve if he is willing to allocate more time and men (to provide better treatment, placing, and curing of the concrete) sufficient to reduce the coefficient of variation from its present value of 15 to 10 percent?

Sketch rough bell-shaped curves for the two conditions.

2.46. The formula for the ultimate moment capacity M of a rectangular (under-reinforced) concrete comes from simple statics; that is,

$$M = AF_Y\left(D - K\frac{AF_Y}{F_c B}\right)$$

in which A is the area of the ductile (elastoplastic) reinforcement, F_Y is the stress at which it yields, D is the depth to the center of gravity of the steel, B is the width of the beam, F_c is the ultimate compressive stress of the concrete, and K is a factor dependent upon the shape of the stress distribution on the concrete portion of the section. All might be treated as independent random variables.

Find the approximate mean and variance of M given:

(a) $m_A = 1.5$ in.^{2} $V_A = 0.05$
(b) $m_{F_y} = 42$ ksi $V_{F_Y} = 0.07$
(c) $m_D = 20$ in. $\sigma_D = 0.3$ in.
(d) $m_B = 10$ in. $\sigma_B = 0.2$ in.
(e) $m_F = 5$ ksi $V_{F_c} = 0.15$
(f) $m_K = 0.6$ $V_K = 0.1$

Which variables "contribute" most significantly to the variance of M? What are the implications to the engineer seeking ecomonical ways to reduce the variance of M?

2.47. Standardized variables, $U = (x - m_X)/\sigma_X$:

(a) Show that

$$E\left[\frac{X - m_X}{\sigma_X}\right] = 0 \quad \text{and} \quad \text{Var}\left[\frac{X - m_X}{\sigma_X}\right] = 1$$

For these reasons the variable $U = (X - m_X)\sigma_X$ is called the *unit, standardized,* or *normalized variable.* Recall that the distribution of U is of exactly the same shape as that of X. Consequently, it is commonly used to facilitate tabulating distribution functions (see Sec. 3.3.1).

(b) Show that the correlation coefficient between any two random variables X and Z is the covariance between their corresponding standardized variables

$$\frac{X - m_X}{\sigma_X} \quad \text{and} \quad \frac{Z - m_Z}{\sigma_Z}$$

2.48. In the determination of the strength of wooden structural members it has been suggested that the estimated strength of full-size pieces X can be found as the product of the clear-wood strength of small, standard specimens Y, times a strength ratio R. After inspection of data, Y and R have been modeled as independent (normal, see Sec. 3.3.1) random variables. Find the mean and variance of X in terms of the corresponding moments of R and Y (see Prob. 2.52).

For construction grade Douglas Fir in bending, $m_R = 0.659$, $\sigma_R = 0.165$, $m_Y = 7480$ psi, and $V_Y = 15$ percent. Evaluate m and σ_x. How many standard deviations below the mean is the present allowable working stress of 1500 psi?

2.49. Uncorrelated variables through a linear transformation (orthogonalization). Consider two random variables X and Y, not necessarily independent.

(a) Find the covariance of Z and W when

$$Z = aX + bY$$
$$W = cX + dY$$

in terms of the moments of X and Y.

(b) Find the values of the coefficients a, b, c, and d such that Z and W will be uncorrelated.

(c) If X and Y are the monthly rainfalls at two neighboring locations, then Z and W, with the coefficient properly chosen, can be considered the uncorrelated rainfalls at two "fictitious" locations. When faced with the problem of creating artificial rainfall records for water-resource systems, Hufschmidt and Fiering [1966] have suggested that the (spatially) uncorrelated records Z and W be generated as independent variables† (a relatively easy task) and that these be properly combined to obtain values X and Y for the actual (correlated) stations. What should the constants be in the equations for X and Y,

$$X = a'Z + b'W$$

$$Y = c'Z + d'W$$

(in terms of the moments of X and Y)? *Hint:* find a', b', c', and d' in terms of a, b, c, and d first, by solving the pair of simulateous equations.

(d) Generate, using random numbers and the method above, a year's record of monthly rainfalls of two stations, each with (for simplicity) independent (in time) monthly flows which have constant means $m = 1$ and standard deviations $\sigma = 0.25$. Assume that owing to the stations' proximity to one another, each pair of monthly rainfalls has a correlation coefficient of 0.7. Assume normal distributions.

Note: The technique is obviously easily extended to n stations. The procedure here is analogous to finding normal modes of vibration.

2.50. The cost of an operation is proportional to the square of the total time required to complete it. Completion time for the first phase is X and the second is Y. X and Y are correlated random variables with moments m_X, m_Y, σ_X, σ_Y, and $\rho_{X,Y}$.
Find

$$E[C] = E[a(X + Y)^2]$$

2.51. Linear transformation. Sketch an arbitrary density function $f_X(x)$ and then sketch density functions for $Y = a + bX$ for the special cases

(a) $a = 0, b = 2$
(b) $a = 0, b = \frac{1}{2}$
(c) $a = 0, b = -2$
(d) $a = +2, b = 1$
(e) $a = -2, b = 1$
(f) $a = 2, b = 2$

The first three cases are particularly important; they represent the effect on changing scale or units. Cases (d) and (e) illustrate the effect of the addition of constants. Describe in words the effects on the PDF of changing units or adding constants.

2.52. Distribution and moments of a product of two random variables. The expression relating one random variable Z to two other random variables:

$$Z = XY$$

is very common in engineering. For example, let X be the demand and Y be the cost per unit of demand.

(a) If X and Y are independent continuous random variables, show that

$$f_Z(z) = \int_{-\infty}^{+\infty} \left| \frac{1}{y} \right| f_X\left(\frac{z}{y}\right) f_Y(y)\, dy$$

and find $f_Z(z)$ if

$$f_X(x) = \frac{2}{9}x \qquad 0 \le x \le 3$$

$$f_Y(y) = \frac{1}{\epsilon} \qquad 0 \le y \le \epsilon$$

---

† If the variables are assumed jointly normally distributed (Sec. 3.6.2), lack of correlation implies independence.

(b) Show that the variance of the product of independent random variables is

$$\text{Var } [XY] = \sigma_X{}^2\sigma_Y{}^2 + m_X{}^2\sigma_Y{}^2 + m_X{}^2\sigma_Y{}^2$$

Hint: Use Eq. (2.4.25): $\text{Var } [XY] = E[X^2Y^2] - [E[XY]]^2$.

(c) Evaluate the variance of Z for the PDF's given above in two ways, using the equation in b and using the PDF of Z computed in a.

2.53. The following technique (algorithm) can be used to generate a sequence† of (pseudo) random numbers for simulation studies or other purposes.

(a) Pick arbitrarily an initial number M_0, which is less than 9999 and not divisible by 2 or 5.

(b) Choose a constant multiplier M of the form

$$M = 200t \pm r$$

where t is any integer and r is any of the values 3, 11, 13, 19, 21, 29, 37, 53, 59, 61, 67, 69, 77, 83, 91. (A good choice of M is a value around 100.)

(c) Compute the product $M_1 = MM_0$. The last four digits form the first "random" number.

(d) Determine successive random numbers by forming the product $M_{i+1} = MM_i$ and retaining the last four digits only.

      (i) Generate by hand a sequence of five random numbers and use the numbers to simulate a sequence of tosses of an unbalanced coin with the probability of "heads" equal to 0.7.

      (ii) Write a computer subroutine to generate such random numbers.

2.54. The following relationships arise in the study of earthquake-resistant design, where Y is ground-motion intensity at the building site, X is the magnitude of an earthquake, and c is related to the distance between the site and center of the earthquake:

$$Y = ce^X$$

If X is exponentially distributed (Sec. 3.2.2),

$$f_X(x) = \lambda e^{-\lambda x} \qquad x \geq 0$$

Show that the cumulative distribution function of Y, $F_Y(y)$, is

$$F_Y(y) = 1 - \left(\frac{y}{c}\right)^{-\lambda} \qquad y \geq c \qquad \cdot$$

Sketch this distribution.

2.55. Specifications are set so that all but a fraction p of the tested specimens will be satisfactory. The number of specimens that have to be inspected before an unsatisfactory one is found is X. The additional number before the second is found is Y. Assume that X and Y are independent and geometrically distributed (as will be shown in Sec. 3.1.3):

$$p_X(x) = (1 - p)^{x-1}p \qquad x = 1, 2, 3, \ldots$$

$$p_Y(y) = (1 - p)^{y-1}p \qquad y = 1, 2, 3, \ldots$$

Find the probability mass function of $Z = X + Y$. What is Z in words?

† This procedure will begin repeating numbers after a sequence of 500. If a different sequence is needed, pick new values of M_0 or M. See "Random Number Generation and Testing" [1959], IBM Reference Manual.

2.56. The amount of water lost from a dam due to evaporation during a summer is proportional to the dam's surface area. The proportionality factor is random, depending as it does on weather conditions. For a particular reservoir with a particular cross section the surface area increases proportionately to the volume of water in storage. The water in storage during the summer (July and August) is the difference between the water made available since the previous September and the water released since that time. Both these volumes are random. Assume that the water released and the water made available during the summer are negligible. Neglecting second-order effects (e.g., the change in surface area due to the evaporation loss), find the mean and variance of the water lost due to evaporation in a summer in terms of the same moments of the various factors mentioned. Assume a lack of correlation among these factors. Define carefully all the variables, constants, moments, relationships, etc., that you use.

2.57. Reconsider the dam-operating-policy problem (Prob. 2.32).

(a) Find the expected value of Z, the water available.

(b) Find the expected value of $Y = g(Z)$, the water released.

2.58. Variability in stream flows makes it difficult to properly design the capacity of new dams. Concern is primarily with providing sufficient storage to avoid water shortage. Early studies focused on the range R, the difference between the maximum value of the impounded water and its minimum value over a period of n years, the design life of the dam. Assuming (unrealistically, but as a first approximation) that the annual "impoundments" are independent with common mean m and variance σ^2, it has been shown† that (for large n)

$$m_R \approx 1.25\sigma n^{\frac{1}{2}}$$

$$\sigma_R \approx 0.272\sigma n^{\frac{1}{2}}$$

(a) Show that the coefficient of variation of R is a constant, independent of the stream characteristics m and σ and of the lifetime n.

(b) A possible dam design rule is to design for a range of value $r_0 = m_R + k\sigma_R$, where k depends on the risk level adopted. For $k = 2$ how many times the mean range is the design value? How many of (the infinite number of) moments of the annual impoundment does the design range depend on? Is design for an infinite lifetime possible? How sensitive to the design lifetime is the design range? (For example, if n is doubled from 50 to 100 years, what is the increase in the design range?) Data suggests that the exponent on n for m_R should be 0.72, which is presumably an influence of correlation.†

2.59. A proposed set of dam-operating policies‡ can be characterized by this (continuity) equation relating the reservoir storage at the end of the $(i + 1)$th time interval S_{i+1} to the storage at the end of the previous interval S_i:

$$S_{i+1} = a(S_i + X_{i+1} - d)$$

---

† W. Feller [1951], The Asymptotic Distribution of the Range of Sums of Independent Random Variables, *Ann. Math. Stat.*, vol. 22, pp. 427–432, September, as reported by M. B Fiering [1967], "Streamflow Synthesis," p. 17, Harvard University Press, Cambridge, Mass. (See Prob. 2.59.)

‡ See Prob. 2.32 for a definition of an operating policy. This problem is based on the work of G. T. Bryant [1961], "Stochastic Theory of Queues Applied to Design of Impounding Reservoirs," Ph.D. dissertation, Harvard University, Cambridge, Mass., as reported by M. B. Fiering [1967], *op. cit.*, p. 17.

in which X_{i+1} is the inflow during the $(i + 1)$th interval and d is the "target" draft (the "desired" amount of release) in any interval. The policy parameter a ($0 \leq a \leq 1$) reflects the importance of carrying over water from one interval to the next. If $a = 0$, $S_{i+1} = 0$ and there is no carryover; if $a = 1$, the policy is the "normal release policy" (Prob. 2.32) (if, as we shall do, one neglects the possibility of the available water in any year being less than d or more than $d + c$, where c is the dam's capacity).

Assume that the inflows are uncorrelated with common mean m_X and variance σ_X^2. Our concern is to calculate the corresponding moments of the storage S and the release or draft Y.

(a) Show that the mean and variance of the reservoir storage in any year (a long time after the initial year) are

$$m_S = \frac{a}{1 - a} (m_X - d)$$

$$\sigma_S^2 = \frac{a^2}{1 - a^2} \sigma_X^2$$

Hint: the S_i's are not independent. Write an expression for S_i in terms of the independent X_i's and take the limit as i becomes large. Check that m_S and σ_S^2 (valid presumably for both S_i *and* S_{i+1} since the moments prove to be independent of i) satisfy the equations found by taking the expectation and variance, respectively, of both sides of the continuity equation. Discuss the design implication of the variance for the "normal release policy," $a = 1$.

(b) Show that the mean and variance of Y, the release in any year (substantially after the first year), are

$$m_Y = m_X$$

$$\sigma_Y^2 = \left(\frac{1 - a}{1 + a}\right) \sigma_X^2$$

Hint: Show first that $Y_{i+1} = (1 - a)(X_{i+1} + S_i) + ad$.

Discuss the implications. If m_Y were any other value, what would the implications be? How do you explain the variance under the limiting cases, $a = 0$ and $a = 1$?

(c) Show that the coefficient of correlation between S_i and S_{i+1} is a. Discuss this result for the limiting cases, in terms of one's ability to predict S_{i+1} given S_i.

(d) In the light of our assumption that we neglect the possibility of the available water being very small or very large, discuss the conditions of target draft and dam capacity for which the results above are valid. Qualitatively, for $a = 1$, how would you expect these results to change if either or both restrictions were dropped? *Hint:* See the sketch in Prob. 2.32. You can easily prove to yourself that the continuity equation we have adopted implies (for $a = 1$) that our operating policy is $y = d$ for *all* z, which may not be feasible.

2.60. The relationship below is often used in water-resources projects to relate short-term benefits U to the amount of water released in any year Y.† (d is the announced or "target" release; see Prob. 2.32.)

(a) Find the annual expected benefits if the annual release Y has a uniform distribution on the interval $d/2$ to $2d$.

† M. B. Fiering [1967], *op. cit.;* see Prob. 2.32.

(b) Find the annual expected benefits when the annual release Y is related to the available water Z by the "normal operating policy," and when Z has the distribution given in Prob. 2.32. *Hint:* If you do not have available the distribution of Y (asked for in part (b) of Prob. 2.32) instead construct first a functional relationship between U and Z.

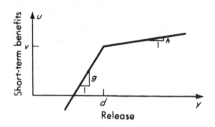

Fig. P2.60

2.61. For a given target release d, the short-term benefits U of a dam release Y are shown in the sketch with Prob. 2.60. If the planned or target release were increased, however, long-term benefits associated with increased downstream development (more irrigation, etc.) could accrue. In other words, the value v associated with a release equal to the target value (see sketch in Prob. 2.60) is itself an increasing function of d, the target release. The combined result can be shown as indicated for two particular target values d_1 and d_2.

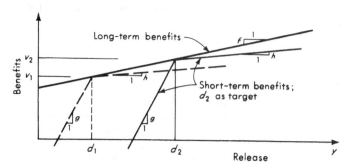

Fig. P2.61

For a given capacity dam, assume that the distribution of Y, the annual release, is independent of d, the target release. The designer's problem is to choose that target release $\tilde{d}$. Show that the value of d that maximizes the expected annual benefits is that value $\tilde{d}$ such that†

$$F_Y(\tilde{d}) = \frac{f - h}{g - h}$$

In short, the optimal target value $\tilde{d}$ is the value that divides the density function of Y into two areas $(f - h)/(g - h)$ to the left and $(g - f)/(g - h)$ to the right.

† This result is attributed by M. B. Fiering [1967], *op. cit.*, to G. T. Bryant, Prob. 2.59.

2.62. Moment generating and characteristic functions. The expectations of two particular functions $g(X)$ of random variables are of great value. They are

$$g(X) = e^{uX}$$

and

$$g(X) = e^{iuX}$$

in which i is the imaginary number $\sqrt{-1}$. The expectations of these functions have special names and properties. The *moment generating function* (mgf) is defined, here for continuous variables, as

$$M(u) = E[e^{uX}] = \int_{-\infty}^{\infty} e^{ux} f_X(x)\, dx$$

and the *characteristic function* (cf) as

$$\mathcal{C}(u) = E[e^{iuX}] = \int_{-\infty}^{\infty} e^{iux} f_X(x)\, dx$$

(a) Show that these functions (integral transforms) can be used as "moment generators." In particular, show that

(i) $E[X] = \dfrac{dM(u)}{du}\bigg|_{u=0}$

(ii) $E[X^2] = \dfrac{d^2 M(u)}{du^2}\bigg|_{u=0}$

(b) Show that

$$E[X^k] = \frac{1}{i^k} \frac{d^k \mathcal{C}(u)}{du^k}\bigg|_{u=0}$$

(c) Show that for independent random variables X_1 and X_2 the moment generating function and characteristic function of $Y = X_1 + X_2$ are

(i) $M_Y(u) = M_{X_1}(u) M_{X_2}(u)$
(ii) $\mathcal{C}_Y(u) = \mathcal{C}_{X_1}(u) \mathcal{C}_{X_2}(u)$

These simple relationships among mgf and cf of independent random variables and their sums explains these functions' great utility in modern probability theory and mathematical statistics.

Although often difficult in practice, it is theoretically possible to transform $\mathcal{C}_Y(u)$ back into $f_Y(y)$. At a minimum the moments of Y are easily made available through $\mathcal{C}_Y(u)$

Hint: Because of the assumptions on convergence, the operations of differentiation and expectation (integration) can be interchanged quite freely.

The student familiar with transform techniques in applied mathematics will recognize these transforms and their relationships when he recalls Eq. (2.3.43). We will not explore these transforms as they might be in the remainder of the text, since most students are not familiar with their use.

(d) Find the moment generating function and characteristic function of the random variable R in the illustration in Sec. 2.3.3:

$$f_R(r) = 2e^{-2r} \qquad r \geq 0$$

Use these functions to find four moments of the variable. *Hint:* In performing the integrations, one can assume $u < 2$. Since after differentiation the functions will only be evaluated at $u = 0$, there is no problem with convergence of the integrals.

2.63. You are performing some computations on the computer which involve a Monte Carlo evaluation of a particular integral. From experience you know that the running time for evaluation of the integral with a variety of data can be described by the negative exponential distribution

$$f_T(t) = \lambda e^{-\lambda t}$$

Total running time of the program can be estimated from the relation

$$Y = (T - 1)^2 + 2$$

(a) If $\lambda = 3$, what is the density function describing the behavior of program running time?

(b) Set up at least two expressions for the expected value of the program running time. At least three are possible.

(c) Use the more convenient expression to evaluate the expected value.

2.64. In the very preliminary planning of some harbor island developments, there was discussion regarding the cost estimates of building four bridges. There had as yet not been a preliminary soil survey in the harbor area, and it was recognized that the bridge costs are highly dependent on soil conditions. Therefore there was great uncertainty in the preliminary cost estimates. A spokesman, however, made this statement; "I recognize the uncertainty in the cost estimate of any bridge. But I am much more confident of our estimate for the total cost of all four bridges because of the likelihood that a high estimate on one will be balanced by a low estimate on another."

Discuss this statement. Use your knowledge, simply, of the means and variances of sums of random variables to support your comments. If your initial intuition lies with that of the spokesman, be sure you resolve in your own mind why it is inconsistent. Compare both the variance and the coefficients of variation of the total cost versus those of an individual cost.

If the four bridge sites are relatively close to one another, soil conditions, although unknown, are probably similar. What implications does this observation have for your analysis?

2.65. A total cost of earthwork on a road construction project will be the total number of cubic yards excavated Y times the contractor's unit bid price P. If the mean and variance of the former are 100,000 yd³ and 10×10^6 (yd³)², and the mean and variance of P are 6 \$/yd³ and 0.25 (\$/yd³)², respectively, find the expected value and variance of the total cost, assuming that Y and P are independent. Compare with the approximate values in Sec. 2.4.4. See Prob. 2.52.

2.66. System reliability. A major application of probability has been in the determination of the reliability of systems made up of components whose reliabilities are known. (This reliability of a component is the probability that it will function properly throughout the period of interest.) Simple block diagrams are helpful in demonstrating how system performance depends on component performance.

(a) *Series system.* For example, if a system will perform only if each and every component proves reliable, then the block diagram will be chainlike:

Fig. P2.66a

If the events C_i = [component i performs satisfactorily] are independent, show in terms of relationships among the events C_i that

$$(1 - p_s) = \prod_{i=1}^{n} (1 - p_i)$$

$$\approx 1 - \sum_{i=1}^{n} p_i \qquad \text{for } \Sigma p_i \ll 1$$

if p_i is the probability of C_i^c and $(1 - p_s)$ is the reliability of the total system.

(b) *Parallel redundant system.* If the system will perform satisfactorily if any one of the components "survives," the block diagram is

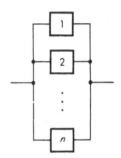

Fig. P2.66*b*

For independent events C_i, show that the system reliability is

$$1 - p_s = 1 - \left(\prod_{i=1}^{n} p_i \right)$$

(c) *Mixed systems.* For a system indicated thus:

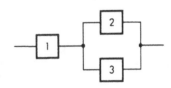

Fig. P2.66*c*

show that the system reliability, $1 - p_s$, is

$$1 - p_s = (1 - p_1)(1 - p_2 p_3)$$

for independent events C_i. Note that at the cost of an additional redundant component, number 3, this system is more reliable than this simpler one.

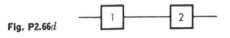

Fig. P2.66*d*

(d) *Examples.* Nuclear power plants, which depend on the functioning of many components, are designed with numerous redundancies. Assume, for simplicity, that during a particular, major, design level of earthquake intensity, the *controlled shutdown* of the reactor depends on the proper functioning of the control system, the cooling system, and the primary containment vessel. Assume that there are three redundant control systems, two redundant cooling systems, and a single, steel primary containment with two critical necessary components A and B. Block model the system and calculate the system reliability with respect to shutdown, in terms of component reliabilities. Assume independence of component performances (at the given earthquake level).

There will be no *major accident* if either the shutdown is controlled or the reinforced-concrete secondary containment vessel performs properly. Model the total system with respect to major accident reliability.

(e) *Component dependence.* Recalculate the system reliability in part (c) if the conditional probability of satisfactory performance of component 3 *given* failure of component 2 is only $1 - p_3^* < 1 - p_3$. (That is, $P[C_{3^c} \mid C_{2^c}] = p_3^* > p_3$.)

Dependence of component performance events is often introduced by the systems *environment* or demand. If a large, random demand on the system is the cause of the failure of component 2, it is likely to cause failure of component 3 also.

Reconsider, for example, part (d), assuming now that the level of the earthquake intensity is uncertain [not a given level as it was in (d)]. Then, given an earthquake occurrence of uncertain level, the reliability of, say, the secondary containment *given* failure of the primary vessel may be substantially smaller than the (marginal) reliability of the secondary containment. In the marginal analysis a whole spectrum of possible earthquakes intensities had to be incorporated in the analysis. Failure of the primary vessel suggests that a large intensity has probably occurred. Conditional on this information, the reliability of the secondary containment must be smaller.

2.67. In evaluating the consolidation settlement of the foundations of new buildings, it is necessary to predict the sustained column loads that are transmitted to the footing. We consider here only the sustained *live* loads.

Let the load on a particular column due to floor i be:

$$L_i = a(B + S_i)$$

and that due to floor j be:

$$L_j = a(B + S_j)$$

in which a is the tributary area. B is a random variable with a mean m_B that is equal to the average unit load over all buildings of this prescribed type of use (e.g., offices), and a variance $\sigma_B{}^2$, that is equal to the variance of *mean* (over the building) building loads from building to building. S_i and S_j are random variables with zero mean representing the spatial variation of load within a *given* building. They both have variance $\sigma_S{}^2$. B and the S's are uncorrelated.

(a) What are the mean and variance of the total (sustained live) load transmitted to a footing by a column supporting n such floors?

(b) For $\sigma_B = 2\sigma_S$ and for $\sigma_B = \frac{1}{2}\sigma_S$, sketch a plot of the coefficient of variation of this total load versus n.

(c) What is the correlation coefficient between L_i and L_j?

(d) What is the "*partial correlation coefficient*" between L_i and L_j, *given* that $B = b_0$? This coefficient is defined in the usual way, except that the expectations are conditional on $B = b_0$. This coefficient appears again in Sec. 4.3.1.

2.68. A storage reservoir is supplied with water at a constant rate k for a period of time Y. Then water is drawn from it at the same rate for a period of time X. X and Y are independent, with distributions

$$f_X(x) = \lambda e^{-\lambda x} \qquad x \geq 0$$
$$f_Y(y) = \nu e^{-\nu y} \qquad y \geq 0$$

(Assume that the reservoir is infinitely large and contains an infinite amount of water so that it cannot run dry or overflow.)

What are the probability density functions of $Z = Y - X$ and of $W = k(Y - X)$, the change of the amount of water in the reservoir after one such cycle of inflow and outflow?

2.69. Hazard functions. If an engineering system is subjected to a random environment, its reliability can be defined in terms of the random variable T, the time to failure, since the reliability of the system during a planned lifetime of t_0 is simply

Reliability $(t_0) = P[\text{no failure before } t_0] = P[T > t_0] = 1 - F(t_0)$

The *hazard function* $h(t)$ is defined such that $h(t)\,dt$ is the probability that the failure will occur in the time interval t to $t + dt$ *given* that no failure occurred prior to time t.

(a) Show that

$$h(t) = \frac{f(t)}{1 - F(t)}$$

Its shape determines, for example, whether a system deteriorates with age or wear (i.e., if $h(t)$ grows with time).

(b) Show that the probability distribution of T, given a ("well-behaved," continuous) hazard function $h(t)$, is

$$F(t) = 1 - e^{-H(t)}$$

in which $H(t) = \int_0^t h(u)\,du$.

(c) Show that if the time to failure has an exponential distribution,

$$f(t) = \nu e^{-\nu t} \qquad t \geq 0$$

then the hazard function is a constant. In Sec. 3.2 we will call such conditions "random" or Poisson failure events, and ν is their average rate of arrival. It is a commonly adopted assumption in reliability analysis.

(d) Find the distribution of the time to failure T of a system which is exposed to two independent kinds of hazard, one due to random occurrences of "rare events" (e.g., earthquakes) and one due to wearout or deterioration (e.g., fatigue). The rare events occur with average annual rate ν. The hazard due to wearout is negligible at time zero, but it grows linearly with time. At time $t = 10$ years it is equal to that due to rare events. Justify why the total hazard function is just the sum of the two hazard functions. What is the reliability of the system if it is desired that it operate for 20 years?

2.70. In simple frame structures (with rigid floors) such as the one shown in the diagram, the total deformation of the top story Y is simply the sum of the deformations of the individual stories X_1 and X_2, acting independently. These variables are uncorrelated and have mean and variance m_1 and σ_1^2, m_2 and σ_2^2, respectively.

(a) Find the mean and variance of Y.

(b) Find the correlation coefficient between Y and X_2. *Discuss* the results in terms of very large relative values of the moments (m_1, m_2, σ_1, and σ_2), for example, σ_2 much larger than σ_1, and vice versa.

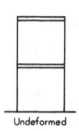 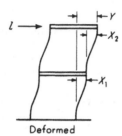

Undeformed   Deformed   **Fig. P2.70**

(c) Find the correlation coefficient between Y and X_2 if X_1 and X_2 are not uncorrelated, but have a positive correlation coefficient ρ, owing, say, to the common source of material and common constructor.

2.71. Tests on full-scale reinforced-concrete-bearing walls indicate that the deflection of such a wall under a given horizontal load is a random variable. The form of the distribution of the variable is

$$f_X(x) = cx^{k-1}e^{-\lambda x} \qquad x \geq 0$$

in which $c = \lambda^k/(k-1)!$ for k integer. It has mean k/λ and variance k/λ^2. Different wall dimensions and different concrete properties will change the values of the parameters k and λ.

In a small two-story building the deflection Y of the roof will be the sum of the deflection of the first-story wall X_1 and the deflection of the second-story wall X_2. X_1 and X_2 are assumed to be independent. Find the probability density function of Y if

$$E[X_1] = 2 \qquad \text{Var }[X_1] = 2$$
$$E[X_2] = 3 \qquad \text{Var }[X_2] = 3$$

2.72. A harbor breakwater is made of massive tanks which are floated into place over a shallow trench scraped out of the harbor floor and then filled with sand. There is concern over the possibility of breakwater sliding under the lateral pressure of a large wave in a major storm. It is difficult to predict the lateral sliding capacity of such a system. What is the reliability (the probability of satisfactory performance) of this system with respect to sliding if the engineer judges the following?

(a) That the sliding resistance has a value 100, 120, or 140 units, with the middle value twice as likely as the low value and twice as likely as the high value.

(b) That the lateral force under the largest wave in the economic lifetime of the breakwater X, has an exponential distribution with parameter $\lambda = 0.02$; that is,

$$f_X(x) = 0.02e^{-0.02x} \qquad x \geq 0$$

The units of sliding resistance and lateral force are the same. Resistance and force are independent.

2.73. In planning a building, the number of elevators is chosen on a basis of balancing initial costs versus the expected delay times of the users. These delays are closely related to the number of stops the elevator makes on a trip. If an elevator runs full (n people) and there are k floors, we want to find the expected number of stops R the elevator makes on any trip. Assuming that the passengers act independently and that any passenger chooses a floor with equal probability $1/k$, show that

$$E[R] = k\left[1 - \left(1 - \frac{1}{k}\right)^n\right]$$

Hint: It is often useful to define "indicator random variables" as follows. Let $X_i = 1$ if the elevator stops at floor i, and 0 if it does not. Then observe that $R = \sum_{i=1}^{k} X_i$. Find the expected value of X_i after finding first the probability that $X_i = 0$.

2.74. The peak annual wind velocity X_i in any year i at a certain site is often assumed to have a distribution of the form

$$F_{X_i}(x) = e^{-(x/u)^{-k}} \qquad x > 0$$

Peak annual wind velocities in different years are independent.

The pseudostatic force on an object subjected to the wind is proportional to the *square* of the wind velocity: force $= c(\text{velocity})^2$. Find the probability density function of Y, the *maximum* force on an object over a period of n years.

2.75. The flooding (peak-flow) potential of a rain storm (defined here to be a ½-day period with more than 2 hr of rainfall) depends on both the total rainfall Y and the duration X. Available data in a particular region suggests that the PDF of X is approximately:

$$f_X(x) = \begin{cases} k(x - 2)^2 & 2 < x \le 7 \text{ hr} \\ k(12 - x)^2 & 7 < x \le 12 \text{ hr} \\ 0 & \text{elsewhere} \end{cases}$$

(a) Find k and sketch this PDF.

(b) The conditional distribution of the total rainfall Y, given that the duration X equals x hr, is uniformly distributed on the interval $\frac{1}{2}x - 1$ to $\frac{1}{2}x + 2$ in.:

$$f_{Y|X}(y,x) = \begin{cases} \frac{1}{3} & \frac{1}{2}x - 1 \le y \le \frac{1}{2}x + 2 \\ 0 & \text{elsewhere} \end{cases}$$

What is the joint distribution of X and Y? Sketch it.

(c) "High" flow rates will occur if the rate of rainfall during the total storm exceeds $\frac{2}{3}$ in./hr. What fraction of storms cause "high" flow rates?

3
Common Probabilistic Models

In the application of probability theory a number of models arise time and time again. In many, diverse, real situations, engineers will often make assumptions about their physical problems which lead to analogous descriptions and hence to mathematically identical forms of the model. Only the numbers, for example, the values of the parameters of a model, and their interpretations differ from application to application. The random variables of interest in these situations have distributions which can be derived and studied independently of the specific application. These distributions have become so common that they have been given proper names and have been tabulated to facilitate their use.

The familiarity of these distributions in fact leads to their frequent adoption simply for reasons of computational expediency. Even though there may exist no argument (no model of the underlying physical mechanism) suggesting that a particular distribution is appropriate, it is often convenient to have a simple mathematical function to describe a variable. The distributions studied in this chapter are the ones most commonly adopted for this empirical use by the engineer. This is simply because they are well known, well tabulated, and easily worked with. In such situations the choice among common distributions is usually based on a comparison between the shape of a histogram of data and the shape of a PDF or PMF of the mathematical distribution. Examples will appear in this chapter, and the technique of comparison will be more fully discussed in Sec. 4.5.†

Such an empirical path is often the only one available to the engineer, but it should be remembered that in these cases the subsequent conclusions, being based on the adopted distribution, are usually dependent to a greater or lesser degree upon its properties. In order to make the conclusions as accurate and as useful as possible, and in order to justify, when necessary, extrapolation beyond available data, it is preferable that the choice of distribution be based whenever possible on an understanding of how the physical situation might have given rise to the distribution. For this reason this chapter is written to give the reader not simply a catalog of common distributions but rather an understanding of at least one mechanism by which each distribution might arise. To the engineer, the existence of such mechanisms, which may describe a physical situation of interest to him, is fundamentally a far more important reason for gaining familiarity with a particular common distribution than, say, the fact that it is a good mathematical approximation of some other distribution, or that it is widely tabulated.

From the pedagogical point of view this chapter has the additional purpose of reinforcing all the basic theory presented in Chap. 2. After the statement of the underlying mechanism, the derivation of the various distributions involved is only an application of those basic principles. In particular, derived distribution techniques will be applied frequently in this chapter.

3.1 MODELS FROM SIMPLE DISCRETE RANDOM TRIALS

Perhaps the single most common basic situation is that where the outcomes of experiments can be separated into two exclusive categories—for example, satisfactory or not, high or low, too fast or not, within specifications or not, etc. The following distributions arise from random variables of interest in this situation.

3.1.1 The Single Trial: The Bernoulli Distribution

We are interested in the simplest kind of experiment, one where the outcomes can be called either a "success" or a "failure"; that is, only

† Alternatively, the choice of distribution form consistent with the engineer's information can be deduced on the basis of the maximum uncertainty or maximum entropy principle. See Tribus [1969].

two (mutually exclusive, collectively exhaustive) events are possible outcomes. Examples include testing a material specimen to see if it meets specifications, observing a vehicle at an intersection to see if it makes a left turn or not, or attempting to load a member to its predicted ultimate capacity and recording its ability or lack of ability to carry the load.

Although at the moment it may hardly seem necessary, one can define a random variable on the experiment so that the variable X takes on one of two numbers—one associated with success, the other with failure. The value of this approach will be demonstrated later. We define, then, the "Bernoulli" random variable X and assign it values. For example,

$X = 1$ if a success is observed

$X = 0$ if a failure is observed

The choice of values 1 and 0 is arbitrary but useful. Clearly, the PMF of X is simply

$$p_X(x) = \begin{cases} p & x = 1 \\ 1 - p & x = 0 \end{cases} \qquad (3.1.1)$$

where p is the probability of success. The random variable has mean

$$E[X] = (1)p + (0)(1 - p) = p \qquad (3.1.2)$$

and variance

$$\begin{aligned} \text{Var}\,[X] &= (1 - p)^2 p + (0 - p)^2 (1 - p) \\ &= (1 - p)p(1 - p + p) \\ &= (1 - p)p \qquad (3.1.3) \end{aligned}$$

As mentioned, another choice of values other than 0,1 might have been made, but this particular choice yields a random variable with an expectation equal to the probability of success p.

Thus, if one has a large stock of items, say high-strength bolts, and he proposes to select and inspect one to determine whether it meets specifications, the mean value of the Bernoulli random variable in such an experiment (with a "success" defined as finding an *un*satisfactory bolt) is p, the proportion of defective bolts.† This direct correspondence between m_X and p will prove useful when in Chap. 4 we discuss methods for estimating the mean value of random variables. Notice that the variance of X is a maximum when p is $\frac{1}{2}$.

† It is assumed that the number of bolts is large and that each is equally likely to be selected. The problem is altered (biased) if the engineer tends to select suspicious looking bolts for testing.

1.1.2 Repeated Trials: The Binomial Distribution

A sequence of simple Bernoulli experiments, when the outcomes of the experiments are mutually independent and when the probability of success remains unchanged, are called *Bernoulli trials*. One might be interested, for example, in observing five successive cars, each of which turns left with probability p; under the assumption that one driver's decision to turn is not affected by the actions of the other four drivers, the set of five cars' turning behaviors represents a sequence of Bernoulli trials. From a batch of concrete one extracts three test cylinders, each with probability p of failing at a stress below the specified strength; these represent Bernoulli trials if the conditional probability that any one will fail remains unchanged given that any of the others has failed. In each of a sequence of 30 years, the occurrence or not of a flood greater than the capacity of a spillway represents a sequence of 30 Bernoulli trials if maximum annual flood magnitudes are independent and if the probability p of an occurrence in any year remains unchanged throughout the 30 years.†

We shall be interested in determining the distributions of various random variables related to the Bernoulli trials mechanism.

Distribution Let us determine the distribution of the total number of successes in n Bernoulli trials, each with probability of success p. Call this random number of successes Y. Consider first a simple case, $n = 3$. There will be no successes (that is, $Y = 0$) in three trials only if all trials lead to failures. This event has probability

$$(1 - p)(1 - p)(1 - p) = (1 - p)^3$$

Any of the following sequences of successes 1 and failures 0 will lead to a total of one success in three trials:

Trial 1	Trial 2	Trial 3
1	0	0
0	1	0
0	0	1

Each sequence is an event with probability of occurrence $p(1 - p)^2$. Therefore the event $Y = 1$ has probability $3p(1 - p)^2$, since the se-

---

† These assumptions are usually made in dam-design studies, but one can argue that, given one large flood, a stream will be altered and hence the conditional likelihood of floods in following years will not be p, or that changing land-use patterns in the watershed might alter p from year to year. To the extent that these and other such effects are important, the model is in error, but still perhaps approximately valid.

quences are mutually exclusive events. Similarly, the mutually exclusive sequences

Trial 1	Trial 2	Trial 3
1	1	0
1	0	1
0	1	1

each occuring with probability $p^2(1 - p)$, lead to $Y = 2$. Hence

$$P[Y = 2] = 3p^2(1 - p)$$

Similarly, $P[Y = 3] = p^3$, since only the sequence 1, 1, 1 corresponds to $Y = 3$. In concise notation,

$$p_Y(y) = \binom{3}{y} p^y (1 - p)^{3-y} \qquad y = 0, 1, 2, 3$$

where $\binom{3}{y}$ indicates the binomial coefficient:

$$\binom{3}{y} = \frac{3!}{y!(3 - y)!}$$

This coefficient is equal to the number of ways that exactly y successes can be found in a sequence of three trials. (Recall that $0! = 1$, by definition.)

If the probability that a contractor's bid will be low is $\frac{1}{3}$ on each of three independent jobs, the probability distribution of Y, the number of successful bids, is

$$p_Y(y) = \begin{cases} (\frac{2}{3})^3 & = \frac{8}{27} & \text{for } y = 0 \\ 3(\frac{1}{3})(\frac{2}{3})^2 & = \frac{12}{27} & 1 \\ 3(\frac{1}{3})^2(\frac{2}{3}) & = \frac{6}{27} & 2 \\ (\frac{1}{3})^3 & = \frac{1}{27} & 3 \end{cases}$$

Contrary to some popular beliefs, it is not certain that the contractor will get a job; the likelihood is almost $\frac{1}{3}$ that he will get no job at all!

In general, if there are n Bernoulli trials, the PMF of the total number of successes Y is given by

$$p_Y(y) = \frac{n!}{y!(n - y)!} p^y (1 - p)^{n-y} \qquad y = 0, 1, 2, \ldots, n \qquad (3.1.4a)$$

$$= \binom{n}{y} p^y (1 - p)^{n-y} \qquad y = 0, 1, 2, \ldots, n \qquad (3.1.4b)$$

It is clear that the parameter n must be integer and the parameter p

must be $0 \leq p \leq 1$.   The binomial coefficient

$$\binom{n}{y} = \frac{n!}{y!(n-y)!} \tag{3.1.5}$$

is the number of ways a sequence of n trials can contain exactly y successes Parzen [1960].

A number of examples of this, the "binomial distribution," are plotted in Fig. 3.1.1.   The shape depends radically on the values of the parameters n and p.   Notice the use of the notation $B(n,p)$ to indicate a binomial distribution with parameters n and p; for example, the random variable Y in the bidding example above is $B(3,\frac{1}{3})$.

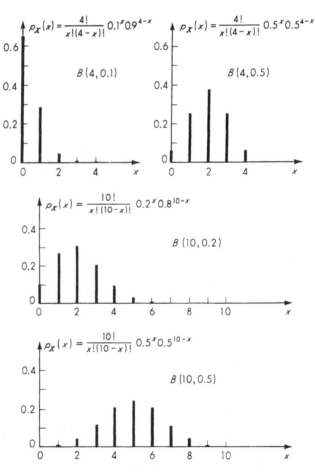

Fig. 3.1.1   Binomial distribution $B(n,p)$.

Formally, the CDF of the binomial distribution is

$$F_Y(y) = P[Y \leq y] = \sum_{u=0}^{y} p_Y(u)$$

$$= \sum_{u=0}^{y} \binom{n}{u} p^u (1-p)^{n-u} \qquad y = 0, 1, \ldots, n \qquad (3.1.6)$$

Thus in the bidding example, the probability that the contractor will receive two or fewer bids is

$$F_Y(2) = \sum_{u=0}^{2} \binom{3}{u} (\tfrac{1}{3})^u (\tfrac{2}{3})^{3-u}$$

$$= \tfrac{8}{27} + \tfrac{12}{27} + \tfrac{6}{27}$$

$$= \tfrac{26}{27}$$

or, more easily,

$$F_Y(2) = 1 - P[Y > 2]$$

$$= 1 - p_Y(3)$$

$$= 1 - \tfrac{1}{27} = \tfrac{26}{27}$$

It is often the case that the probabilities of complementary events are more easily computed than those of the events themselves. As another example, the probability that the contractor gets at least one bid is

$$P[Y \geq 1] = 1 - P[Y < 1] = 1 - p_Y(0)$$

$$= 1 - \tfrac{8}{27} = \tfrac{19}{27}$$

Moments   The mean and variance of a binomial random variable Y are easily determined, and the exercise permits us to demonstrate some of the techniques of dealing with sums which many readers find unfamiliar. Thus,

$$E[Y] = \sum_{y=-\infty}^{\infty} y p_Y(y)$$

$$= \sum_{y=0}^{n} y \binom{n}{y} p^y (1-p)^{n-y}$$

Since the first term is zero, it can be dropped. The y cancels a term in $y!$:

$$E[Y] = \sum_{y=1}^{n} \frac{n!}{(y-1)!(n-y)!} p^y (1-p)^{n-y}$$

Bringing np in front of the sum,

$$E[Y] = np \sum_{y=1}^{n} \frac{(n-1)!}{(y-1)!(n-y)!} p^{y-1}(1-p)^{n-y}$$

Letting $u = y - 1$,

$$E[Y] = np \sum_{u=0}^{n-1} \frac{(n-1)!}{u!(n-1-u)!} p^{u}(1-p)^{n-1-u}$$

Notice now that the sum is over all the elements of a $B(n-1, p)$ mass function and hence equals 1, yielding

$$E[Y] = np \tag{3.1.7}$$

This is an example of a most common method of approach in probability problems, namely, manipulating integrals and sums into integrals and sums over recognizable PDF or PMF's, which are known to equal unity. In a similar manner, we can find

$$\text{Var}\,[Y] = E[Y^2] - E^2[Y]$$
$$= (n^2 p^2 - np^2 + np) - (n^2 p^2)$$

or

$$\text{Var}\,[Y] = np(1-p) \tag{3.1.8}$$

Y as a sum   Notice that the total number of successes in n trials Y, can be interpreted as the sum

$$Y = X_1 + X_2 + X_3 + \cdots + X_n$$

of n independent identically distributed Bernoulli random variables; thus $X_i = 1$ if there is a success on the ith trial, and $X_i = 0$ if there is a failure. As such, the results above could have been more easily obtained as

$$E[Y] = \sum_{i=1}^{n} E[X_i] = nE[X_i] = np$$

$$\text{Var}\,[Y] = \sum_{i=1}^{n} \text{Var}\,[X_i] = n\,\text{Var}\,[X_i] = np(1-p)$$

using Eqs. (3.1.2) and (3.1.3). The techniques of derived distributions (Sec. 2.3) could also have been used to find the distribution of Y from those of X_i. The interpretation of a binomial variable as the sum of Bernoulli variables also explains why the sum of two binomial random variables, $B(n_1,p)$ and $B(n_2,p)$, also has a binomial distribution, $B(n_1 + n_2, p)$, as long as p remains constant. This fact is easily verified using the techniques of Sec. 2.3.

The binomial distribution is tabulated,[†] but its evaluation, which is time-consuming for large n, can frequently be carried out approximately using more readily available tables of the Poisson distribution (Sec. 3.2.1) or the normal distribution (Sec. 3.3.1); Prob. 3.18 explains these approximations. The latter approximation is justified in Sec. 3.3.1 in the light of the binomial random variable's interpretation as the sum of n Bernoulli random variables.

3.1.3 Repeated Trials: The Geometric and Negative Binomial Distributions

Rather than asking the question, "How many successes will occur in a fixed number of repeated Bernoulli trials?" the engineer may alternatively be interested in the question, "At what trial will the *first* success occur?" For example, how many bolts will be tested before a defective one is encountered if the proportion defective is p? When will the first critical flood occur if the probability that it will occur in any year is p?

Geometric distribution   Assuming independence of the trials and a constant value of p, the distribution of N, the number of trials to the first success, is easily found. The first success will occur on the nth trial if and only if (1) the first $n - 1$ are failures, which occurs with probability $(1 - p)^{n-1}$; and (2) the nth trial is a success, which occurs with probability p, or

$$P[N = n] = p_N(n) = (1 - p)^{n-1}p \qquad n = 1, 2, 3, \ldots \qquad (3.1.9)$$

Notice that N may, conceptually at least, take on any value from one to infinity.[‡]

This distribution is called the *geometric*[§] distribution with parameter p. Symbolically we say that N is $G(p)$. A plot of a geometric distribution appears in Fig. 3.1.2.

The cumulative distribution function of the geometric distribution is

$$F_N(n) = \sum_{j=1}^{n} p_N(j) = \sum_{j=1}^{n} (1 - p)^{j-1}p$$

$$= p\left(\frac{(1 - p)^n - 1}{(1 - p) - 1}\right)$$

$$= 1 - (1 - p)^n \qquad\qquad n = 1, 2, 3, \ldots \qquad (3.1.10)$$

[†] For example, see National Bureau of Standards [1950] or other references listed at the end of Chap. 3.

[‡] But what is the probability that $N = \infty$?

[§] The geometric PMF is also commonly defined as $p(1 - p)^j$, $j = 0, 1, 2, \ldots$, which is the distribution of $J = N - 1$, the number of trials *before* the first success or the number of failures between successes. We have seen, however, that such linear transformations do not alter the basic shape or type of the distribution (Sec. 2.3.1).

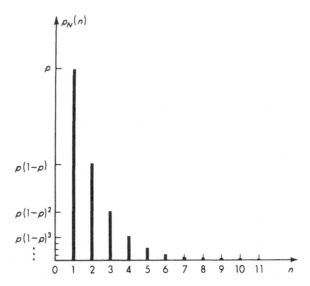

Fig. 3.1.2   Geometric distribution $G(p)$.

where use is made of the familiar formula for the sum of a geometric progression.   This result could have been obtained directly by observing that the probability that $N \leq n$ is simply the probability that there is at least one occurrence in n trials, or

$$1 - P[\text{no occurrences in } n \text{ trials}] = 1 - (1 - p)^n$$

The geometric distribution also follows from a totally different mechanism (see Sec. 3.5.3).

Moments of the geometric   The first moment of a geometric random variable is found by substitution and a number of algebraic manipulations similar to those used in the binomial case.   The reader can show that

$$E[N] = \sum_{n=1}^{\infty} np(1 - p)^{n-1}$$

$$= \frac{1}{p} \tag{3.1.11}$$

In words, the average number of trials to the first occurrence is the reciprocal of the probability of occurrence on any one trial.   For example, if the proportion of defective bolts is 0.1, on the average, 10 bolts would be tested before a defective would be found, assuming the conditions of independent Bernoulli trials hold.

The variance of N can easily be shown to be

$$\text{Var } [N] = E[N^2] - E^2[N]$$

$$= \frac{1 - p}{p^2} \tag{3.1.12}$$

Negative binomial distribution   We have just determined the answer to the question, "On which trial will the first success occur?"   Next consider the more general question, "On which trial will the kth success occur?" We are dealing now with a random variable, say W_k, which is the sum of random variables $N_1, N_2, \ldots, N_k$, where N_i is the number of trials between the $(i - 1)$th and ith successes.   Thus

$$W_k = N_1 + N_2 + N_3 + \cdots + N_k$$

Because of the assumed independence of the Bernoulli trials, it is clear that the N_i $(i = 1,2, \ldots k)$ are mutually independent random variables, each with a common geometric distribution with parameter p.   As a result of this observation, we can obtain partial information about W_k, namely, its moments, very easily.   The mean and variance of W_k can be written down immediately as

$$m_{W_k} = \sum_{i=1}^{k} m_{N_i} = \frac{k}{p} \tag{3.1.13}$$

$$\sigma_{W_k}^2 = \sum_{i=1}^{k} \sigma_{N_i}^2 = \frac{k(1 - p)}{p^2} \tag{3.1.14}$$

The distribution of W can be found using the methods discussed in Sec. 2.3.   For example, for $k = 2$,

$$W_2 = N_1 + N_2$$

Writing Eq. (2.3.43) for the discrete case, we have

$$p_{W_2}(w) = \sum_{n=1}^{w} p_{N_1}(n) p_{N_2}(w - n)$$

which states that the probability that $W_2 = w$ is the probability that $N_1 = n$ and $N_2 = w - n$, summed over all values of n up to w.   Notice that when $n = w$ in the sum, we get $p_{N_2}(0)$, which is zero; hence

$$p_{W_2}(w) = \sum_{n=1}^{w-1} (1 - p)^{n-1} p (1 - p)^{w-n-1} p$$

$$= p^2 (1 - p)^{w-2} \sum_{n=1}^{w-1} 1$$

or, simply,

$$p_{W_2}(w) = (w - 1)(1 - p)^{w-2}p^2 \qquad w = 2, 3, \ldots \qquad (3.1.15)$$

This result could have been arrived at directly by arguing that the second success will be achieved at the wth trial only if there are $w - 2$ failures, a success on the wth trial, and one success in any one of the $w - 1$ preceding trials.

With this result for W_2, one could find the distribution of W_3, knowing $W_3 = W_2 + N_3$. In turn the distribution for any k could be arrived at. If this exercise is carried out, a pattern emerges immediately, and one can conclude that

$$p_{W_k}(w) = \binom{w - 1}{k - 1}(1 - p)^{w-k}p^k \qquad w = k, k + 1, \ldots \qquad (3.1.16)$$

The result is reasonable if one argues that the kth success occurs on wth trial only if there are exactly $k - 1$ successes in the preceding $w - 1$ trials and there is a success on the wth trial. The probability of exactly $k - 1$ successes in $w - 1$ trials we know from our study of binomial random variables to be $\binom{w - 1}{k - 1}(1 - p)^{w-k}p^{k-1}$.

This distribution is known as the *Pascal*, or *negative binomial, distribution* with parameters k and p, denoted here $NB(k,p)$.† It has been widely used in studies of accidents,‡ cosmic rays, learning processes, and exposure to diseases. It should not be surprising, owing to its possible interpretation as the number of trials to the $(k_1 + k_2)$th event, that a random variable which is the sum of a $NB(k_1,p)$ and a $NB(k_2,p)$ pair of (independent) random variables is itself negatively binomially distributed, $NB(k_1 + k_2, p)$.

Illustration: Turn lanes If a left-turn lane at a traffic light has a capacity of three cars, what is the probability that the lane will not be sufficiently large to hold all the drivers who want to turn left in a string of six cars delayed by a red signal? In the mean, 30 percent of all cars want to turn left. The desired probability is simply the probability that the number of trials to the fourth success (left-turning car) is less than 6. Letting W_4 equal that number, W_4 has a negative

† Again the reader should be cautioned that other forms of this distribution are commonly found, in particular, that of $Z_k = W_k - k$, the number of failures before the kth success: $p_{Z_k}(z) = p^k(1 - p)^z \binom{z + k - 1}{k - 1}$, $z = 0, 1, 2, \ldots$. This form conveniently is always defined on $z = 0, 1, 2, \ldots$, and not $k, k + 1, \ldots$, as the distribution of W_k is.

‡ See Prob. 3.47 for an illustration in which the negative binomial distribution arises from a totally different mechanism.

binomial distribution. Therefore,

$$P[W_4 \leq 6] = F_{W_4}(6) = \sum_{w=4}^{6} \binom{w-1}{4-1} (1 - 0.3)^{w-4}(0.3)^4$$

$$= (0.3)^4 + 4(1 - 0.3)(0.3)^4 + 10(1 - 0.3)^2(0.3)^4$$

$$= 0.07$$

Alternatively, one could have deduced this answer by asking for the probability of four, five, or six successes in six Beroulli trials and used the PMF of the appropriate binomial random variable.

A more realistic question might be, "What is the probability that the left-turn lane capacity will be insufficient when the number of red-signal-delayed cars is unspecified?" This number is itself a random variable, say Z, with a probability mass function, $p_Z(z)$, $z = 0, 1, 2, \ldots$. Then the probability of the event A, that the lane is inadequate, is found by considering all possible values of Z:

$$P[A] = \sum_{z=0}^{\infty} P[A \mid Z = z]p_Z(z)$$

$$= \sum_{z=0}^{\infty} F_{W_4}(z)p_Z(z) \qquad\qquad\qquad (3.1.17a)$$

or, since $F_{W_4}(z) = 0$ for $z < 4$,

$$P[A] = \sum_{z=4}^{\infty} F_{W_4}(z)p_Z(z) \qquad\qquad\qquad (3.1.17b)$$

A logical choice for the mass function of Z will be considered in the following section, Sec. 3.2.1.

Illustration: Design values and return periods In the design of civil engineering systems which must withstand the effects of rare events such as large floods or high winds, it is necessary to consider the risks involved for any particular choice of design capacity. Given a design, the engineer can usually estimate the largest magnitude of the rare event which the design can withstand, e.g., the maximum flow possible through a spillway or the maximum wind velocity which a structure can resist. The engineer then seeks (from past data, say) an estimate of the probability p that in a given time period, usually a year, this critical magnitude will be exceeded.

If, as is commonly assumed, the magnitude of the maximum annual flow rates in a river or the maximum annual wind velocities are independent, and if p remains constant from year to year, then the successive years represent independent Bernoulli trials. A magnitude greater than the critical value is denoted a success. With the knowledge of the distributions in this section the engineer is now in a position to answer a number of critical questions.

Let us assume, for example, that $p = 0.02$; that is, there is 1 chance in 50 that a flood greater than the critical value will occur in any particular year. What is the probability that at least one large flood will take place during the 30-year economic lifetime of a proposed flood-control system? Let X equal the number of large floods in 30 years. Then X is $B(30,0.02)$, and

$$P[X \geq 1] = 1 - p_X(0) = 1 - \binom{30}{0}(0.02)^0(0.98)^{30} = 1 - (0.98)^{30}$$

Using the familiar binomial theorem,

$$P[X \geq 1] = 1 - (0.98)^{30} = 1 - (1 - 0.02)^{30}$$

$$= 1 - \sum_{i=0}^{30}\binom{30}{i}(0.02)^i$$

$$= 1 - \left[1 - 30(0.02) + \frac{(30)(29)}{2}(0.02)^2 \right.$$

$$\left. - \frac{(30)(29)(28)}{(2)(3)}(0.02)^3 + \cdots \right]$$

$$= 0.6 - 0.174 + 0.03248 - \cdots$$

$$\approx 0.46$$

If this risk of at least one critical flood is considered too great relative to the consequences, the engineer might increase the design capacity such that the magnitude of the critical flood would only be exceeded with probability 0.01 in any one year. Then X is $B(30,0.01)$, and

$$P[X \geq 1] = 1 - p_X(0) \approx 0.26$$

The engineer seeks, of course, to weigh increased initial cost of the system versus the decreased risk of incurring the damage associated with a failure of the system to contain the largest flood.

The number of years N to the first occurrence of the critical flood is a random variable with a geometric distribution, $G(0.01)$, in the latter case. The probability that it is greater than 10 years is

$$P[N > 10] = 1 - F_N(10)$$

$$= 1 - \{1 - (1 - p)^{10}\} = (1 - p)^{10}$$

$$\approx 0.92$$

What is the probability that $N > 30$? Clearly it is just equal to the probability that there are no floods in 30 years, i.e., that $X = 0$, where X is $B(30,0.01)$. Here, using a previous result,

$$P[N > 30] = P[X = 0] = 1 - P[X \geq 1]$$

$$\approx 1 - 0.26$$

$$= 0.74$$

Average Return Periods   The expected value of N is simply

$$m_N = \frac{1}{p} = \frac{1}{0.01} = 100$$

This is the average number of trials (years) to the first flood of magnitude greater than the critical flood. In civil engineering this is referred to as the *average return period* or simply the *return period*. The critical magnitude is often referred to as the "m_N-year flood," here the "100-year flood." The term is somewhat unfortunate, since its use has led the layman to conclude that there will be 100 years between such floods when in fact the probability of such a flood in any year remains 0.01 independently of the occurrence of such a flood in the previous or a recent year (at least according to the engineer's model).

The probability that there will be no floods greater than the m-year flood in m years is, since X is then $B[m, 1/m]$,

$$P[X = 0] = (1 - 1/m)^m = \sum_{i=0}^{m} \binom{m}{i} \left(\frac{1}{m}\right)^i$$

$$= 1 - m\left(\frac{1}{m}\right) + \frac{m(m-1)}{2}\left(\frac{1}{m}\right)^2 - \frac{m(m-1)(m-2)}{(2)(3)}\left(\frac{1}{m}\right)^3 \cdots$$

$$\approx 1 - \frac{u}{1!} + \frac{u^2}{2!} - \frac{u^3}{3!} \cdots \approx e^{-u}$$

where $u = m(1/m) = 1$; hence, *for large m,*

$$P[X = 0] \approx e^{-1} = 0.368$$

That is, the likelihood that one or more m-year events will occur in m years is approximately $1 - e^{-1} = 0.632$. Thus a system "designed for the m-year flood" will be inadequate with a probability of about 2/3 at least once during a period of m years.

Cost Optimization What is the optimum design to minimize total expected cost? Assume that the cost c associated with a system failure is independent of the flood magnitude, and that, in the range of designs of interest, the cost is a constant I plus an incremental cost whose logarithm is proportional to the mean return period m_N of the design demand.† For an economic design life of 30 years,

$$E[\text{total cost}] = I + \text{antilog}\,(bm_N) + E[\text{cost due to failures}]$$

$$= I + e^{bm_N} + \sum_{x=0}^{30} (xc)P[x \text{ failures occur}]$$

Since the cost of failure is cX,

$$E[\text{total cost}] = I + e^{bm_N} + E[cX] = I + e^{bm_N} + cE[X]$$

Assume that the system itself remains in use with an unaltered capacity after any failure. Then, if X is the number of failures in 30 years, it is distributed

† For reasons which will become evident in Sec. 3.3.3, it is commonly assumed by hydrological engineers that the logarithm of the magnitude of the m-year flood is proportional to m. Thus the assumption above implies that the incremental cost is proportional to the magnitude of the design flood.

$B(30, 1/m_N)$ (assuming no more than one failure per year).   The mean of X is $30/m_N$.   The expected cost, given a value of m_N, is

$$E[\text{total cost}] = I + e^{bm_N} + c\,\frac{30}{m_N}$$

The design flood magnitude $\tilde{m}_N$, which minimizes this expected total cost, is found by setting the derivative of total cost equal to zero:

$$\frac{dE[\text{total cost}]}{dm_N} = be^{b\tilde{m}_N} - \frac{30c}{\tilde{m}_n{}^2} = 0$$

$$(\tilde{m}_N)^2 e^{b\tilde{m}_N} = \frac{30c}{b}$$

This equation can be solved by trial and error for given values of b and c. For example, for $b = 0.1$ year^{-1} and $c = \$200,000$, $\tilde{m}_N = 90$ years.   Thus the designer should provide a system with a capacity sufficient to withstand the demand which is exceeded with probability $1/90 = 0.011$ in any year.   In Sec. 3.3, we will encounter distributions commonly used to describe annual maximum floods from which such design magnitudes could be easily obtained, once the design return period is fixed.

3.1.4 Summary

The basic model discussed here is that of Bernoulli trials.   They are a sequence of *independent* experiments, the outcome of any one of which can be classified as either *success or failure* (or, numerically, 0 or 1). The probability of success p remains *constant* from trial to trial.

  If this model holds, then:

1. Y, the total number of successes in n trials, has a *binomial distribution:*

$$p_Y(y) = \binom{n}{y} p^y (1 - p)^{n-y} \qquad y = 0, 1, \ldots, n$$

   with mean np and variance $np(1 - p)$.
2. N, the trial number at which the first success occurs, has a *geometric distribution:*

$$p_N(n) = p(1 - p)^{n-1} \qquad n = 1, 2, \ldots$$

   with mean $1/p$ and variance $(1 - p)/p^2$.   The former is called the mean return period.
3. W_k, the trial number at which the kth success will occur, has a *negative binomial* distribution:

$$p_{W_k}(w) = \binom{w - 1}{k - 1} (1 - p)^{w-k} p^k \qquad w = k, k + 1, \ldots$$

   with mean k/p and variance $k(1 - p)/p^2$.

3.2 MODELS FROM RANDOM OCCURRENCES

In many situations it is not possible to identify individual discrete trials at which events (or successes) might occur, but it is known that the number of such trials is many. The models in this section arise out of consideration of situations where the number of possible trials is infinitely large. Examples include situations where events can occur at any instant over an interval of time or at any location along the length of a line or on the area of a surface.

3.2.1 Counting Events; The Poisson Distribution

A derivation of the distribution Suppose that a traffic engineer is interested in the total number of vehicles which arrive at a specific location in a fixed interval of time, t sec long. If he knows that the probability of a vehicle occurring in any second is a small number p, (and if he assumes that the probability of two or more vehicles in any one second is negligible), then the total number of vehicles X in the $n = t$ (assumed) independent trials is binomial, $B(n,p)$:

$$p_X(x) = \binom{n}{x} p^x (1 - p)^{n-x} \qquad x = 0, 1, 2, \ldots, n$$

Consider what happens as the engineer takes smaller and smaller time durations to represent the individual trials. The number of trials n increases and the probability p of success on any one trial decreases, but the expected number events in the total interval must remain constant at pn. Call this constant ν and consider the PMF of X in the limit as the trial duration shrinks to zero, such that

$$n \to \infty$$
$$p \to 0$$
$$np = \nu$$

Substituting for $p = \nu/n$ in the PMF of X and rearranging,

$$
\begin{aligned}
p_X(x) &= \frac{n!}{x!(n-x)!} \left(\frac{\nu}{n}\right)^x \left(1 - \frac{\nu}{n}\right)^{n-x} \\
&= \frac{\nu^x}{x!} \left(1 - \frac{\nu}{n}\right)^n \frac{n!}{(n-x)!} \frac{1}{n^x(1 - \nu/n)^x} \\
&= \frac{\nu^x}{x!} \left(1 - \frac{\nu}{n}\right)^n \left\{ \frac{n(n-1)(n-2) \cdots (n-x+1)}{[n(1 - \nu/n)]^x} \right\}
\end{aligned}
$$

The term in braces has x terms in the numerator and x terms in the denominator. For large n each of these terms is very nearly equal to n; hence in the limit, as n goes to infinity, the term in braces is simply

n^x/n^x or 1. The term $(1 - \nu/n)^n$ is known to equal $e^{-\nu}$ in the limit. Hence the PMF of X is

$$p_X(x) = \frac{\nu^x e^{-\nu}}{x!} \qquad x = 0, 1, 2, \ldots, \infty \qquad (3.2.1)$$

Moments This extremely useful distribution is known as the *Poisson distribution*, denoted here $P(\nu)$. Notice it contains but a single parameter ν, compared with the two required for the binomial distribution, $B(n,p)$. Its mean and variance are both equal to this parameter. Following the same steps used to find the mean of the binomial distribution,

$$E[X] = \sum_{x=0}^{\infty} x \frac{\nu^x e^{-\nu}}{x!}$$

$$= \nu \sum_{x=1}^{\infty} \frac{\nu^{x-1} e^{-\nu}}{(x-1)!}$$

Letting $y = x - 1$,

$$E[X] = \nu \sum_{y=0}^{\infty} \frac{\nu^y e^{-\nu}}{y!}$$

$$m_X = \nu \qquad (3.2.2)$$

since the sum is now simply the sum over a Poisson PMF. A similar calculation shows that also

$$\sigma_X{}^2 = \nu \qquad (3.2.3)$$

Plots of Poisson distributions are displayed in Fig. 3.2.1.† Notice the fading of skew as ν increases.

Also, consideration of the derivation should make it clear that the sum of two Poisson random variables with parameters ν_1 and ν_2 must again be a Poisson random variable with parameters $\nu = \nu_1 + \nu_2$. (How might this fact be verified?) Distributions with the peculiar and valuable property that the sum of independent random variables with the distribution has the same distribution are said to be "regenerative." The binomial and negative binomial distributions are, recall, regenerative only on the condition that the parameter p is the same for all the distributions.

Poisson process It is clear from the derivation that if a time interval of a different duration, say $2t$, is of interest, then the number of trials at any stage in the limit would be twice as great and the parameter of the

† Use of Table A.2 to simplify computation of the CDF will be discussed in Sec. 3.4.2.

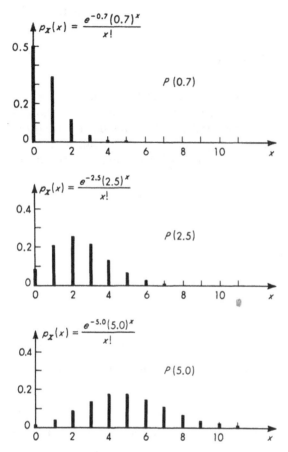

Fig. 3.2.1 Poisson distribution $P(\nu)$.

resulting Poisson distribution would be 2ν.   In such cases, the parameter of the Poisson distribution can be written advantageously as λt rather than ν:

$$p_X(x) = \frac{(\lambda t)^x e^{-\lambda t}}{x!} \qquad x = 0, 1, 2, \ldots \infty \qquad (3.2.4)$$

This form of the Poisson distribution is suggestive of its association with the *Poisson process*. A stochastic process is a random function of time (usually).   In this case we are interested in a stochastic process $X(t)$, whose value at any time t is the (random) number of arrivals or incidents which have occurred since time $t = 0$.   Just as samples of a random variable X are numbers, $x_1,\ x_2,\ \ldots$, so observations of a

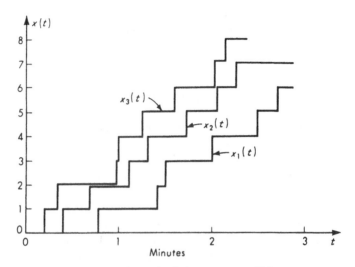

Fig. 3.2.2 Sample functions of a Poisson process $X(t)$.

random process $X(t)$ are *sample functions* of time, $x_1(t)$, $x_2(t)$, . . . , as shown in Fig. 3.2.2.

The samples shown there are of a Poisson process which is counting the number of vehicle arrivals versus time. Other examples of stochastic processes include wave forces versus time, total accumulated rainfall versus time, and strength of soil versus depth. We will encounter other cases of stochastic processes in Secs. 3.6 and 3.7 and in Chap. 6.

At any fixed value of the (time) parameter t, say $t = t_0$, the value $X(t_0)$ of a stochastic process is a simple random variable, with an appropriate distribution $p_{X(t_0)}(x)$, which has an appropriate mean $m_{X(t_0)}$, variance $\sigma^2_{X(t_0)}$, etc.† In general, this distribution, mean, and variance are functions of time. In addition, the joint behavior of two (or more) values, say $X(t_0)$ and $X(t_1)$, of a stochastic process is governed by a joint probability law. Typically one might be interested in studying the conditional distribution of a future value $X(t_1)$, given an observation at the present time $X(t_0)$, in order to "predict" that future value (e.g., see Sec. 3.6.2).

An elementary result of the study of stochastic processes‡ is that the distribution of the random variable $X(t_0)$ is the Poisson distribution with parameter λt_0 if the stochastic process $X(t)$ is a Poisson process with

† In the study of stochastic processes the more common notation is $p_x(t)$, $m(t)$, $\sigma^2(t)$, etc.

‡ A recommended text is E. Parzen [1962], "Stochastic Processes," Holden-Day, Inc., Publisher, San Francisco.

parameter λ. To be a Poisson process, the underlying physical mechanism generating the arrivals or incidents must satisfy the following important assumptions:

1. *Stationarity.* The probability of an incident in a *short* interval of time t to $t + h$ is approximately λh, for any t.
2. *Nonmultiplicity.* The probability of two or more events in a short interval of time is negligible compared to λh (i.e., it is of smaller order than λh).
3. *Independence.* The number of incidents in any interval of time is independent of the number in any other (nonoverlapping) interval of time.

The close analogy to the assumptions underlying discrete Bernoulli trials (Sec. 3.1) is evident. Therefore, the observed convergence of the binomial distribution to the Poisson distribution is to be expected.

In short, if we are observing a Poisson process, the distribution of $X(t)$ at any t is the Poisson distribution [Eq. (3.2.4)], with mean $m_{X(t)} = \lambda t$ [from Eq. (3.2.2)]. Therefore, the parameter λ is usually referred to as the *average rate* (of arrival) of the Poisson process.

The basic mechanism from which the Poisson process arises, namely, independent incidents occurring along a continuous (time) axis with a constant average rate of occurrence, suggests why it is often referred to as the "model of random events" (or random arrivals). It has been successfully employed to describe such diverse problems as the occurrences of storms (Borgman [1963]), major floods (Shane and Lynn [1964]), overloads on structures (Freudenthal, Garrelts, and Shinozuka [1966]), and, distributed in space rather than time, flaws in materials and particles of aggregate in surrounding matrices of material.† It is also widely employed in other fields of engineering to describe the arrival of telephone calls at a central exchange and the demands upon service facilities.

The Poisson process is often used by traffic engineers to model the flow of vehicles past a point when traffic is freely flowing and not dense (e.g., Greenshields and Weida [1952], Haight [1963]). For this reason the Poisson distribution describes well the number of vehicles which arrive at an intersection during a given time interval, say a cycle of a traffic light. The distribution logically might have been used in the illustrations dealing with traffic lights in Secs. 2.3 and 3.1.

Illustration: Left-turn lane and the model of random selection   If, in the illustration in Sec. 3.1.2 involving left-turning cars at a traffic light, the number Z of cars arriving during one cycle is Poisson-distributed with parameter $\nu = 6$, the

† See Prob. 3.14.

probability that less than four cars will arrive is

$$P[Z < 4] = F_Z(3)$$

$$= \sum_{z=0}^{3} \frac{e^{-6}6^z}{z!}$$

$$= e^{-6}\left(1 + 6 + \frac{6^2}{2} + \frac{6^3}{6}\right) = 61e^{-6}$$

$$= 0.152$$

If this event occurs, there is no chance that the left-turn lane capacity will be exceeded. The probability of the event A, that the lane will be inadequate on any cycle (assuming there are no cars remaining from a previous cycle), is given in Eq. (3.1.17b):

$$P[A] = \sum_{z=4}^{\infty} F_{W_4}(z)p_Z(z)$$

$$= \sum_{z=4}^{\infty} \left[\sum_{w=4}^{z} \binom{w-1}{4-1} p^4(1-p)^{w-4} \right] \frac{e^{-\nu}\nu^z}{z!} \tag{3.2.5}$$

where p is the proportion of cars turning left. This result could be evaluated as it stands for any values of p and ν, but it is more informative to reason to another, simpler form.

If the probability is p that any particular car will turn left, then we can consider directly the arrival only of those cars which desire to turn left. Returning to the derivation of the Poisson distribution from the binomial, one can see that, since such left-turning cars arrive with a probability† p times the probability that any car arrives, the number X of these left-turning cars is also Poisson-distributed, but with parameter $p\nu$.‡ Hence the probability that the left-turn lane is inadequate is simply the probability that X is greater than or equal to four, or§:

$$P[A] = P[X \geq 4] = 1 - F_X(3) = 1 - \sum_{x=0}^{3} e^{-p\nu} \frac{(p\nu)^x}{x!}$$

For $\nu = 6$ and $p = 0.3$, $p\nu = 1.8$ and

$$P[A] = 1 - e^{-1.8}\left(1 + 1.8 + \frac{(1.8)^2}{2} + \frac{(1.8)^3}{6}\right)$$

$$= 1 - .859 = 0.141$$

---

† Do not confuse this p, the (constant) proportion of left-turning cars, with the "shrinking" p used briefly in the derivation as the probability of a car arriving in any one short duration trial.

‡ Alternatively, in the assumptions of the Poisson process, the probability of occurrence in t to $t + h$ becomes $p\lambda h$.

§ The equivalence of the two expressions for $P[A]$ could also be shown formally by proper mathematical operations.

If this number is considered by the engineer to be too large, he might consider increasing the capacity of the lane in order to reduce the likelihood of inadequate performance.

It is important to realize the generality of the result used here. The implication is that if a random variable Z is Poisson-distributed, then so too is the random variable X, which is derived by (independently) selecting only with probability p each of the incidents counted by Z; that is if Z is $P(\nu)$, then X is $P'(p\nu)$. More formally, the distribution of X is found as

$$p_X(x) = P[X = x] = \sum_{z=x}^{\infty} P[X = x \mid Z = z]p_Z(z)$$

The conditional term is simply the probability of observing x successes in z Bernoulli trials; thus

$$p_X(x) = \sum_{z=x}^{\infty} \binom{z}{x} p^x(1-p)^{z-x} \frac{e^{-\nu}\nu^z}{z!} \tag{3.2.6}$$

which, upon changing the variable to $u = z - x$, reduces to

$$p_X(x) = \frac{e^{-p\nu}(p\nu)^x}{x!} \tag{3.2.7}$$

Examples of application of this result might include X being the number of hurricane-caused floods greater than critical magnitude when Z is the total number of hurricane arrivals, or X being the number of vehicles recorded by a defective counting device when Z is the total number of vehicles passing in a given interval. The latter example was encountered in Sec. 2.2.2 when the joint distribution of X and Z was investigated.

3.2.2 Time Between Events: The Exponential Distribution

The traffic engineer observing a traffic stream is often concerned with the length of the time interval between vehicle arrivals at a point. If an interval is too short, for example, it will cause a car attempting to cross or merge with the traffic stream to remain stationary or to interrupt the stream. Let us seek the distribution for this time between arrivals under the conditions describing traffic flow used in the preceding section, namely, that the vehicles follow a Poisson arrival process with average arrival rate λ.

Distribution and moments   If we denote by random variable T the time to the *first* arrival, then the probability that T exceeds some value t is equal to the probability that *no* events occur in that time interval of length t. The former probability is $1 - F_T(t)$. The latter probability is $p_X(0)$, the probability that a Poisson random variable X with parameter λt is zero.   Substituting into Eq. (3.2.4),

$$1 - F_T(t) = \frac{(\lambda t)^0 e^{-\lambda t}}{0!} = e^{-\lambda t} \qquad t \geq 0$$

Therefore,

$$F_T(t) = 1 - e^{-\lambda t} \qquad t \geq 0 \tag{3.2.8}$$

while

$$f_T(t) = \frac{dF_T(t)}{dt} = \lambda e^{-\lambda t} \qquad t \geq 0 \tag{3.2.9}$$

This defines the "exponential" distribution wh ch we shall denote $EX(\lambda)$. It describes the time to the first occurrence of a Poisson event.   Therefore it is a continuous analog of the geometric distribution [Eq. (3.1.3)]. But, owing to the stationarity and independence properties of the Poisson process, $e^{-\lambda t}$ is the probability of no events in any interval of time of length t, whether or not it begins at time 0.   If we use the arrival time of the nth event as the beginning of the time interval, then $e^{-\lambda t}$ is the probability that the time to the $(n + 1)$th event is greater than t.   In short, the *interarrival times* of a Poisson process are independent and exponentially distributed.

The mean of the exponential distribution is

$$E[T] = \int_0^\infty t\lambda e^{-\lambda t}\, dt$$

Letting $u = \lambda t$,

$$E[T] = \frac{1}{\lambda} \int_0^\infty u e^{-u}\, du = \frac{1}{\lambda} \left[e^{-u}(-u - 1) \right] \Big|_0^\infty$$

$$m_T = \frac{1}{\lambda} \tag{3.2.10}$$

Recall that in the previous section, λ was found to be the average number of events per unit time, while here $1/\lambda$ is revealed as the *average time between arrivals*.

In a similar manner,

$$\sigma_T{}^2 = \frac{1}{\lambda^2} \tag{3.2.11}$$

Notice that the coefficient of variation of T is unity for any value of the parameter λ.

The exponential distribution is plotted in Fig. 3.2.3 as a function of λt, the ratio of t to the mean interarrival time.   The distribution of

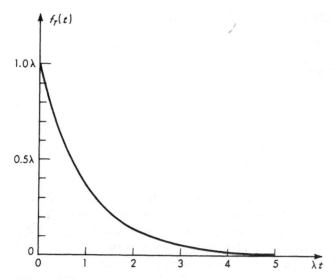

Fig. 3.2.3   Exponential distribution $EX(\lambda)$.

the sum of two independent exponential random variables with different parameters α and β was discussed in Sec. 2.3.2.

Memoryless property   The Poisson process is often said to be "memoryless," meaning that future behavior is independent of its present or past behavior.   This memoryless character of the Poisson arrivals and of the exponential distribution is best understood by determining the conditional distribution of T given that $T > t_0$, that is, the distribution of the time between arrivals given that no arrivals occurred before t_0:

$$F_{T|[T>t_0]}(t) = P[T \leq t \mid T > t_0]$$

$$= \frac{P[(T \leq t) \cap (T > t_0)]}{P[T > t_0]}$$

For t less than t_0, the numerator is zero; for $t \geq t_0$ it is simply equal to $P[t_0 < T \leq t]$.   Thus

$$F_{T|[T>t_0]}(t) = \frac{F_T(t) - F_T(t_0)}{1 - F_T(t_0)} = \frac{(1 - e^{-\lambda t}) - (1 - e^{-\lambda t_0})}{e^{-\lambda t_0}}$$

$$= \frac{e^{-\lambda t_0} - e^{-\lambda t}}{e^{-\lambda t_0}} = 1 - e^{-\lambda(t-t_0)} \qquad t \geq t_0 \qquad (3.2.12)$$

$$f_{T|[T>t_0]}(t) = \lambda e^{-\lambda(t-t_0)} \qquad t \geq t_0 \qquad (3.2.13)$$

Or if time τ is reckoned from t_0, $\tau = t - t_0$,

$$f_{T|[T>t_0]}(t_0 + \tau) = \lambda e^{-\lambda \tau} \qquad \tau \geq 0 \qquad (3.2.14)$$

In words, failure to observe an event up to t_0 does not alter one's prediction of the length of time (from t_0) before an event will occur. The future is not influenced by the past if events are Poisson arrivals. An implication is that any choice of the time origin is satisfactory for the Poisson process.

Applications The very tractable exponential distribution is widely adopted in practice. The close association between the Poisson distribution and the exponential distribution, both arising out of a mechanism of events arriving independently and "at random," suggests that the exponential is applicable to the description of interarrival times (or distances, in the case of flaw distribution, for example) in those situations mentioned in Sec. 3.2.1 in which the number of events in a fixed interval is Poisson distributed. In addition, observed data is often suggestive of an exponential distribution even when the assumptions of a Poisson-arrivals mechanism may not seem wholly appropriate. Times between vehicle arrivals, for example, are influenced (for small values at least) by effects such as minimum spacings between vehicles and "platooning" of vehicles behind a slower vehicle, implying a lack of independence between event arrivals in two neighboring short intervals. Nonetheless, experience† has indicated that in many circumstances the adoption of the exponential distribution of vehicle interarrival times is reasonable. In studies of the lengths of the lifetimes in service of mechanical and electrical components, the analytical tractability of the exponential distribution has led to its wide adoption, even though gradual wearout of such components would suggest that the risk of failure in intervals of equal length would not be constant in time. This time dependence is, of course, in contradiction to the stationarity assumption in the Poisson process model, which predicts that the exponential distribution will describe interarrival times. In short, the exponential distribution, like many others we will encounter, is often adopted simply as a convenient representation of a phenomenon when no more than the shape of the observed data or the analytical tractability of the exponential function seem to suggest it.

3.2.3 Time to the kth Event: The Gamma Distribution

Distribution and moments As in the case with discrete trials, it is also of interest to ask for the distribution of the time X_k to the kth arrival of a

† See, for example, data displayed and discussed in Sec. 4.4.

Poisson process. Now, the times between arrivals, T_i, $i = 1, 2, \ldots, k$, are independent and have exponential distributions with common parameter λ. X_k is the sum $T_1 + T_2 + \cdots + T_k$. Its distribution follows from repeated application of the convolution integral, Eq. (2.3.43).[†] For any $k = 1, 2, 3, \ldots$,

$$f_{X_k}(x) = \frac{\lambda(\lambda x)^{k-1}e^{-\lambda x}}{(k-1)!} \qquad x \geq 0 \tag{3.2.15}$$

We say X is *gamma-distributed* with parameters k and λ, $G(k,\lambda)$[‡] if its density function has the form in Eq. (3.2.15). By integration or, more simply, by consideration of X as the sum of k independent exponentially distributed random variables, if X is $G(k,\lambda)$, then

$$m_X = \frac{k}{\lambda} \tag{3.2.16}$$

and

$$\sigma_X{}^2 = \frac{k}{\lambda^2} \tag{3.2.17}$$

In fact, the gamma distribution is more broadly defined than is implied by its derivation as the distribution of the sum of k independently, identically distributed exponential random variables. More generally the parameter k need not be integer-valued[§] when the gamma distribution is written

$$f_X(x) = \frac{\lambda(\lambda x)^{k-1}e^{-\lambda x}}{\Gamma(k)} \qquad x \geq 0 \tag{3.2.18}$$

the only restrictions being $\lambda > 0$ and $k > 0$. The gamma function $\Gamma(k)$ (from which the distribution gets its name) is equal to $(k-1)!$ if k is an integer, but more generally is defined by the definite integral

$$\Gamma(k) = \int_0^\infty e^{-u}u^{k-1}\,du \tag{3.2.19}$$

The integral arises here as a constant needed to normalize the function to a proper density function. The gamma function is widely tabulated,

---

[†] The results of the waiting-time illustration in Sec. 2.3.2, which dealt with the sum of two exponentially distributed random variables, cannot be used directly for $X_2 = T_1 + T_2$ because for $\alpha = \beta$ those results reduce to an indeterminate result, zero over zero. The reader can retrace that illustration and easily find what step must be changed for this case.

[‡] This notation should not cause confusion with the geometric $G(p)$, which has only one parameter.

[§] For integer-valued k, the gamma distribution is sometimes referred to as the *Erlang distribution*. The gamma distribution is also called the *Pearson Type-III distribution*.

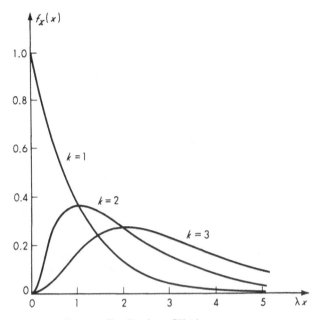

Fig. 3.2.4 Gamma distributions $G(k,\lambda)$.

as is the "incomplete gamma function":

$$\Gamma(k,x) = \int_0^x e^{-u}u^{k-1}\,du \tag{3.2.20}$$

which can be used to evaluate the cumulative distribution function $F_X(x)$:†

$$F_X(x) = \int_0^x f_X(x)\,dx = \frac{\Gamma(k,\lambda x)}{\Gamma(k)} \tag{3.2.21}$$

The equations given above for the mean and variance of X hold for noninteger k as well.

The shape of the gamma density function, Fig. 3.2.4, indicates why it is widely used in engineering applications. Like observed data from many phenomena it is limited to positive values and is skewed to the right. Notice that λ can be interpreted as a scaling parameter and k as a shape parameter of the distribution. The skewness coefficient is

$$\gamma_1 = \frac{E[(X - m_X)^3]}{\sigma_X{}^3} = 2k^{-\frac{1}{2}} \tag{3.2.22}$$

† The use of Table A-2 to evaluate the CDF of the gamma distribution will be discussed in Sec. 3.4.2.

Owing to this shape and its convenient mathematical form (rather than to any belief that it arose from a fundamental, underlying mechanism where the time to the kth occurrence of some event was critical), the gamma distribution has been used by civil engineers to describe such varied phenomena as maximum stream flows (Markovic [1965]), the yield strength of reinforced concrete members (Tichy and Vorlicek [1965]), and the depth of monthly precipitation (Whitcomb [1940]).

In addition to its ability to describe observed data, the gamma distribution is, of course, important as the time to the occurrence of the kth [or the time between the nth and $(n + k)$th occurrence] in a Poisson process.† Applications include the time to the arrival of a fixed number of vehicles, and the time to the failure of a system designed to accept a fixed number of overloads before failing, when the vehicles and the overloads are Poisson arrivals. The sum of exponentials interpretation explains (at least for integer k) why the gamma is regenerative under fixed λ; that is, if X is $G(k_1,\lambda)$, Y is $G(k_2,\lambda)$, and $Z = X + Y$, then Z is $G(k_1 + k_2, \lambda)$ if X and Y are independent. The result is true for noninteger k as well.

Illustration: Maximum flows   Based on a histogram of data of maximum annual river flows in the Weldon River at Mill Grove, Missouri, for the years 1930 to 1960, the distribution was considered as a model by Markovic [1965]. The parameters were estimated‡ as $k = 1.727$ and $\lambda = 0.00672$ $(\text{cfs})^{-1}$.

The mean is

$$m_X = \frac{k}{\lambda} = \frac{1.727}{0.00672} = 256.7 \text{ cfs}$$

$$\sigma_X = \frac{\sqrt{k}}{\lambda} = 190.0 \text{ cfs}$$

The probability that the maximum flow is less than 400 cfs in any year is ($\lambda 400 = 2.70$):

$$F_X(400) = \frac{\Gamma(1.727,2.70)}{\Gamma(1.727)}$$

$$= \frac{0.71}{0.914} = 0.78$$

These values were taken from tables of the gamma function. In Sec. 3.4 more convenient tables will be introduced exploiting a simple relationship between the gamma and the widely tabulated "chi-square distribution."

† As such it bears the same relationship to the discrete negative binomial as the exponential does to the geometric, the negative binomial and geometric distributions being associated with discrete trials rather than a continuous time axis.
‡ Parameter estimation is treated in Sec. 4.1.

3.2.4 Summary

In a Poisson stochastic process incidents occur "at random" along a
time (or other parameter) axis. If the conditions of *stationarity, nonmultiplicity, and independence of nonoverlapping intervals* hold, then for
an *average arrival rate* λ,

1. For any interval of time of length t, the number of events X which
   occur has a *Poisson distribution* with parameter $\nu = \lambda t$:

$$p_X(x) = \frac{\nu^x e^{-\nu}}{x!} \qquad x = 0, 1, 2, \ldots$$

   The mean and variance are both equal to ν.
2. The distribution of time T between incidents has an *exponential
   distribution* with parameter λ:

$$f_T(t) = \lambda e^{-\lambda t} \qquad t \geq 0$$

   with mean $1/\lambda$ and variance $1/\lambda^2$.
3. The time S_k between k incidents has a *gamma distribution:*

$$f_{S_k}(s) = \frac{\lambda(\lambda s)^{k-1} e^{-\lambda s}}{(k - 1)!} \qquad s \geq 0$$

   with mean k/λ and variance k/λ^2. The distribution is also defined
   for noninteger k when $(k - 1)!$ should be replaced by $\Gamma(k)$.

3.3 MODELS FROM LIMITING CASES

The single most important class of models is that in which the model
arises as a limit to an argument about the relationship between the
phenomenon of interest and its (many) "causes." The uncertainty in
a physical variable may be the result of the combined effects of many
contributing causes, each difficult to isolate and observe. In several
important situations, if we know the mechanism by which the individual
causes affect the variable of interest, we can determine a model (or
distribution) for the latter variable without studying in detail the
individual effects. In particular, we need not know the distributions
of the causes. Three important cases will be considered—that where the
individual causes are additive, that where they are multiplicative, and
that where their extremes are critical.

3.3.1 The Model of Sums: The Normal Distribution

Convergence of the shape of the distribution of sums We shall introduce
this case through an example. The total length of an item may be
made up of the sum of the lengths of a number of similar individual

parts. Examples are the total length of a line of cars in a queue, the total error in a long line surveyed in 100-ft increments, or the total time spent repeating a number of identical operations in a construction schedule.

Suppose, as a specific example, that the total length of several pieces is desired, and each has been measured on an accurate rule and recorded to the nearest even number of units (for example, 248 mm). It may be reasonable to assume that each "error," that is, each difference between the recorded length and the true length, is in this case equally likely to lie anywhere between plus and minus one unit (e.g., the true length is between 247 and 249 mm).

Let us define a random variable:

X_i = the measurement error in the ith piece

Then by our argument this variable has a constant (or uniform, Sec. 3.4.1) density function

$$f_{X_i}(x) = \begin{cases} \frac{1}{2} & -1 \le x \le 1 \\ 0 & \text{elsewhere} \end{cases}$$

Consider first the total error in the combined length of two pieces, that is, the difference between the sum of their recorded lengths and the sum of their true lengths. Let

Y_2 = error in total length of two pieces

   = $X_1 + X_2$

Assuming that the two measurement errors are unrelated, i.e., that X_1 and X_2 are independent random variables, we can determine the density function of Y_2 using Eq. (2.3.43). The following result will be found:

$$f_{Y_2}(y) = \begin{cases} \frac{1}{4}(2 - |y|) & -2 \le y \le 2 \\ 0 & \text{elsewhere} \end{cases}$$

The distributions of X_i (or Y_1) and Y_2 are shown in Fig. 3.3.1.

The density function of $Y_3 = X_1 + X_2 + X_3$ is found in the same manner:

$$f_{Y_3}(y) = \begin{cases} \frac{1}{16}(3 - |y|)^2 & 1 \le |y| \le 3 \\ \frac{3}{8} - \frac{1}{8}y^2 & |y| \le 1 \\ 0 & \text{elsewhere} \end{cases}$$

Similarly, we find too

$$f_{Y_4}(y) = \begin{cases} \frac{1}{96}(4 - |y|)^3 & 2 \le |y| \le 4 \\ \frac{1}{3} + \frac{y^2}{32}(|y| - 4) & 0 \le |y| \le 2 \\ 0 & \text{elsewhere} \end{cases}$$

These results are sketched in Fig. 3.3.1, along with the corresponding cumulative distribution functions. Note the scale changes.

Looking at the sequence of curves, it is evident that the distribution of the sum of a number of uniformly distributed random variables quickly takes on a bell-shaped form. By proper adjustment of the parameter c in each case,† a density function of the form

$$f_{Y_i}(y) = ke^{-cy^2} \qquad -\infty \leq y \leq \infty \qquad (3.3.1)$$

will closely approximate the density functions of the successive Y_i. Such curves are shown in Fig. 3.3.1 as dotted lines. The approximations are successively better as more pieces are considered, and the approximation is always better around the mean, $y = 0$, than in the tails. This very important "double exponential" density function is called the *normal or gaussian distribution*.

Central limit theorem   The ability of a curve of this shape to approximate the distribution of the sum of a number of uniformly distributed random variables is not coincidental. It is, in fact, one of the most important results of probability theory that:

> *Under very general conditions, as the number of variables in the sum becomes large, the distribution of the sum of random variables will approach the normal distribution.*

Several of the phrases in this rather loose statement of the *central limit theorem* deserve elaboration. Some idea of the "very general conditions" can be obtained by considering some special cases. The theorem‡ holds for most physically meaningful§ random variables: (1) if the variables involved are independent and identically distributed; (2) if the variables are independent, but not identically distributed (provided that each individual variable has a small effect on the sum); or (3) if the variables

---

† Clearly the constant k will also have to be adjusted at each step to insure that the area under the curve is unity.

‡ This theorem will not be proved in this text. The student is asked to demonstrate another special case in Prob. 3.18. All standard, intermediate-level probability references (e.g., Parzen [1960]) prove the theorem for the special case where the variables are independent and identically distributed, usually using characteristic functions (Prob. 2.62). For more lenient conditions see Parzen [1960], Fisz [1963], or advanced probability texts.

§ For example, the variances must be finite in case 1; it is sufficient in case 2 that the variables are all bounded (that is, $P[|X_i| \leq a] = 1$ for all i and some a short of infinity); in case 3 it is sufficient that the variables are bounded and the dependence does not lead to pathological cycling among a small number of values.

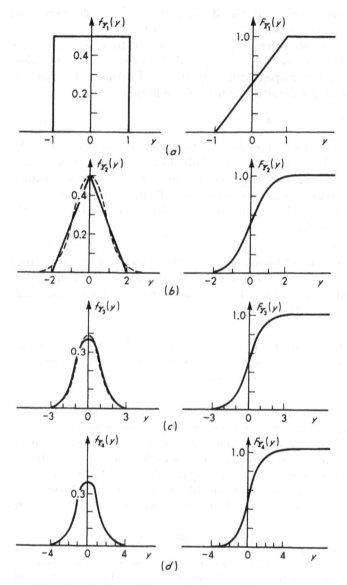

Fig. 3.3.1 Distributions of the error in the total length of one, two, three, or four pieces. (a) Distribution of Y_1; (b) distribution of Y_2; (c) distribution of Y_3; (d) distribution of Y_4.

are not independent at all, but jointly distributed such that correlation is effectively zero between any variable and all but a limited number of others. The word "approach" in the theorem's statement would be interpreted by a mathematician as "converge to," but an engineer will read "be approximated by." The question of how many is "large" depends, as in any such approximation, on what accuracy is demanded, but in this case it depends too on the shape of the distributions of the random variables being summed (highly skewed distributions are relatively slow to converge). The degree of dependence is also a factor. For example, if all variables were perfectly linearly dependent on the first, $X_i = a_i + b_i X_1$, then the distribution of the sum of any number of them would have the same shape as the distribution of the first, since the sum $Y_n = \sum_{i=1}^{n} X_i$ could always be written $Y_n = c + dX_1$.

The important fact is that, even if the number of variables involved is only moderately large, as long as no one variable dominates and as long as the variables are not highly dependent, the distribution of their sum will be very near normal. The immense practical importance of the normal distribution lies in the fact that this statement of the central limit theorem can be made *without* exact knowledge (1) of the marginal distributions of the contributing random variables, (2) of their number, or (3) of their joint distribution. Since the random variation in many phenomena arises from a number of additive variations, it is not surprising that histograms approximating this distribution are frequently observed in nature and that this distribution is frequently adopted as a model in practice. In fact, owing to its analytic tractability and to the familiarity of many engineers with the distribution, the normal model is very often used in practice when there is no reason to believe that an additive physical mechanism exists. For all these reasons the distribution deserves special attention.

Parameters: Mean and variance   First, it should be pointed out that the distributions need not be centered at the origin. Consider then in general the shifted case

$$f_X(x) = ke^{-c(x-m)^2} \qquad -\infty \leq x \leq \infty \tag{3.3.2}$$

The distance to the center of the distribution is labeled m, for, by the symmetry of the distribution, it is its mean. We can determine the normalizing constant k by integration:

$$1 = \int_{-\infty}^{\infty} ke^{-c(x-m)^2} \, dx$$

We find

$$1 = \frac{k \sqrt{\pi}}{\sqrt{c}}$$

Hence

$$k = \frac{\sqrt{c}}{\sqrt{\pi}}$$

or

$$f_X(x) = \frac{\sqrt{c}}{\sqrt{\pi}} e^{-c(x-m)^2} \qquad -\infty \leq x \leq \infty \tag{3.3.3}$$

Let us next determine the variance of the normal distribution.

$$\text{Var}\,[X] = \frac{\sqrt{c}}{\sqrt{\pi}} \int_{-\infty}^{\infty} (x - m)^2 e^{-c(x-m)^2}\,dx$$

A change of variable and integration by parts yield

$$\text{Var}\,[X] = \frac{1}{2c}$$

$$\sigma_X = \frac{1}{\sqrt{2c}}$$

Noting that

$$\sqrt{c} = \frac{1}{\sigma_X \sqrt{2}}$$

we replace c as a parameter of the normal distribution and write the density function in the form

$$f_X(x) = \frac{1}{\sigma_X \sqrt{2\pi}} \exp\left[-\frac{1}{2}\left(\frac{x - m_X}{\sigma_X}\right)^2\right] \qquad -\infty \leq x \leq \infty \tag{3.3.4}$$

In this, its usual form, the mean and standard deviation (or variance) are used as parameters of the distribution, and we would write "X is $N(m,\sigma^2)$." The effect of changes in m and σ is shown in Fig. 3.3.2. The normal curves plotted in dotted lines in Fig. 3.3.1 are those with mean and standard deviation equal to the corresponding moments of the true distributions.

Using normal tables   The normal distribution is widely tabulated in the literature.[†]   This job is simplified by tabulating only the standardized

[†] See National Bureau of Standards [1953] or, in less detail, any standard mathematical tables.   A small table appears in Table A.1.

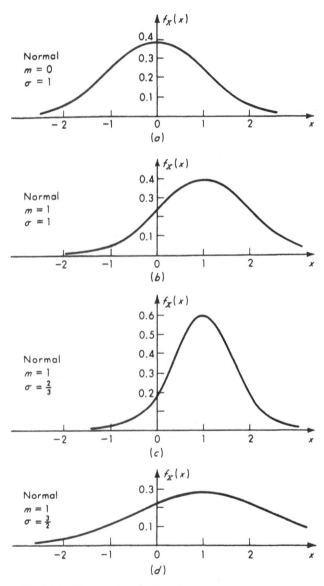

Fig. 3.3.2 Normal density functions.

variable

$$U = \frac{X - m_X}{\sigma_X} \tag{3.3.5}$$

This variable has mean 0 and standard deviation 1 (as the reader was asked to show in Prob. 2.47). The density of the standardized normal random variable is consequently $N(0,1)$ or

$$f_U(u) = \frac{1}{\sqrt{2\pi}} e^{-\frac{1}{2}u^2} \qquad -\infty \leq u \leq \infty \tag{3.3.6}$$

which is shown in Fig. 3.3.2a. The desired density function, that of $X = m_X + \sigma_X U$, is found using Eq. (2.3.15):

$$f_X(x) = \frac{1}{\sigma_X} f_U\left(\frac{x - m_X}{\sigma_X}\right) \qquad -\infty \leq x \leq \infty \tag{3.3.7}$$

For example, the value of the density function in Fig. 3.3.2d at $x = 1$ is

$$f_X(1) = \frac{2}{3} f_U\left(\frac{1-1}{\frac{2}{2}}\right) = \frac{2}{3} f_U(0) = \frac{2}{3}(0.3989) = 0.2660$$

Tables give only half the range of u, $u \geq 0$, owing to the symmetry of the PDF.

To determine the probability that a normal random variable lies in any interval, the integral of $f_X(x)$ over the interval is required; alternatively of course, the difference between the two values of the cumulative distribution function will yield this information. There is no simple expression for the CDF of the normal distribution, but it has been evaluated numerically and tabulated, again for the standardized random variable. In general

$$F_X(x) = P[X \leq x] = P\left[U \leq \frac{x - m_X}{\sigma_X}\right]$$

$$= F_U\left(\frac{x - m_X}{\sigma_X}\right) = F_U(u)$$

$$= \frac{1}{\sqrt{2\pi}} \int_{-\infty}^{u} e^{-\frac{1}{2}v^2}\, dv \qquad -\infty \leq u \leq \infty \tag{3.3.8}$$

in which $u = (x - m_X)/\sigma_X$. Note that u can be interpreted as the number of standard deviations by which x differs from the mean. Tables yield values of $F_U(u)$. Because of the symmetry of the PDF, tables give only half of the range of u, usually for $u \geq 0$.

For example, the probability that a normal random variable X, with mean 4000 and standard deviation 400, will be less than 4800 is:

$$F_X(x) = \frac{1}{400} \frac{1}{\sqrt{2\pi}} \int_{-\infty}^{4800} \exp\left[-\frac{1}{2}\left(\frac{x - 4000}{400}\right)^2\right] dx$$

$$= F_U\left(\frac{4800 - 4000}{400}\right)$$

$$= F_U(2)$$

$$= \frac{1}{\sqrt{2\pi}} \int_{-\infty}^{2} e^{-\frac{1}{2}u^2} du$$

$$= 0.9772$$

Concrete compressive strength is commonly assumed to be normally distributed, and a particular mix might have parameters near the values used here.

The symmetry of the distribution permits one to find values for negative arguments. A quick sketch should convince the reader that for the standardized normal variable,

$$F_U(-u) = 1 - F_U(u) \tag{3.3.9}$$

Consequently, for the variable just cited:

$$P[X \leq 3200] = F_X(3200)$$

$$= F_U\left(\frac{3200 - 4000}{400}\right)$$

$$= F_U(-2)$$

$$= 1 - F_U(2)$$

$$= 1 - 0.9772$$

$$= 0.0228$$

In other tables $1 - F_U(u)$ is tabulated for $u \geq 0$, while still other tables give $1 - 2F_U(-u)$, $u \geq 0$.

The last form is related to the common situation in the study of errors or tolerances where it is desired to know the probability that a normal random variable will fall within k standard deviations of its mean:

$$P[m - k\sigma \leq X \leq m + k\sigma] = P\left[-k \leq \frac{X - m}{\sigma} \leq +k\right]$$

$$= F_U(k) - F_U(-k)$$

$$= 1 - F_U(-k) - F_U(-k)$$

$$= 1 - 2F_U(-k) \tag{3.3.10}$$

If one recalls the discussion in Sec. 2.4.1, he will recognize this value as the probability of a normal random variable remaining within the "k-sigma" bounds. In fact, the "rules of thumb" stated there are approximately those values one obtains for the probabilities of a normal variable. Therefore they are appropriate if the distribution is approximately bell-shaped.

Higher moments The symmetry of the normal distribution about its mean implies that all odd-order central moments (and the skewness coefficient) are zero. The even-order moments must be related to only the mean and standard deviation because these two moments serve as the parameters of the family of normal distributions. It can be shown by integration that the even-order moments are given by

$$\mu_X^{(n)} = E[(X - m_X)^n]$$

$$= \frac{n!}{2^{n/2}(n/2)!} \sigma^n \qquad n = 2, 4, \ldots \tag{3.3.11}$$

Note that

$$\mu_X^{(4)} = \frac{4!}{2^2 2!} \sigma^4 = 3\sigma^4 \tag{3.3.12}$$

Hence the coefficient of kurtosis [Eq. (2.4.19)] is

$$\gamma_2 = \frac{\mu_X^{(4)}}{\sigma^4} = \frac{3\sigma^4}{\sigma^4} = 3 \tag{3.3.13}$$

It is not uncommon to compare the coefficient of kurtosis of a random variable with this "standard" value of 3, that of the normal distribution. If a distribution has $\gamma_2 > 3$, it is said to be "flatter" than standard. In fact, one often sees defined the *coefficient of excess*, $\gamma_2 - 3$. Positive coefficients of excess indicate distributions flatter than the normal, negative, more peaked than normal.

Distribution of the sum of normal variables Since by the central limit theorem the sum of many variables tends to be normally distributed, it should be expected that the sum of two independent random variables each normally distributed would also be normally distributed. That is, in fact, the case. The proof is a direct application of Eq. (2.3.45), but will not be demonstrated here. Accepting the statement, we conclude that the distribution of $Z = X + Y$ when X is $N(m_X, \sigma_X^2)$ and Y is $N(m_Y, \sigma_Y^2)$ is also normal if X and Y are independent. We can use the results of Sec. 2.4.3 to determine the parameters of Z:

$$m_Z = m_X + m_Y \tag{3.3.14}$$

$$\sigma_Z = \sqrt{\sigma_X^2 + \sigma_Y^2} \tag{3.3.15}$$

Thus Z is $N(m_X + m_Y, \sigma_X^2 + \sigma_Y^2)$.

If we are interested in $Z = X_1 + X_2 + \cdots + X_n$, where the X_i are independent, normally distributed random variables, we can conclude, by considering pairs of the variables in sequence, that, in general, the sum of independent normally distributed random variables is normally distributed with a mean equal to the sum of their means and a variance equal to the sum of their variances. A generalization of this statement for nonindependent, jointly normally distributed random variables will be found in Sec. 3.6.2.

Illustration: Footing loads For example, suppose that a foundation engineer wanting to estimate the long-term settlement of a footing states that the total sustained load on the footing is the sum of the dead load of the structure and the load imposed by furniture and the occupancy loads. Since each is the sum of many relatively small weights, the engineer adopts the assumption that the dead load X and sustained occupancy load Y are normally distributed. Unable to see any important correlation between them, he also decides to treat X and Y as independent. Data from numerous buildings of a similar type suggest to him that

$m_X = 100$ kips

$\sigma_X = 10$ kips

$m_Y = 40$ kips

$\sigma_Y = 10$ kips

The distribution of the total sustained load, $Z = X + Y$, then is also normally distributed with mean

$m_Z = 100 + 40 = 140$ kips

and

$\sigma_Z = \sqrt{10^2 + 10^2} = 10\sqrt{2} = 14.1$ kips

A reasonable design load might be that load which will be exceeded only with probability 5 percent. This is the value z^* such that

$1 - F_Z(z^*) = 5$ percent

or

$F_Z(z^*) = 95$ percent

In terms of U, the tabulated $N(0,1)$ variable, we find $u^* = (z^* - m_Z)/\sigma_Z$ such that

$F_U(u^*) = 95$ percent

From tables,

$u^* = 1.65$

implying that

$z^* = m_Z + 1.65\sigma_Z$

$= 140 + (1.65)(14.1) = 163$ kips

Applications It can be said without qualification that the normal distribution is the single most used model in applied probability theory. We have seen that a sound reason may exist for its adoption as a description of certain natural phenomena. Namely, it can be expected to represent those variables which arise as the sum of a number of random effects, no one of which dominates the total. As a consequence the normal model has been used fruitfully to describe the error in measurements, to represent deviations from specified values of manufactured products and constructed items that result from a number of pieces and/or operations each of which may add some deviation to the total, and to model the capacity of a system which fails only after saturation of a number of parallel, redundant (random capacity) components has taken place. Specific examples of the last case include the capacity of a road which is the sum of the capacities of its lanes, the strength of a collapse mechanism of a ductile, elastoplastic frame, which is a sum of constants times the yield moments of certain joints, and the deflection of an elastic material which is the sum of the deflections of a number of small elementary volumes. The normal model was adopted in Fig. 2.2.5 to represent the distribution of total annual runoff (given that it was not zero) based on the assumption that this runoff was the sum of a number of individual daily contributions.

The normal distribution is also often adopted as a convenient approximation to other distributions which are less widely tabulated. Those cases in which the normal distribution is a good numerical approximation can usually be anticipated by considering how this distribution arises and then invoking the central limit theorem. The gamma distribution with parameters λ and integer k, for example, was shown in Sec. 3.2.3 to be the distribution of the sum of k exponentially distributed random variables. For k large (integer or not) the normal distribution will closely approximate the gamma. Figure 3.2.4 shows that the approximation is not good for $k \leq 3$ but that it rapidly improves as k increases.

Similarly, discrete random variables such as the binomial or Poisson variables can be considered to represent the sum of a number of "zero-one" random variables and can be expected, by the central limit theorem, to be approximated well by the continuous, normal random variable for proper values of their parameters. In particular, if the approximation is to be good for the binomial, n should be large and p not near zero or one, and for the Poisson, ν should be large. Figures 3.1.1 and 3.2.1 verify these conclusions. The ability to use the normal approximation may be particularly appreciated in such discrete cases where the evaluation of factorials of large numbers is involved. Many references treat these approximations in some detail (Hald [1952], Parzen [1960], Feller [1957], etc.)

Although certain corrections associated with representing discrete by continuous variables are available to increase the accuracy of the approximation,† it is usually sufficient simply to equate first and second moments of the two distributions (discrete or continuous) in order to achieve a satisfactory approximation. For example, the probability that X, a gamma-distributed random variable, $G(12.5,3.16)$, is less than 3 is approximately equal to the probability that a normally distributed random variable, say X^*, with the same mean, $m_{X^*} = m_X = 12.5/3.16 = 3.95$, and same standard deviation, $\sigma_{X^*} = \sigma_X = \sqrt{12.5/(3.16)^2} = 1.12$, is less than 3. That is, X^* is $N(3.95,(1.12)^2)$. Thus

$$
\begin{aligned}
F_X(3) &\approx F_{X^*}(3) \\
&= F_U\left(\frac{3 - 3.95}{1.12}\right) = F_U(-0.85) \\
&= 0.1977
\end{aligned}
\tag{3.3.16}
$$

U is, as above, $N(0,1)$. The exact value of $F_X(3)$ can be found using tables of the incomplete gamma function or of the χ^2 distribution (Sec. 3.4). This value is 0.2000.

Whether the normal model is adopted following a physical argument or as an approximation to other distributions, it should be noted that its validity may break down outside the region about its mean. Tails of the distribution are much more sensitive to errors in the model formulation than the central region. Recall, for example, the original illustration, Fig. 3.3.1. To say that the sum of four uniformly distributed random errors is normally distributed is a valuable approximation in the range close to the mean, but it clearly falls down in the tails, where the normal distribution predicts a small, but nonzero probability of occurrence of errors larger than four units. The limits on the argument of a normal density function are theoretically plus and minus infinity, yet it may still be useful in practice to assume that some variable, such as load, weight, or time, which is physically limited to nonnegative values, is normally distributed. Caution must be urged in drawing conclusions

† This is discussed in Prob. 3.18.

from models based on such assumptions, if, as is often the case in civil engineering, it is just these extreme values (large or small) which are of most concern. In short, the engineer must never forget the range of validity of his model nor the accuracy to which he can meaningfully use it. This statement holds true for all probabilistic models, as well as for deterministic formulations of engineering problems, although it is too frequently ignored in both.

The ease with which one can work with the normal distribution,[†] its many tables, and its well-known properties causes its adoption as a model in many situations when little or no physical justification exists. Frequently it is used simply because an observed histogram is roughly bell-shaped and approximately symmetrical. In this case the reason for the choice is simply mathematical convenience, with a continuous, easily defined curve replacing empirical data. Even with little or no data, the normal distribution is often adopted as a "not-unreasonable" model. The American Concrete Institute specifications for concrete strength are based on the normal distribution because it seems to fit observed data. The U.S. Bureau of Public Roads, in initial development of statistical quality-control techniques for highway construction, has adopted the normal assumption for material and construction properties, often when no information is yet available. Such uses of the normal distribution serve a valid computational and operational purpose, but the engineer should be quick to question highly refined conclusions or high-order accuracy leading from such premises.[‡] In particular, results highly dependent upon the tail probabilities in or beyond the region where little data has been observed should be suspect unless there is a valid physical reason to expect the model to hold.

3.3.2 The Model of Products: The Lognormal Distribution

Multiplicative models While the normal distribution arose from the sum of many small effects, it is desirable also to consider the distribution of a phenomenon which arises as the result of a multiplicative mechanism acting on a number of factors. An example of such a mechanism occurs in breakage processes, such as in the crushing of aggregate or transport of sediments in streams. The final size of a particle results from a number of collisions of particles of many sizes traveling at different velocities. Each collision reduces the particle by a random proportion of its size at the time (Epstein [1947]). Therefore, the size Y_n of a randomly chosen particle after the nth collision is the product of Y_{n-1} (its

† In addition to the qualities mentioned, it will be seen that the normal model underlies most of the work in applied statistics (Chap. 4).
‡ For a discussion of this point with respect to reinforced concrete design specifications see Granholm [1965], Chap. 16.

size prior to that collision) and W_n (the random reduction factor). Extending this argument back through previous collisions,

$$Y_n = Y_{n-1}W_n = Y_{n-2}W_{n-1}W_n = \cdots$$
$$= Y_0 W_1 W_2 \cdots W_n \quad (3.3.17)$$

Also, a number of physical systems can be characterized by a mechanism which dictates that the increment $Y_n - Y_{n-1}$ in the response of the system when it is subjected to a random impulse of input Z_n is proportional to the present value of the response Y_{n-1}. Formally,

$$Y_n - Y_{n-1} = Z_n Y_{n-1} \qquad (3.3.18)$$

or

$$Y_n = Y_{n-1}(1 + Z_n) = Y_{n-2}(1 + Z_{n-1})(1 + Z_n) = \cdots$$
$$= Y_0(1 + Z_1)(1 + Z_2) \cdots (1 + Z_n)$$

Letting $W_i = 1 + Z_i$, this function is in the same multiplicative form as the breakage model above. The growth of certain economic systems may follow this model.

Finally, the fatigue mechanism in materials has been described as follows (Fruedenthal [1951]†). The internal damage Y_n, after n cycles of loading, is

$$Y_n = g(Y_{n-1})W_n \qquad (3.3.19)$$

In this expression, W_n is the internal stress state resulting from the nth load application, subject to variation because of the internal differences in materials at the microscopic level. If, as a first approximation, $g(Y_{n-1})$ is taken equal to $c_{n-1}Y_{n-1}$,

$$Y_n = Y_0 W_1 W_2 \cdots W_n \qquad (3.3.20)$$

Distribution   In all these cases the variable of interest Y is expressed as the product of a large number of variables, each of which is, in itself, difficult to study and describe. In many cases something can be said, however, of the distribution of Y. Take natural logarithms‡ of both sides of any of the equations above. The result is of the form

$$\ln Y = \ln Y_0 + \ln W_1 + \ln W_2 + \cdots + \ln W_n \qquad (3.3.21)$$

Since the W_i are random variables, the functions $\ln W_i$ are also random variables (Sec. 2.3.1). Calling upon the central limit theorem, one

---

† It should be stated that its author has since proposed more appropriate probabilistic models of this phenomenon (see Sec. 3.3.3).

‡ Logarithms could be taken to other bases. Results will differ only by an awkward constant, since, for example, $\log_{10} Y = 0.4343 \ln Y$. If logs to the base 10 are used, the right-hand sides of Eqs. (3.3.24) and (3.3.25) should be multiplied by 0.4343.

may predict that the sum of a number of these variables will be approximately normally distributed. In this case, then, we expect $\ln Y$ to be normally distributed. Let

$$X = \ln Y \tag{3.3.22}$$

Our problem is, knowing that X is normally distributed, to determine the distribution of Y or

$$Y = e^X \tag{3.3.23}$$

A random variable Y whose logarithms are normally distributed is said to have the *logarithmiconormal or lognormal* distribution. Its form is easily determined.

In Sec. 2.3.1 we found that the density function in such a case of one-to-one, monotonic transformation is [Eq. (2.3.13)]

$$f_Y(y) = \left| \frac{dg^{-1}}{dy} (y) \right| f_X(g^{-1}(y))$$

Here

$$Y = g(X) = e^X$$
$$X = g^{-1}(Y) = \ln Y$$
$$\frac{dg^{-1}(y)}{dy} = \frac{1}{y}$$

and X is normally distributed:

$$f_X(x) = \frac{1}{\sigma_X \sqrt{2\pi}} \exp\left[-\frac{1}{2}\left(\frac{x - m_X}{\sigma_X} \right)^2 \right] \qquad -\infty \le x \le \infty$$

Therefore, upon substituting,

$$f_Y(y) = \frac{1}{y\sigma_X \sqrt{2\pi}} \exp\left[-\frac{1}{2}\left(\frac{\ln y - m_X}{\sigma_X} \right)^2 \right] \qquad y \ge 0 \tag{3.3.24}$$

The random variable Y is lognormally distributed, whereas its logarithm X is normally distributed. The range of Y is zero to infinity, whereas that for X is minus to plus infinity. If $Y = 1$, $X = 0$, and if $Y > 1$, X is positive. In the range of $0 \le Y \le 1$, X is in the range minus infinity to zero, since the logarithm of a number between zero and unity is negative. Y cannot have negative values, since the logarithm of a negative number is not defined.

Common parameters In Eq. (3.3.24), for the PDF of Y, the parameters used, σ_X and m_X, are the mean and standard deviation of X or $\ln Y$, *not* of Y.   A somewhat more natural pair of parameters involving the median of Y is available.   The *median* $\breve{m}$ of a random variable was defined in Sec. 2.3.1 as that value below which one-half of the probability mass lies.   That is,

$$F(\breve{m}) = 0.5$$

When X equals $\ln Y$,

$$0.5 = P[Y \leq \breve{m}_Y] = P[X \leq \ln \breve{m}_Y] = F_X(\ln \breve{m}_Y)$$

Thus,

$$\ln \breve{m}_Y = \breve{m}_X$$

For the normal and other symmetrical distributions, the median equals the mean: $\breve{m}_X = m_X$.   Consequently,

$$0.5 = F_X(\breve{m}_X) = F_X(m_X)$$

Thus,

$$\ln \breve{m}_Y = m_X$$

Substituting into Eq. (3.3.24) for $\ln y - m_X$,

$$\ln y - m_X = \ln y - \ln \breve{m}_Y = \ln \frac{y}{\breve{m}_Y}$$

And, writing σ_X as $\sigma_{\ln Y}$, we obtain the most common form of the lognormal PDF:

$$f_Y(y) = \frac{1}{y \sqrt{2\pi}\, \sigma_{\ln Y}} \exp\left\{ -\frac{1}{2}\left[\frac{1}{\sigma_{\ln Y}} \ln\left(\frac{y}{\breve{m}_Y}\right) \right]^2 \right\}$$

$$y \geq 0 \quad (3.3.25)$$

If y has this density function, we say Y is $LN(\breve{m}_Y, \sigma_{\ln Y})$, where $\breve{m}_Y$ and $\sigma_{\ln Y}$ are the two parameters of the distribution.

Using normal tables   In terms of the PDF of a standardized $N(0,1)$ variable U,

$$f_Y(y) = \frac{1}{y\sigma_{\ln Y}} f_U\left(\frac{\ln (y/\breve{m}_Y)}{\sigma_{\ln Y}} \right) \qquad (3.3.26)$$

That is,

$$f_Y(y) = \frac{1}{y\sigma_{\ln Y}} f_U(u) \qquad (3.3.27)$$

where

$$u = \frac{1}{\sigma_{\ln Y}} \ln \frac{y}{\tilde{m}_Y} \tag{3.3.28}$$

The function $f_U(u)$ is widely tabulated.

The CDF of Y is also most easily evaluated using a table of the normal distribution, since

$$F_Y(y) = P[Y \leq y] = P[\ln Y \leq \ln y]$$
$$= P[X \leq \ln y] = F_X(\ln y)$$

Because X is $N(m_X, \sigma_X{}^2)$ or $N(\ln \tilde{m}_Y, \sigma_{\ln Y}^2)$,

$$F_Y(y) = F_U \left(\frac{\ln y - \ln \tilde{m}_Y}{\sigma_{\ln Y}} \right)$$

$$= F_U \left(\frac{\ln (y/\tilde{m}_Y)}{\sigma_{\ln Y}} \right)$$

$$= F_U(u) \tag{3.3.29}$$

with u given above in Eq. (3.3.28).

Moments   Through experimental evidence one usually has available the moments of the random variable Y. To determine the parameters of the PDF and CDF, it then becomes necessary to compute $\tilde{m}_Y$ and the standard deviation of X or $\ln Y$ from these moments. To determine the equations to do this, we seek first the moments of Y in terms of its parameters. By integration, it is easily shown that the moments of Y are

$$E[Y^r] = \int_0^\infty y^r f_Y(y) \, dy$$
$$= (\tilde{m}_Y)^r \exp \left(\tfrac{1}{2} r^2 \sigma_{\ln Y}^2 \right) \tag{3.3.30}$$

In particular,

$$m_Y = \tilde{m}_Y \exp \left(\tfrac{1}{2} \sigma_{\ln Y}^2 \right) \tag{3.3.31}$$

and

$$\sigma_Y{}^2 = E[Y^2] - m_Y{}^2 = \tilde{m}_Y{}^2 \exp \left(2\sigma_{\ln Y}^2 \right) - \tilde{m}_Y{}^2 \exp \left(\sigma_{\ln Y}^2 \right)$$
$$= m_Y{}^2 \left[\exp \left(\sigma_{\ln Y}^2 \right) - 1 \right] \tag{3.3.32}$$

Thus

$$V_Y{}^2 = \exp \left(\sigma_{\ln Y}^2 \right) - 1 \tag{3.3.33}$$

It follows that the desired relationships for the parameters in terms of the moments are

$$\tilde{m}_Y = m_Y \exp \left(-\tfrac{1}{2} \sigma_{\ln Y}^2 \right) \tag{3.3.34}$$

$$\sigma_{\ln Y}^2 = \ln \left(V_Y{}^2 + 1 \right) \tag{3.3.35}$$

It is also important to note that the mean of $\ln Y$ or X is

$$m_{\ln Y} = \ln \check{m}_Y = \ln [m_Y \exp (-\tfrac{1}{2}\sigma_{\ln Y}^2)]$$
$$= \ln m_Y - \tfrac{1}{2}\sigma_{\ln Y}^2 \qquad (3.3.36)$$

The mean of $\ln Y$ is *not* equal to $\ln m_Y$; that is, the mean of the log is *not* the log of the mean, illustrating again Eq. (2.4.29).

Illustration: Fatigue life   From a number of tests on rotating beams of A285 steel, the number of cycles Y to fatigue failure were concluded to be lognormally distributed with an estimated mean of 430,000 cycles and a standard deviation of 215,000 cycles (when tested to $\pm 36,400$ psi). Let us plot $f_Y(y)$ and $F_Y(y)$. We are given from data

$$m_Y = 430,000$$
$$\sigma_Y = 215,000$$
$$V_Y = \frac{215,000}{430,000} = 0.5$$

Using the previous results,

$$\sigma_{\ln Y}^2 = \ln (V_Y^2 + 1) = \ln (0.25 + 1) = 0.223$$
$$\sigma_{\ln Y} = 0.472$$

If V_Y is less than 0.2, one can assume that $\sigma_{\ln Y}^2 \approx V_Y^2$ with less than 2 percent error in $\sigma_{\ln Y}^2$. This follows from an expansion of $\ln (V_Y^2 + 1)$ as

$$\sigma_{\ln Y}^2 = \ln (V_Y^2 + 1) = V_Y^2 - \frac{1}{2} V_Y^4 + \frac{1}{3} V_Y^6 \cdots \approx V_Y^2 \qquad (3.3.37)$$

for $V_Y < 0.2$.

The other parameter of the lognormal distribution is

$$\check{m}_Y = m_Y \exp (-\tfrac{1}{2}\sigma_{\ln Y}^2)$$
$$= 430,000e^{-0.223/2} = 430,000e^{-0.111} = 385,000 \text{ cycles}$$

Again, if V_Y is less than 0.2, a good approximation is $\ln \check{m}_Y \approx \ln m_Y - \tfrac{1}{2}V_Y^2$. These two approximations can also be derived using the technique described in Sec. 2.4.4.

In this example, Eqs. (3.3.25) and (3.3.26) become

$$f_Y(y) = \frac{1}{0.472 \sqrt{2\pi}\, y} \exp \left[-\frac{1}{2}\left(\frac{1}{0.472} \ln \frac{y}{385,000} \right)^2 \right] \qquad y \geq 0$$

$$f_Y(y) = \frac{1}{0.472y} f_U \left(\frac{\ln (y/385,000)}{0.472} \right) \qquad y \geq 0$$

while the CDF is

$$F_Y(y) = F_U \left(\frac{\ln (y/385,000)}{0.472} \right) \qquad y \geq 0$$

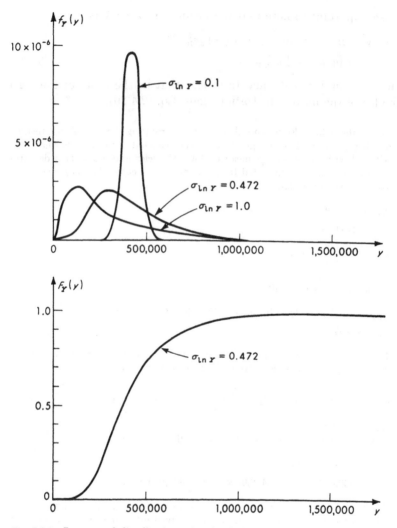

Fig. 3.3.3 Lognormal distributions showing influence of $\sigma_{\ln y}$.

To plot these functions, one might pick y at a number of selected locations and evaluate. It is somewhat easier, however, to select values of the standardized normal variable u, locate $f_U(u)$ and $F_U(u)$ in the tables, calculate the corresponding values of y, and finally compute $f_Y(y)$. The functions $f_Y(y)$ and $F_Y(y)$ are shown graphed in Fig. 3.3.3. Also shown in Fig. 3.3.3 are plots of lognormal distributions with the same means but with coefficients of variation equal to 1.32 and 0.1. In these cases the parameters of the distributions are $\sigma_{\ln Y}$, equal to 1.0 and 0.1, and $\breve{m}_Y$, equal to 260,000 and 429,800, respectively.

Skewness and small V approximations Compared to a normal distribution, the most salient characteristic of the lognormal is its skewed
shape. Through Eq. (3.3.30) the third central moment and thence the
skewness coefficient of the lognormal distribution can be computed to be

$$\gamma_1 = \frac{E[(Y - m_Y)^3]}{\sigma_Y{}^3} = 3V_Y + V_Y{}^3 \tag{3.3.38}$$

Through Eqs. (3.3.38) and (3.3.35) it is clear that the coefficient
$\sigma^2_{\ln Y}$ and the skewness of the distribution depend only on the coefficient
of variation of Y. The ratio of the mean to the median [Eq. (3.3.31)]
also depends only on this factor. Furthermore, for small values of V
(less than about 0.2), $\sigma_{\ln Y}$ and γ_1 are very nearly linear in V [Eqs.
(3.3.37) and (3.3.38)], and the mean of the logarithm of Y is approximately equal to the logarithm of the mean of Y. The distribution of Y
is approximately normal for small V (Sec. 2.4.4).

Distribution of a product of lognormals The lognormal distribution does
not have the additive regenerative property; that is, the distribution
of a random variable which is the sum of two lognormally distributed
random variables is no longer lognormal. But the distribution is rare
in possessing a multiplicative type of regeneration ability. If

$$Z = Y_1 Y_2 \cdots Y_n \tag{3.3.39}$$

and the Y_i are independent and all lognormally distributed with parameters $\tilde{m}_{Y_i}$ and $\sigma_{\ln Y_i}$, then Z is also lognormally distributed with

$$m_{\ln Z} = \sum_{i=1}^{n} m_{\ln Y_i} \tag{3.3.40}$$

and

$$\sigma^2_{\ln Z} = \sum_{i=1}^{n} \sigma^2_{\ln Y_i} \tag{3.3.41}$$

By Eq. (3.3.36),

$$m_{\ln Y_i} = \ln \tilde{m}_{Y_i}$$

Hence

$$m_{\ln Z} = \ln (\tilde{m}_{Y_1} \tilde{m}_{Y_2} \cdots \tilde{m}_{Y_n})$$

and

$$\tilde{m}_Z = \tilde{m}_{Y_1} \tilde{m}_{Y_2} \cdots \tilde{m}_{Y_n} \tag{3.3.42}$$

That Z is lognormally distributed follows [after logarithms of both sides of Eq. (3.3.39) have been taken] from the fact established in Sec. 3.3.1 that the distribution of the sum of independent normally distributed random variables is normally distributed. What is more, this argument also implies that under the same conditions on the Y_i the random variable W:

$$W = Y_1{}^{a_1} Y_2{}^{a_2} \cdots Y_n{}^{a_n} \tag{3.3.43}$$

is also lognormally distributed with easily determined constants.

Applications The lognormal probability law has a long history in civil engineering. It was adopted early in the statistical studies of hydrological data (Hazen [1914]) and of fatigue failures. It seems to have been adopted originally only because the observed data were found to be skewed, and better fit was obtained using this simple transformation of the familiar normal distribution. This skewed quality,† not uncommon in many kinds of data, plus the fact that the distribution avoids the nonzero probability (however small) of negative values associated with the normal model, have combined to make this distribution remain one commonly used in civil engineering practice. It is particularly frequently encountered in hydrology studies (for example, Chow [1954], Beard [1953], and Beard [1962]) to model daily stream flow, flood peak discharges, annual floods, and annual, monthly, and daily rainfall. Chow [1954] argues that the hydrological event is the result of the joint action of many hydrological and geographical factors which can be expressed in the mathematical form

$$Y = W_1 W_2 \cdots W_n$$

where n is large. This form is typical of those which we have seen to lead to the lognormal distribution.

Lomnitz [1964] has used the multiplicative model to describe the distribution of earthquake magnitudes. He found the lognormal distribution to fit both the magnitudes and the interarrival times between earthquakes.

The distribution has also been found to describe the strength of elementary volumes of plastic materials (Johnson [1953]), the distribution of small particle sizes (Kottler [1950]), and the yield stress in steel reinforcing bars (Fruedenthal [1948]). A thorough treatment of the distribution is available in Aithchison and Brown [1957].

---

† The lognormal distribution can be shifted and the skew reversed from right-hand to left-hand skew by simple linear transformations, Sec. 3.5.1.

3.3.3 The Model of Extremes: The Extreme Value Distributions

In civil engineering applications concern often lies with the largest or smallest of a number of random variables. Success or failure of a system may rest solely on its ability to function under the maximum demand or load to which it is subjected, not simply the typical values. Floods, winds, and floor loadings are all variables whose largest value in a sequence may be critical to a civil engineering system. The capacity, too, of a system may depend only on extremes, for example, on the strength of the weakest of many elementary components.

If the key variable Y is the maximum[†] of n random variables $X_1, X_2, \ldots, X_n$, then the probability

$$F_Y(y) = P[Y \leq y]$$
$$= P[\text{all } n \text{ of the } X_i \leq y]$$

If the X_i are independent,

$$F_Y(y) = P[X_1 \leq y]P[X_2 \leq y] \cdots P[X_n \leq y]$$
$$= F_{X_1}(y)F_{X_2}(y) \cdots F_{X_n}(y) \qquad (3.3.44)$$

In the special case where all the X_i are identically distributed with CDF $F_X(x)$:

$$F_Y(y) = [F_X(y)]^n \qquad (3.3.45)$$

If the X_i in the last case are continuous random variables[‡] with common density function $f_X(x)$,

$$f_Y(y) = \frac{d}{dy} F_Y(y) = n[F_X(y)]^{n-1}f_X(y) \qquad (3.3.46)$$

Knowing from past experience the distribution of X_i, the flood in any, say the ith, year, one might need the distribution of Y, the largest flood in 50 years, this being the design lifetime of a proposed water resource system. Equations (3.3.45) and (3.3.46) define the distribution of Y if the X_i can be assumed to be (1) mutually independent and (2) identically distributed. Related to these two assumptions are questions of long (multiyear) weather cycles, meteorological trends, changes (natural or man-made) in the drainage basin, etc. It may be possible to estimate the magnitude of these effects, if they exist, and, after alteration of the distributions of the X_i, make use of the more general form Eq. (3.3.44).

---

† In Prob. 3.46 we consider the distribution of $Y^{(i)}$ the ith largest of n independent random variables. In this section our concern is only with the maximum or minimum, that is, $i = 1$ or $i = n$.

‡ The reader is asked in Prob. 3.29 to show the analogous result for discrete random variables.

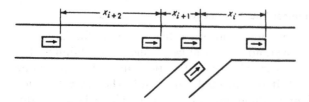

Fig. 3.3.4 Merging-traffic illustration. X values are times between vehicles.

Illustration: Merging lengths An engineer is studying how drivers merge into freely flowing traffic (Fig. 3.3.4). A certain class of drivers will merge only if the time X_i between passing cars is at least y sec. (The value of y will depend upon the class of drivers being studied.) If he assumes that the traffic is following a Poisson arrival process, then the X_i's are independent, exponentially distributed random variables with parameter λ (Sec. 3.2). What is the probability that after n cars have passed, a driver will not have been able to merge?

Consider the maximum Y of the n times between cars, $X_1, X_2, \ldots, X_n$. Then there will have been no merge if Y is less than y. For any particular value of y.

$$P[\text{no merge}] = F_Y(y) = F_X^n(y) = (1 - e^{-\lambda y})^n$$

The density function of Y, the maximum of times, is simply

$$f_Y(y) = n\lambda(1 - e^{-\lambda y})^{n-1}e^{-\lambda y} \qquad y \geq 0$$

which is compared in Fig. 3.3.5 with the distribution of X_i for $n = 2$ and 10. Note that the maximum of say 10 values is likely to be (i.e., its mode is) in the right-hand tail of the distribution of X_i.

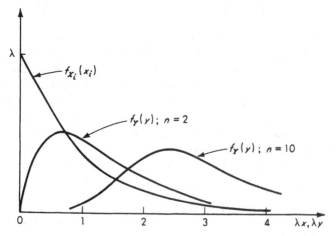

Fig. 3.3.5 Comparison of PDF of interarrival times X_i with that of maximum Y of n times.

Asymptotic distributions   If the conditions of independence and common distribution hold among the X_i, then in a number of cases of great practical importance the shape of the distribution of Y is relatively insensitive to the exact shape of the distribution of the X_i.   In these cases, limiting forms (as n grows) of the distribution of Y can be found, which can be expected to describe the behavior of that random variable even when the exact shape of the distribution of the X_i is not known precisely.   In this sense this situation is not unlike those already encountered in the previous two sections on the normal and lognormal distributions.   When dealing with extreme values, however, no single limiting distribution exists.   The limiting distribution depends, obviously, on whether largest or smallest values are of interest, and also on the general way in which the appropriate tail of the underlying distribution, that of the X_i, behaves. Three specific cases, which have been studied in some detail (Gumbel [1958]) and applied widely, will be mentioned here.

Type I: Distribution of largest value   The so-called Type I limiting distribution arises under the following circumstances.   We wish to know the limiting distribution of the *largest* of n values of X_i, as n gets large. Suppose that it is known only that the distribution of the x_i is unlimited in the positive direction and that the upper tail falls off "in an exponential manner"; that is, suppose that, in the upper tail, at least, the common CDF of the X_i may be written in the form

$$F_X(x) = 1 - e^{-g(x)} \qquad\qquad (3.3.47)$$

with $g(x)$ an increasing function of x.   For example, if $g(x) = \lambda x$, this is the familiar negative exponential distribution (Sec. 3.2.2).   The normal and gamma distributions are also of this general type.

We have the following conclusion† for the distribution Y, the largest of *many* independent random variables with a common exponential type of upper-tail distribution:

$$F_Y(y) = \exp\left[-e^{-\alpha(y-u)}\right] \qquad -\infty \leq y \leq \infty \qquad (3.3.48)$$

and

$$f_Y(y) = \alpha \exp\left[-\alpha(y - u) - e^{-\alpha(y-u)}\right] \qquad\qquad (3.3.49)$$

The parameters α and u would be estimated from observed data in such cases.   The result is asymptotic, being approximately true for any large value of n.   For example, the engineer may argue that the annual maximum gust velocity has such an extreme-value distribution because it represents the largest of some unknown, but large number of gusts

† See Appendix B for a simplified derivation of this widely quoted result.

in a year (Court [1953]).   A random variable with this, *the Type I asymp-*
totic extreme value distribution of largest values, will be said to be $EX_{I,L}$ (u,α)
distributed.   It is also called the *Gumbel distribution*, or simply the
extreme-value distribution, when there is no opportunity for confusion
with other extreme-value distributions.

Setting the derivative of $f_Y(y)$ equal to zero will reveal that u is
in fact the mode of the distribution.   The parameter α is a measure of
dispersion.   Integration leads to the following expressions for the mean,
variance, standard deviation, and third central moment of Y:

$$m_Y = u + \frac{\gamma}{\alpha} \approx u + \frac{0.577}{\alpha} \tag{3.3.50}$$

(γ is "Euler's constant.")   Also

$$\sigma_Y{}^2 = \frac{\pi^2}{6\alpha^2} \approx \frac{1.645}{\alpha^2} \tag{3.3.51}$$

$$\sigma_Y = \frac{\pi}{\sqrt{6}\,\alpha} \approx \frac{1.282}{\alpha} \tag{3.3.52}$$

The skewness coefficient is

$$\gamma_1 = 1.1396 \tag{3.3.53}$$

The distribution has a positive skewness coefficient which is independent
of the parameters α and u.

The distribution is sketched in Fig. 3.3.6 in terms of a *reduced*

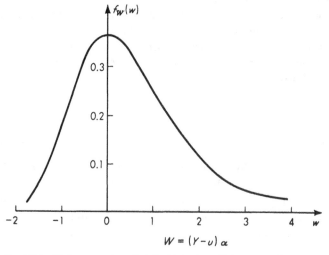

Fig. 3.3.6   Extreme-value distribution: type I, largest value.

variate W,†

$$W = (Y - u)\alpha \tag{3.3.54}$$

This variable has $u = 0$ and $\alpha = 1$.

$$F_W(w) = e^{-e^{-w}} \tag{3.3.55}$$

Table A.5 gives the distribution tabulated in terms of this reduced variate.

The tables are used as those of the normal distribution are, since clearly

$$f_Y(y) = \alpha f_W((y - u)\alpha) \tag{3.3.56}$$

$$F_Y(y) = F_W((y - u)\alpha) \tag{3.3.57}$$

The use of the tables will be illustrated shortly.

Type I: Smallest value By a symmetrical argument one can find the distribution of the smallest of *many* independent variables with a common unlimited distribution with an exponential-like *lower* tail. Let Z be this minimum value; then the asymptotic (large n) distribution of Z is

$$F_Z(z) = 1 - \exp\left(-e^{\alpha(z-u)}\right) \qquad -\infty \le z \le \infty \tag{3.3.58}$$

$$f_Z(z) = \alpha \exp\left[\alpha(z - u) - e^{\alpha(z-u)}\right] \qquad -\infty \le z \le \infty \tag{3.3.59}$$

with

$$m_Z = u - \frac{\gamma}{\alpha} \tag{3.3.60}$$

$$\sigma_Z = \frac{\pi}{\sqrt{6}\,\alpha} \tag{3.3.61}$$

$$\gamma_1 = -1.1396 \tag{3.3.62}$$

The table for the distribution of largest values (Table A.5) may also be used for smallest values owing to the antisymmetry of the pair of distributions. In terms of the tabulated, reduced variate W [Eq. (3.3.54)]

$$
\begin{aligned}
F_Z(z) &= P[Z \le z] \\
&= P[W \ge -(z - u)\alpha] = 1 - F_W(-(z - u)\alpha) \\
&\qquad\qquad\qquad\qquad -\infty \le z \le \infty
\end{aligned} \tag{3.3.63}
$$

and

$$f_Z(z) = \alpha f_W(-(z - u)\alpha) \qquad -\infty \le z \le \infty \tag{3.3.64}$$

† This reduced variate is clearly analogous to the standardized variate

$$U = \frac{(X - m_X)}{\sigma_X}$$

used in tabulating and plotting the normal distribution.

Applications The Type I asymptotic extreme-value distribution has been used by a number of investigators to describe the strength of brittle materials (e.g., Johnson [1953]). The basic argument here is that a specimen of brittle material fails when the weakest of one of many microscopic, "elementary" volumes fails;[†] hence the distribution of smallest values is appropriate. The model is also commonly adopted to describe hydrological phenomena such as the maximum daily flow in a year or the annual peak hourly discharge during a flood (e.g., Chow [1952]), the reasoning being that the values are the maximum of many (365 in the first example) random variables. One might justly question whether the daily flows are independent or identically distributed, and certainly the daily flows cannot take on negative values. Nonetheless, the Type I distribution has proved useful to describe observed phenomena and seems to have some physical justification. Numerous other applications are referenced and discussed in Gumbel [1954a, 1958]. In Chap. 4 a special graph paper will be introduced that facilitates applications of this model.

As with the lognormal and normal models, having hypothesized the functional form or shape of the distribution, one would proceed to estimate parameters α and u (for example, from estimates of m and σ). With these in hand one can make probability statements of the kind, "A flood of magnitude y will be exceeded with probability p in any given year," or, "The strength which will, with 98 percent reliability, be exceeded is z."

Illustration: Peak annual flow As an example, assume that an engineer has good estimates of the mean and variance of the peak annual flow in a small stream and that he suspects that the distribution is of the extreme-value type because the annual peak flow is the largest of many daily peak flows. He desires to plot the PDF of the assumed model in order to compare it with a histogram of the observed data. The mean and variance are given[‡]:

$m_Y = 100$ cfs

$\sigma_Y{}^2 = 2500$

$\sigma_Y = 50$ cfs

[†] Of particular interest in such studies is the effect of the size of the specimen on the parameters (for example, u and α) of the distribution of breaking stress. The theory provides a hypothesis regarding this relationship: larger specimens have more elementary volumes and hence a larger n. The mode and mean of the smallest-value Type I distribution decrease with n, predicting the often observed size-effect phenomenon, i.e., larger specimens of brittle materials failing at lower average stresses than smaller specimens. See Appendix B and Prob. 3.28.

[‡] These estimates might be based on 50 observations (50 years) of the annual peak flows. Data on daily peak flows are not necessary.

Solving for the parameters α and u,

$$\alpha = \frac{1.282}{\sigma_Y} = \frac{1.282}{50} = 0.0256$$

$$\frac{1}{\alpha} = 39.1 \text{ cfs}$$

$$u = m_Y - \frac{0.577}{\alpha}$$

$$u = 100 - (0.577)(39.1) = 77.5 \text{ cfs}$$

Now the engineer can compute, for example, the probability that the peak flow in a particular year will exceed, say, 200 cfs.

$$P[Y \geq 200] = 1 - F_Y(200)$$
$$= 1 - \exp\left[-e^{-0.0256(200-77.5)}\right] = 0.043$$

In order to plot the distribution of Y, one should read from the table (for various values of the reduced variate w) the values of $f_W(w)$ and $F_W(w)$. The corresponding values of y and $f_Y(y)$ are easily found. The reduced variate is

$$W = (Y - u)\alpha$$

Hence, a value w of W corresponds to a value of Y equal to

$$y = u + \frac{w}{\alpha}$$

From Table A.5, we find:

w	$f_W(w)$	$F_W(w) = F_Y(y)$	$y = u + w/\alpha$	$f_Y(y) = \alpha f_W(w)$
-2	0.0046	0.0006	-0.7	0.0001
-1	0.1794	0.0660	38.40	0.0046
-0.5	0.3170	0.1923	57.95	0.0081
$+1$	0.2546	0.6922	116.60	0.0065
$+1.5$	0.1785	0.8000	136.15	0.0048
$+2$	0.1182	0.8734	155.70	0.0030
$+2.5$	0.0756	0.9212	175.25	0.0019
$+3.0$	0.0474	0.9514	194.80	0.0012

The PDF of Y is plotted in Fig. 3.3.7a.

Illustration: Minimum annual flows To illustrate the Type I distribution of smallest values, consider a distribution with the same mean and variance as used above for largest values. The distribution might now be the model of the minimum annual flow in a large stream. With

$$m_Z = 100 \text{ cfs}$$

$$\sigma_Z = 50 \text{ cfs}$$

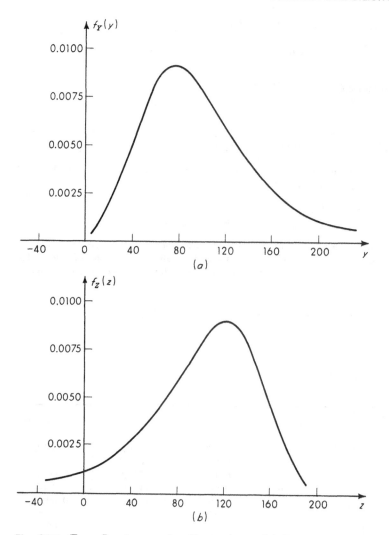

Fig. 3.3.7 Type I extreme-value illustrations. (a) Largest values, $m = 100$, $\sigma = 50$; (b) smallest values, $m = 100$, $\sigma = 50$.

the parameters are

$$\alpha = \frac{1.282}{\sigma_Z} = \frac{1.282}{50} = 0.0256$$

$$\frac{1}{\alpha} = 39.1 \text{ cfs}$$

$$u = m_Z + \frac{0.577}{\alpha} = 100 + (0.577)(39.1) = 122.5 \text{ cfs}$$

To plot the PDF of Z, values of the PDF of W at a number of values of its argument, $w = 1.0$, for example, are tabulated and the corresponding value of z found. For $w = 1.0$, $f_W(w) = 0.2546$, and solving $w = -\alpha(z - u) = 1.0$, we find the corresponding z equal to $u - 1.0/\alpha = 122.5 - 39.1 = 83.4$. At $z = 83.4$, $f_Z(83.4) = \alpha f_W(1.0) = (0.0256)(0.2546) = 0.0065$. The entire PDF is plotted in Fig. 3.3.7b. The indicated high probability of negative values suggests that the model is not a good choice if the physical variable Z is necessarily positive, as, say, a stream flow must be.

Type II distribution   The Type II extreme-value distribution also arises as the limiting distribution of the largest value of many independent identically distributed random variables. In this case each of the underlying variables has a distribution limited on the left at zero, but unlimited to the right in the tail of interest. This tail falls off such that the CDF of the X_i is of the form

$$F_X(x) = 1 - \beta \left(\frac{1}{x}\right)^k \qquad x \geq 0 \tag{3.3.65}$$

The asymptotic distribution of Y, the largest of many X_i, is of the form

$$F_Y(y) = e^{-(u/y)^k} \qquad y \geq 0 \tag{3.3.66}$$

$$f_Y(y) = \frac{k}{u}\left(\frac{u}{y}\right)^{k+1} e^{-(u/y)^k} \qquad y \geq 0 \tag{3.3.67}$$

We say Y is $\mathrm{EX}_{II,L}(u,k)$.

Interestingly, the moments of order l of Y do not exist for $l \geq k$, but if $l < k$, integration yields

$$E[Y^l] = u^l \Gamma\left(1 - \frac{l}{k}\right) \tag{3.3.68}$$

Consequently

$$m_Y = u\Gamma\left(1 - \frac{1}{k}\right) \qquad k > 1 \tag{3.3.69}$$

$$\sigma_Y{}^2 = u^2 \left[\Gamma\left(1 - \frac{2}{k}\right) - \Gamma^2\left(1 - \frac{1}{k}\right)\right] \qquad k > 2 \tag{3.3.70}$$

Thus

$$V_Y{}^2 = \frac{\Gamma(1 - 2/k)}{\Gamma^2(1 - 1/k)} - 1 \qquad k > 2 \tag{3.3.71}$$

This function of k only is plotted in Fig. 3.3.8, permitting one to find the parameter k if V is known from data.

The relationship between this distribution and the Type I distribution is identical to that between the lognormal and normal distributions. Using the technique of derived distributions, we can show that if Y has

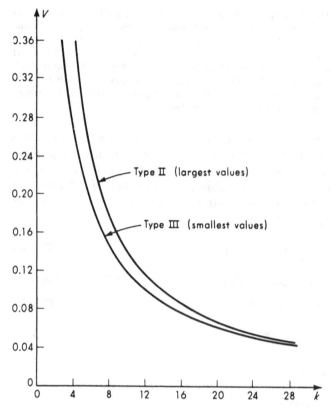

Fig. 3.3.8 Extreme-value distributions. Coefficient of variation versus parameter k.

Type II distribution with parameters u and k, then $Z = \ln Y$ has the Type I distribution [Eqs. (3.3.48) and (3.3.49)] with parameters $u^0 = \ln u$ and $\alpha = k$. This fact can be utilized so that Table A.5 (and the special graph paper to be introduced in Chap. 4) can be used for Type II distributions as well as for Type I distributions. The relationship implies that

$$F_Y(y) = F_Z(\ln y) \tag{3.3.72}$$

$$f_Y(y) = \frac{1}{y} f_Z(\ln y) \tag{3.3.73}$$

where Y is $\mathrm{EX}_{\mathrm{II},L}(u,k)$ and Z is $\mathrm{EX}_{\mathrm{I},L}(\ln u,k)$. Thus in terms of the tabulated variable W, which is $\mathrm{EX}_{\mathrm{I},L}(0,1)$,

$$F_Y(y) = F_W((\ln y - \ln u)k) \tag{3.3.74}$$

$$f_Y(y) = \frac{k}{y} f_W((\ln y - \ln u)k) \tag{3.3.75}$$

A corresponding distribution of smallest values can also be found, but it lacks common practical interest.

This Type II distribution of largest values has been used (following Thom [1960]) to represent annual maximum winds, owing to the physically meaningful limitation of the X_i to positive values and owing to the interpretation of the annual maximum wind velocity as being the largest of daily values, or perhaps the largest of many storms' maxima, which are in turn the maxima of many gusts' velocities. The distribution has also been employed to model meteorological and hydrological phenomena. (See references in Gumbel [1958].)

Illustration: Maximum annual wind velocity   In Boston, Massachusetts, the measured data suggest that the mean and standard deviation of the maximum annual wind velocity are about

$$m_Y = 55 \text{ mph}$$

$$\sigma_Y = 12.8 \text{ mph}$$

As mentioned, such velocities are suspected to follow the Type II extreme value distribution (although apparently no studies have been made to verify that the underlying distributions have the required shape in their upper tails).

To find the parameters, we find first the coefficient of variation:

$$V = \frac{12.8}{55} = 0.23$$

And from Fig. 3.3.8,

$$k = 6.5$$

while, from Eq. (3.3.69),

$$u = \frac{55}{\Gamma(1 - 1/6.5)}$$

$$= \frac{55}{\Gamma(0.846)}$$

Standard mathematical tables yield $\Gamma(0.846) = 1.12$.   Therefore

$$u = \frac{55}{1.12} = 49.4 \text{ mph}$$

With the parameters known, probabilities can be calculated. For example, it has been recommended[†] that most structures be designed for the 50-year wind (see the illustration in Sec. 3.1). The velocity y, which will be exceeded with probability $1/50 = 0.02$ in any year, is found by setting the complementary CDF of Y equal to 0.02:

$$1 - F_Y(y) = 0.02$$

$$e^{-(49.4/y)^{6.5}} = 0.98$$

† By the ASCE Committee on Wind Forces [1961].

Solving for y,

$y = 91$ mph

Alternatively, using Table A.5, we find that

$F_W(4.0) = 0.98$

Using Eq. (3.3.74), this implies that

$(\ln y - \ln 49.4)6.5 = 4.0$

or

$y = 91$ mph

As indicated, this table can also be used to calculate values of $f_Y(y)$. For example,

$$f_Y(60) = \frac{6.5}{60} f_W((\ln 60 - \ln 49.4)6.5)$$

$$= \frac{6.5}{60} f_W (1.35)$$

$$= \frac{6.5}{60} (0.77) = 0.084$$

With several values, the PDF of Y can be sketched as in Fig. 3.3.9.

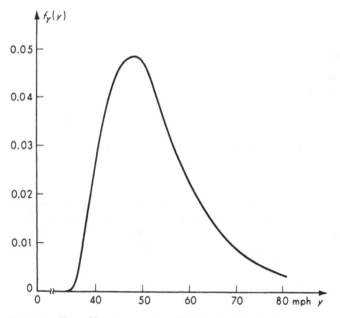

Fig. 3.3.9 Type II extreme-value-distribution illustration.

Type III distribution   The final asymptotic extreme-value distribution of interest is the Type III distribution, which arises from underlying distributions *limited in the tail of interest*. For example, if largest values are of interest and if the density function of the X_i falls off to some maximum value of w in a manner such that the following form of the CDF holds† near w

$$F_X(x) = 1 - c(w - x)^k \qquad x \le w, \, k > 0 \tag{3.3.76}$$

then the distribution of Y, the largest of many X_i, is of the form:

$$F_Y(y) = \exp\left[-\left(\frac{w - y}{w - u}\right)^k\right] \qquad y \le w \tag{3.3.77}$$

$$f_Y(y) = \frac{k}{w - u}\left(\frac{w - y}{w - u}\right)^{k-1} \exp\left[-\left(\frac{w - y}{w - u}\right)^k\right]$$

$$y \le w \tag{3.3.78}$$

Since most useful applications of this model deal with smallest values, we shall restrict the remainder of the discussion to this case. We assume that the left-hand tail of the PDF of the X_i rises from zero for values of $x \ge \epsilon$ such that the CDF has the following form near $x = \epsilon$:

$$F_X(x) = c(x - \epsilon)^k \qquad x \ge \epsilon \tag{3.3.79}$$

where ϵ is a lower limit of x. For example, the gamma distribution is known to be of this form (with $\epsilon = 0$). If the X_i are independent and identically distributed, the distribution of Z, the smallest or minimum of many values, is

$$F_Z(z) = 1 - \exp\left[-\left(\frac{z - \epsilon}{u - \epsilon}\right)^k\right] \qquad z \ge \epsilon \tag{3.3.80}$$

$$f_Z(z) = \frac{k}{u - \epsilon}\left(\frac{z - \epsilon}{u - \epsilon}\right)^{k-1} \exp\left[-\left(\frac{z - \epsilon}{u - \epsilon}\right)^k\right] \qquad z \ge \epsilon \tag{3.3.81}$$

The moments are:

$$m_Z = \epsilon + (u - \epsilon)\Gamma\left(1 + \frac{1}{k}\right) \tag{3.3.82}$$

$$\sigma_Z^2 = (u - \epsilon)^2\left[\Gamma\left(1 + \frac{2}{k}\right) - \Gamma^2\left(1 + \frac{1}{k}\right)\right] \tag{3.3.83}$$

$$E[(Z - \epsilon)^l] = (u - \epsilon)^l\Gamma\left(1 + \frac{l}{k}\right) \tag{3.3.84}$$

† Examples include the uniform distribution where $k = 1$, or a triangular PDF, where $k = 2$.

In many cases the primary purpose of experimental investigations may be to determine the lower limit ϵ, a parameter of this distribution. For example, Weibull [1939] has studied material strength in tension and fatigue, ϵ representing the lower bounds on strength and on number of cycles before which no failures occur, respectively. Freudenthal ([1951], [1956]) has given physical reasons why this model might hold in both cases. (Weibull's justifications for its adoption were strictly empirical.) The distribution, with somewhat modified forms of the parameters, is often referred to as the *Weibull distribution* and is commonly employed in statistical reliability studies of the lifetimes of components and systems (see, for example, Lloyd and Lipow [1962].)

In other cases, it may be reasonable to set $\epsilon = 0$ simply from professional knowledge of the phenomenon of interest. Droughts have been studied in this manner (Gumbel [1954b]). If ϵ equals zero, the form of the distribution simplifies greatly, and we have

$$F_Z(z) = 1 - e^{-(z/u)^k} \qquad z \geq 0 \tag{3.3.85}$$

with

$$m_Z = u\Gamma\left(1 + \frac{1}{k}\right) \tag{3.3.86}$$

$$\sigma_Z^2 = u^2\left[\Gamma\left(1 + \frac{2}{k}\right) - \Gamma^2\left(1 + \frac{1}{k}\right)\right] \tag{3.3.87}$$

$$V_Z^2 = \frac{\Gamma(1 + 2/k)}{\Gamma^2(1 + 1/k)} - 1 \tag{3.3.88}$$

This last relationship is also graphed in Fig. 3.3.8.

In any case, a logarithmic transformation again permits the use of the tables of Type I. For, if Z is distributed following the Type III distribution of smallest values with parameters ϵ, u, and k [denoted $EX_{III,S}(\epsilon,u,k)$], then $X = \ln(Z - \epsilon)$ has the Type I distribution of smallest values [Eqs (3.3.58) and (3.3.59)] with parameters $u_0 = \ln(u - \epsilon)$ and $\alpha = k$; that is, X is $EX_{I,S}[\ln(u - \epsilon),k]$. Thus

$$F_Z(z) = F_X(\ln(z - \epsilon))$$

$$= 1 - F_W(-k[\ln(z - \epsilon) - \ln(u - \epsilon)]) \qquad z \geq \epsilon \tag{3.3.89}$$

where W is the tabulated (Table A.5) $EX_{I,L}(0,1)$ variable. Consequently,

$$f_Z(z) = \frac{1}{z - \epsilon} f_X(\ln(z - \epsilon))$$

$$= \frac{k}{z - \epsilon} f_W\left(-k \ln \frac{z - \epsilon}{u - \epsilon}\right) \qquad z > \epsilon \tag{3.3.90}$$

The illustration of the Type II distribution shows the steps in treating numerical examples.

General comments   Some general comments concerning all three limiting extreme-value distributions are in order.† The results for the asymptotic distributions of extremes are not so powerful as those following from the central limit theorem for additive mechanisms. As we have seen, more assumptions must be made as to the general form of the underlying distributions. It may be difficult to determine which of the three forms [Eqs. (3.3.47) (3.3.65), or (3.3.76)] applies. Nor are these three forms exhaustive; distributions exist whose largest-value distribution does not converge to one of the three types given here. The effects of lack of independence and lack of identical distributions in the underlying variables are not so well understood as for the central limit theorem, but analogous conclusions (see Sec. 3.3.1) are known to hold (Juncosa [1948] and Watson [1954]). The convergence of the exact distribution [Eq. (3.3.44)] toward the asymptotic distribution may not be "fast" (Gumbel [1958]). Nonetheless, these distributions provide a valuable link between observed data of extreme events and mathematical models which aid the engineer in evaluating past results and predicting future outcomes.

3.3.4 Summary

In this section three general families of models have been studied. They share the common feature that they all arise when one is interested in a variable that is a consequence of many other factors whose individual behaviors are not well understood.

The normal or gaussian probability law

$$f_X(x) = \frac{1}{\sigma \sqrt{2\pi}} \exp\left[-\frac{1}{2}\left(\frac{x-m}{\sigma}\right)^2 \right]$$

is the distribution of the sum of many random factors.

The lognormal distribution

$$f_Y(y) = \frac{1}{y\sigma_{\ln Y} \sqrt{2\pi}} \exp\left\{ -\frac{1}{2}\left[\frac{\ln (y/\check{m})}{\sigma_{\ln Y}}\right]^2 \right\}$$

---

† As is true of virtually all the distributions studied in this chapter, other models may lead to distributions of the same or nearly the same analytical form. For example, if one asks what is the distribution of Y, the largest of a random number N of random variables, $X_1, X_2, \ldots, X_N$, when the X_i are exponentially distributed and N is Poisson-distributed, the distribution looks much like the Type I extreme value distribution (even if N is expected to be quite small). Examples will be found in the problems. This and a similar argument leading to the Type II distribution arise in the study of earthquake forces on structures (Cornell [1968]).

describes the probabilistic behavior of the product of many random factors.

The extreme-value distributions are the distributions of the largest (or smallest) of many random variables. The particular form depends on the general analytical form of the corresponding tail of the distribution of these underlying variables.

3.4 ADDITIONAL COMMON DISTRIBUTIONS†

In this section a number of other commonly encountered distributions are presented. Their utility stems in some cases from their ability to represent conveniently certain empirically observed characteristics of variables, such as both upper and lower limits. In other cases physical interpretations may be involved, some of which are considered here. Several of the distributions find their most frequent application in statistics, as will be discussed in Chap. 4.

3.4.1 The Equally Likely Model: The Rectangular or Uniform Distribution

If a random variable X is equally likely to take on any value in the interval 0 to 1, its density function is constant over that range

$$f_X(x) = 1 \qquad 0 \leq x \leq 1 \tag{3.4.1}$$

The shape of this density function is simply that of a rectangle, which gives it the name *rectangular* or *uniform distribution*. Its cumulative distribution function is triangular:

$$F_X(x) = \begin{cases} 0 & x < 0 \\ x & 0 \leq x \leq 1 \\ 1 & x > 1 \end{cases} \tag{3.4.2}$$

The mean and variance of the rectangular distribution are, by inspection,

$$m_X = \tfrac{1}{2} \tag{3.4.3}$$

$$\sigma_X{}^2 = \tfrac{1}{12} \tag{3.4.4}$$

Generalized slightly to an arbitrary range, say a to b, the density function becomes

$$f_X(x) = \begin{cases} \dfrac{1}{b - a} & a \leq x \leq b \\ \text{elsewhere} & 0 \end{cases} \tag{3.4.5}$$

† This section can be read very selectively at first reading.

when we denote it $R(a,b)$. The mean and variance become

$$m_X = a + \frac{b-a}{2} \qquad (3.4.6)$$

$$\sigma_X{}^2 = \frac{(b-a)^2}{12} \qquad (3.4.7)$$

The distribution is commonly applied in situations where there is no reason to give other than equal likelihoods to the possible values of a random variable.† For example, the direction from which the earthquake shock waves may approach a proposed structure might be considered, in absence of information to the contrary, to be uniformly distributed on 0 to 360°. With respect to the signal's cycle, the instant of arrival of an arbitrary vehicle at a traffic signal might be considered to be "at random" in the interval 0 to t_0, where t_0 is the duration of the signal's cycle; and, if properly generated, a "random number"‡ can be considered to be uniformly distributed on the interval 0 to 0.999 . . . .

3.4.2 The Beta Distribution

Distribution on 0 to 1 Although it can be shown to arise from the consideration of various underlying mechanisms,§ the beta distribution, $BT(r,t)$ is presented here simply as a very flexible distribution for use in describing empirical data. Like the simple rectangular distribution, it is limited in its basic form to the range $0 \le x \le 1$. For r and $t - r$ positive, the beta distribution is

$$f_X(x) = \frac{1}{B} x^{r-1}(1-x)^{t-r-1} \qquad 0 \le x \le 1 \qquad (3.4.8)$$

where the normalizing constant is

$$B = \frac{(r-1)!(t-r-1)!}{(t-1)!} \qquad (3.4.9)$$

if r and $t - r$ are integer-valued, or

$$B = \frac{\Gamma(r)\Gamma(t-r)}{\Gamma(t)} \qquad (3.4.10)$$

if r and $t - r$ are not restricted to being integer-valued.¶

---

† This is perhaps the most widely accepted interpretation of the too-common phrase "at random."
‡ The use of random numbers was discussed in Sec. 2.3.3 and Prob. 2.53.
§ See, for example, Pieruschka [1963], page 86; Gumbel [1958]; or Thomas [1948].
¶ B as a function of r and t is known as the beta function; hence the name of this distribution.

The mean of the beta distribution is

$$m_X = \frac{r}{t} \tag{3.4.11}$$

while

$$\sigma_X{}^2 = \frac{m_X(1 - m_X)}{t + 1} = \frac{r(t - r)}{t^2(t + 1)} \tag{3.4.12}$$

The skewness coefficient is

$$\gamma_1 = \frac{1}{\sigma_X{}^3} \frac{r}{t} \left[\frac{(r + 2)(r + 1)}{(t + 2)(t + 1)} - \frac{3r(r + 1)}{t(t + 1)} + \frac{2r^2}{t^2} \right] \tag{3.4.13}$$

Shapes The great value of the beta distribution lies in the wide variety of shapes it can be made to assume simply by varying the parameters. It contains as special cases the rectangular distribution ($r = 1$, $t = 2$) and triangular densities ($t = 3$ and $r = 1$ or 2). It is symmetrical about $x = 0.5$ if $r = \frac{1}{2}t$. The beta distribution is skewed right if $r < \frac{1}{2}t$, and left if $r > \frac{1}{2}t$. It is U-shaped if $r < 1$ and $t \leq 2r$; it is J-shaped if $r < 1$ and $t > r + 1$ or if $r > 1$ and $t < r + 1$. It is unimodal and bell-shaped (generally skewed) if $r > 1$ and $t > r + 1$, with the mode at $x = (r - 1)/t - 2$. Concentration about a fixed mode position increases as the values of the parameters are increased. Interchange of parameters, $r' = t - r$ and $t' = t$, yields mirror images. Many of these relationships are displayed in Fig. 3.4.1.

Tables For integer-valued t and r one can use tables of the binomial distribution to evaluate $f_X(x)$ or $F_X(x)$. As a function of y, n, and p the distribution of a binomial random variable Y is

$$p_Y(y) = \frac{n!}{y!(n - y)!} p^y(1 - p)^{n-y}$$

which, when compared to the beta distribution

$$f_X(x) = \frac{(t - 1)!}{(r - 1)!(t - r - 1)!} x^{r-1}(1 - x)^{t-r-1} \tag{3.4.14}$$

suggests why

$$f_X(x) = (t - 1)p_Y(y) \tag{3.4.15}$$

if $p_Y(y)$ is evaluated for $n = t - 2$ and $y = r - 1$ for various values of $p = x$. For example, if it is desired to find $f_X(x)$ when X is BT(1,4), then it equals $3p_Y(0)$ for Y-distributed B(2,x). In particular,

$$f_X(0.4) = 3p_Y(0) = 3 \binom{2}{0} (0.4)^0(0.6)^2 = 3(0.36) = 1.08$$

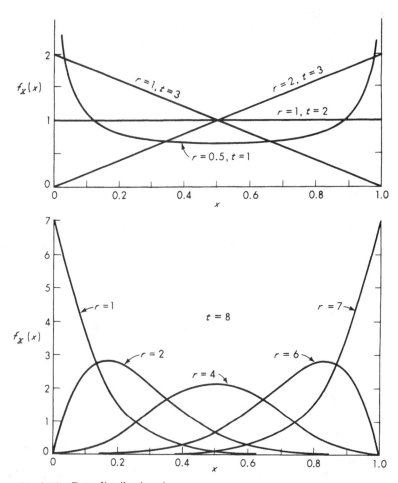

Fig. 3.4.1 Beta-distribution shapes.

in which $0.36 = P[Y = 0]$ when Y is B(2,0.4).   The equality is verified by substituting values of r, t, x, and m, p, y in the respective PDF and PMF of X and Y.

Similarly, one can verify that

$$F_X(x) = 1 - F_Y(r - 1) \tag{3.4.16}$$

where X is BT(r,t) and Y is B$(t - 2, x)$.

For noninteger r and t, interpolation is possible, or more general tables must be used (e.g., Benjamin and Nelson [1965] or Thompson [1941]).   A small table for some noninteger values of r and t is reprinted as Table A.4.

Beta distribution on a to b   Although variables of interest may be distributed on the interval 0 to 1, it is common to want to use other limits on the range of the variable, say a to b.   In this case, the beta distribution $BT(a,b,r,t)$ generalizes to

$$f_Y(y) = \frac{1}{B(b-a)^{t-1}} (y-a)^{r-1}(b-y)^{t-r-1} \qquad a \le y \le b \quad (3.4.17)$$

when

$$m_Y = a + \frac{r}{t}(b-a) \tag{3.4.18}$$

$$\sigma_Y^2 = (b-a)^2 \frac{r(t-r)}{t^2(t+1)} \tag{3.4.19}$$

It is apparent that a simple linear relationship exists between Y, a $BT(a,b,r,t)$ variable, and X, a $BT(0,1,r,t)$ [or $BT(r,t)$] variable.   In particular,

$$Y = a + (b-a)X \tag{3.4.20}$$

Consequently to find CDF and PDF of Y from the tabulated values of X, we use the familiar results for linear relationships (Sec. 2.3.1):

$$F_Y(y) = F_X\left(\frac{y-a}{b-a}\right) \tag{3.4.21}$$

$$f_Y(y) = \frac{1}{b-a} f_X\left(\frac{y-a}{b-a}\right) \tag{3.4.22}$$

Illustration: Wind direction   For planning runway directions or for studying dispersions of air pollutants from power plants, both the "prevailing" wind direction and the random variations in the direction are critical.   The wind direction Y at the Boston airport measured in degrees from north has mean and standard deviation of about

$$m_Y = 205°$$

$$\sigma_Y = 99.7°$$

These are the averages of many observations at many different hours, days, and seasons.   The direction is of course limited between 0 and 360°.   Thus, assuming that a beta distribution is appropriate,

$$a = 0$$

$$b = 360°$$

$$\frac{r}{t} 360 = m_Y = 205°$$

$$(360)^2 \frac{r(t-r)}{t^2(t+1)} = \sigma_Y^2 = 9950$$

Solving for r and t,

$$r = 1.25 \qquad t = 2.2$$

Thus Y is BT$(0,360,1.25,2.2)$, and its density function is

$$f_Y(y) = \frac{1}{360} f_X\left(\frac{y}{360}\right)$$

where X is BT$(1.25,2.2)$. The tabulated density BT$(1.2,2)$ is sketched in Fig. 3.4.2. A histogram of observed wind directions is sketched to the same scale. Although the comparison with data is not bad, the shape of this beta distribution is not satisfactory physically. The sharp discontinuity between the values of f_Y at 0 and 360°, which are both the same direction, north, is an unrealistic but perhaps not important feature of the model.

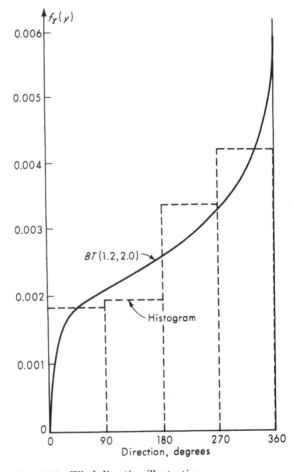

Fig. 3.4.2 Wind-direction illustration.

3.4.3   Some Normal Related Distributions: Chi-square, Chi, t, and F

A number of important distributions arise as a consequence of operations on normal distributions. They are most commonly encountered in statistical applications (Chap. 4), but their derivations are simply problems of probability theory of the type studied in Sec. 2.3.

The chi-square distribution   The most important of these normal-related distributions describes the distribution of Y_ν, the sum of the squares of ν independent normal random variables. Consider the case first of only one normal variable

$$Y_1 = X_1{}^2$$

where X_1 is $N(0,1)$. Using the techniques of Sec. 2.3,

$$F_{Y_1}(y) = P[Y_1 \le y] = P[-\sqrt{y} \le X \le +\sqrt{y}]$$
$$= F_X(\sqrt{y}) - F_X(-\sqrt{y}) = F_X(\sqrt{y}) - [1 - F_X(\sqrt{y})]$$
$$= 2F_X(\sqrt{y}) - 1 \tag{3.4.23}$$

Therefore,

$$f_{Y_1}(y) = \frac{dF_{Y_1}(y)}{dy} = 2(\tfrac{1}{2})y^{-\frac{1}{2}}f_X(\sqrt{y})$$
$$= \frac{1}{\sqrt{2\pi y}}\, e^{-y/2} \qquad y \ge 0 \tag{3.4.24}$$

The distribution of Y_ν could be found by successive convolutions (Sec. 2.3.2) of the distribution of Y_1, but it is simpler to recognize, comparing Eqs. (3.2.18) and (3.4.24), that $f_{Y_1}(y)$ is gamma-distributed with $\lambda = \tfrac{1}{2}$ and $k = \tfrac{1}{2}$ (since $\Gamma(\tfrac{1}{2}) = \sqrt{\pi}$). But recall (Sec. 3.2.3) that the sum of ν independent $G(k,\lambda)$ gamma variables (with common λ) is gamma-distributed with $k_\nu = \Sigma k_i = \nu k$; that is, the sum is $G(\nu k,\lambda)$. It is therefore clear that Y_ν, the sum of ν squared independent normal random variables, is $G(\nu/2,\tfrac{1}{2})$, with PDF [Eq. (3.2.18)]

$$f_Y(y) = \frac{\tfrac{1}{2}(y/2)^{(\nu/2)-1}e^{-y/2}}{\Gamma(\nu/2)} \qquad y \le 0 \tag{3.4.25}$$

$$m_{Y_\nu} = \nu \tag{3.4.26}$$

$$\sigma_{Y_\nu}{}^2 = 2\nu \tag{3.4.27}$$

This distribution is called the *chi-square or χ^2 distribution*,† with parameter ν or with ν "degrees of freedom," and is denoted here as $\mathrm{CH}(\nu)$. From its derivation it is clear that the distribution is regenerative;

† Or chi-squared distribution.

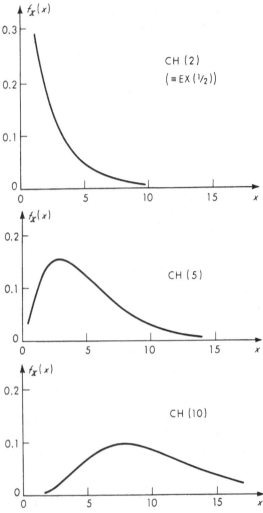

Fig. 3.4.3 χ^2 distributions. (Adapted from R. S. Burington and D. C. May, "Handbook of Probability and Statistics," Handbook Publishers, Inc., Sandusky, Ohio, 1953.)

the distribution of $Y_{\nu_1} + Y_{\nu_2}$ is $CH(\nu_1 + \nu_2)$. It is plotted in Fig. 3.4.3. These plots and the central limit theorem (Sec. 3.3.1) suggest that the χ^2 distribution can be closely approximated by a $N(\nu,2\nu)$ distribution for larger values of ν.

χ^2, **gamma, and Poisson tables**   The cumulative χ^2 distribution is found tabulated in Appendix A.2. Values of the cumulative distribution of a

χ^2 random variable can be read directly. For example, from the table, the probability that a χ^2 random variable with $\nu = 6$ degrees of freedom is less than 1.1 is

$$F(1.1) = 1 - 0.98154 = 0.01846$$

The table also provides the means to evaluate the CDF of a gamma-distributed random variable. As shown above, if Z is $G(k,\lambda)$, then $Y = 2\lambda Z$ is CH($2k$) (if $2k$ is integer). Hence the CDF of Z is

$$
\begin{aligned}
F_Z(z) &= P[Z \leq z] = P[2\lambda Z \leq 2\lambda z] \\
&= P[Y \leq 2\lambda z] \\
&= F_Y(2\lambda z)
\end{aligned}
\tag{3.4.28}
$$

where F_Y is tabulated. For example, if Z is $G(1.5,2)$, i.e., gamma with $k = 1.5$ and $\lambda = 2$, then

$$
\begin{aligned}
F_Z(1.0) &= F_Y((2)(2)(1.0)) \\
&= F_Y(4.0)
\end{aligned}
$$

where Y is CH(3), i.e., χ^2-distribution with $\nu = 2k$ degrees of freedom. Looking in the tables of the χ^2 distribution with $\nu = 3$, $F_Y(4.0)$ is found to be

$$F_Z(1.0) = F_Y(4.0) = 1 - 0.26146 = 0.73854$$

The cumulative distribution of a discrete Poisson random variable X with mean (parameter) m can also be evaluated using this table.

Thinking in terms of a Poisson process with average rate λ, X is the number of arrivals in time $t = m/\lambda$ (since $m = \lambda t$, Sec. 3.2.1). But the probability that x *or fewer* arrivals occur in time t is equal to the probability that Z, the waiting time to the $(x + 1)$th event, is greater than t:

$$F_X(x) = 1 - F_Z(t) = 1 - F_Z\left(\frac{m}{\lambda}\right)$$

But Z is gamma-distributed with parameters $k = x + 1$ and λ; therefore, from above $F_Z(t) = F_Y(2\lambda t)$ in which Y is a χ^2 random variable with $\nu = 2k = 2(x + 1)$ degrees of freedom. Finally, substituting $t = m/\lambda$,

$$F_X(x) = 1 - F_Y(2m)$$

where F_Y is tabulated. For example, if X is Poisson-distributed with mean $m = 1.3$, then the probability that $X \leq 3$ is

$$F_X(3) = 1 - F_Y((2)(1.3)) = 1 - F_Y(2.6)$$

where Y is CH(8); that is, Y is χ^2-distributed with $\nu = 2(3 + 1) = 8$.

From Table A.2 with $\nu = 8$,

$$F_X(3) = 1 - F_Y(2.6) = 0.95691$$

The chi distribution   If the χ^2 distribution represents the sum of the squares of two, say, independent $N(0,1)$ random variables, and these latter two variables are, for example, the errors in the N–S and in the E–W directions in a surveying measurement, then the distribution of the square root of this sum of squares is of interest as the distribution of the actual amplitude of the error, regardless of direction.   If Y is $CH(\nu)$, and $X = (+) \sqrt{Y}$,

$$
\begin{aligned}
F_X(x) = P[X \le x] &= P[\sqrt{Y} \le x] \\
&= P[Y \le x^2] \\
&= F_Y(x^2)
\end{aligned}
\tag{3.4.29}
$$

which indicates how one may use χ^2 tables to evaluate the CDF of X. Also,

$$
\begin{aligned}
f_X(x) = \frac{dF_X(x)}{dx} &= 2x f_Y(x^2) \\
&= \frac{x(\tfrac{1}{2}x^2)^{\nu/2-1}e^{-\frac{1}{2}x^2}}{\Gamma(\nu/2)} \qquad x \ge 0
\end{aligned}
\tag{3.4.30}
$$

This distribution is known as the *chi or χ distribution*.   Its mean is

$$m_X = \sqrt{2}\,\frac{\Gamma((\nu+1)/2)}{\Gamma(\nu/2)}
\tag{3.4.31}$$

Since

$$\sigma_X{}^2 = E[X^2] - m_X{}^2$$

and since

$$E[X^2] = E[Y]$$

it follows from Eq. (3.4.26) for the mean of Y that

$$
\begin{aligned}
\sigma_X{}^2 &= \nu - m_X{}^2 \\
&= \nu - 2\frac{\Gamma^2((\nu+1)/2)}{\Gamma^2(\nu/2)}
\end{aligned}
\tag{3.4.32}
$$

We shall denote the χ distribution by $CHI(\nu)$.

The t and F distributions   Two other distributions that are related to normal random variables should be mentioned here.   These are the *Student's t distribution* and the *F distribution*.

The former, denoted here $T(\nu)$, is the distribution of the ratio of X, a $N(0,1)$ random variable, to $Y/\sqrt{\nu}$, where Y is an independent χ-distributed random variable, $\mathrm{CHI}(\nu)$. That is, $Z = X\sqrt{\nu}/Y$ is a normal random variable divided by the square root of the sum of ν squared, independent, standardized normal random variables (the sum being normalized first by its mean.) The density of Z can be derived to be (for $\nu > 2$)

$$f_Z(z) = \frac{1}{\sqrt{\nu\pi}} \frac{\Gamma((\nu+1)/2)}{\Gamma(\nu/2)} \left(1 + \frac{z^2}{\nu}\right)^{-(\nu+1)/2}$$

$$-\infty \leq z \leq \infty \quad (3.4.33)$$

The t distribution is symmetrical about zero, having almost normal-looking shapes for large values of ν (Fig. 3.4.4). Its mean is zero and its variance is

$$\sigma_Z{}^2 = \frac{\nu}{\nu-2} \quad (3.4.34)$$

The cumulative t distribution is tabulated (Table A.3), but it is clear from Fig. 3.4.4 that for reasonably large values of ν the distribution can be approximated by a $N(0, \nu/(\nu-2))$ distribution. The t distribution is found almost solely in statistical applications (Sec. 4.1), but it has been employed to describe empirically the strength of structural assemblies by Freudenthal, Garrelts, and Shinozuka [1966].

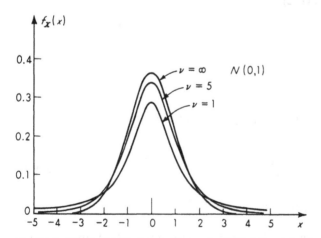

Fig. 3.4.4 The t distribution. (From A. Hald, "Statistical Theory with Engineering Applications," John Wiley and Sons, New York, 1952.)

The final normal related distribution of interest is the F *distribu-tion*. If X and Y are independent χ^2 random variables with degrees of ν_1 and ν_2, respectively, then the F distribution is the distribution of the ratio of

$$Z = \frac{X/\nu_1}{Y/\nu_2} \tag{3.4.35}$$

In words, the ratio of the sum of the squares of ν_1 $N(0,1)$ random variables to the sum of the squares of ν_2 $N(0,1)$ random variables (the sums being divided by their means) can be determined to be

$$f_Z(z) = \frac{\Gamma((\nu_1 + \nu_2)/2)z^{(\nu_1-2)/2}(1 + z)^{-(\nu_1+\nu_2)/2}}{\Gamma(\nu_1/2)\Gamma(\nu_2/2)} \qquad z \geq 0 \tag{3.4.36}$$

We shall denote the F distribution by $\mathrm{F}(\nu_1,\nu_2)$, being careful to avoid confusion with the joint CDF of a pair of random variables. Its mean is

$$m_Z = \frac{\nu_2}{\nu_2 - 2} \qquad \text{for } \nu_2 > 2 \tag{3.4.37}$$

and its variance is

$$\sigma_Z{}^2 = \frac{2\nu_2{}^2(\nu_1 + \nu_2 - 2)}{\nu_1(\nu_2 - 2)^2(\nu_2 - 4)} \qquad \text{for } \nu_2 > 4 \tag{3.4.38}$$

The F distribution is widely tabulated; Table A.6 gives only the value of z which is exceeded with probability 5 percent. The shape of the density function is similar to that of the χ^2 distribution which it approaches as ν_2 grows. This distribution also finds its primary application in statistical procedures (Chap. 4).

Illustration: Random vibrations If a simple linear 1-degree-of-freedom oscillator like that pictured in Fig. 3.4.5c is acted on by a forcing function $F(t)$, its dynamic response $X(t)$ (from rest at time zero) is well known from mechanics to be

$$X(t) = \frac{1}{\omega} \int_0^t F(u) \sin{(\omega(t - u))} \, du \tag{3.4.39}$$

where ω is the natural (circular) frequency of the oscillator and $F(t)$ is measured in units of force per unit mass. A similar form of the convolution (or Duhamel) integral holds if $F(t)$ is a ground acceleration. A number of engineers (see, for example, Rodriguez [1956]) have proposed that the dynamic response of structures to earthquakes might be represented by treating the ground accelera-tion as a sequence of evenly spaced impulses F_i, $i = 1, 2, \ldots$, of random magnitude (assumed for simplicity here to be independent and identically distributed) (Fig. 3.4.5a). Similar approaches have been taken to represent the response of towers to wind (Chiu [1964]). In such cases, the integral

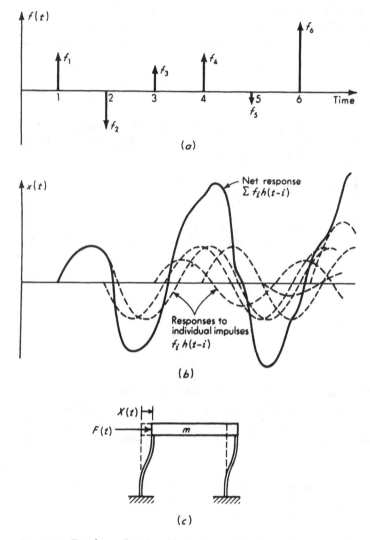

Fig. 3.4.5 Random-vibration illustration. (*a*) Typical sample of forcing function; (*b*) dynamic response to typical sample of forcing function; (*c*) simple oscillator representation of a structure.

reduces to a sum (Fig. 3.4.5*b*) which we can write as

$$X_t = \sum_{i=0}^{t} F_i h(t - i) \tag{3.4.40}$$

where $h(t - i)$ is the response of the structure $t - i$ time units (say seconds) after an impulse of unit magnitude strikes the structure.

For our purposes here it is sufficient to recognize simply that the response X_t is the sum of a number of random variables, $F_i h(t - i)$, and under many circumstances might, by the central limit theorem, be assumed to be normally distributed, $N(0,\sigma_t{}^2)$. The mean response is assumed to be zero, a reasonable assumption if the earthquake accelerations are symmetrically distributed about zero or if we are dealing with dynamic response to fluctuations in the wind (about a mean velocity to which response is static). The variance is proportional to the assumed common variance of the F_i [Eq. (2.4.81)]. In a similar fashion the velocity of the system $\dot{X}_t = dX/dt$ can be represented as

$$\dot{X}_t = \sum_{i=0}^{t} F_i g(t - i) \tag{3.4.41}$$

where $g(t - i)$ is the derivative of $h(t - i)$. $\dot{X}_t$, too, might often be expected to be normally distributed. It can be considered to have zero mean, and its variance is approximately $\omega^2 \sigma_t{}^2$. A somewhat surprising result of this theory of random vibrations is that under many practical circumstances, after the initial starting conditions have been damped out, X_t and $\dot{X}_t$ are independent random variables.†

Energy If we seek the distribution of the energy E_t in the system at time t, it is simply the sum of the kinetic energy of the mass and the potential energy stored in the spring:

$$E_t = \tfrac{1}{2}m\dot{X}_t{}^2 + \tfrac{1}{2}kX_t{}^2 \tag{3.4.42}$$

in which m is the mass and k the spring constant of the oscillator. Then

$$E_t = \tfrac{1}{2}k\left(\frac{m}{k}\dot{X}_t{}^2 + X_t{}^2\right)$$

or, since $\omega = \sqrt{k/m}$,

$$E_t = \tfrac{1}{2}k\left(\frac{1}{\omega^2}\dot{X}_t{}^2 + X_t{}^2\right)$$

Multiplying and dividing by $\sigma_t{}^2$,

$$E_t = \tfrac{1}{2}k\sigma_t{}^2\left[\left(\frac{1}{\omega\sigma_t}\dot{X}_t\right)^2 + \left(\frac{1}{\sigma_t}X_t\right)^2\right]$$

or

$$\frac{2}{k\sigma_t{}^2}E_t = \left(\frac{1}{\omega\sigma_t}\dot{X}_t\right)^2 + \left(\frac{1}{\sigma_t}X_t\right)^2 \tag{3.4.43}$$

Since X_t is $N(0,\sigma_t{}^2)$ and $\dot{X}_t$ is $N(0,\omega^2\sigma_t{}^2)$, the variables on the right-hand side are the squares of independent $N(0,1)$ random variables; hence $2E_t/k\sigma_t{}^2$ is χ^2-distributed with $\nu = 2$. Given information regarding the intensity or gustiness of the forcing function, which can be interpreted in terms of the variance of

† Strictly, this is true if X_t and $\dot{X}_t$ are jointly normally distributed (Sec. 3.6.2) and if the response is "stationary," i.e., in a condition analogous to a steady state in nonprobabilistic vibration theory. In this case, σ_t is independent of time.

the F_i and subsequently the σ_t^2, one can compute the probability that the energy in the system at time t will exceed any critical value, since

$$P[E_t > u] = 1 - P[E_t \leq u] = 1 - P\left[\frac{2}{k\sigma_t^2} E_t \leq \frac{2u}{k\sigma_t^2}\right]$$

$$= 1 - F_Y\left(\frac{2u}{k\sigma_t^2}\right)$$

where Y is CH(2). Recall that a CH(2) distribution is equivalent to a gamma distribution, $G(1,\frac{1}{2})$, which in turn is simply an exponential distribution with parameter $\frac{1}{2}$, EX$(\frac{1}{2})$. Thus, from Eq. (3.2.8),

$$P[E_t > u] = e^{-\frac{1}{2}(2u/k\sigma_t^2)}$$

$$= e^{-u/k\sigma_t^2}$$

In short, the energy in the system at any time t is exponentially distributed, EX$(1/k\sigma_t^2)$.

Amplitude A second response characteristic of interest is the amplitude of vibration

$$A_t = \sqrt{\frac{1}{\omega^2} \dot{X}_t^2 + X_t^2} \tag{3.4.45}$$

It is shown in elementary dynamics that if the system were in free vibration, A_t would be the maximum value of X_t; consequently it is significant as an indicator of the maximum displacement (and hence strains and stresses) which the system will undergo. By an argument similar to that above

$$\frac{1}{\sigma_t} A_t = \sqrt{\left(\frac{1}{\omega\sigma_t} \dot{X}\right)^2 + \left(\frac{1}{\sigma_t} X\right)^2} \tag{3.4.46}$$

is the square root of the sum of the squares of two independent $N(0,1)$ random variables and hence is χ-distributed with 2 degrees of freedom. Letting $Z = (1/\sigma_t)A_t$, we know that Z is CHI(2):

$$f_Z(z) = ze^{-\frac{1}{2}z^2} \qquad z \geq 0$$

A_t is linearly related to Z:

$$f_{A_t}(a) = \frac{ae^{-\frac{1}{2}(a/\sigma_t)^2}}{\sigma_t^2} \qquad a \geq 0 \tag{3.4.47}$$

This special case of the χ distribution is usually referred to as the *Rayleigh distribution*. Its shape is plotted in Fig. 3.4.6. As mentioned earlier, the χ^2 tables can be used to evaluate χ CDF's.

$$F_{A_t}(a) = P[A_t \leq a] = P\left[Z \leq \frac{a}{\sigma_t}\right]$$

$$= F_Y\left(\frac{a^2}{\sigma_t^2}\right) \tag{3.4.48}$$

where Z is CHI(2) and Y is CH(2) or EX$(\frac{1}{2})$. Thus

$$1 - F_{A_t}(a) = e^{-\frac{1}{2}(a/\sigma_t)^2} \tag{3.4.49}$$

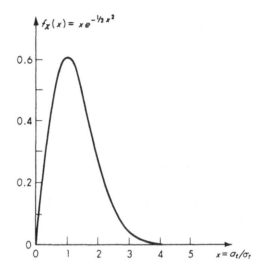

Fig. 3.4.6 Rayleigh or χ (with $\nu = 2$) distribution.

which could have been determined by integrating $f_{A_t}(a)$. Evaluation of Eq. (3.4.49) for a critical value of a (say the value of displacement at which cracking of plaster or damage to windows will occur), will determine approximately the likelihood of this event at or near time t.†

3.4.4 Summary

Several common models are catalogued in this section. They were two limited distributions:

1. The uniform distribution,

$$f_X(x) = \frac{1}{b - a} \qquad a \leq x \leq b$$

and

2. The beta distribution,

$$f_X(x) = \frac{1}{B(b - a)^{t-1}} (x - a)^{r-1}(b - x)^{t-r-1} \qquad a \leq x \leq b$$

and four distributions of functions of independent normal random variables:

3. The chi-square (χ^2) distribution, which is the distribution of sum of ν squared normal variables.

† For purposes of simplifying the illustrations of the applications of these distributions, a great deal has been left unsaid or not well said about a number of important points in random vibrations. The interested reader is referred to Crandall and Mark [1963] for a more complete discussion.

4. The chi (χ) distribution, which is the square root of the sum of the squares of ν normal variables.
5. The t distribution, the ratio of a $N(0,1)$ variable to the square root of the sum of ν squared $N(0,1)$ variables divided by ν.
6. The F distribution, the ratio of the sum of ν_1 squared $N(0,1)$ variables (divided by ν_1) to the sum of ν_2 squared $N(0,1)$ variables (divided by ν_2).

3.5 MODIFIED DISTRIBUTIONS†

A number of modifications to the more common distributions presented in the earlier sections of this chapter are frequently adopted in order to relate the mathematical model more closely to the physical problem or simply to provide a better empirical fit to observed data.

3.5.1 Shifted and Transformed Distributions

Shifts The simplest modification to a distribution is simply a shift of the origin. For example, the exponential interarrival time between vehicles has been found to be a satisfactory model even in fairly dense traffic if it is shifted an amount t_0 to account for the impossibility of vehicles arriving within time intervals on the order of the length of a vehicle divided by its velocity. Thus the engineer would assume $T - t_0$ to be exponentially distributed, implying

$$f_T(t) = \lambda e^{-\lambda(t-t_0)} \qquad t \geq t_0 \tag{3.5.1}$$

$$F_T(t) = \begin{cases} 0 & t \leq t_0 \\ 1 - e^{-\lambda(t-t_0)} & t \geq t_0 \end{cases}$$

which is plotted‡ in Fig. 3.5.1.

A similar shift has been used by Chow and Ramaseshan [1965] with the lognormal distribution to describe hydrologic data. The shifted gamma has been used by Vorlicek and Suchy [1958] to describe the strength of concrete and steel and by Markovic [1965] to describe maximum annual flows. In most such cases the shift was introduced not because the author had reason to believe that there was a natural lower bound, but rather simply to gain an additional parameter (three rather than two in the case of the lognormal and gamma distributions) to insure a better fit to the data. Markovic [1965], in fact, compared (using a

---

† This section can be omitted on first reading.
‡ Notice that in the particular case of the exponential distribution this shifted distribution is identical to the conditional distribution of T given that T is greater than t_0 (see Sec. 3.2.1).

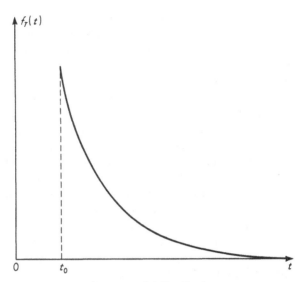

Fig. 3.5.1 Shifted exponential distribution.

method to be discussed in Sec. 4.5) the closeness of the fit produced by
the two- and three-parameter gamma and lognormal distributions to
many different rivers' records of annual maximum flows.   Notice that
the parameters w and ϵ in the Type III asymptotic extreme-value distributions (Sec. 3.3.3) can, in fact, be considered as simply shift parameters
added to basically two-parameter distributions.

The mean of the shifted distribution is simply the original mean
plus the shift.   All *central* moments such as the variance are unchanged.

The simple shift is, of course, no more than a special case of the
linear transformation

$$Y = b(X - x_0) \tag{3.5.2}$$

which was seen in Sec. 2.3.1 to leave the shape of the distribution
unchanged but to cause a shift to the right (if $x_0 > 0$) by x_0 and a scale
expansion (if $b > 1$) by a factor b.   The density of Y if X is a continuous
random variable was found to be simply

$$f_Y(y) = \frac{1}{|b|} f_X\left(\frac{y}{b} + x_0\right) \tag{3.5.3}$$

Such linear transformations have been used to reduce distributions (e.g.,
the normal, Sec. 3.3.1, or the gamma, Sec. 3.2.2 and 3.4.3) to standardized
forms more easily tabulated.

Transformations   More general transformations are commonly used if professional knowledge so indicates or if better fits to observed data result.   Examples of the former case included the lognormal distribution (Sec. 3.3.2), where, if a multiplicative mechanism is indicated, it may be reasonable to assume that the logarithm of the variable of interest is normally distributed.   Examples of the latter case, transformations for convenience, include the assumption by Stidd [1953] and Beals [1954] that X^a is normally distributed where X is the total precipitation during a fixed period.†   In general, of course, the greater the number of independent parameters available for estimating from data (Chap. 4) the better the fit that can be expected.

Many of the distributions given special names in the previous sections of this chapter can be recognized as no more than cases where some function of the variable has one of the other distributions.   In many cases they were derived this way.   For example, if the distribution of X is assumed to be χ^2-distributed, then the distribution of $\sqrt{X}$ is χ-distributed.   Notice, too, that if

$$\left(\frac{X - a}{b}\right)^c$$

is assumed to be exponential with parameter $\lambda = 1$, one is led to the conclusion that X has the Type III asymptotic extreme-value distribution, since

$$F_X(x) = 1 - e^{-[(x-a)/b]^c}$$

Similarly, if $e^{[-a(X-b)]}$ is assumed to be exponentially distributed, X has the Type I extreme-value distribution.   Knowledge of such relationships can often simplify problem solutions.

3.5.2   Truncated and Censored Distributions

Truncated distributions   In certain situations tails of distributions may be "lost" owing, say, to an inspection procedure which leads to the elimination of all those members of the population of manufactured items or mixed materials, for example, with values less than a particular value x_0.   Such a distribution is said to be *truncated* below x_0.   The resulting distribution might then have a shape like that in Fig. 3.5.2 if the distribution of the original population was normal, for example.

If the original population had PDF $f_X(x)$ and CDF $F_X(x)$ and if the variable of interest Y has been truncated below x_0, its PDF is zero

† Stidd [1953] found that $a = \frac{1}{3}$ gives good fits to daily, monthly, and annual precipitation, while Beals [1954] found better results with $a = \frac{1}{4}$ for daily precipitation.

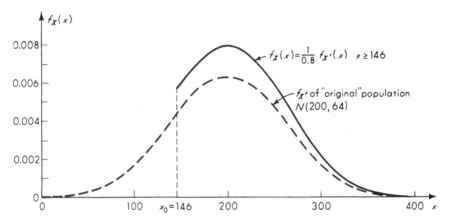

Fig. 3.5.2 The truncated normal distribution.

up to x_0 and $f_X(x)$ is renormalized† for $x \geq x_0$:

$$f_Y(y) = \begin{cases} 0 & y < x_0 \\ k f_X(y) & y \geq x_0 \end{cases} \qquad (3.5.4)$$

where $k = 1/[1 - F_X(x_0)]$. Similar results hold if the upper tail is truncated or if both tails are truncated.

Truncated distributions have been employed by Fiering [1962] (the normal) to describe hydrologic data, by Das [1955] (the gamma) to describe daily precipitation, and by Freudenthal, Garrelts, and Shinozuka [1966] (the t distribution) to describe the ultimate capacity of structural members.

Censored distributions *Censored* distributions differ slightly in application in that the total population is present but the exact frequencies are known only up to (or beyond) a certain value x_0. For example, in fatigue testing it is not uncommon at low stress levels to stop the test of long-lived members short of failure at an arbitrary, high, number, say 10^6 cycles, simply to free the apparatus for further testing. The observed failures and the unfailed specimens represent samples from a population characterized by the parent or underlying population up to the cutoff value x_0, but of unspecified magnitude (other than greater than x_0) beyond this cutoff point. When, as in many engineering applications, the nature of the underlying distribution is being determined from observed data, it is wasteful and improper to neglect the information that a certain proportion of the observations exceeded x_0 in value. In many cases the distribution is assumed to have the probability or pro-

† In fact, $f_Y(Y)$ is the conditional distribution of X given that $X \geq x_0$.

portion above x_0 lumped at x_0. If the underlying distribution is continuous, the censored distribution is mixed with a spike at x_0. See, for example, the pump-capacity illustration in Sec. 2.4.2. Such questions are treated in Hald [1952].

3.5.3 Compound Distributions

Compound distributions arise when a parameter of the distribution of a random variable is itself treated as a random variable.

Illustration: Vehicle delays during construction operations For example, assuming vehicle arrivals on a rural road are Poisson events (Sec. 3.2.1), closing the road for a construction operation for t min will lead to the delay of X vehicles. X has the Poisson distribution with parameter λt, where λ is the average number of cars arriving per minute:

$$p_X(x) = e^{-\lambda t} \frac{(\lambda t)^x}{x!} \qquad x = 0, 1, 2, \ldots \tag{3.5.5}$$

Suppose now that, since it may be caused by one of a number of different operations, the duration of the closed period is a random variable with a probability mass function $p_T(t)$ that is related directly to the relative likelihood that a particular operation taking a particular time causes the closure. Then, considering all possible values of t, each with its relative likelihood, the number of cars stopped by an arbitrary unspecified type of closure has distribution

$$p_X(x) = \sum_t e^{-\lambda t} \frac{(\lambda t)^x}{x!} p_T(t) \tag{3.5.6}$$

If each of the operations takes a random amount of time, it is reasonable to treat T as a continuous random variable when the previous equation becomes

$$p_X(x) = \int_0^\infty e^{-\lambda t} \frac{(\lambda t)^x}{x!} f_T(t) \, dt \tag{3.5.7}$$

For example, if T is exponentially distributed with parameter α,

$$p_X(x) = \int_0^\infty e^{-\lambda t} \frac{(\lambda t)^x}{x!} \alpha e^{-\alpha t} \, dt$$

$$= \frac{\alpha}{x!} \int_0^\infty e^{-(\alpha+\lambda)t} \left(\frac{\lambda}{\alpha+\lambda}\right)^x (\alpha + \lambda)^x t^x \, dt$$

Letting $u = (\alpha + \lambda)t$,

$$p_X(x) = \frac{\alpha}{(\alpha+\lambda)} \left(\frac{\lambda}{\alpha+\lambda}\right)^x \frac{1}{x!} \int_0^\infty e^{-u} u^x \, du$$

The integral is by definition $\Gamma(x + 1)$, which for integer x is $x!$. Thus

$$p_X(x) = \frac{\alpha}{\alpha+\lambda} \left(\frac{\lambda}{\alpha+\lambda}\right)^x$$

$$= \frac{\alpha}{\alpha+\lambda} \left(1 - \frac{\alpha}{\alpha+\lambda}\right)^x \qquad x = 0, 1, 2, \ldots \tag{3.5.8}$$

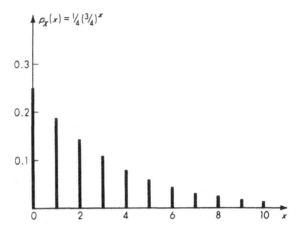

Fig. 3.5.3 Geometric distribution arising from vehicle-delay illustration.

which coincidently is simply the geometric distribution† (Sec. 3.1.2) with parameter $\alpha/(\alpha + \lambda)$. For example, if the mean arrival rate is $\lambda = 3$ cars per minute and the average delay is $1/\alpha = 1$ min, $\alpha/(\alpha + \lambda) = \frac{1}{4}$. The mean number of cars stopped is $4 - 1 = 3$, which is intuitively satisfactory; the distribution of X is plotted in Fig. 3.5.3. Notice that the geometric distribution does *not* arise here from the argument about the number of Bernoulli trials preceding the first success, which was used in Sec. 3.1.3. The illustration was included here to emphasize that totally different underlying mechanisms may lead to the same distribution. Therefore one cannot infer conclusively that a particular mechanism is acting if observed data appear to follow a particular distribution.

General form   In general, if a random variable X has CDF $F_X(x)$ which depends upon a parameter θ [and which therefore might be written $F_X(x;\theta)$], then we can find the CDF of another random variable Y when Θ is a random variable‡ as follows:

$$F_Y(y) = \sum_{\theta} F_X(x;\theta) p_\Theta(\theta) \tag{3.5.9}$$

or

$$F_Y(y) = \int_{-\infty}^{\infty} F_X(x;\theta) f_\Theta(\theta) \, d\theta \tag{3.5.10}$$

† See footnote page 228.
‡ In Chap. 6 this assumption will also be made in the context of treating statistical problems, i.e., problems where the engineer does not know precisely the values of the parameters of his models, but he has data to aid him in estimating them or in making decisions dependent on them. Note that where feasible we distinguish between a random variable Θ and a particular value θ, in a manner analogous to X and x.

depending on whether Θ is discrete or continuous. If Y is discrete, which it will be if X is, its PMF can be found by replacing $F_X(x)$ by $p_X(x)$. Similarly, as was done in the previous illustration, the PDF of Y can be found by putting the PDF of X in place of its CDF in Eqs. (3.5.9) and (3.5.10). The distribution of Y is just the weighted average of the distributions of X for each value θ.

Illustration: Uncertain supplier of material The simple case when the parameter has a discrete distribution, Eq. (3.5.9), is simply a formal way of treating situations where the engineer may be faced with uncertainty as to which of several distributions is applicable. Then the values of $p_\Theta(\theta)$ can be interpreted as the engineer's weights or degree-of-belief assignments (see Sec. 2.1.2) as to the true distribution. For illustration, a job's material may be furnished by one of two manufacturers, each producing material with his own characteristic distribution, the two being different in means, variances, and, possibly, shapes. Knowledge of these distributions and the relative likelihoods of receiving the material from each manufacturer will yield by Eq. (3.5.9) a material property distribution which accounts for uncertainty in the supplier. For example, if two contractors just meet ACI specifications† for 4000-psi concrete, but one produces material with a coefficient of variation $V_1 = 10$ percent, while the other's concrete has $V_2 = 15$ percent, then the probability that the supplied strength will be less than 3500 psi is (assuming, as ACI does, normal distributions)

$$F_Y(3500) = 0.2F_{X_1}(3500) + 0.8F_{X_2}(3500)$$

if 0.2 and 0.8 are the respective likelihoods that contractors 1 and 2 will get

† See Prob. 2.45. Contractor 1 will have $m_1 = 4600$ and $\sigma_1 = 460$, while contractor 2 will have $m_2 = 5000$ and $\sigma_2 = 750$.

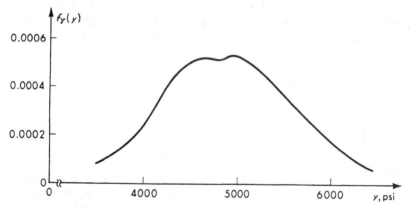

Fig. 3.5.4 Bimodal PDF from illustration involving uncertain supplier of materials.

the job. In this case, letting U be $N(0,1)$,

$$F_Y(3500) = (0.2)F_U\left(\frac{3500 - 4600}{460}\right) + (0.8)F_U\left(\frac{3500 - 5000}{750}\right)$$

$$= (0.2)(0.0224) + (0.8)(0.054)$$

$$= 0.048 \tag{3.5.11}$$

It should be clear that Y is not normally distributed, but, in fact, is a bimodal distribution having peaks near 4600 and 5000 psi:

$$f_Y(y) = 0.2f_{x_1}(y) + 0.8f_{x_2}(y) \tag{3.5.12}$$

This density function is shown in Fig. 3.5.4.

Moments   Although the distributions resulting from compounding or "mixing" may be difficult to determine analytically, owing to the integration or summation involved, the moments are more easily found.   It is readily shown through substituting into their definitions, Eq. (2.4.16), that the moments about the origin are simply

$$E[Y^n] = m_Y{}^{(n)} = \sum_\theta m_X{}^{(n)}(\theta)p_\Theta(\theta) \tag{3.5.13}$$

or

$$E[Y^n] = m_Y{}^{(n)} = \int_{-\infty}^{\infty} m_X{}^{(n)}(\theta)f_\Theta(\theta)\,d\theta \tag{3.5.14}$$

where $m_X{}^{(n)}(\theta)$ is the nth moment of X given that the parameter of the distribution of X is θ.   In particular,[†]

$$m_Y = \sum_\theta m_X(\theta)p_\Theta(\theta) \tag{3.5.15}$$

The mean of Y is just the weighted average of the means of X for the various values of θ.   In terms of m_Y and $m_Y{}^{(2)} = E[Y^2]$, the variance we know to be

$$\sigma_Y{}^2 = E[Y^2] - m_Y{}^2$$

$$= m_Y{}^{(2)} - m_Y{}^2$$

Substitution and manipulation let the variance of Y be written

$$\sigma_Y{}^2 = \sum_\theta \sigma_X{}^2(\theta)p_\Theta(\theta) + \sum_\theta (m_X(\theta) - m_Y)^2 p_\Theta(\theta) \tag{3.5.16}$$

which makes it clear the variance of Y is *not* the weighted average of the variances of X for the various values of θ.   Engineers will see an

---

[†] The interpretation in terms of conditional probabilities and conditional expectations leads one to these same results, written in Sec. 2.4.3 as $E[Y] = E[E[Y \mid \Theta]]$, the inner expectation being with respect to Y, the outer with respect to Θ.

analogy with the parallel-axis theorem for obtaining the moment of inertia of an area made up of several small areas.†

Similar relationships hold for the first and second moments of Y when the distribution of Θ is continuous.

Illustration: Total rainfall　As a final illustration of compound distributions, recall the illustration in Sec. 2.3.3. There we sought the distribution of the total annual rainfall T, when the number of rainy weeks N is Poisson-distributed with mean $\nu = 20$ and the rain in a rainy week is exponentially distributed with parameter $\lambda = 2$ in.$^{-1}$ (or mean $\frac{1}{2}$ in.). In that section a Monte Carlo technique was used to solve the problem. Using the notions of compound distributions, we can find the distribution of T directly.

If the number of rainy weeks were known, the distribution of T would be gamma, $G(n,\lambda)$, since the sum of n (assumed independent and identically distributed) exponential random variables is gamma (Sec. 3.2.3).

$$f_T(t;n) = \frac{\lambda e^{-\lambda t}}{\Gamma(n)}(\lambda t)^{n-1} \qquad t \geq 0 \tag{3.5.17}$$

if $n > 0$. (If $n = 0$, $T = 0$.) The parameter n of this gamma distribution is Poisson-distributed‡; hence, for $t > 0$,

$$f_T(t) = \sum_{n=1}^{\infty} \frac{\lambda e^{-\lambda t}}{\Gamma(n)}(\lambda t)^{n-1}\frac{e^{-\nu}\nu^n}{n!} \qquad t > 0 \tag{3.5.18}$$

The distribution of T is a mixed distribution. Equation (3.5.18) gives the continuous part. At $t = 0$, we must include a spike, a finite probability of exactly zero rainfall, if there are no rainy weeks. The magnitude of this spike is simply $P[N = 0] = e^{-\nu}$.

Simplifying Eq. (3.5.18) and letting $u = n - 1$,

$$f_T(t) = \lambda e^{-\lambda t-\nu}\sum_{u=0}^{\infty}\frac{(\lambda t)^u \nu^{u+1}}{u!\Gamma(u+2)} \qquad t > 0 \tag{3.5.19}$$

This density could be evaluated numerically§ for a set of values of t. The distribution of T is sketched in Fig. 3.5.5 for $\nu = 20$ and $\lambda = 2$. Note that the continuous part is approximately normal, as should be expected. This is clearly one reasonable form for a model of the problem and data discussed in Sec. 2.2.1. The mean and variance of T are, by Eqs. (3.5.15) and (3.5.16)

† In the notation of Sec. 2.4.3, $\text{Var}[Y] = E[\text{Var}[Y\mid\Theta]] + \text{Var}[E[Y\mid\Theta]]$.
‡ Recall that we are neglecting the error in the model associated with N taking on values greater than 52 weeks.
§ Alternatively, it happens that the sum of this series is closely related to a widely tabulated function, the modified Bessel function of order 1 (see for example, National Bureau of Standards [1964]). Denoting this function I_1, the function Eq. (3.5.19), becomes

$$f_T(t) = \lambda e^{-\lambda t-\nu}\sqrt{\frac{\nu}{\lambda t}}I_1(2\sqrt{\lambda t\nu}) \qquad t > 0$$

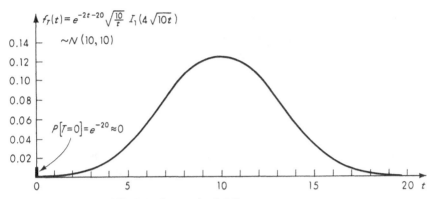

Fig. 3.5.5 Mixed PDF of total annual rainfall.

$$m_T = E[T] = \sum_{n=0}^{\infty} m(n)p_N(n)$$

$$= \sum_{n=0}^{\infty} \frac{n}{\lambda} p_N(n) = \frac{1}{\lambda} E[N] = \frac{\nu}{\lambda} = 10 \text{ in.}$$

$$\sigma^2_T = \sum_{n=0}^{\infty} \frac{n}{\lambda^2} p_N(n) + \sum_{n=0}^{\infty} \left(\frac{n}{\lambda} - \frac{\nu}{\lambda}\right)^2 p_N(n)$$

$$= \frac{1}{\lambda^2} E[N] + \frac{1}{\lambda^2} \text{Var}[N] = \frac{2\nu}{\lambda^2} = 10 \text{ in.}^2$$

$$\sigma_T = 3.16 \text{ in.}$$

Recall that the mean and variance of T were also calculated in Sec. 2.4.3 by an alternate approach.

3.5.4 Summary

Common distributions are often modified in various ways to extend their usefulness in applications. These modifications include:

1. *Transformations:*
   A function $Y = g(X)$ of the variable of interest is assumed to have a particular common distribution. A frequent case is $Y = X - a$, which leads to a *shift* of the common distribution.
2. *Tail modifications:*
   The probability mass in a tail (or tails) of the common distribution may either be removed (truncated) or lumped at the cutoff point (censored).

3. *Compounding:*

A new distribution may be formed by treating θ, the parameter(s) of a common distribution, as a random variable. In this case the distribution becomes a weighted distribution

$$F_Y(y) = \int_{-\infty}^{\infty} F_X(x;\theta)f_\Theta(\theta) \, d\theta$$

3.6 MULTIVARIATE MODELS

Clearly, many different multivariate models can be formed easily from the common univariate models discussed in the previous sections of this chapter. For example, in Sec. 2.2.2, a Poisson marginal and binomial conditional distribution led to a joint distribution of the number of arriving vehicles and the number of counted vehicles when the counter was defective. Also, reexamination of the models formed from compound distributions will make it clear that the integrand in such an equation as [see Eq. (3.5.10)]

$$f_Y(y) = \int_{-\infty}^{\infty} f_X(x;\theta)f_\Theta(\theta) \, d\theta \tag{3.6.1}$$

is simply the joint density function of Y and Θ since it can be interpreted as the product of the conditional distribution of Y given Θ and the marginal distribution of Θ (Eq. 2.2.45a). In the annual-rainfall illustration of Sec. 3.5.3, for example, extracting the function from under the summation sign in Eq. (3.5.18) yields the joint distribution of the total rainfall T and the total number of rainy weeks N:

$$f_{T,N}(t,n) = \begin{cases} \lambda e^{-\lambda t} \dfrac{(\lambda t)^{n-1}}{\Gamma(n)} \dfrac{e^{-\nu}\nu^n}{n!} & t > 0, \; n = 1, 2, 3, \ldots \\ e^{-\nu} & n = 0 \end{cases} \tag{3.6.2}$$

which is continuous in t and discrete in n. Examples of this type will not be studied in this section as they represent only special applications of the ideas in Sec. 3.5. Two multivariate distributions do deserve separate treatment, however, if only because of their prevalence in applications. One of these, the multinomial distribution, is discrete; the other, the multivariate normal distribution, is continuous.

3.6.1 Counting Multiple Events: The Multinomial Distribution

In Sec. 3.1.2 we discussed the binomial distribution as describing the total number of "successes" X observed in a sequence of n independent Bernoulli trials. The Bernoulli trial was described as one where either one of only two events—success or failure—occurred. The binomial distribution was found to be

$$p_X(x) = \frac{n!}{x!(n-x)!} \, p^x(1-p)^{n-x} \qquad x = 0, 1 \ 2, \ldots, n \qquad (3.6.3)$$

where p is the (constant) probability of success at each trial. Notice that we could have counted both successes X_1 and failures $X_2 (= n - X_1)$ and derived a joint distribution of X_1 and X_2 simply:

$$p_{X_1,X_2}(x_1,x_2) = \begin{cases} \dfrac{n!}{x_1!x_2!} \, p^{x_1}(1-p)^{x_2} & \begin{array}{l} \text{for } x_1 \text{ and } x_2 \text{ such} \\ \text{that } x_1 + x_2 = n \end{array} \\ 0 & \text{elsewhere} \end{cases} \qquad (3.6.4)$$

This distribution is zero except on those pairs of nonnegative integers x_1 and x_2 whose sum is n.

This form of the binomial distribution is unnecessarily clumsy, but it is suggestive of the desired generalization, the multinomial distribution. Consider a sequence of n independent trials in which not two but k mutually exclusive, collectively exhaustive outcomes are possible on each trial, with probabilities $p_1, p_2, p_3, \ldots, p_k$, respectively $(p_1 + p_2 + \cdots + p_k = 1)$. Then the joint probability mass function of $X_1, X_2, \ldots, X_k$, the total numbers of outcomes of each kind, is the *multinomial distribution*

$$p_{X_1,X_2,\ldots,X_k}(x_1, x_2, \ldots, x_k)$$

$$= \begin{cases} \dfrac{n!}{x_1!x_2! \cdots x_k!} \, p_1^{x_1} p_2^{x_2} \cdots p_k^{x_k} & \begin{array}{l} \text{for } x_1, x_2, \ldots, x_k \text{ such} \\ \text{that } x_1 + x_2 + \\ \qquad\qquad \cdots + x_k = n \end{array} \\ 0 & \text{elsewhere} \end{cases} \qquad (3.6.5)$$

For the special (binomial) case, $k = 2$, $p_1 = p$, $p_2 = 1 - p_1$, this equation reduces to Eq. (3.6.4). The coefficient $n!/x_1!x_2! \cdots x_k!$ can be interpreted as a normalizing factor or as a generalization of the binomial coefficient determining the number of different ways the combination of x_1 occurrences of event 1, x_2 occurrences of event 2, etc., could occur in a sequence of n trials.

Illustration: Vehicle turns In the mean, 20 percent of all cars arriving at an intersection turn left, 30 percent turn right, and the remainder go straight. Of a sequence of four cars, what is the joint distribution of X_1, the number of left-turning cars, X_2, the number of right-turning cars, and X_3, the number of cars going straight?

$$p_{X_1,X_2,X_3}(x_1,x_2,x_3) = \frac{4!}{x_1!x_2!x_3!} \, (0.2)^{x_1}(0.3)^{x_2}(0.5)^{x_3}$$

for $x_1, x_2,$ and x_3 such that $x_1 + x_2 + x_3 = 4$.

In particular the probability that one car will turn right, one left, and two not at all is

$$p_{X_1,X_2,X_3}(1,1,2) = \frac{4 \cdot 3 \cdot 2 \cdot 1}{1 \cdot 1 \cdot 2} (0.2)(0.3)(0.5)^2 = 0.18$$

Since the marginal distribution of any X_i is clearly binomial with parameters p_i and n, it follows immediately that

$$E[X_i] = np_i \qquad\qquad\qquad (3.6.6)$$

$$\text{Var}\,[X_i] = np_i(1 - p_i) \qquad\qquad (3.6.7)$$

With somewhat more effort it can be shown that

$$\text{Cov}\,[X_i,X_j] = -np_ip_j \qquad\qquad (3.6.8)$$

What is the correlation between X_1 and X_2 in the binomial $(k = 2)$ case, Eq. (3.6.4)? It must be unity in absolute value (Sec. 2.4.3), since there is a linear functional relationship between X_1 and X_2: $X_2 = n - X_1$.

3.6.2 The Multivariate Normal Distribution

For reasons analogous to those discussed in Sec. 3.3.1, the joint or multivariate normal distribution is the most commonly adopted model when joint behavior of two or more variables is of interest. It can usually be expected as the result of additive mechanisms, many other distributions converge to it, and it is mathematically convenient. In multivariate situations the last of these factors is often of utmost importance. Many powerful results can be shown to hold for the normal case which do not hold generally. We shall discuss in detail only the bivariate case; higher dimensions offer few new notions and demand matrix techniques for efficient treatment.†

Standardized form Recall the basic form of the univariate normal distribution $N(0,1)$:

$$f_U(u) = \frac{1}{\sqrt{2\pi}}\, e^{-u^2/2} \qquad -\infty < u < \infty \qquad (3.6.9)$$

The analogous bivariate distribution replaces u^2 by a quadratic form in two variables u and v,

$$f_{U,V}(u,v) = k \exp\left[-\frac{1}{2(1 - \rho^2)} (u^2 - 2\rho uv + v^2) \right]$$

$$-\infty \le u \le \infty, \ -\infty \le v \le \infty \qquad (3.6.10)$$

† See any of a number of intermediate or advanced probability texts, e.g., Feller [1966].

where ρ is the correlation coefficient and k is a normalizing factor found by setting an integral over $f_{U,V}(u,v)$ equal to 1:

$$k = \frac{1}{2\pi \sqrt{1 - \rho^2}} \tag{3.6.11}$$

In this standardized form the bivariate normal is a distribution with one parameter ρ. It is tabulated in a number of references.† U and V have means 0 and standard deviations 1.

General form   More generally, interest lies in variables X and Y with means m_X and m_Y, standard deviations σ_X and σ_Y, and covariance $\rho\sigma_X\sigma_Y$. As with the univariate case, U and V represent the standardized variables

$$U = \frac{X - m_X}{\sigma_X} \tag{3.6.12}$$

$$V = \frac{Y - m_Y}{\sigma_Y} \tag{3.6.13}$$

In its expanded form the bivariate normal has five parameters, m_X, m_Y, σ_X, σ_Y, and ρ:

$$f_{X,Y}(x,y) = \frac{1}{2\pi\sigma_X\sigma_Y \sqrt{1 - \rho^2}} \exp\left\{ -\frac{1}{2(1 - \rho^2)} \left[\left(\frac{x - m_X}{\sigma_X}\right)^2 \right.\right.$$
$$\left.\left. - 2\rho \frac{(x - m_X)(y - m_Y)}{\sigma_X\sigma_Y} + \left(\frac{y - m_Y}{\sigma_Y}\right)^2 \right] \right\}$$
$$-\infty \le x \le \infty, \; -\infty \le y \le \infty \tag{3.6.14}$$

The distribution is sketched in Fig. 3.6.1.

Marginal and conditional distributions   The marginal density of X is found by integrating over y [Eq. (2.2.43)]

$$f_X(x) = \int_{-\infty}^{\infty} f_{X,Y}(x,y) \, dy$$

which becomes simply

$$f_X(x) = \frac{1}{\sqrt{2\pi}\sigma_X} e^{-\frac{1}{2}[(x-m_X)/\sigma_X]^2} \qquad -\infty \le x \le \infty \tag{3.6.15}$$

which is the normal distribution $N(m_X, \sigma_X^2)$. A similar result holds for Y. Thus the marginal distributions of the bivariate normal are normal. Unfortunately the converse does not necessarily hold; it is possible to have normal random variables whose joint density is not bivariate

† See, for example, National Bureau of Standards [1959].

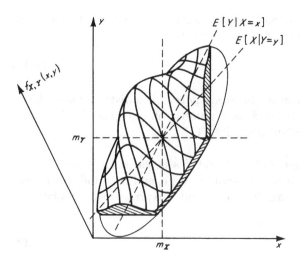

Fig. 3.6.1 The bivariate normal distribution. (From
P. G. Hoel, "Introduction to Mathematical Statistics,"
John Wiley and Sons, New York, 1962.)

normal.[†] In practice, the assumption of *marginal* normality may be
well motivated (or, at least, compared to data to verify a reasonable
correspondence between shapes), but the further assumption of *joint*
normality it is often simply an act of faith.

Notice in Eq. (3.6.14) that, if the jointly normal random variables
are uncorrelated, i.e., if $\rho = 0$, their joint distribution factors into the
product of their marginals [Eq. (3.6.15)], implying that they are also
independent. Recall (Sec. 2.4.3) that lack of correlation does *not*, in
general, imply independence, only lack of *linear* dependence.[‡]

The conditional distribution of X given Y, say, can also be found
to be normal with parameters which will be given in an illustration to
follow.

Sum of joint normals   It is an easily (but tediously) verified property of
two jointly distributed normal random variables X and Y that their
sum Z is also normally distributed. This fact is a generalization of the

† See Feller [1966], page 69, for examples.
‡ The implication is that jointly normal random variables are, if dependent at all,
only linearly dependent. The wide application of jointly normal random variables,
particularly in statistical applications, often stems from the analytical simplifications
offered by this "linearity" property. In engineering mechanics, of course, a similar
statement holds; linear problems, systems, equations, etc., being more easily solved,
are more widely applied, either by appropriate system design or as approximations
to truly nonlinear systems.

statement in Sec. 3.3.1 that *independent* normally distributed random variables have a normally distributed sum. The mean and variance of $Z = X + Y$ are found easily from the relationships given in Sec. 2.4.3, Eqs. (2.4.81a,) and (2.4.82):

$$m_Z = m_X + m_Y \tag{3.6.16}$$

$$\text{Var}[Z] = \text{Var}[X] + \text{Var}[Y] + 2\,\text{Cov}[X,Y] \tag{3.6.17}$$

or

$$\sigma_Z{}^2 = \sigma_X{}^2 + \sigma_Y{}^2 + 2\rho_{X,Y}\sigma_X\sigma_Y \tag{3.6.18}$$

Multivariate normal   The multivariate normal distribution, a joint distribution of n variables, is an extension of the bivariate normal (Feller [1966]).   Its parameters are the marginal means and variances of all n variables plus a set of correlation coefficients $n(n - 1)/2$ in number) between all pairs of variables.   The distribution is not conveniently treated without recourse to matrix notation, but multidimensional generalizations to all the properties encountered here still hold; e.g., marginal and conditional distributions remain normal, and if

$$Z = X_1 + X_2 + \cdots + X_n$$

where the X_i have a *multivariate normal distribution*, then Z is normally distributed with mean and variance given by Eqs. (2.4.81a,) and (2.4.81b).

Illustration: Stream-flow prediction   The hydrological engineer often assumes that the watershed area and tributaries leading into a stream form a linear system. Thus he may calculate the flow in the stream $X(t)$ (in cfs. say) in response to any rainstorm which has a given intensity $Y(t)$ (in units inches per hour, say), if he has found, analytically and/or experimentally, the system's "unit hydrograph" $h(t)$, i.e., the response of the system at time t to a unit impulse of rain at time 0.   Such a unit hydrograph appears in Fig. 3.6.2.   Owing to the assumed linearity, superposition is valid and the response $X(t)$ becomes

$$X(t) = \int_0^t Y(\tau)h(t - \tau)\,d\tau \tag{3.6.19}$$

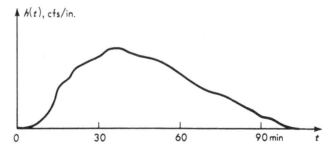

Fig. 3.6.2   A typical small-basin unit hydrograph.

where t is measured from the beginning of the storm. This equation, of course, is of the same convolution-integral form as Eq. (3.4.39) which described the dynamic response of a linear structural system to a time dependent input. Following arguments analogous to those in the dynamic response illustration in Sec. 3.4.3, it is to be expected that X_1, the stream flow at time t_1, and X_2, the flow at a later time t_2, will each be approximately normally distributed random variables if the rainfall intensity $Y(t)$ (in inches per minute) is of a random nature. Although more difficult to justify, it is not unreasonable to assume that X_1 and X_2 are bivariate normal,† Eq. (3.6.14), with X_1 taking the place of X and X_2 of Y.

Let us consider the problem of predicting or forecasting the flow at time t_2 or X_2. The parameters m_{X_1}, m_{X_2}, σ_{X_1}, σ_{X_2}, and ρ are presumably known, computed from properties of a stochastic description of rainstorms and a system response function $h(t)$. Methods to accomplish these computations are not our immediate concern.‡

Without further information our ability to "predict" X_2 is limited to the (normal) marginal distribution of X_2, $N(m_{X_2}, \sigma_{X_2}{}^2)$. Forced to give a single number as the "predicted value of X_2," the most reasonable choice is m_{X_2}, but the uncertainty involved in this particular prediction is measured by the marginal variance of X_2 or

$$\sigma_{X_2}{}^2 = E[(X_2 - m_{X_2})^2] \tag{3.6.20}$$

which can also be interpreted here as the expected value of the squared "error" of prediction, $X_2 - m_{X_2}$. (See also the discussion in Sec. 2.4.3 on prediction.) Let us next investigate the value of the information offered by an observation of the stream flow at some earlier time t_1. That is, we are given the information that X_1 was observed to be some number, say x_1 cfs.

We first determine the conditional distribution of X_2 given $X_1 = x_1$. By definition, Eq. (2.2.45a),

$$f_{X_2|X_1}(x_2, x_1) = \frac{f_{X_1, X_2}(x_1, x_2)}{f_{X_1}(x_1)} \tag{3.6.21}$$

which after substitution and reduction becomes simply

$$f_{X_2|X_1}(x_2, x_1) = \frac{1}{\sqrt{2\pi}\sigma_2'} e^{-\frac{1}{2}[(x_2 - m_2')/\sigma_2']^2} \qquad -\infty \leq x_2 \leq \infty,$$

$$-\infty \leq x_1 \leq \infty \tag{3.6.22}$$

where

$$m_2' = m_{X_2} + \rho \frac{\sigma_{X_2}}{\sigma_{X_1}} (x_1 - m_{X_1}) \tag{3.6.23}$$

and

$$\sigma_2' = \sqrt{(1 - \rho^2)}\sigma_{X_2} \tag{3.6.24}$$

† We are skirting about the edges of random processes, i.e., time-dependent random phenomena. (Parzen [1962].) In that language, we might (more strongly still) assume $X(t)$ to be a "normal process."
‡ See, for example, Chow [1964], Pattison [1964] and Eagleson, Mejia, and March [1965].

In words, the conditional PDF of X_2 given $X_1 = x_1$ is normal with mean m_2' and standard deviation σ_2'. That is,

$$m_2' = E[X_2 \mid X_1 = x_1] = m_{X_2|X_1} = m_{X_2} + \rho \frac{\sigma_{X_2}}{\sigma_{X_1}}(x_1 - m_{X_1}) \qquad (3.6.25)$$

$$(\sigma_2')^2 = \mathrm{Var}\,[X_2 \mid X_1 = x_1] = (1 - \rho^2)\sigma_{X_2}^2 \qquad (3.6.26)$$

Still attempting to predict X_2, we are now in a state of information defined by this *conditional* (normal) density function. That is, X_2 is still a random variable unknown to us in value, but, forced to predict its value, our logical predicted value is now m_{X_2} altered by an amount $\rho(\sigma_{X_2}/\sigma_{X_1})(x_1 - m_{X_1})$. If x_1, the observed value of X_1, was higher than predicted (m_{X_1}), then we would now expect X_2 to be higher than we previously predicted if the correlation is positive (Fig. 3.6.1). If correlation is negative, the conditional predicted value of X_2 will be lower than the marginal value.

The dispersion of the conditional distribution (or, alternatively interpreted, the expected value of the squared error of prediction) is

$$E[(X_2 - m_2')^2 \mid X_1 = x_1] = (\sigma_2')^2 = (1 - \rho^2)\sigma_{X_2}^2 \qquad (3.6.27)$$

which is smaller than or equal to $\sigma_{X_2}^2$, the marginal variance or expected squared error. If there is any correlation between the two flows, observation of one will reduce the uncertainty in predicting the other.

For example, a storm of total magnitude 3 in.† will, owing to variation in intensity throughout its duration (of random length) lead to a random input to the watershed system. The engineer's knowledge of the nature of such variations and of the watershed of interest has led to estimates $m_{X_1} = 100$ cfs, $\sigma_{X_1} = 29$ cfs, $m_{X_2} = 120$ cfs, $\sigma_{X_2} = 30$ cfs, and $\rho = -0.5$. These are the parameters of a bivariate normal distribution relating X_1 and X_2, the runoff flows at a particular location in a stream at times $t_1 = \frac{1}{2}$ hr and $t_2 = 1$ hr after the beginning of the storm.‡

If flows in excess of 160 cfs are anticipated, it is necessary to initiate a sequence of operations required to open a diversion gate in a small hydraulic installation. For a "3-in." storm, the likelihood of such a flow at the later time t_2 is only

$$P[X_2 \geq 160] = 1 - F_{X_2}(160)$$

$$= 1 - F_U\left(\frac{160 - 120}{30}\right) = 1 - F_U(1.33)$$

$$= 0.091$$

But if the engineer, needing time to carry out the operations, makes a measurement at the earlier time and finds the flow to be $x_1 = 70$ cfs (or lower than the expected values of 100 cfs), then the conditional distribution of X_2, the later

---

† It would clearly be possible to treat this problem without giving a specific value to the total magnitude. The total magnitude could be considered to be a random variable. Knowing the parameters of the joint distribution of X_1 and X_2 for each value of the storm total magnitude, one could use the techniques of compound distributions (Sec. 3.5.3), which are easily generalized to multivariate distributions.

‡ The negative correlation results from the fact that, for the same total rainfall, larger early flows imply smaller later flows and vice versa. The storm may be short and intense, as in a thunderstorm, or longer and less intense.

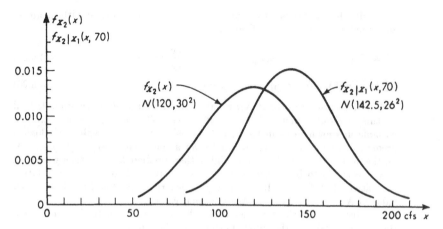

Fig. 3.6.3 The marginal and a conditional distribution of X_2.

flow, is normal with parameters

$$m_2' = 120 + (-0.5)(^{3}\!\%_0)(70 - 100) = 120 + 22.5 = 142.5 \text{ cfs}$$

$$\sigma_2' = \sqrt{(1 - (-.5)^2)}\,(30) = 0.87(30) = 26 \text{ cfs}$$

Conditional on this new information, the risk of X_2 exceeding the critical flow is now

$$P[X_2 \geq 160/X_1 = 70] = 1 - F_U\left(\frac{160 - 142.5}{26}\right)$$

$$= 1 - F_U(0.67)$$

$$= 0.25$$

or nearly three times as great as before. If the consequences of failing to open the gates versus those of opening it unnecessarily are of appropriate magnitudes, this change in the high-flow likelihood could alter the engineer's decision to take action. The marginal and conditional distributions of X_2 are plotted in Fig. 3.6.3.

   In words, if X_1 and X_2 are bivariate normal and correlated (positively, say) and if observation of X_1 yields an observation greater than the expected (predicted) value m_1, it is logical to alter the prediction of X_2 upward [Eq. (3.6.23)]. The amount of change should be proportional to the amount X_1 deviated from its mean, to the degree of correlation between X_2 and X_1 and to the relative dispersion of the two variables. The added information contained in the observation of X_1 is reflected in a more peaked conditional distribution of X_2. This distribution has a smaller variance (σ_2'),[2] a reduced mean square error of prediction; the fractional reduction in mean square error is proportional to the square of the correlation coefficient. Perfect correlation (or linear functional dependence: $X_2 = a + bX_1$) implies that X_2 can be predicted exactly, that is, that the expected squared error in estimating X_2 can be completely eliminated by an observation of X_1. Complete lack of correlation (or independence in this bivariate normal case) implies that ρ is equal to zero; the predicted value of X_2 and the mean square error are left unaltered by the unhelpful information that $X_1 = x_1$.

Many of the problems of engineering experimentation and testing can be interpreted in the light of the engineer's observing one random variable to predict another. Measurements of asphalt content, ultimate compressive stress of concrete, or percentage of optimum compacted soil density are usually made, not because they are of central interest, but because they are easily performed tests on materials properties which are correlated with critical, difficult-to-measure properties such as pavement strength, concrete durability, and embankment stability. The higher the correlation, the better is the test in predicting the adequacy of the key variable.

3.6.3 Summary

In this section two important multivariate distributions were considered. The multinomial distribution is a generalization of the binomial distribution. At each of n independent trial, any of k different events might take place.

The multivariate normal (or gaussian) distribution is a generalization of the univariate normal distribution. This commonly applied distribution is extremely tractable. The marginal and conditional distributions, and the distributions of linear combinations of multivariate normal variables, remain normal. In particular the conditional distribution of Y given $X = x$, when X and Y are bivariate normal, is normal with mean and variance

$$m_{Y|X} = m_Y + \rho \frac{\sigma_Y}{\sigma_X} (x - m_X)$$

$$\sigma^2_{Y|X} = (1 - \rho^2)\sigma_Y{}^2$$

3.7 MARKOV CHAINS†

In practice, the assumption of stochastic independence is frequently made. It is a fundamental assumption, for example, in the Bernoulli trials and Poisson-process models studied in Secs. 3.1 and 3.2. In many applications, however, significant dependence exists between the succes-

† Subsequent sections of the text are not dependent on this section.

sive trials or years or steps that can be identified in the physical process. This stochastic dependence may be quite general, e.g., the conditional distribution of the amount of rainfall tomorrow may depend in varying degrees on the amounts of rainfall on a number of preceding days. On the other hand, this stochastic dependence may be relatively simple; the conditional distribution of the total rainfall next month may depend only on the amount of rainfall observed this month, but not on that in previous months. This last illustration is an example of *Markov* dependence, which we shall consider in this section. Initially, in Sec. 3.7.1, we shall make some necessary definitions and consider some examples of simple processes exhibiting the Markov property. In Sec. 3.7.2 we shall treat the simplest case in some detail, and in Sec. 3.7.3, the results for more general models will be introduced and illustrated.

3.7.1 Simple Markov Chains

Definition and illustration of the Markov property We want to study phenomena whose probabilistic behavior can be characterized by Markov dependence. The clearest example is the running sum of the number of heads encountered in n *independent* flips of a coin. Denote this sum by† $X(n)$ for $n = 1, 2, \ldots$. Suppose we have observed the entire history of the coin-flipping process from the first to nth flips, finding $X(1) = x_1$, $X(2) = x_2, \ldots, X(n) = x_n$. Then the probability that the *next* value of the process, $X(n + 1)$, will equal any particular value x_{n+1}, *given this entire history*, is written in general as

$$P[X(n + 1) = x_{n+1} \mid (X(1) = x_1) \cap (X(2) = x_2) \cap \cdots$$
$$\cap (X(n) = x_n)]$$

But for this coin-flipping process this conditional probability is equal to simply

$$P[X(n + 1) = x_{n+1} \mid (X(1) = x_1) \cap (X(2) = x_2) \cap \cdots$$
$$\cap (X(n) = x_n)] = P[X(n + 1) = x_{n+1} \mid X(n) = x_n] \quad (3.7.1)$$

The conditional probability is functionally independent of the values of process prior to the present value. The probabilities of how many heads we shall have observed after the next flip depend only on how many we have now, and *not* by the path by which we got to this present number.

---

† Recall that the same kind of notation was used for another stochastic process, the Poisson counting process in Sec. 3.2.1. $X(t)$, for $t \geq 0$, was the total number of incidents in the time interval 0 to t. There the "time" parameter t was continuous; in this section we shall consider only a discrete time parameter, $n = 1, 2, \ldots$.

In this simple case, if p is the probability of a head on any flip:

$$P[X(n + 1) = x_{n+1} \mid X(n) = x_n]$$
$$= \begin{cases} p & \text{for } x_{n+1} = x_n + 1 \\ 1 - p & \text{for } x_{n+1} = x_n \\ 0 & \text{otherwise} \end{cases} \quad (3.7.2)$$

Illustration: Soil testing If a series of compaction tests are being carried out on a particular lift of soil during a large embankment construction, we may assume that each is an independent sample of the material. Record is kept of the number of individual specimens which meet specifications. Let $X(n)$ equal the total number of specimens found satisfactory at the completion of n tests. After four tests, the recorded values were

$X(1) = 1$

$X(2) = 2$

$X(3) = 2$

$X(4) = 3$

(Clearly, the third specimen was not satisfactory.) The next or fifth specimen will be either satisfactory or not, implying that the value of the process at step 5, that is, $X(5)$, will be either 4 or 3, respectively. The conditional probabilities of these two values of $X(5)$ are p and $1 - p$, respectively, p being the probability that any specimen is satisfactory. But this would also be true if the past recorded values had been 0, 1, 2, 3 or 1, 1, 2, 3 or 1, 2, 3, 3, that is, no matter which previous specimen was unsatisfactory. In short, the conditional probabilities of the values $X(5)$ depend only on the fact that the present value, $X(4)$, is 3:

$P[X(5) = 4 \mid X(4) = 3] = p$

$P[X(5) = 3 \mid X(4) = 3] = 1 - p$

A stochastic process which has the property defined by Eq. (3.7.1) is said to be a *Markov process*. Such a process is said to be *memoryless;* the future behavior depends on only its present state and not on its past history.

The Markov property can be appreciated better by considering two *non-Markov* processes in the coin-flipping phenomenon. Using stochastic-process notation, we can also treat the *individual* coin flips as a process. Let $Y(n) =$ outcome of the nth flip ($= 0$ or 1 for tail or head, say). This Bernoulli process has *no* trial-to-trial dependence:

$$P[Y(n + 1) = y_{n+1} \mid (Y(1) = y_1) \cap (Y(2) = y_2) \cap \cdots$$
$$\cap (Y(n) = y_n)] = P[Y(n + 1) = y_{n+1}] = p \quad (3.7.3)$$

As discussed, the total-number-of-heads process $X(n)$ has the *one-step* or Markov dependence defined in Eq. (3.7.1). Still other processes associated with coin flipping can be defined which depend on a part or the entire *past history* of the process. For example, define $Z(n)$ = number of heads in the *last two tosses*. In this case, the probability of the next, or $(n + 1)$th, value of the process is influenced by more than just the present, or nth, value of the process. For example, in both of the following cases the nth value of the process is 1, but the probability of the next value being 1 depends on what the value was at step $n - 1$:

$$P[Z(n + 1) = 1 \mid (Z(n) = 1) \cap (Z(n - 1) = 0)] = 1 - p \quad (3.7.4)$$
$$P[Z(n + 1) = 1 \mid (Z(n) = 1) \cap (Z(n - 1) = 2)] = p \quad (3.7.5)$$

Here knowledge of the previous, $(n - 1)$th, value (or state) of the process permits one to deduce that in the first case, the nth flip was a head, while in the second case the nth flip was a tail. This information in turn influences the probability that $Z(n + 1)$ will equal 1. The Markov property [Eq. (3.7.1)] does not hold in this case, and so the process $Z(n)$ is *not* a Markov process.

Typical observed "time" histories of these three stochastic processes, $X(t)$, $Y(t)$, and $Z(t)$, are shown in Fig. 3.7.1. These observations are called *sample functions* or *realizations* of the process.

The word "state" is frequently and conveniently used in discussing Markov processes. If the total number of heads is five at the eighth flip, i.e., if $X(8) = 5$, it would be said that the process is in "state" 5 at "time" 8. Although the set of possible states (i.e., the sample space) can be either continuous or discrete, we shall restrict our attention to *discrete-state processes*, called *Markov chains* (or *random walks*). In this case it is possible to index or number the states, typically 0, 1, 2, . . . , to a finite or infinite number. Therefore we shall usually write a statement such as $P[X(n) = i]$ rather than $P[X(n) = x]$ and read "the probability that the process is 'in state i' at step (or 'time') n," rather than "the probability that the value of the process at step n is x."

Examples Engineering processes which have been or might be modeled as Markov processes are common. Some examples are:

1. The total number of subway system users with destination station B, which have entered at station A, observed each 1 min from 5:30 P.M. to 5:45 P.M.
2. The number of vehicles waiting in a toll-booth queue, the process being observed (not at fixed intervals of time but) just after a car either enters the queue or leaves the toll booth.
3. The state of deterioration of the wearing surface of a bridge (e.g., state 0 = smooth; state 1 = fair; state 2 = rough; state 3 = unacceptable) observed each year after its placement.

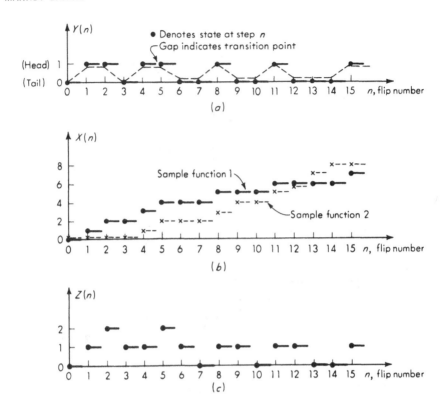

Fig. 3.7.1 Typical sample functions from coin tossing. (a) A typical function of $Y(n)$, the outcome of the nth flip; note the different ways of sketching the sample functions. (b) Two typical sample functions of the process $X(n)$, the total number of heads at the nth flip; sample function 1 corresponds to observations shown in (a). (c) A typical sample function of the process $Z(n)$, the number of heads in previous two flips; function corresponds to $Y(n)$ observation in (a). Define $Z(n) = X(n)$ for $n = 0$ and $n = 1$.

Note that in the last case, the states are not naturally quantitative, the state numbers simply index the various states. Typical sample functions of these three processes might appear as shown in Fig. 3.7.2.

If truly continuous-state spaces are approximated by a discrete set of states, still other examples are possible: the amount of water in a dam observed each month after the inflows and releases for the month have occurred, or the total distance downstream that a particular sediment particle has moved after n impacts by other particles.

The Markov property In each application one must ask if the Markov property [Eq. (3.7.1)] is a reasonable assumption. In words, does the

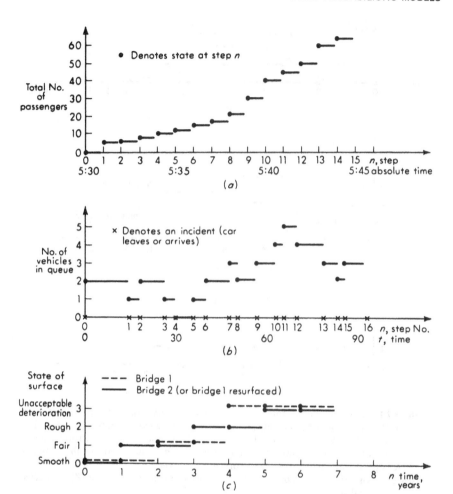

Fig. 3.7.2 Typical sample functions of processes that can be modeled as Markov chains. (*a*) A typical sample function of a passenger-arriving process; note that average rate of arrival may depend on time. (*b*) A typical sample function of a toll-booth queue; note that the discrete steps occur at random times. (*c*) Two typical sample functions of a bridge-wearing-surface process; note that the process may remain in the same state for two or more steps.

future behavior depend only on the present state, and not on the past states? Considering some of the illustrations above, for example, the process of total passengers will be Markov if the numbers arriving in each minute are mutually independent; the bridge-surface-condition process has the Markov property if, roughly, the "damage rate" depends only on the present degree of damage and possibly the age n, but not on, for example, how long the surface spent in state 1; and the dam-contents process will be Markov if, for instance, the monthly flows are independent and the amount of water released depends on the contents (and possibly the date n). In the last case, the process would *not* be Markov if there were correlation between monthly inflows. Positive correlation would mean that low flows are likely to follow low flows. This correlation would imply that the low dam contents *this month* are more likely to be followed by low contents *next month* if the contents were also low last month than if they were not. This is a dependence on the past which is inconsistent with the Markov property. Like independence, the Markov assumption must in practice be made in the basis of professional judgment. The advantage of adopting the assumption is its relative analytical simplicity.

Typical questions We shall be interested in asking a number of kinds of questions about Markov chains. For example, what is the probability that the process is in state i at time n? The reader should immediately know the answer to this question in the case of the total-number-of-heads process $X(n)$ (see Sec. 3.1.2). Or given that the process (e.g., the dam) is now in state i (e.g., half full), we might ask what is the probability that k steps (e.g., months) from now it will be in state j (e.g., full)? Symbolically, calling the present time n, we seek

$$P[X(n + k) = j \mid X(n) = i] \tag{3.7.6}$$

We might also ask, when appropriate, what is the average fraction of the steps (e.g., average proportion of the time) that the process (e.g., the toll-booth queue) spends in state i (e.g., empty). Of the bridge-wearing surface-deterioration process, we might ask what is the probability distribution of the time to the unacceptable state (i.e., the time to "first passage" into state 3)?

Parameters To describe the probabilistic behavior of a Markov chain, we need to have two sets of information. These represent the parameters of the process.

1. First, we must know the state in which the system originates at "time" 0, or, more generally, *a distribution on the initial states:*

$$q_i(0) = P[X(0) = i] \qquad \text{for all } i \tag{3.7.7}$$

Note the introduction of a shorthand notation $q_i(n)$ for the probability that the process is in state i at time (or step) n, in general:

$$q_i(n) = P[X(n) = i] \tag{3.7.8}$$

2. Second, we need the transition probabilities $p_{ij}(n)$. Each of these is the probability that the process will be in state j at time n given that it was in state i at the previous step:

$$p_{ij}(n) = P[X(n) = j \mid X(n-1) = i] \qquad \text{for all } i, j \text{ pairs} \qquad (3.7.9)$$

In different words, the transition probability $p_{ij}(n)$ is the probability that the process will "jump" into state j at the nth step if it is in state i at the $(n-1)$th step. In general this probability is a function of "time" or the step number n. If not, the process is said to be "*homogeneous*" in time, in which case we write simply

$$p_{ij}(n) = p_{ij} \qquad \text{for all } n \qquad (3.7.10)$$

In the previous examples, we can discern immediately some characteristics of their initial-state probabilities $q_i(0)$ and transition probabilities $p_{ij}(n)$. For example, the total-number-of-passengers process starts with certainty in state 0, by definition, and its transition probabilities are such that $p_{ij}(n) = 0$ for $j < i$, since the process is a cumulative one. The transition probabilities could be calculated if the distribution of number of passengers arriving in each minute were known. This might be the same for all minutes over the 15-min interval of interest, implying that the chain is homogeneous, i.e., that the $p_{ij}(n)$ are independent of time p_{ij}.

The process describing the number of vehicles waiting in a toll queue, on the other hand, might start in most any state. The transition probabilities are restricted to permitting jumps to the next lower or next higher state only, since a transition occurs only when a vehicle is added or subtracted from the queue. Transitions to other states have zero probability. Again it might be reasonable to assume that the transition probabilities are time-independent. It can be shown, for example, that, if the vehicle arrivals follow a Poisson processes with arrival rate λ, and if the service time taken at the toll booth is exponentially distributed with mean $1/\mu$ (i.e., the mean service rate is μ), then the queuing process is Markov[†] with transition probabilities dependent in a simple and intuitive manner on these "rates" (Parzen [1962]):

$$p_{i,i-1} = \frac{\mu}{\mu + \lambda} \qquad (3.7.11)$$

$$p_{i,i+1} = \frac{\lambda}{\mu + \lambda} \qquad (3.7.12)$$

A bridge-surface deterioration process with four states (0, 1, 2, 3) denoting advancing states of deterioration would probably start in state

---

[†] The Markov property follows from the memoryless character of the Poisson process (Sec. 3.2.1).

0 at the time of opening of a new surface. In any year the surface condition could potentially advance to any of the higher numbered states of deterioration, or it might remain as it was. It must remain in the final state.† The implication is that $p_{33}(n) = 1$. Such a state (i.e., with $p_{ii}(n) = 1$ for all n) is said to be a *trapping state*, since once there the process cannot leave it. Since the bridge surface cannot improve, the process can never make a transition to a lower state, for example, $p_{21}(n) = 0$ for all n. The values of the other transition probabilities could presumably be estimated from maintenance-department records, depending on the type of surface and type of traffic conditions. If absolute age of the material were influential, the probabilities $p_{ij}(n)$ would depend on n and a complete set of $4 \times 4 = 16$ transition probabilities would be needed for each time n.

Transition-probability matrix Transition probabilities are conveniently displayed in square arrays or *matrices*, which we denote $\mathbf{\Pi}(n)$ (or simply $\mathbf{\Pi}$ for homogeneous chains). For example, in the second year, $n = 2$, the transition probabilities for the bridge surface would appear as:

$$
\mathbf{\Pi}(2) = \begin{array}{c} \\ \text{State left} \begin{cases} (0) \\ (1) \\ (2) \\ (3) \end{cases} \end{array} \overset{\overset{\text{State entered}}{\overbrace{\begin{array}{cccc} (0) & (1) & (2) & (3) \end{array}}}}{\begin{bmatrix} p_{00}(2) & p_{01}(2) & p_{02}(2) & p_{03}(2) \\ 0 & p_{11}(2) & p_{12}(2) & p_{13}(2) \\ 0 & 0 & p_{22}(2) & p_{23}(2) \\ 0 & 0 & 0 & 1 \end{bmatrix}} \tag{3.7.13}
$$

For example $p_{00}(2)$ is the probability of remaining in state 0 during the second year given that in the previous year the state is 0, and $p_{01}(2)$ is the probability of a transition leaving state 0 and entering state 1 in the second year, *given* that it is in state 0 in year 1. Since the surface cannot improve, $p_{10}(2) = 0$, the probability of a transition from state 1 to state 0 in the second year is zero. This accounts for all the zero elements. For the trapping state, $p_{33}(2) = 1$ because the pavement can only remain in state 3. The state number designations on the margins of the matrix will usually be omitted.

In the case of the toll-booth queue the array of (homogeneous) transition probabilities would have this structure:

$$
\mathbf{\Pi} = \begin{bmatrix} 0 & 1 & 0 & 0 & 0 & \cdots \\ p & 0 & 1-p & 0 & 0 & \cdots \\ 0 & p & 0 & 1-p & 0 & \cdots \\ 0 & 0 & p & 0 & 1-p & \cdots \\ \multicolumn{6}{c}{\cdots \cdots \cdots \cdots \cdots \cdots \cdots} \end{bmatrix} \tag{3.7.14}
$$

† That is, it must remain in the final state, at least until resurfacing, when the process would begin anew at time "zero" in state zero.

Recall that a transition takes place when a new car arrives or an existing car is serviced. This matrix is constructed by noting that if zero cars are in the queue, the only possible transition is for a car to arrive: $p_{00} = 0$ and $p_{01} = 1$. With one vehicle in the queue the probability of its being serviced and the queue length becoming zero is p, i.e., $p_{10} = p$. The only other possible event involves the addition of a vehicle to the queue. Thus $p_{12} = 1 - p$, since p_{10} and p_{12} describe the only events that can occur at a transition. Note again that the transitions do not take place at fixed time spacings but at random times with the addition or subtraction of a vehicle from the queue. Note too that in this case there is no logical upper limit to the set of states. Such a process is called an *infinite chain*.

Since at each step *any* Markov process must either remain in the same state or make a transition to some other state, and since these are mutually exclusive events, it is obvious that the sum of the probabilities in any *row* of the array of transition probabilities must equal unity. We can use this observation to calculate one of the elements in a row if all the others are known. This was done in the preceding example.

In the next section we shall demonstrate typical calculations on Markov chains using the simplest possible example, that is, a two-state homogeneous chain. In Sec. 3.7.3 we shall see more general results and illustrations.

3.7.2 Two-state Homogeneous Chains

The simplest example of a Markov chain is one which has but two states, 0 and 1, and one which is homogeneous, i.e., its transition probabilities p_{ij} are independent of time. This case will serve to introduce the analysis of Markov chains, while keeping the algebra simple. The illustration is one of a system with random demand, both the existence and the duration of the demand being probabilistic.

We shall consider the occupancy history of a "slot" in a transit system proposed for congested central cities. In the proposal, a continuous moving "belt" of "slots" (each slot designed for a single passenger) passes along the transit route. The route has a large number of stations or stops, at two- or three-block intervals. At these stations each individual slot or compartment slows down. If it is full, its passenger may exit if he wishes; if the slot enters the station empty or becomes empty, a waiting user, if any is present, will fill it. We want to follow the history of a *single, particular* slot as it passes from station to station.

The process $X(n)$, $n = 0, 1, 2, \ldots$, will take on value 0 or 1 at step n depending on whether the slot is empty or occupied between the nth station and the $(n + 1)$th station; that is, the slot is "observed" after it *leaves* station n.

The transition probabilities can be calculated from the following assumptions. At any station there is a probability p that one or more persons will be waiting to use the slot if it is empty. For simplicity (i.e., to make the process homogeneous) we assume that this probability is the same at all stations. Again for simplicity we shall assume that if there is an individual in the slot as it enters the station, he will get off with probability p', independently of when he got on, and independently of whether another user is waiting.

Under these independence assumptions about the transit users' behavior, the process $X(n)$ is a *Markov* process. The Markov condition, Eq. (3.7.1), is satisfied, since, in words, the probability that the slot will be full, say, as it leaves the next station depends upon whether it is full or empty as it leaves the present station, but not on the history of how it got to its present condition.†

The parameters of the process are:

1. *Initial-state probabilities:*

   For illustration, we assume arbitrarily that the slot starts out empty (i.e., it is empty when *entering* station 1) with certainty; that is,

$$q_0(0) = 1 \tag{3.7.15}$$

$$q_1(0) = 0 \tag{3.7.16}$$

2. *Transition probabilities:*

   (a) If the car leaves a station empty, it will leave the next full only if a passenger is waiting there. Therefore:

$$p_{01} = p \tag{3.7.17}$$

and, since p_{01} plus p_{00} must equal 1,

$$p_{00} = 1 - p_{01} = 1 - p \tag{3.7.18}$$

   (b) If the car leaves one station full, it will leave the next empty only if that passenger gets off *and* none is waiting. This event happens with probability $p'(1 - p)$. Calling this product r, we

---

† The reader might ask himself if any of the simplifying assumptions about the behavior of the transit users can be relaxed without losing the Markov property and its subsequent analytical advantages. Notice that the present assumption about passenger behavior implies a geometric distribution (Sec. 3.1.3) on trip length. This may not be too unrealistic. It is interesting to note, too, that the route could be a *circuit* of, say, c stations; "time" steps n, $n + c$, $n + 2c$, etc. would represent, physically, leaving the same station at different cycles.

have

$$p_{10} = p'(1 - p) = r$$
$$p_{11} = 1 - p_{10} = 1 - r \tag{3.7.19}$$

The transition-probability matrix is

$$\mathbf{\Pi} = \begin{bmatrix} 1 - p & p \\ r & 1 - r \end{bmatrix} \tag{3.7.20}$$

The marginal probabilities of state   We seek first the probability that the process will be in any state i at any time n.   For example, we might want the probability that the slot is full on leaving the second station. We find the probability

$$q_i(n) = P[X(n) = i] \tag{3.7.21}$$

by working frontward one step at a time.   For example, to find $q_i(2)$, consider first that

$$q_1(1) = q_1(0)p_{11} + q_0(0)p_{01}$$

and $\tag{3.7.22}$

$$q_0(1) = q_1(0)p_{10} + q_0(0)p_{00}$$

The first equation says, in words, that the probability that the process is in state 1 at time 1 is the probability that it was in state 1 at the previous step and stayed there *plus* the probability that it was previously in state 0 and then it jumped to state 1.   Knowing these probabilities for step 1, we can find $q_1(2)$, since, by the same argument,

$$q_1(2) = q_1(1)p_{11} + q_0(1)p_{01}$$

Then, substituting, we obtain

$$q_1(2) = q_1(0)p_{11}p_{11} + q_0(0)p_{01}p_{11}$$
$$+ q_1(0)p_{10}p_{01} + q_0(0)p_{00}p_{01} \tag{3.7.23}$$

Each term represents a path by which the process might have ended up in state 1 at time 2.   The third term, for example, is the probability that the slot, say, was initially full, then empty, and then full again.

For the case of a two-state homogeneous chain, one can show by induction that for any† n:

$$q_1(n) = \left[q_1(0) - \frac{p_{01}}{p_{01} + p_{10}} \right] (p_{00} + p_{11} - 1)^n + \frac{p_{01}}{p_{01} + p_{10}} \tag{3.7.24}$$

The probability of being in state 0 at step n or $q_0(n)$ is, of course, just

† This result does not hold for the uninteresting cases: $p_{00} = p_{11} = 1$ or $p_{00} = p_{11} = 0$.

$1 - q_1(n)$. For example, in our case, the probability that the slot will be full after n stations is

$$q_1(n) = \left(0 - \frac{p}{p+r}\right)(1 - p - r)^n + \frac{p}{p+r}$$

$$= \frac{p}{p+r} - \frac{p}{p+r}(1 - p - r)^n \qquad (3.7.25)$$

and

$$q_0(n) = 1 - q_1(n) = \frac{r}{p+r} + \frac{p}{p+r}(1 - p - r)^n$$

If $r = 0.4$ and $p = 0.5$, then the probability that the slot will be empty as it leaves station 4 (given, recall, the initial condition that it *entered* station 1 empty) is

$$q_0(4) = \frac{0.4}{0.9} + \frac{0.5}{0.9}(1 - 0.5 - 0.4)^4 = 0.444 \qquad (3.7.26)$$

This result accounts, of course, for all the possible sequences of empty and full conditions that start empty and end up empty on leaving station 4.

The m-step transition probabilities   More generally we might seek the probability that the process will be in state j m steps later *given* that it is in state i at time n. For example, we might ask, given that the slot was full on leaving station 8, what is the probability that it will be full two stations later, i.e., after station 10. These are called *m-step transition probabilities* $p_{ij}^{(m)}$; the simple transition probability p_{ij} represents the $m = 1$ or one-step case. In general,

$$p_{ij}^{(m)} = P[X(n + m) = j | X(n) = i] \qquad (3.7.27)$$

Owing to the homogeneity in "time" of the process, the argument used above for a start from the initial state still holds, now translated n steps in time. Now, too, we are given that the process is with certainty in a specific state at time n. Therefore, for $n = 8$, $m = 2$, $i = 0$, and $j = 1$, the two-step transition probability $p_{01}^{(2)}$ can be found from Eq. (3.7.23), letting $q_1(0) = 0$ and $q_0(0) = 1$:

$$p_{01}^{(2)} = p_{01}p_{11} + p_{00}p_{01}$$

This is a result we could also arrive at by a simple direct argument, since each term represents a path by which the process starts empty at step n and ends up full two steps later.

To generalize to any m we can make use of Eq. (3.7.24) in a similar way and conclude that the m-step transition probabilities are, given a start in state 1 at time n:

$$p_{11}^{(m)} = \left(1 - \frac{p_{01}}{p_{01} + p_{10}}\right)(p_{00} + p_{11} - 1)^m + \frac{p_{01}}{p_{01} + p_{10}} \qquad (3.7.28)$$

and, given a start in state zero at time n,

$$p_{00}^{(m)} = 1 - p_{01}^{(m)}$$

$$= 1 - \left[0 - \left(\frac{p_{01}}{p_{01} + p_{10}}\right)\right](p_{00} + p_{11} - 1)^m - \frac{p_{01}}{p_{01} + p_{10}} \qquad (3.7.29)$$

The other two m-step transition probabilities, $p_{10}^{(m)}$ and $p_{01}^{(m)}$, are simply $1 - p_{11}^{(m)}$ and $1 - p_{00}^{(m)}$.

For example, if we ask for the probability that a slot will be empty four stations after leaving station n empty, we ask for

$$p_{00}^{(4)} = 1 + \frac{p}{p + r}(1 - p - r)^4 - \frac{p}{p + r}$$

If $p = 0.5$ and $r = 0.4$,

$$p_{00}^{(4)} = 1 + \frac{0.5}{0.9}(1 - 0.5 - 0.4)^4 - \frac{0.5}{0.9} = 0.444 \qquad (3.7.30)$$

This is, of course, exactly equal to $q_0(4)$ [Eq. (3.7.26)], since we assumed there that the initial state was empty, and since the m-step transition probabilities are independent of the time n for homogeneous Markov chains.

Steady-state probabilities   In many cases the initial state probabilities are not well known, and it is meaningful to ask, "Do the probabilities of state $q_i(n)$ take on values which are independent of the initial state?" Can one give the probability that a slot will be full at some future time without knowing whether it was initially full or empty?   In different words, does the process reach a probabilistic steady-state or equilibrium condition, and is this condition independent of its starting position? Here, steady state does *not* suggest that the process is fixed in a given state, but rather that it is moving (stochastically) among the states in a manner which is uninfluenced by "starting transients."   The obvious analogy is with the dynamic response of simple deterministic mechanical systems to periodic forcing functions.

The answer to these questions for the simple homogeneous chain we are considering is affirmative.   Notice that as n grows in Eq. (3.7.24) for $q_1(n)$, the first term grows smaller and smaller in absolute value,†

---

† This is true because the absolute value of $(p_{00} + p_{11} - 1)$ is less than 1 unless we have a degenerate case where p_{00} and p_{11} both equal 0 or 1.

leaving only the second term:

$$q_1(n) \approx \frac{p_{01}}{p_{01} + p_{10}} \qquad \text{for large } n \qquad (3.7.31)$$

and

$$q_0(n) = 1 - p_1(n) \approx \frac{p_{10}}{p_{01} + p_{10}} \qquad \text{for large } n \qquad (3.7.32)$$

These results are independent of the initial-state probabilities and of the time n. The heuristic interpretation of the first result is that the probability that the process is in state 1 at *any* time (after the process has been going for some time) is equal to the ratio of the "rate" at which the process enters the state (p_{01}) compared to the total rate at which the process enters all states $(p_{01} + p_{10})$.

These long-run or steady-state probabilities will be denoted q_i^*. Formally, they are the limits

$$q_i^* = \lim_{n \to \infty} q_i(n) \qquad (3.7.33)$$

The probability q_i^* can be interpreted as either:

1. The probability that the process will be in state i at any point in time (a sufficiently long time after the process starts), or
2. The *average fraction*† *of the time* (i.e., the average fraction of the number of steps) that the process is in state i.

In our illustration, the steady-state probabilities are

$$q_1^* = \frac{p_{01}}{p_{01} + p_{10}} = \frac{p}{p + r} \qquad (3.7.34)$$

$$q_0^* = \frac{p_{10}}{p_{01} + p_{10}} = \frac{r}{p + r} \qquad (3.7.35)$$

With specific numerical values for p and r, we can study the rate at which the transit system process leaves the transient initial-state–dependent condition and approaches the steady-state condition. For example, for $r = 0.4$ and $p = 0.5$ the following values for $q_0(n)$ are obtained, depending on whether the slot begins in the empty or full state:

n	0	1	2	3	4	5	$\cdots$	∞
Initially empty	1	0.5	0.450	0.445	0.444	0.444	$\cdots$	$0.444 = q_0(n)$
Initially full	0	0.4	0.440	0.444	0.444	0.444	$\cdots$	$0.444 = q_0(n)$

† This result will not be proved here. See Parzen [1962].

No matter what the initial state, the probability of being empty is $r/(p + r) = 0.444$ within four stations. If the probability of a passenger's getting off is smaller, say $r = 0.18$, and if the probability of a passenger's waiting is smaller, say $p = 0.3$, then the state condition is approached more slowly[†]:

n	0	1	2	3	4	5	6	$\cdots$	∞
Initially empty	1	0.7	0.544	0.463	0.421	0.399	0.387	$\cdots$	$0.375 = q_0(n)$
Initially full	0	0.18	0.274	0.322	0.348	0.361	0.368	$\cdots$	$0.375 = q_0(n)$

Convergence is slower, but eventually the probability of the slot being empty is $r/(p + r) = 0.375$, no matter whether it was initially empty or full.

We interpret this last result, $p_0^* = 0.375$, as either the probability that the slot will be empty on leaving any particular station or as the average number of stations from which the slot will depart empty.

3.7.3 Multistate Markov Chains

The results of the previous section will be stated in this section for multistate Markov chains with homogeneous or nonhomogeneous transition probabilities. The derivations and interpretations of these results are analogous to those for the simple two-state homogeneous chain; therefore they will not be discussed in detail. Illustrations will follow.

m-Step transition probabilities; homogeneous chains $p_{ij}^{(m)}$ If the process has r possible states and if the transition probabilities are independent of time, the array of transition probabilities p_{ij} can be written in this form:

$$\mathbf{\Pi} = \begin{bmatrix} p_{11} & p_{12} & p_{13} & \cdots & p_{1r} \\ p_{21} & p_{22} & p_{23} & \cdots & p_{2r} \\ \cdots\cdots\cdots\cdots\cdots\cdots \\ p_{r1} & p_{r2} & p_{r3} & \cdots & p_{rr} \end{bmatrix} \tag{3.7.36}$$

The m-step transition probabilities $p_{ij}^{(m)}$ can be found by working frontward one step at a time. The one-step transition probabilities are simply

$$p_{ij}^{(1)} = p_{ij}$$

[†] If both p and r are large enough (such that $p(1 + r) > 1$), then $q_0(n)$ will *oscillate* about q_0^*, converging to it at a slower or faster rate, depending on the specific values of p and r.

The two-step transition probabilities can be found from the one-step probabilities:

$$p_{ij}^{(2)} = p_{i1}^{(1)}p_{1j} + p_{i2}^{(1)}p_{2j} + p_{i3}^{(1)}p_{3j} + \cdots + p_{ir}^{(1)}p_{rj}$$

$$= \sum_{k=1}^{r} p_{ik}^{(1)}p_{kj} \tag{3.7.37}$$

This equation states simply that the probability that the process is in state j, two steps after being state in i, is the sum over all states of the probabilities that it went to state k in the first step and then passed from state k to state j on the second step.

Knowing the two-step probabilities, we can find the three-step transition probabilities as

$$p_{ij}^{(3)} = \sum_{k=1}^{r} p_{ik}^{(2)}p_{kj} \tag{3.7.38}$$

Continuing, we can finally write†

$$p_{ij}^{(m)} = \sum_{k=1}^{r} p_{ik}^{(m-1)}p_{kj} \tag{3.7.39}$$

For a homogeneous Markov chain the m-step transition probabilities are independent of time n. Thus for any n, $p_{ij}^{(m)}$ is the probability that the process will be in state j, m steps later, given that it is in state i at step n.

Marginal state probabilities $q_j(n)$; homogeneous chains In particular, the m-step transition probabilities hold for $n = 0$, the initial time. Therefore, knowing the initial probabilities $q_i(0)$, we can easily find $q_j(n)$, the unconditional probability that the process is in any particular state j at any particular time n. It equals simply‡

$$q_j(n) = q_1(0)p_{1j}^{(n)} + q_2(0)p_{2j}^{(n)} + \cdots + q_r(0)p_{rj}^{(n)}$$

$$= \sum_{i=1}^{r} q_i(0)p_{ij}^{(n)} \tag{3.7.40}$$

---

† The reader familiar with simple matrix algebra will recognize that Eq. (3.7.37) defines the operation of the multiplication of two matrices, $\Pi^{(2)} = \Pi^{(1)}\Pi$, in which, in general, $\Pi^{(m)}$ is the matrix of m-step transition probabilities $p_{ij}^{(m)}$. Equation (3.7.39) can be written concisely in these terms as $\Pi^{(m)} = \Pi^{(m-1)}\Pi$. Substituting $\Pi^{(m-2)}\Pi$ for $\Pi^{(m-1)}$, we obtain finally simply $\Pi^{(m)} = \Pi\Pi\Pi \cdots \Pi$, the product of the matrix Π by itself m times, or the "mth power" of Π, which we denote simply Π^m. We shall continue to show the concise matrix representation of Markov-chain results in footnotes.

‡ In matrix form we find the *vector* of state probabilities $\mathbf{q}(n)$ as

$$\mathbf{q}(n) = \mathbf{q}(0)\Pi^{(n)} = \mathbf{q}(0)\Pi^n$$

Marginal and m-step transition probabilities; nonhomogeneous chains
If a process is not homogeneous, its transition probabilities $p_{ij}(n)$ depend
on time n. In this case the m-step transition probabilities, denoted
$p_{ij}^{(m)}(n)$, also will depend on the absolute time n at which the process
is given to be in state i. Otherwise the results are similar. Eq. (3.7.37)
becomes

$$p_{ij}^{(2)}(n) = p_{i1}^{(1)}(n)p_{1j}(n + 2) + p_{i2}^{(1)}(n)p_{2j}(n + 2)$$
$$+ \cdots + p_{ir}^{(1)}(n)p_{rj}(n + 2)$$
$$= \sum_{k=1}^{r} p_{ik}^{(1)}(n)p_{kj}(n + 2) \tag{3.7.41}$$

$p_{ik}^{(1)}(n)$ is, of course, simply equal to $p_{ik}(n + 1)$. Equation (3.7.39)
becomes, for nonhomogeneous chains,

$$p_{ij}^{(m)}(n) = \sum_{k=1}^{r} p_{ik}^{(m-1)}(n)p_{kj}(n + m) \tag{3.7.42}$$

which we interpret as saying the probability that a process will be in
state j at time $n + m$ given that it was in state i at time n is the sum over
all states, $k = 1, 2, \ldots, r$, of the probabilities that the process goes to
state k in $m - 1$ steps and then passes to state j on the mth step.

The marginal probabilities of state $q_j(n)$ for a nonhomogeneous
chain are†

$$q_j(n) = \sum_{i=1}^{r} q_i(0)p_{ij}^{(n)}(0) \tag{3.7.43}$$

Equilibrium probabilities; homogeneous chains   For chains whose transition probabilities are independent of time, we often want to ask if the
marginal probabilities of state $q_i(n)$ become, in sufficient time, independent of the initial-state probabilities and independent of time n. In
many practical cases such limiting or *equilibrium* or *stationary probabilities* q_i^* do exist. It is beyond the scope of this text to define formally
the conditions under which these limiting probabilities exist. Roughly,
if communication or passage to and from all pairs of states is possible
(not necessarily in one step, but perhaps through other states), then the
probabilities of state do in time become independent of the initial con-

---

† In matrix form, where $\boldsymbol{\Pi}^{(m)}(n)$ is the array of $p_{ij}^{(m)}(n)$ probabilities:

$$\boldsymbol{\Pi}^{(m)}(n) = \boldsymbol{\Pi}(n + 1)\boldsymbol{\Pi}(n + 2) \cdots \boldsymbol{\Pi}(n + m)$$

and

$$\mathbf{q}(n) = \mathbf{q}(0)\boldsymbol{\Pi}^{(n)}(0)$$

in which $\mathbf{q}(n)$ is the column vector of unconditional probabilities of state at time n.

ditions.   This condition excludes, for example, chains which have closed or trapping states (i.e., states which cannot be left because $p_{ii} = 1$) or closed *groups* of states.

If the stationary probabilities exist, they can be found—through lengthy computation—by simply finding the m-step transition probabilities $p_{ij}{}^{(m)}$ for large enough m.   In time these probabilities will be the same for all m and all starting states i; that is, all the rows in the matrix of $p_{ij}{}^{(m)}$'s will be identical.†

On the other hand, we can find the probabilities directly if we observe that *when* the equilibrium condition exists, the probability of being in state i at step n is *the same* as that at step $n + 1$, or

$$q_i(n + 1) = q_i(n) = q_i^*   \tag{3.7.44}$$

but, $q_i(n + 1)$ is related to $q_i(n)$ through the transition probabilities [e.g., Eq. (3.7.40)]

$$q_i(n + 1) = \sum_{k=1}^{r} q_k(n) p_{ki}$$

Therefore when the equilibrium condition holds, the r stationary probabilities are related through a set of r equations:

$$q_i^* = \sum_{k=1}^{r} q_k^* p_{ki}   \qquad i = 1, 2, \ldots, r   \tag{3.7.45}$$

We must also include the equation‡

$$\sum_{i=1}^{r} q_i^* = 1   \tag{3.7.46}$$

Solution of these simultaneous linear equations§ will yield the stationary probabilities.   An illustration will follow.

Again, as in the two-state case, these probabilities can be interpreted as probabilities of state in the long run or as the average fraction of time spent in a particular state.

Illustration: Bridge-wearing surface deterioration   In Sec. 3.7.1, we introduced a Markov-chain example concerned with the deterioration of a bridge-wearing surface with time, measured from the year it is applied.   The states represent

† In matrix form $\mathbf{\Pi}^{(m)} = \mathbf{\Pi}^m$ becomes:

$$\begin{bmatrix} q_1^* & q_2^* & \cdots & q_r^* \\ \cdots\cdots\cdots\cdots\cdots \\ q_1^* & q_2^* & \cdots & q_r^* \end{bmatrix} \quad \text{for large } m$$

‡ Since the trival solution $q_1^* = q_2^* = \cdots = q_r^* = 0$ satisfies the set of equations, Eq. (3.7.45).

§ In matrix form we must solve $\mathbf{q}^* = \mathbf{q}^* \mathbf{\Pi}$, together with Eq. (3.7.46).

successively rougher surfaces. Due to annual variations in traffic and weather, the transitions to states of higher deterioration are probabilistic. Since deterioration enhances roughness growth by impounding water and by increasing vehicle vibrations, the transition probabilities depend on the state. Owing to aging of the surface material (e.g., embrittlement, hydration, freezing, and thawing), the conditional probabilities of making a particular transition in any year change with time. Assume that the following (fictitious) sets of transition matrices hold for a particular bridge [Eq. (3.7.13)]:

$$\mathbf{\Pi}(1) = \begin{bmatrix} 0.7 & 0.2 & 0.1 & 0 \\ 0 & 0.5 & 0.3 & 0.2 \\ 0 & 0 & 0.3 & 0.7 \\ 0 & 0 & 0 & 1 \end{bmatrix} \qquad \mathbf{\Pi}(2) = \begin{bmatrix} 0.6 & 0.3 & 0.1 & 0 \\ 0 & 0.4 & 0.4 & 0.2 \\ 0 & 0 & 0.2 & 0.8 \\ 0 & 0 & 0 & 1 \end{bmatrix}$$

$$\mathbf{\Pi}(3) = \begin{bmatrix} 0.5 & 0.3 & 0.2 & 0 \\ 0 & 0.3 & 0.4 & 0.3 \\ 0 & 0 & 0.1 & 0.9 \\ 0 & 0 & 0 & 1 \end{bmatrix}$$

Assume for simplicity that $\mathbf{\Pi}(4)$, $\mathbf{\Pi}(5)$, etc., are the same as $\mathbf{\Pi}(3)$. The initial-state probability vector

$$q_i(0) = \begin{cases} 0.9 & i = 0, \text{ smooth} \\ 0.1 & i = 1, \text{ fair} \\ 0 & i = 2, \text{ rough} \\ 0 & i = 3, \text{ unacceptable} \end{cases}$$

reflects the fact that the surface is only under design, not in place, and that construction control may not yield a smooth surface.

Let us find $p_{ij}{}^{(m)}(0)$, the m-step transition probabilities given that the process is in state i at time $n = 0$, using Eqs. (3.7.41) and (3.7.42).

For one year or step, $m = 1$:

$$p_{ij}{}^{(1)}(0) = p_{ij}(1)$$

For example, $p_{00}{}^{(1)}(0) = p_{00}(1) = 0.7$ and $p_{01}{}^{(1)}(0) = p_{01}(1) = 0.2$.

For two years or steps, $m = 2$ [(Eq. (3.7.41)]:

$$p_{ij}{}^{(2)}(0) = \sum_{k=0}^{3} p_{ik}{}^{(1)}(0) p_{kj}(2)$$

For a start in state $i = 0$:

$$p_{00}{}^{(2)}(0) = p_{00}{}^{(1)}(0) p_{00}(2) + p_{01}{}^{(1)}(0) p_{10}(2) + \cdots$$
$$= (0.7)(0.6) + (0.2)(0) + 0 + 0 = 0.42$$

$$p_{01}{}^{(2)}(0) = p_{00}{}^{(1)}(0) p_{01}(2) + p_{01}{}^{(1)}(0) p_{11}(2) + p_{02}{}^{(1)}(0) p_{21}(2) + \cdots$$
$$= (0.7)(0.3) + (0.2)(0.4) + (0.1)(0) + 0 = 0.29$$

$$p_{02}{}^{(2)}(0) = p_{00}{}^{(1)}(0) p_{02}(2) + p_{01}{}^{(1)}(0) p_{12}(2) + p_{02}{}^{(1)}(0) p_{22}(2) + p_{03}{}^{(1)}(0) p_{23}(2)$$
$$= (0.7)(0.1) + (0.2)(0.4) + (0.1)(0.2) + 0 = 0.17$$

$$p_{03}{}^{(2)}(0) = 1 - [p_{00}{}^{(2)}(0) + p_{01}{}^{(2)}(0) + p_{02}{}^{(2)}(0)]$$
$$= 1 - (0.42 + 0.29 + 0.17) = 0.12$$

Similarly, given a start in state $i = 1$:

$$p_{10}^{(2)}(0) = 0$$

$$p_{11}^{(2)}(0) = (0.5)(0.4) = 0.20$$

$$p_{12}^{(2)}(0) = (0.5)(0.4) + (0.3)(0.2) = 0.26$$

$$p_{13}^{(2)}(0) = (0.5)(0.2) + (0.3)(0.8) + (0.2)(1)$$

$$= 1 - (0.20 + 0.26) = 0.54$$

In this case we are not interested in chains which begin in states 2 or 3 at time zero, since we know from the given initial probability vector that this will not happen. These results say, for example, that a surface starting out in only fair condition ($i = 1$) is about 11 times $(0.54/0.05)$ more likely to be in an unacceptable condition ($j = 3$) in 2 years than a surface starting out smooth ($i = 0$).

For three years or steps, $m = 3$ [Eq. (3.7.42)]:

$$p_{ij}^{(3)}(0) = \sum_{k=0}^{3} p_{ik}^{(2)}(0)p_{kj}(3)$$

For example,

$$p_{02}^{(3)}(0) = p_{00}^{(2)}(0)p_{02}(3) + p_{01}^{(2)}(0)p_{12}(3) + p_{02}^{(2)}(0)p_{22}(3) + p_{03}^{(2)}(0)p_{32}(3)$$

$$= (0.42)(0.2) + (0.29)(0.4) + (0.17)(0.1) + 0$$

$$= 0.207$$

Similarly, we can find all the elements of interest in the array $p_{ij}^{(3)}(0)$

$$\begin{matrix} 0 \\ 1 \\ 2 \\ 3 \end{matrix} \begin{bmatrix} 0.210 & 0.213 & 0.207 & 0.370 \\ 0 & 0.060 & 0.106 & 0.834 \\ & \text{Not of interest} \\ & \text{in this case} \end{bmatrix}$$

Element $p_{03}^{(3)}(0)$ equals 0.353, which says that there is a probability of this magnitude that a surface which was initially smooth will be unacceptable 3 years later.

In a similar manner we could find, say, $p_{03}^{(3)}(2)$, which would give the probability that a surface which was still "smooth" in the second year would be unacceptable 3 years later.

The $p_{ij}^{(m)}(0)$ probabilities can be used to find the $q_j(n)$, that is, the unconditional or marginal probabilities that the surface will be in any state in any year. Using Eq. (3.7.43)

$$q_j(3) = \sum_{i=0}^{3} q_i(0)p_{ij}^{(3)}(0) \qquad \text{for } n = 3$$

We find that the probability that the bridge surface is smooth 3 years after construction is

$$q_0(3) = q_0(0)p_{00}^{(3)}(0) + q_1(0)p_{10}^{(3)}(0) + q_2(0)p_{20}^{(3)}(0) + q_3(0)p_{30}^{(3)}(0)$$

$$= (0.9)(0.210) + (0.1)0 + 0 + 0$$

$$= 0.189$$

The probabilities of other states after year 3 are

$$q_1(3) = (0.9)(0.213) + (0.1)(0.060) = 0.198$$
$$q_2(3) = (0.9)(0.207) + (0.1)(0.106) = 0.197$$
$$q_3(3) = (0.9)(0.370) + (0.1)(0.834) = 0.416$$

Time to first passage. Finally, the engineer might want to know the distribution of the random variable T, the time from construction until the surface becomes unacceptable. (This would be the distribution of time *between* resurfacings if the model holds for resurfaced bridges as well.) Clearly, since the process cannot leave state 3, the probability that T is *less* than or equal to any value t is the probability that the process is in state 3 at that time

$$F_T(t) = P[T \le t] = q_3(t) \tag{3.7.47}$$

Thus

$$F_T(1) = q_3(1) = (0.9)(0) + (0.1)(0.2) = 0.02$$
$$F_T(2) = q_3(2) = (0.9)(0.12) + (0.1)(0.54) = 0.114$$
$$F_T(3) = q_3(3) = 0.416$$

. . . .

The PMF is

$$p_T(1) = F_T(1) = 0.02$$
$$p_T(2) = F_T(2) - F_T(1) = 0.114 - 0.02 = 0.094$$
$$p_T(3) = F_T(3) - F_T(2) = 0.416 - 0.114 = 0.302$$

. . . .

The result in Eq. (3.7.47) is true *only* because state 3 is a trapping state. But if the distribution of the "time to first passage" into any state (state i) is desired, it can be found by simply modifying all the transition matrices to make state i a trapping state (i.e., make $p_{ii}(n) = 1$, for all n). Then (different) probabilities $q_i(t)$ will be obtained which can be interpreted as the CDF of the time to first entrance into that state. These times are generalizations of the geometrically distributed time to first success in a Bernoulli process.

Illustration: Dam storage   The stochastic behavior of dam storage has been thoroughly analyzed as a Markov process (Moran [1959]). We shall consider a simple example of a small irrigation storage dam. Every year n, the total amount of water flowing into the dam is a random quantity Y_n with the discretized probability distribution shown in Fig. 3.7.3a. The unit used is one-third of the capacity of the dam. The intent of the regulating agency is to provide two units of water per year to neighboring agricultural lands, if possible. Without the dam for storage it is clear that in 40 percent of the years this much water would not be available, but in 30 percent of the years excess water would arrive at the dam site. With the dam, the agency will release in each year two units of water *if* this is feasible with the *total* water available, that is, with the water from the year's inflow plus that stored or carried over from

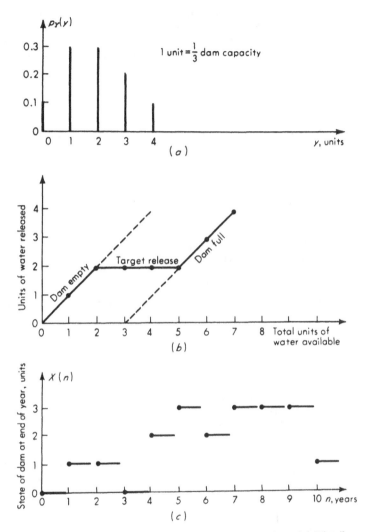

Fig. 3.7.3 Storage-dam illustration of a Markov chain. (a) Distribution of annual inflows; (b) "operating-policy" curve; (c) typical sample function or history of dam.

the previous year (if any). If water in excess of the "target release" of two units plus the storage capacity of the dam (three units) is available, the excess will be released also. This rule of action is summarized in the operating-policy curve Fig. 3.7.3b. (See also Prob. 2.32.)

To study the behavior over time of the dam under this policy, we shall "observe" the water remaining in the dam at the end of each year. If we assume that the annual inflows are stochastically independent, then the model is a

Markov chain; the probabilities of future states of the process depend only on the present state, not the past sequence of states. We denote the process $X(n)$. The process is homogeneous. A sample function is shown in Fig. 3.7.3c.

With the information available the transition-probability matrix is easily assembled, since

$$p_{ij} = P[X(n+1) = j \mid X(n) = i]$$

$$= P[\text{storage } (n) + \text{inflow } (n+1) - \text{release } (n+1) = j \mid \text{storage } (n) = i]$$

For example if the dam is empty, it will be empty the next year if two or fewer units flow in (since they will all be released):

$$p_{00} = P[Y_{n+1} \le 2] = 0.7$$

If there is one unit in storage this year, there will be none next year if zero or one unit flows in:

$$p_{10} = P[Y_{n+1} \le 1] = 0.4$$

and there will be one next year if two flow in (yielding a total of three units available, two of which are released):

$$p_{11} = P[Y_{n+1} = 2] = 0.3$$

If the dam is full this year, it will remain full if two or more units flow in:

$$p_{33} = P[Y_{n+1} \ge 2] = 0.6$$

In this way the entire transition matrix can be constructed:

$$
\mathbf{\Pi} = \begin{array}{c} \\ \\ \text{State left} \\ \\ \end{array}
\begin{array}{c}
\\
(0) \\ (1) \\ (2) \\ (3)
\end{array}
\overset{\text{State entered}}{
\begin{array}{cccc}
(0) & (1) & (2) & (3) \\
\left[\begin{array}{cccc}
0.7 & 0.2 & 0.1 & 0 \\
0.4 & 0.3 & 0.2 & 0.1 \\
0.1 & 0.3 & 0.3 & 0.3 \\
0 & 0.1 & 0.3 & 0.6
\end{array}\right]
\end{array}}
$$

Although we will not use this information in this illustration, one might reasonably assume that initially, say just after construction, the dam is empty with certainty:

$$q_i(0) = \begin{cases} 1 & i = 0 \\ 0 & i = 1 \\ 0 & i = 2 \\ 0 & i = 3 \end{cases}$$

As with the nonhomogeneous bridge-surface illustration, we could answer any number of questions about the likelihoods of being in state i at time n or of being in state j m years after leaving state i at time n, etc. In this case we will seek the answers to only two questions however. They are:

1. What is the long-run fraction of time that the dam will be full, $\frac{2}{3}$ full, $\frac{1}{3}$ full, and empty when observed at the end of each year?
2. What is the distribution of the time at which the dam is first empty given that it is full at present?

Equilibrium probabilities To answer the first question we must find the stationary probabilities of state q_i^*. There is no question about their existance in this chain, since there is free communication to and from all states. (Note, however, that since p_{03} is zero, access to state 3 from state 0 is only through some other state.) Applying Eqs. (3.7.45) and (3.7.46), we have the following set of equations solve:

$$q_0^* = q_0^* p_{00} + q_1^* p_{10} + q_2^* p_{20} + q_3^* p_{30}$$
$$q_1^* = q_0^* p_{01} + q_1^* p_{11} + q_2^* p_{21} + q_3^* p_{31}$$
$$q_2^* = q_0^* p_{02} + q_1^* p_{12} + q_2^* p_{22} + q_3^* p_{32}$$
$$q_3^* = q_0^* p_{03} + q_1^* p_{13} + q_2^* p_{23} + q_3^* p_{33}$$

and

$$q_0^* + q_1^* + q_2^* + q_3^* = 1$$

Substituting for the p_{ij} and rearranging, we have

$$(0.7 - 1)q_0^* + 0.4q_1^* + 0.1q_2^* = 0$$
$$0.2q_0^* + (0.3 - 1)q_1^* + 0.3q_2^* + 0.1q_3^* = 0$$
$$0.1q_0^* + 0.2q_1^* + (0.3 - 1)q_2^* + 0.3q_3^* = 0$$
$$0.1q_1^* + 0.3q_2^* + (0.6 - 1)q_3^* = 0$$
$$q_0^* + q_1^* + q_2^* + q_3^* = 1$$

The solution is

$$q_i^* = \begin{cases} 0.52 & i = 0 \\ 0.38 & i = 1 \\ 0.05 & i = 2 \\ 0.05 & i = 3 \end{cases}$$

After the startup transients have lost influence, there is a probability of 5 percent that the dam will be full at (any) time n. In different words, on the average, the dam will be full 5 percent of the times it is observed (i.e., after inflows and releases for the year). More than half the time it will be empty. Only about 1 time in 5 that it is not empty will it be more than $\frac{1}{3}$ full.

First-passage times Recall the argument in the previous illustration. To calculate the distribution of T, the first-passage time to the empty state given that the dam is now full, we need only to find the m-step transition probabilities *after* changing the empty state to a trapping state. The modified transition matrix is

$$\Pi_{\text{modified}} = \begin{bmatrix} 1 & 0 & 0 & 0 \\ 0.4 & 0.3 & 0.2 & 0.1 \\ 0.1 & 0.3 & 0.3 & 0.3 \\ 0 & 0.1 & 0.3 & 0.6 \end{bmatrix}$$

Then

$$F_T(t) = P[T \leq t] = P[\text{process enters state 0 at some time prior to or at}$$
$$\text{step } t \mid \text{now in state 3}]$$
$$= P[\text{process is in ``trapping state'' 0 at time } t \mid \text{now in state 3}]$$
$$= P[\text{process is in ``trapping state'' 0 } t \text{ steps after being in}$$
$$\text{state 3}]$$
$$= p_{30}{}^{(t)}$$

Using Eqs. (3.7.37) to (3.7.39),

$$F_T(1) = p_{30}{}^{(1)} = 0$$

The dam cannot be empty only 1 year later.   Next,

$$F_T(2) = p_{30}{}^{(2)} = p_{30}{}^{(1)}p_{00} + p_{31}{}^{(1)}p_{10} + p_{32}{}^{(1)}p_{20} + p_{33}{}^{(1)}p_{30}$$
$$= (0)(1) + (0.1)(0.4) + (0.3)(0.1) + (0.6)(0) = 0.07$$

In words, the dam will be empty 2 years after being full only if it was first $\frac{1}{3}$ full and then empty, or first $\frac{2}{3}$ full and then empty.   Continuing,

$$F_T(3) = p_{30}{}^{(3)} = p_{30}{}^{(2)}p_{00} + p_{31}{}^{(2)}p_{10} + p_{32}{}^{(2)}p_{20} + p_{33}{}^{(2)}p_{30}$$
$$= (0.07)1 + (0.18)(0.4) + (0.29)(0.1) + (0.46)(0) = 0.17$$

There is a probability 0.171 that the dam will have been empty on or before the third year.   But there was a probability of 0.07 that it would be empty by the second year; therefore, there is a probability of $0.17 - 0.07 = 0.10$ that it will be empty for the *first* time 3 years after it was full.   Continuing, we find the following sequence:

$F_T(1) = 0$	$p_T(1) = 0$
$F_T(2) = 0.07$	$p_T(2) = 0.07$
$F_T(3) = 0.17$	$p_T(3) = 0.10$
$F_T(4) = 0.27$	$p_T(4) = 0.10$
$F_T(5) = 0.36$	$p_T(5) = 0.09$
$F_T(6) = 0.44$	$p_T(6) = 0.08$
$F_T(7) = 0.51$	$p_T(7) = 0.07$
$F_T(8) = 0.57$	$p_T(8) = 0.06$
$F_T(9) = 0.62$	$p_T(9) = 0.05$
$F_T(10) = 0.66$	$p_T(10) = 0.04$

. . . .

This PMF is plotted in Fig. 3.7.4.
   In Prob. 3.45, the results of this illustration are incorporated into an economic decision analysis with respect to the construction (or not) of the dam.

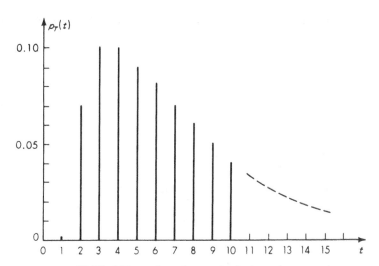

Fig. 3.7.4 PMF of T, number of years to first empty year, given an initially full dam.

3.7.4 Summary

Markov chains $X(n)$ are simple stochastic processes typified by a set of discrete states, $i = 1, 2, \ldots, r$, by a set of discrete observation points in time, $n = 0, 1, 2, \ldots$, and by the memoryless Markov property:

$$P[X(n + 1) = j \mid (X(0) = x_0) \cap (X(1) = x_1) \cap \cdots$$
$$\cap (X(n - 1) = x_{n-1}) \cap (X(n) = i)]$$
$$= P[X(n + 1) = j \mid X(n) = i)] = p_{ij}(n + 1)$$

For the behavior of the process to be completely defined, the initial-state probabilities,

$$q_i(0) = P[X(0) = i]$$

and the transition probabilities $p_{ij}(n)$, for all times $n = 1, 2, \ldots$, must be known. In a homogeneous chain, the $p_{ij}(n)$ are independent of time and are denoted simply p_{ij}.

The (unconditional) probability $q_j(n)$ that the process is in state j at time n is found to be

$$q_j(n) = \sum_{i=1}^{r} q_i(0) p_{ij}{}^{(n)}(0)$$

in which the m-step transition probabilities $p_{ij}{}^{(m)}(n)$ are defined as the probability that the process will be in state j m steps after being in state i at time n. These (conditional) probabilities are found recursively from the equation:

$$p_{ij}^{(m)}(n) = \sum_{k=1}^{r} p_{ik}^{(m-1)}(n)p_{kj}(n+m)$$

For homogeneous chains the m-step transition probabilities, denoted $p_{ij}^{(m)}$, are independent of n. They are found from

$$p_{ij}^{(m)} = \sum_{k=1}^{r} p_{ik}^{(m-1)}p_{kj}$$

In certain cases homogeneous chains have stationary (or steady-state or equilibrium) probabilities of state q_i^*. If they exist, they can be found by solving r simultaneous equations

$$q_i^* = \sum_{k=1}^{r} q_k^* p_{ki} \qquad \text{for } i = 1, 2, \ldots, r$$

with the condition

$$\Sigma q_i^* = 1$$

3.8 SUMMARY FOR CHAPTER 3

In this chapter a variety of the most common probabilistic models of physical phenomena have been presented. Many practical problems fit successfully the assumptions made about the underlying mechanisms considered here. Study of such physical problems is simplified by the availability of many tractable analytical results for these general classes of models.

The basic underlying mechanisms considered were Bernoulli trials in Sec. 3.1; the Poisson arrival process in Sec. 3.2; additive, multiplicative, and extreme-value mechanisms in Secs. 3.3 and 3.6; multinominal trials in Sec. 3.6; and the Markov process in Sec. 3.7. A number of commonly adopted distributions and a number of distributions related to the normal distribution appeared in Sec. 3.4. Various modifications (shifting, truncating, compounding, etc.) of distributions were presented in Sec. 3.5.

In many practical situations the distributions appearing in this chapter are adopted empirically without a supporting physical model. The justifications are "fit" to data and analytical tractability.

In the next chapter we shall see how observed data is used to estimate the values of the parameters of the model, no matter for what reason it was adopted. We shall also find there methods for verifying the model in light of the observed data. In Chap. 6 these models will appear again in decision analyses.

REFERENCES

General

Feller, W. [1957]: "An Introduction to Probability Theory and Its Applications," vol. 1, 2d ed., John Wiley & Sons, Inc., New York.

Fisz, M. [1963]: "Probability Theory and Mathematical Statistics," 3d ed., John Wiley & Sons, Inc., New York.

Gumbel, E. J. [1958]: "Statistics of Extremes," Columbia University Press, New York.

Hahn, G. J. and S. S. Shapiro [1967]: "Statistical Models in Engineering," John Wiley & Sons, Inc., New York.

Hald, A. [1952]: "Statistical Theory with Engineering Applications," John Wiley & Sons, Inc., New York.

Parzen, E. [1960]: "Modern Probability Theory and Its Applications," John Wiley & Sons, Inc., New York.

—— [1962]: "Stochastic Processes," Holden-Day, Inc., Publisher, San Francisco.

Specific text references

Aitchison, J. and J. A. C. Brown [1957]: "The Lognormal Distribution," Cambridge University Press, Cambridge, England.

Allen, C. R., P. St. Amand, C. F. Richter, and J. M. Nordquist [1965]: Relationship between Seismicity and Geologic Structure in the Southern California Region, *Bull. Seismological Soc. Am.*, vol. 55, pp. 753–97.

Altouney, E. G. [1963]: The Role of Uncertainties in the Economic Evaluation of Water-Resources Projects, Stanford University Program in Engineering-Economic Planning, Stanford, Calif.

ASCE Task Committee on Wind Force Final Report [1961]: *ASCE Trans.* vol. 126, part II, p. 1124.

Beard, L. R. [1954]: Estimation of Flood Probabilities, *Proc. ASCE*, vol. 80, sep. no. 438.

Beard, L. R. [1962]: "Statistical Methods in Hydrology," U.S. Army Engineer District, Corps of Engineers, Sacramento, Calif.

Berrettoni, J. N. [1962]: Practical Applications of the Weibull Distribution, *Proc. 16th Ann. Conv. Am. Soc. Quality Control.*

Borgman, L. E. [1963]: Risk Criteria, *J. Waterways Harbors Div., Proc. ASCE*, WW3, vol. 89, August, pp. 1–35.

Chiu, A. N. L., D. A. Sawyer, and L. E. Grinter [1964]: Vibration of Towers as Related to Wind Pulses, *J. Structural Div., Proc. ASCE*, ST 5, vol. 90, October.

Chow, V. T. [1951]: A General Formula for Hydrologic Frequency Analysis, *Trans. Am. Geophys. Union*, vol. 32, p. 231.

Chow, V. T. (ed.) [1964]: "Handbook of Applied Hydrology," McGraw-Hill Book Company, New York.

Chow, V. T. [1954]: The Log-Probability Law and Its Engineering Applications, *Proc. ASCE*, vol. 80, sep. no. 536.

Cornell, C. A. [1968]: Engineering Seismic Risk Analysis, *Bull. Seismological Soc. Am.*, vol. 58, pp. 1583–1606.

Cornell, C. A. [1964]: Stochastic Process Models in Structural Engineering, *Tech. Rept. No. 34*, Department of Civil Engineering, Stanford University, Stanford, Calif.

Court, A. [1959]: Wind Extremes as Design Factors, *J. Franklin Inst.*, vol. 256, p. 39.

Crandall, S. H. and W. D. Mark [1963]: "Random Vibration in Mechanical Systems," Academic Press, Inc., New York.

Eagleson, P. S., R. Mejia, and F. March [1965]: The Computation of Optimum Realizable Unit Hydrographs from Rainfall and Runoff Data, *M.I.T. Dept. Civil Eng. Res. Rept.* R65–39.

Epstein, B. [1947]: The Mathematical Description of Certain Breakage Mechanisms Leading to the Logarithmico-Normal Distribution, *J. Franklin Inst.*, vol. 244, pp. 471–477.

Feller, W. [1966]: "An Introduction to Probability Theory and Its Applications," vol. 2, John Wiley & Sons, Inc., New York.

Fiering, M. B. [1962]: Queueing Theory and Simulation in Reservoir Design, *Trans. ASCE*, vol. 127, part I, pp. 1114–1144.

Freudenthal, A. M. [1951]: Planning and Interpretation of Fatigue Tests, *ASTM Symp. Statis. Aspects Fatigue, Spec. Tech. Pub. no. 121.*

Freudenthal, A. M. [1948]: Reflections on Standard Specifications for Structural Design, *Trans. ASCE*, vol. 113, p. 269.

Freudenthal, A. M., J. M. Garrelts, and M. Shinozuka [1966]: The Analysis of Structural Safety, *J. Structural Div., Proc. ASCE*, ST1, vol. 92, February, pp. 267–325.

Freudenthal, A. M. and E. J. Gumbel [1956]: Physical and Statistical Aspects of Fatigue, *Advan. Appl. Mech.*, vol. 4.

Goodman, L. E., E. Rosenblueth, and N. M. Newmark [1955]: Aseismic Design of Elastic Structures, *Trans. ASCE*, vol. 120, p. 782.

Granholm, Hjalmar [1965]: "A General Flexural Theory of Reinforced Concrete," John Wiley & Sons, Inc., New York.

Greenshields, B. D. and F. M. Weida [1952]: "Statistics with Applications to Highway Traffic Analysis," The Eno Foundation for Highway Control, New Haven, Conn.

Greenwood, M. and G. U. Yule [1920]: An Inquiry into . . . Repeated Accidents, *J. Roy. Statist. Soc.* vol. 83, p. 255.

Gumbel, E. J. [1954a]: Statistical Theory of Extremes and Some Practical Applications, National Bureau of Standards Applied Mathematics Series, 33.

Gumbel, E. J. [1954b]: Statistical Theory of Droughts, *Proc. ASCE*, vol. 80, sep. no. 439.

Haight, F. A. [1963]: "Mathematical Theories of Traffic Flow," Academic Press, Inc., New York.

Hazen, Allen [1914]: Discussion on Flood Flows, *ASCE Trans.*, vol. 77, p. 664.

Hoel, P. G. [1962]: "Introduction to Mathematical Statistics," 3d ed., John Wiley & Sons, Inc., New York.

Johnson, A. I. [1953]: Strength, Safety, and Economical Dimensions of Structures, *Roy. Inst. Technol., Div. Bldg. Statics Structural Eng. (Stockholm), Bull. 12.*

Juncosa, M. L. [1948]: On the Distribution of the Minimum of a Sequence of Mutually Independent Random Variables, *Duke Univ. Math. J.*, vol. 16.

Kottler, F. [1950]: The Distribution of Particle Sizes, *J. Franklin Inst.*, vol. 250, pp. 339–356 and 419–441.

Lloyd, D. K. and M. Lipow [1962]: "Reliability: Management, Methods, and Mathematics," Prentice-Hall, Inc., Englewood Cliffs, N.J.

Loeve, M. [1950]: Fundamental Limit Theorems of Probability Theory, *Ann. Math. Statist.*, vol. 21, pp. 321–337.

Lomnitz, C. [1964]: Estimation Problems in Earthquake Series, *Tectnophysics*, vol. 2, pp. 193–203.

Markovic, R. D. [1965]: Probability Function of Best Fit to Distributions of Annual Precipitation and Runoff, *Colorado State Univ. Hydrol. Paper*, no. 8, August.

Moran, P. A. P. [1959]: "The Theory of Storage," Methuen & Co., Ltd., London.

Pattison, A. [1964]: Synthesis of Rainfall Data, *Stanford Univ. Dept. Civil Eng. Tech. Rept. No. 40*, July.

Pieruschka, E. [1963]: "Principles of Reliability," Prentice-Hall, Inc., Englewood Cliffs, N.J.

Press, H. [1949]: The Application of the Statistical Theory of Extreme Values to Gustload Problems, *NACA Tech. Note 1926*, Rept. 991, Washington, D.C.

Rascon, O. A. [1967]: Stochastic Model to Fatigue, *J. Engineering Mechanics Div., Proc. ASCE*, EM3, vol. 93, June, pp. 147–156.

Robertson, L. [1967]: A Study on the Wind Effects on the Seattle First National Bank Building, *Proc. Design Seminar Bldgs. Earthquakes Wind, ASCE Structural Eng. Conf.*, Seattle, Wash., May.

Rodriguez, M. [1956]: Aseismic Design of Simple Plastic Steel Structures on Firm Ground, *Proc. 1st World Conf. Earthquake Eng.*, Berkeley, Calif., 1956.

Shane, R. M. and W. R. Lynn [1964]: Mathematical Model for Flood Risk Evaluation, *J. Hydraulics Div., Proc. ASCE*, HY6, vol. 90, November, pp. 1–20.

Thom, H. C. S. [1960]: Distributions of Extreme Winds in the United States, *J. Structural Div., Proc. ASCE*, ST4, vol. 86, April.

Thomas, H. A. [1948]: Frequency of Minor Floods, *J. Boston Soc. Civil Engrs.*, vol. 34, pp. 425–442.

Tichy, M. and M. Vorlicek [1965]: Safety of Reinforced Concrete Framed Structures, *Proc. Intern. Symp. Flexural Mech. Reinforced Concrete (Miami)*, November.

Vorlicek, M. [1963]: The Effect of the Extent of Stressed Zone Upon the Strength of Material, *Acta Tech. Csav*, no. 2, pp. 149–176.

Vorlicek, M. and J. Suchy [1958]: Discussion of Synopsis of First Progress Report of Committee on Factors of Safety, *J. Structural Div., Proc. ASCE*, ST1, vol. 84, paper 1522, January, pp. 59–64.

Watson, G. S. [1954]: Extreme Values in Samples from Dependent Stationary Stochastic Processes, *Am. Math. Statist.*, vol. 23.

Weibull, W. [1939]: A Statistical Theory of the Strength of Materials, *Proc. Roy. Swedish Inst. Eng. Res.*, no. 159.

Whitcomb, M. [1940]: A Statistical Study of Rainfall, M. S. Thesis, Department of Meteorology, Massachusetts Institute of Technology, Cambridge, Mass.

Tables

Benjamin, J. R. and M. F. Nelson [1966]: Tables of the Standardized Beta Distribution, *Stanford Univ. Dept. Civil Eng. Tech. Rept. No. 59*, January.

Hald, A. [1952]: "Statistical Tables and Formulas," John Wiley & Sons, Inc., New York.

Harvard Computation Laboratory [1966]: "Tables of the Cumulative Binomial Probability Distribution," Harvard University Press, Cambridge, Mass.

National Bureau of Standards [1964]: "Handbook of Mathematical Functions," Applied Mathematics Series 55, Washington, D.C.

National Bureau of Standards [1953]: "Probability Tables for the Analysis of Extreme-Value Data," Applied Mathematics Series 22, Washington, D.C.

National Bureau of Standards [1950]: "Tables of the Binomial Probability Distribution," Applied Mathematics Series 6, Washington, D.C.

National Bureau of Standards [1959]: "Tables of the Bivariate Normal Probability Distribution and Related Functions," Applied Mathematics Series 50, Washington, D.C.

National Bureau of Standards [1954]: "Tables of the Error Function and Its Derivative," Applied Mathematics Series 41, Washington, D.C.

National Bureau of Standards [1953]: "Tables of Normal Probability Functions," Applied Mathematics Series 23, Washington, D.C.

Pearson, E. S. and H. O. Hartley [1966]: "Biometrika Tables for Statisticians," Cambridge University Press, Cambridge, England.

Thompson, C. M. [1941]: Percentage Points of the Incomplete Beta-Function, *Biometrika*, vol. 32, part II, pp. 168–181, October.

PROBLEMS

3.1. If the probability of finding a net profit on any one job is a constant p, what is the probability of observing all successes or all failures in a sequence of m jobs (Bernoulli trials)? If $p = 0.1$, plot the PMF and CDF of zero, one, two, three, four, five profits in five jobs. What is the probability of at least two profits? Two or fewer profits?

3.2. Four widely spaced piles are to be driven either until bedrock is reached or until under test loads of 10 tons further penetration is less than 6 in. The engineer supervising this activity estimates that for any pile, the chances of hitting bedrock are 3 to 1 and if he hits bedrock with any pile he has no better knowledge than his original estimate for any subsequent pile. If hitting bedrock is called a success S, and everything else is failure F, one possible outcome of the experiment is the sequence $SSFS$.

    (a) List every possible sequence.

    (b) How many sequences contain exactly two successes?

    (c) What is the probability of each of these sequences?

    (d) What is the probability of exactly two successes? Also calculate directly from the appropriate PMF.

    (e) Sketch the PMF and the CDF of the number of successes.

Note in part (d) that the coefficient is simply the number of ways in which a sequence of length n can be rearranged when it has k successes.

3.3. In a transportation system having the network shown below, a sequence of passenger arrivals occurs at terminal 3. It is known that the arrivals' desired destinations are independently distributed with probability:

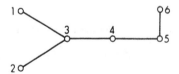

Fig. P3.3

$$p_i = P[\text{arrival has destination } i] = 0.2 \qquad i = 1, 2, 4, 5, 6$$

    (a) What is the PMF of the number of arrivals before the first appearance of a passenger wanting to go to destination 6?

    (b) Suppose there is a vehicle waiting with five remaining seats with the operating rule that it will leave for station 4 when it is full or at the latest after the next seven arrivals. What is the probability the seats will be filled if station-4 travelers are the only ones accepted?

    (c) Suppose the vehicle described in part (b) will accept travelers to terminals 4, 5, or 6 instead of just 4. What now is the probability it will be full after seven more arrivals (or before)?

3.4. If, on the average, 30 percent of all cars entering a particular intersection from the north make left turns, what is the distribution of X, the number out of six successive entering vehicles that will make left turns? (Assume that the turning decision by a driver is independent of actions of other drivers.) What is the probability that four of six cars will turn left? What is the probability that four or more of six will turn left?

3.5 "N-Year" design magnitudes. A system (e.g., a dam or a dike) is said to be designed for the N-year flood if it has a capacity which will be exceeded only by a flood equal to or greater than the N-year flood. The magnitude of the N-year flood is that which is exceeded with probability $1/N$ in any given year. Assume that successive annual floods are independent.

(a) What is the probability that exactly one flood equal to or in excess of the "50-year flood" will occur in a 50-year period?

(b) What is the probability that exactly three floods will equal or exceed this 50-year flood in 50 years?

(c) What is the probability one or more floods will equal or exceed the 50-year flood in 50 years? *Hint:* this is most easily calculated as $1 - P[\text{no floods exceed}]$.

(d) If an agency designs each of 20 independent systems—i.e., systems at widely scattered locations—for its particular 500-year flood, what is the distribution of the *number of systems* which will fail at least once within the first 50 years after their construction? Assume that $(1 - x)^n \approx 1 - xn$ if $xn \ll 1$. *Hint:* what is the probability that any one system will not fail in 50 years?

(e) In 1958 the 50-year flood was estimated to be a particular size. In the next 10 years, two floods were observed in excess of that size. If the original estimate was correct, what is the probability of such an observation? Such a rare event may be so unlikely that the engineer prefers to believe (i.e., act as if) his original estimate was wrong. In Chaps. 4 to 6 we shall see that such calculations are an important part of decision making.

3.6. The Richter magnitude of an earthquake—given that one has occurred—has been hypothesized to be exponentially distributed.[†] In Southern California the value of the parameter of this distribution has been estimated to be 2.35. What is the probability that any given earthquake will be larger than 6.3, the magnitude of the disastrous 1933 Long Beach earthquake?

In the Southern California region there is on the average one earthquake per year with magnitude equal to or greater than 6.1. What is the probability of an earthquake in this area of magnitude greater than 7.7, "a truly great earthquake," in any given year? (The great Alaskan earthquake of 1964 had a magnitude of 8.4.)

3.7. "Partial-duration" flows—flows greater than some arbitrary base value—have been treated as random variables with a shifted exponential distribution by Shane and Lynn [1964]. From 23 years of records during which 56 flows in excess of 5500 cfs were observed in the South Chickamauga Creek near Chickamauga, Tennessee, the following distribution was found adequate[‡]

$$f_X(x) = (\tfrac{1}{6500})e^{-(x-5500)/6500} \qquad x \geq 5500$$

where X is the peak-flow magnitude given a flow in excess of 5500 occurred.

[†] The hypothesis has been verified experimentally at least for magnitudes between about 0 and 8.0. (Problem adapted from C. R. Allen, P. St. Amand, C. F. Richter, and J. M. Nordquist [1965].)

[‡] The authors justify the choice on the observation that the upper tails of many distributions are exponential-like.

Sketch the distribution and compute the probability that a flow in excess of 20,000 cfs will be observed given that the flow exceeds the base value of 5500 cfs. (Incidentally, 5 of the 56 observed flows were in excess of 20,000.)

3.8. Our interest is in how people use urban multimode public transportation systems and in whether the service might be improved by changing the routes. In the morning, during peak demand, an express bus now travels from suburb A to terminal B, where it discharges all its passengers. It then returns to suburb A, empty, to repeat the trip. Some of these discharged passengers may shop or work near terminal B, while others transfer to another system, the subway. Some of the passengers who go on the subway get off at terminal C in order to walk some distance to work at a large industrial center Q.

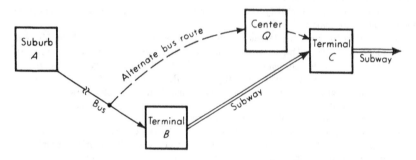

Fig. P3.8

Data suggest that on the average a fraction p of the persons on the bus from suburb A now transfer to the subway at terminal B. Of these persons on the average a fraction q get off at terminal C to work at center Q.

(a) What is the probability distribution of X, the number of persons who transfer from a particular express bus to the subway at terminal B? The bus arrives full, carrying n passengers. Individual passengers behave independently of one another at all stations.

(b) Let Y equal the number of persons on this same bus who get on the subway and get off at terminal C to walk to Q. They also act independently of one another. *Show that* the joint probability distribution of X and Y is

$$p_{X,Y}(x,y) = \frac{n!}{(n-x)!\,y!\,(x-y)!}\, p^x q^y (1-p)^{n-x}(1-q)^{x-y}$$

over a certain region of the x,y plane and zero elsewhere. *What is that region?*

(c) The system might be more efficient if the express bus avoided the traffic congestion around terminal B by transferring the passengers who want to stay near that terminal to a local bus at the outskirts of that congestion. The original express bus could then bypass the terminal B congestion and continue with the remaining passengers, those who would have used the subway. This bus would stop first at industrial center Q, and then at terminal C, before returning. With the available information we could study the relative advantages of such a revision to the system. Here we answer only two questions:

(i) What is the probability that exactly x persons stay on the bus (i.e., do not go to terminal B) and that *all* these people get off at center Q? (Notice that no matter what the value of x, in this case the bus could return directly to suburb A without continuing all the way to terminal C.)

(ii) What is the probability that the bus could return directly to suburb A (either directly from Q or directly from terminal B) without having to go to terminal C?

3.9. Consider a water-storage tank containing some large constant number n of water molecules. Water is entering and leaving through neighboring pipes at the same rates, m molecules per second. We are interested in the "mixing" going on within the tank. We shall measure this by considering the "residence time" T that any molecule spends in the tank (a number which can be measured by injecting a radioactive tracer molecule and waiting for it to exit). If all the water passed through the tank in a perfectly uniform way (e.g., if the tank were simply a long pipe) and hence with no mixing, each molecule would spend exactly n/m sec in the tank. Suppose on the other extreme that, owing to great turbulence, all molecules within the tank have equal likelihood $k \, \Delta t$ of leaving in any small interval of time Δt. Show that under the latter circumstances the residence time T is a random variable exponentially distributed with parameter $\lambda = m/n$.

Hint: Show that $G(t + \Delta t) = G(t)(1 - (m/n) \, \Delta t)$ and solve for $G(t) = 1 - F(t)$ as $\Delta t \to 0$. (Strictly we should deal with discrete time increments being equal to the time required for one molecule to pass from the tank. In this case T would be geometric.)

Suggest an easily estimated parameter which might be used as a measure of the degree of mixing (e.g., the standard deviation or the coefficient of variation of T) and discuss it (e.g., its probable range and interpretation of the end points on the range).

3.10. On the average five accidents per 10,000 passing vehicles have been observed on a particular piece of road. What is the probability that, in the next interval of 10,000 passages, between two and eight accidents will be observed? (Assume that the number of trials—possible accident situations—is sufficiently large that a Poisson approximation is valid.)

Suppose a new side road entering on this road has been added and it is desired to study its influence on safety: that is, has the average accident rate increased since this road was added? This question will be studied by observing the number of accidents in the next interval of 10,000 cars. The problem is that that number is random, and even if there has been no degradation of safety (i.e., no increase in average accident rate), there is a small probability that a large number of accidents will occur in the next interval.

Suppose in this case nine accidents were observed in the 10,000 car interval after the side-road installation. Has the average rate increased or was this just a rare occurrence of a large number of accidents with the unchanged *average* rate of 5? In practice this question is often answered by saying, "If the observation represents a 'quite rare' event (say with probability of occurrence less than 5 percent) for an unchanged average rate, then 'probably' the rate is not unchanged but increased." Is the probability of nine or more accidents with an unchanged average rate 5, small enough (<5 percent) that the engineer might conclude that the new side road has increased the average accident rate? As will be discussed in Chaps. 4 and 5, more important than his conclusion is the action implied by the conclusion; that is, should he install a traffic light or not? Consideration of actions and consequences will lead to the replacement of arbitrary criteria with regard to what is "rare enough."

3.11.† In a harbor structure exposed to the sea, cracks form along the length of tension members in the reinforced-concrete trusses. Crack occurrence along the length of the member follows a (spatial) Poisson process with mean rate of 2/ft. The members are 20 ft long. A crack exposes the reinforcing steel to corrosion at that point. At any point in time the amount of corrosion and hence the strength loss at any crack is approximately exponentially distributed. At present the mean loss is 1 kip. The remaining strength of any member is its initial strength, 100 kips, minus the *largest* strength loss along the member.

(a) Find the distribution of the remaining strength of any member.

(b) What is the probability that any particular member will fail when subjected to an applied force of 90 kips?

(c) What is the distribution of the strength of a (statically determinate) truss that will fail if any of 10 such (independent) tension members fail? (Neglect the possibility that failure of the truss will be caused by anything other than failure of one of these 10 members.) A unit load causes a unit force in each member.

3.12.‡ Solve for the answer sought in the left-turn traffic illustration (Secs. 3.1.3 and 3.2.1), that is, $P[A] = P[$the lane capacity is inadequate$]$, using the interpretation of the gamma distribution as the time to the kth arrival of Poisson events. That is,

$$P[A] = P[X \leq t]$$

where X is $G(4,\lambda)$ and $\lambda t = p\nu = (0.3)(6) = 1.8$.

Notice that the result must depend only on the value of the product λt and not on specific values of λ and t.

3.13. A traffic signal is said to "fail" if it does not provide a sufficiently long green signal for the accumulated cars to enter the intersection. It has been measured that a maximum of six vehicles can enter an intersection in the 20 sec after a light changes. If a light is set with a cycle of 20-sec red and 20-sec green (ignore the yellow phase) and the average traffic flow is 400 vehicles per hour in the direction of interest, what is the probability that the signal will fail during any particular cycle? (Assume that failure occurs if seven or more vehicles arrive during the cycle. Assume that the arrivals are Poisson.)

Notice that the first of the preceding two assumptions is not strictly correct since the failure on the cycle following a failure is more likely owing to the leftover vehicles. If the preceding cycle has not failed, find the probability that the next two cycles in a row will fail? What is the conditional probability of failure of the second given that the first failed?

The failure probability can be reduced by increasing cycle times, but this also increases the likelihood that cars will be unnecessarily delayed. A properly designed signal balances these two effects.

3.14. Idealize the structure of a composite material such as concrete (rock aggregate in cement paste) as randomly located spheres in a volume of the surrounding matrix. An important question in experimental work is, "How large need a specimen of this material be before one can be reasonably certain that the number of spheres in the specimen is 'representative' (i.e., very nearly the average value) so that the specimen may be treated as homogeneous and with the same properties as a large volume of the composite material?"

Under assumptions similar to those used in Sec. 3.2 with respect to vehicle distances or times but now applied to a volume v, the PMF of the number of spheres

† This problem was suggested by Ove Ditlevsen of Denmark.
‡ Adapted from Greenshields and Weida [1952], page 199.

in the volume X is Poisson-distributed with mean λv. Find an expression for the coefficient of variation of X versus volume. C. B. Brown has proposed a standard of homogeneity based on choosing the specimen volume so that the probability is about 0.683 that X is within a region $\pm k m_X$, say $\pm 0.1 m_X$, about m_X. Show (use a normal approximation to the Poisson) that for any value k, to satisfy this standard one must choose the volume such that the coefficient of variation equals k.

What is the interpretation of λ?

3.15. The distribution of X, the intensity of ground motion at a site, given that an earthquake larger in magnitude than m_0, say 5, occurs in some neighboring region, is $F_X(x)$. Show that the CDF of Y, the maximum intensity at the site in a period of time t, is

$$F_Y(y) = e^{-\nu t[1 - F_X(y)]}$$

if the average number of earthquakes (greater than m_0) in the region is ν per unit time and if these occurrences may be treated as Poisson arrivals.

3.16. In Prob. 2.54, for large enough x (namely, $x \geq c_1 + c_2 m_0 - c_3 \ln d$) the distribution of ground motion intensity X at a site given an earthquake on a neighboring fault was of the form

$$F_X(x) = 1 - c e^{-\lambda x}$$

(a) Using the assumptions and results of Prob. 3.15, what is the distribution of Y for values larger than this same number (that is, $y \geq c_1 + c_2 m_0 - c_3 \ln d$)?

(b) For this same case, show that the intensity which will be exceeded only with probability p during the lifetime t_0 of a dam or structure is (assuming it is larger than $c_1 + c_2 m_0 - c_3 \ln d$) for small p.

$$y_p \approx \frac{1}{\lambda} \ln \left(\frac{\nu t_0 c}{p} \right)$$

(c) For this case, show that for large values of n, the n-year intensity (assuming it is larger than $c_1 + c_2 m_0 - c_3 \ln d$) is simply

$$y_n \approx \frac{1}{\lambda} \ln (\nu c n)$$

(when ν is in average number *per year*).

3.17. Assume that the number X of large hurricanes which strike along a given coastline in one year has PMF $p_X(x)$, $x = 0, 1, 2 \ \ldots$.

(a) What is the distribution of Y, the number of years out of n years in which one or more hurricanes are observed along the coastline? (The number of occurrences is independent from year to year.)

(b) A reasonable assumption for $p_X(x)$ may be (Sec. 3.2.1)

$$p_X(x) = \frac{e^{-\nu} \nu^x}{x!} \qquad x = 0, 1, 2, \ \ldots$$

Evaluate the mean and variance of Y for $\nu = 0.2$ and $n = 30$.

3.18. Normal approximation to the binomial distribution. For large values of n, the binomial distribution CDF,

$$F_X(x) = \sum_{i=0}^{x} \binom{n}{i} p^i (1 - p)^{n-i}$$

can be tedious to evaluate. The interpretation of the binomial random variables as the sum of n independent Bernoulli random variables suggests that for large values of n the random variable should be approximately normally distributed.†

That is, if Y is a normal random variable with the same mean np and variance $np(1 - p)$ then

$$F_X(x) \approx F_Y(x)$$

(a) Find the exact value of $F_X(5)$ for X having a B(10,0.5) distribution, and compare this value with $F_Y(5)$ when Y is normal with the same mean and variance.

(b) The sketch below suggests why, when a continuous distribution is being used to approximate a discrete distribution, a correction should be applied. Namely, if the distribution is defined on the integers,

$$F_X(x) \approx F_Y(x + 0.5)$$

Compare $F_Y(5.5)$ with $F_X(5)$ computed above.

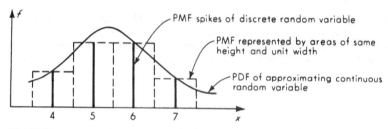

Fig. P3.18

3.19. The observed annual maximum runoff X in the Weldon River at Mill Grove, Missouri, has been modeled (see Markovic [1965]) by a normal distribution with parameters $m = 256.7$ cfs and $\sigma = 191$ cfs, by a lognormal distribution with $\ln \tilde{m} = 5.228$ and $\sigma_{\ln X} = 0.840$, or by the gamma distribution with $\lambda = 0.00676$ (cfs)$^{-1}$ and $k = 1.727$. Find the probability that $X \geq 400$ cfs predicted by each model.

3.20. The expectations of the reciprocals of powers of a random variable arise frequently. For example, the mean and variance of the safety factor (SF) for independent demand D and capacity C involve two such moments:

$$E[\text{SF}] = E\left[\frac{C}{D}\right] = E[C]E\left[\frac{1}{D}\right]$$

$$E[(\text{SF})^2] = E[C^2]E\left[\frac{1}{D^2}\right]$$

† Also, for good approximations neither np or $n(1 - p)$ should be small in order to avoid very skewed distributions. For these cases the Poisson distribution offers a good approximation to the binomial (Sec. 3.2.1).

(a) Show that for a normally distributed random variable the following expansion holds (if $V_X < 1.0$):

$$E\left[\frac{1}{X^r}\right] = \frac{1}{m_X{}^r}\left[1 + \frac{r(r+1)}{2!}V_X{}^2\frac{\mu_X{}^{(2)}}{\sigma_X{}^2}\right.$$
$$\left. + \frac{r(r+1)(r+2)(r+3)}{4!}V_X{}^4\frac{\mu_X{}^{(4)}}{\sigma_X{}^4} + \cdots\right]$$
$$= \frac{1}{m_X{}^r}\left[1 + \frac{r(r+1)}{2!}V_X{}^2 + \frac{r(r+1)(r+2)(r+3)}{2^{\frac{1}{2}}(\frac{4}{2})!}V_X{}^4 + \cdots\right]$$

Hint: Substitute for X: $X = m_X + \sigma_X U = m_X + m_X V_X U$ in which U is the standardized normal variable.

(b) For $V_X = 0.2$ and 0.5, find numerical approximations for $E[1/X]$ and Var $[1/X]$ such that the error in the first ignored term is less than 5 percent of the approximate value. Let $m_X = 1$.

3.21. The dynamic response $X(m)$ at "time" m of a simple linear structure to a sequence of impulses $Y(n)$, $n = 1, 2, \ldots, m$, can be described by the sum

$$X(m) = \frac{1}{p}\sum_{n=0}^{m} Y(n)\sin\left[p(m - n)\right]$$

(in which p is the natural circular frequency of the system—a function of the mass and stiffness of the structure—and the $Y(n)$ are measured in units of impulse per unit of mass of the system). A typical sample of $Y(n)$ might appear as shown in the sketch. Assume that the $Y(n)$ are independent identically distributed random

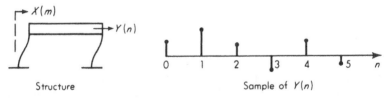

Structure   Sample of $Y(n)$

Fig. P3.21

variables, each distributed like a random variable Y. Such a model has been proposed to model earthquake ground acceleration.†

(a) (i) Find an expression for $E[X(m)]$ in terms of the moments of Y of the form

$$E[X(m)] = (\text{term involving } Y) \times (\text{term involving sin functions})$$

(ii) What is this result for the case when

$$p_Y(y) = \begin{cases} \frac{1}{2} & y = +1 \\ \frac{1}{2} & y = -1 \\ 0 & \text{otherwise} \end{cases}$$

(Do *not* try to evaluate the second term! You do not have enough information.)

(b) (i) Find Var $[X(m)]$ in a form similar to that in (a) above.

† See, for example, Goodman, Rosenblueth, and Newmark [1955].

(ii) What is this result when p_Y is as in (a), part (ii)?

(c) (i) If m is large (say >50) what is a reasonable approximate, continuous model for the distribution of $X(m)$?   Why?

(ii) Under this assumption, what are the symmetrical limits within which $X(m)$ will be found with probability 0.945 if the distribution of Y is as given in (a), part (ii.)?

(d) Since structural interest is not so much in the direction as the magnitude of $X(m)$, find the density function of the absolute value of $X(m)$, $|X(m)|$, using the approximate model adopted in (c), part (i), and the distribution of Y given in (a), part (ii).

3.22. If the applied thickness T of a layer of paving material is normally distributed with mean 3 in. and standard deviation 0.2 in., find

(a) $f_T(2.5)$
(b) $P[T \leq 2.5]$
(c) $P[|T - m_T| \leq .2]$
(d) $P[|T - m_T| \leq .2 \mid T \geq 2]$

3.23. Consider the simple model shown of a flexible (long-span) structure resting on two footings.   The foundation engineer's concern is in the relative deflection of the

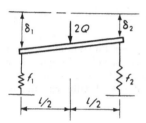

Fig. P3.23

ends of the structure:

$$\Delta = \delta_1 - \delta_2 = Qf_1 - Qf_2$$

Assume that flexibilities (the reciprocals of the stiffnesses) f_1 and f_2 of the two soil and footing systems are jointly distributed random variables with common mean m and common variance σ^2, and with covariance $\rho\sigma^2$, where ρ is the correlation coefficient, a measure of the dependence between flexibilities, and hence a measure of lateral homogeneity of the soil.   Find

(a) The expected value of Δ.

(b) The variance and standard deviation of Δ.   (Note the relationship between variance and the soil homogeneity measure ρ.   It is unlikely that ρ would ever be negative.)

(c) If f_1 and f_2 are jointly normally distributed (bivariate normal), with $Q = 1000$ kips, $\sigma = 0.002$ in./kip, and $\rho = 0.9$, find

(i) $P[|\Delta| \geq 2 \text{ in.}]$
(ii) $E[|\Delta|]$

(d) Repeat (c), both (i) and (ii), for $\rho = 0.1$, representing soil with little homogeneity.

3.24. Find and sketch the probability density function of vertical support reaction Y of a large radome during a windstorm when

$$Y = 3000 - 100X$$

where X is a measure of the magnitude of the wind causing the uplift. X is lognormally distributed with parameters $\breve{m}_X$ and $\sigma_{\ln X}$.

3.25. A structural engineer believes the yield stress S in mild steel reinforcing bars to be lognormally distributed with parameters $\breve{m}_S = 40.0$ ksi and $\sigma_{\ln S} = 0.1$, while the bar areas A of nominal $\frac{9}{8}$-in.-diameter bars ("no. 9" bars) are lognormally distributed with mean 1.0 in.^{2} and standard deviation 0.05 in.^{2}

Find the distribution of the yield force $F = SA$, assuming independence of S and A. Sketch this density function and find $P[F \le 35$ kips] and the value f_{allow} such that

$$P[F \le f_{\text{allow}}] = 0.01$$

3.26. The observed annual peak runoff X in the Weldon River at Mill Grove, Missouri, has been modeled (Markovic [1965]) by a shifted lognormal distribution in which $Y = X - 12.8$ cfs is lognormally distributed with parameters $\ln m_Y = 5.031$ and $\sigma_{\ln Y} = 0.933$. Find the probability that $X \ge 400$ cfs. Compare these results with those in Prob. 3.19. Markovic also considered a shifted gamma distribution, but the estimated (see Sec. 4.1) shift parameter was negative, implying that negative flows were possible.

3.27. The annual maximum rate of discharge of a particular river is thought to have the Type I extreme-value distribution with mean 10,000 cfs and standard deviation 3,000 cfs. Compute

$$P[\text{annual maximum discharge} \ge 15,000]$$

Find an expression for the CDF of the river's maximum discharge rate over the 20-year lifetime of an anticipated flood-control project. Assume that the individual annual maxima are mutually independent random variables. What is the probability that the 20-year maximum flow will exceed 15,000 cfs?

3.28. The Type I extreme-value distribution of largest values is commonly used to describe annual maximum floods. In the Sec. 3.1 illustration, m-year floods were discussed. Show that for large m the magnitude x_m of the m-year flood is approximately equal to

$$x_m \approx u + \frac{1}{\alpha} \ln m$$

if the annual floods are $\text{EX}_{\text{I},L}(u,\alpha)$. The implication is that the design magnitude grows like the logarithm of the return period. This fact was used in the economic study accompanying the illustration.

The Type II distribution of largest values has also been found to be applicable in some such design situations. Show that in this case

$$x_m \approx u m^{-1/k}$$

3.29. Distribution of maximum of n discrete random variables. Show that the probability mass function of Y_n, the largest of n independent discrete random variables

X_i, each with PMF $p_X(x)$ and CDF $F_X(x)$, is (letting $F_X^*(x) = P[X < x]$)

$$p_{Y_n}(y) = \sum_{k=1,2,\ldots}^{n} \binom{n}{k} [p_X(y)]^k [F_X^*(y)]^{n-k}$$

$$= [p_X(y) + F_X^*(y)]^n - [F_X^*(y)]^n = [F_X(y)]^n - [F_X^*(y)]^n$$

Find the PMF of Y, the largest of 10 independent demands each of which has PMF

$$p_X(x) = \begin{cases} 0.8 & x = 10 \\ 0.15 & x = 15 \\ 0.05 & x = 20 \end{cases}$$

[Note that the probability of the event that a specific X_i equals y and at the same time the other X_i are less than or equal to y is $p_X(y)[F_X(y)]^{n-1}$. Why does $p_{Y_n}(y)$ *not* equal $np_X(y)[F_X(y)]^{n-1}$, which "looks like" Eq. (3.3.46)?]

3.30. It has been argued that distribution of an arbitrary gust velocity is exponential and hence a rapid convergence to the Type I asymptotic distribution of the maximum gust velocity in a thunderstorm should be expected. Press [1949] has verified this experimentally and has observed the extreme gusts in many (nearly 500) similar thunderstorms. He found a mean extreme gust of 15.6 ft/sec and a standard deviation of 6.2 ft/sec. Find the parameters of the Type I distribution and the probability that the maximum gust velocity in such a storm will exceed 60 mph.

3.31. In a design study for a tall building (L. Robertson [1967]), it was found that the distribution of the peak wind-induced displacement D of a particular building in an arbitrary 20-min interval could be approximated by a "Weibull" distribution:

$$F_D(d) = 1 - e^{-(cd)^k}$$

in which $c = 46.0$ and $k = 0.68$. This distribution is of the exponential family.

(a) Find the asymptotic extreme-value distribution for the annual maximum peak displacement Y, the maximum of 20,000 such peak displacements. (In 1 year there are about 26,000 20-min intervals. The reduction to 20,000 is a crude attempt to account for the lack of independence among the 20-min peaks. The results are not too sensitive to the choice of the number of intervals.)

(b) Find the 50-year peak displacement.

3.32. Of the vehicles entering a particular interchange from the south, 40 percent, on the average, want to go to the west, 10 percent to the east, and the remainder straight through. What is the probability that of eight entering cars, half will want to go west and half straight? What is the distribution of the west-bound number X_1, the east-bound number X_2, and the north-bound number X_3 out of eight vehicles entering from the south? What is the marginal joint distribution of X_1 and X_3? What is the marginal distribution of the number of cars (out of eight) turning to the west? What is the probability that more than half will want to turn to the west (i.e., to the left)?

Such questions are of concern to the interchange designer who must balance cost of interchange versus cost of traffic delays. (Assume independence among driver desires throughout.)

3.33.† Assume that the probability density of flaw size in specimens of brittle material is exponential and that there are very many such flaws in any specimen. What is

† Adapted from Gumbel [1954].

the form of the distribution of the size of the largest flaw X? Find the form of the distribution of the breaking strength S if S is a linear function of the size of the largest flaw X:

$$S = s_0 - cX$$

in which s_0 is the strength of a flawless specimen.

3.34. The force F at which a type of steel reinforcing bar yields has been found to be lognormally distributed with parameters $m_{\ln F} = \ln \breve{m}_F$ and $\sigma_{\ln F}$. The cross-sectional area A of these bars is also lognormally distributed with parameters $m_{\ln A}$ and $\sigma_{\ln A}$.

What is the distribution of Y, the stress F/A at which a bar will yield, if F and A are independent? (Note that $Y = FA^{-1}$ is of the form $Y = F^a A^b$.)

3.35. Airplanes arriving at a certain airport are assumed to follow a Poisson arrival process with average rate λ. An air-control official comes each day and observes the arrivals of $m + 1$ planes.

(a) Show that the distribution of Y_m, the *smallest* of the m interarrival times, is exponential.

(b) At the end of each 5-day week (excluding weekends), the official calculates the average of the smallest (of m) interarrival times observed that week.

$$Z = \frac{Y_m{}^{\text{Mon}} + Y_m{}^{\text{Tues}} + Y_m{}^{\text{Wed}} + Y_m{}^{\text{Thurs}} + Y_m{}^{\text{Fri}}}{5}$$

What is the distribution of Z?

3.36. A three-city intercity transportation system uses small two-passenger vehicles. A car is available at each city for each of the other cities in the network and is dispatched only when it is full of passengers all wanting to go to the same city. At

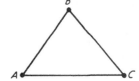

Fig. P3.36

city C, the possible states of the system are defined as

	No. of persons waiting	
State No.	In car for city A	In car for city B
1	0	0
2	0	1
3	1	0
4	1	1

when the system is observed just after the arrival of a traveler (and after a departure if he fills a car); that is, the process parameter is not explicitly time but rather arrival number.

(a) If the probability is $\frac{1}{3}$ that an arriving passenger wants to go to city A, what is the one-step transition matrix of this (assumed) Markov chain?

(b) What does the Markov assumption imply is assumed about the desired destination of passengers? Why?

(c) Write the simultaneous equations which must be solved to find the steady-state probabilities q_1^*, q_2^*, q_3^*, and q_4^*.

3.37. A garage with m spaces has, at time $t = 0$, X spaces filled, where X is a random variable with PMF

$$p_X(x) = \frac{q(1 - q)^x}{1 - (1 - q)^{m+1}} \qquad x = 0, 1, \ldots, m$$

in which q is a parameter between 0 and 1. In the next t hr, $N(t)$ vehicles will arrive. They can be assumed to follow the laws of a Poisson counting process with mean arrival rate λ. $N(t)$ is independent of X. Assume that no cars leave.

(a) Find the distribution of the number of vehicles in the garage at time t.

(b) For $q = 0.75$, $m = 5$, $\lambda = 20/\text{hr}$, find the probability that the garage will be full at $t = 0.25$ hr.

3.38. Consider the transition matrix below of a homogeneous Markov chain:

$$\begin{bmatrix} 0.6 & 0.2 & 0.2 \\ 0.2 & 0.6 & 0.2 \\ 0 & 0 & 1 \end{bmatrix}$$

(a) Given that the process starts in state 1 at time 0 what is the probability that at time 2 it will be in state 3?

(b) In the long run what is the probability that the process will be in state 1, in state 2, in state 3?

(c) Describe some real-world situation for which the Markov chain above might be a crude model. Note particularly the last row of the transition matrix. Comment briefly on why and under what circumstances your Markov-chain model might fail to be a good or even approximate representation of the physical problem you consider.

3.39. If a system with a randomly distributed capacity C is subjected to an independent randomly distributed demand D, their distributions plotted on the same axis might appear as shown. The reliability R of the system is by definition

$$R = P[D \leq C] = \int_0^\infty \int_0^c f_D(d)f_C(c) \, dd \, dc$$

which equals

$$R = \int_0^\infty F_D(c)f_C(c) \, dc$$

or which equals

$$R = \int_0^\infty f_D(d)[1 - F_C(d)] \, dd$$

In general these calculations require numerical integration. In some cases the reliability is easily calculated using the safety margin $M = C - D$, since

$$R = P[M > 0]$$

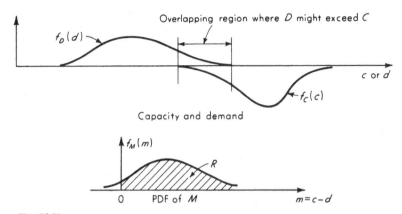

Fig. P3.39

If C and D are both independent and normally distributed, what is the distribution of M? Show that in this case

$$R = 1 - F_U\left(-\frac{1}{V_M}\right)$$

in which F_U is the standardized normal distribution and V_M is the coefficient of variation of M

$$V_M = \frac{\sigma_M}{m_M}$$

Find an expression for R in terms of F_U, m_D, m_C, σ_D, and σ_C. For $m_D = 10$, $\sigma_D = 2$, and $\sigma_C = 1$, make a sketch of R versus the specified (or estimated) capacity m_C (for $10 \le m_C \le 20$). It is suggested that semilog paper be used with $1 - R$, not R, on the logarithmic scale.

3.40. Consider again Prob. 3.39. The overturning moment on a retaining wall is lognormally distributed with median 100 kip-ft and coefficient of variation 0.15. The resisting moment of the proposed design is lognormally distributed with median 200 kip-ft and coefficient of variation 0.1. Find the reliability of this design. *Hint:* the safety factor $F = C/D = CD^{-1}$ is lognormally distributed, since $\ln F = \ln C - \ln D$ is normally distributed. Note that $R = P[F \ge 1]$.

3.41. What is the distribution of the cross-sectional area of a reinforced-concrete column which is

(a) A square with equal sides, each equal to a random variable Z with a lognormal distribution with the following parameters: median $\check{m}$ and standard deviation of its logarithm $\sigma_{\ln z}$?

(b) A "random square" with sides Z_1 and Z_2, two identically distributed independent random variables each distributed like Z in part (a) above?

(c) Which model of the column area implies a larger dispersion? Note that model (a) is the same as (b) if Z_1 and Z_2 are assumed to be not independent but perfectly correlated, that is, $Z_1 = Z_2$.

3.42. A proposed traffic-control system includes a new type of left-turn lane, monitored by instruments. Drivers are permitted to turn left *at any time* that a gap in opposing traffic permits. (The probability of a gap large enough for two or more cars to turn· is negligible.) But, if at any time the capacity (two) of the storage lane is exceeded by the arrival of a third car, a signal will stop all other traffic and permit all three cars to turn left. Assume that this can be done *instantaneously* upon arrival of the third car so that its arrival and the departure of all three vehicles coincide.

(a) Set up the transition matrix for a Markov-chain model of this lane. A state corresponds to the number of vehicles in the lane. What states need to be included if the process will be observed just after the arrival of either a left-turning vehicle or a gap in the opposing traffic? Assume that the occurrences of gaps in the opposing traffic and the occurrences of left-turning northbound vehicles are independent Poisson processes with *equal* average arrival rates. You need this asumption *only* to be able to state (i) that the lane-storage process is a Markov process and (ii) that the probability that a left-turning car has arrived, given that either a gap has arrived or a left-turning car has arrived, is $\frac{1}{2}$.

(b) Calculate the long-run fraction of the observations that the lane will be in the full state.

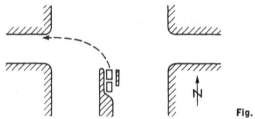

Fig. P3.42

3.43. Four independent regions of equal seismic activity have been selected as sites for a system of four instruments designed to record strong earthquakes. In *each* region strong earthquakes are Poisson arrivals with the same average arrival rate λ per year. What is the probability that after t years exactly one strong-motion earthquake will have been recorded by the total system of four instruments?

3.44. Investigate the relative safety with respect to stability of two earth embankments of equal length s by finding the probability, in each case, that the overturning moment $(e)(w)$ exceeds the resisting moment, $r(X_A l_A s + X_B l_B s)$ or $r X_C l_C s$, respectively, in which the X's are the shear strengths of the homogeneous soils, in lb/ft² and the l's are the lengths over which these stresses act ($l_A = 0.3 l_C$; $l_B = 0.7 l_C$). The X's all

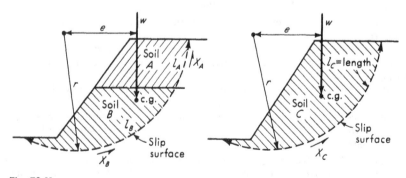

Fig. P3.44

have means $1.5(ew/srl_C)$ and coefficients of variation of 20 percent. X_C is normally distributed. X_A and X_B are jointly normally distributed with correlation coefficient 0.5.

3.45. Reconsider the dam-storage illustration in Sec. 3.7.

(a) The engineer would like to determine the benefits of his dam by comparing the long-run fraction of the time the release will meet (or exceed) the target value of two units and $P[Y_n \leq 1] = 0.4$, the fraction of time that the water would be insufficient if no dam were built.

Can you set up a *Markov* chain $Z(n)$, $n = 0, 1, \ldots$, in which $Z(n) =$ number of units released in year n? Why not?

(b) Find the probability asked for in (a) by considering the (long-run) probabilities of state of the dam process $X(n)$ in Sec. 3.7 and the probability distribution of Y_n, the annual inflow.

(c) If the initial cost of the dam (and its operation less recreational and flood benefits) reduces to an equivalent annual cost of \$100,000 and if the loss to agriculture is \$1 million in the event that no water is available in any particular year and \$100,000 if only one unit is available, should the dam be built? Base the decision on long-run expected annual costs of building the dam or doing nothing. (This decision analysis does not account for the possible economic influences of sustained periods of low flows beyond the annual cost mentioned.)

3.46. The ith largest and smallest variables. Consider n independent identically distributed continuous random variables, $X_1, X_2, \ldots, X_n$. Consider next these same variables rearranged in increasing order and denoted $Y^{(1)}, Y^{(2)}, \ldots, Y^{(n)}$, for example, $Y^{(1)} = \min [X_1, X_2, \ldots, X_n]$ and $Y^{(n)} = \max [X_1, X_2, \ldots, X_n]$. Then the cumulative distribution of $Y^{(i)}$ is

$$F_{Y^{(i)}}(y) = P[\text{at least } i \text{ of the } X\text{'s are less than or equal to } y]$$

$$= \sum_{j=i}^{n} \binom{n}{j} [F_X(y)]^j [1 - F_X(y)]^{n-j}$$

(a) Interpret this result in words.

(b) Consider a system with n components each with independent capacities, $X_1, X_2, \ldots, X_n$, The system will fail if the capacities of any j components are exceeded. Find the cumulative distribution of the jth smallest capacity, i.e.. the distribution of the system capacity.

(c) The lifetimes of individual street-lighting units on a major intersection have exponential distributions with common parameter λ. The illumination level falls below acceptable safety standards when any j of the n units are out. At that point a maintenance cost minimization program calls for the replacement of all n units. What is the distribution of the time between system replacements?

3.47. Assume that among heavy construction workers of a particular bracket of age, experience, job type, job conditions, etc., the number of accidents in which each is involved in an interval of time of length t is Poisson-distributed with parameter $\nu = \lambda t$ (i.e., the accident "rate" is λ). If one considers now all heavy construction workers, the parameter ν can be considered to be a random variable with relative likelihoods depending upon the fraction of all workers associated with any particular rate λ. Assume that ν can be treated as a continuous random variable N with a gamma distribution $G(k,\mu)$:

$$f_N(n) = \frac{\mu(\mu n)^{k-1} e^{-\mu n}}{(k - 1)!} \qquad n \geq 0$$

(a) Show that X, the number of accidents in which an arbitrary construction worker will be involved in an interval of length t, has distribution of the negative binomial form. [Strictly, $Y = X + k$ is $NB(k, \mu/(1 + \mu))$.]

(b) If 414 workers were observed for 3 months and the following table of number of accidents x for each worker was obtained:

x No. of accidents	n No. of workers
0	296
1	74
2	26
3	8
4	4
5	4
6–8	2
	Total 414

Estimate the parameters k and μ by equating the sample mean and variance to the model mean and variance. Compute and compare the observed frequency of each value of x with the model predicted probability that $X = x$. (In Chap. 4 we shall formalize such parameter estimation and comparison procedures.) (Adapted from Greenwood and Yule [1920].)

3.48. An engineer concerned with cumulative damage due to fatigue anticipates monitoring the deformation (actually the strain) versus time of a main structural member in a room at a newspaper press where the rolls of blank paper are stored prior to being placed on the presses. The (discretized) states are strains; the time (observation) points are each day just after the delivery of the rolls. Therefore the strains are assumed to represent *peaks* in the strain-time process. The loads are random in time, but the load-strain diagram of the member has been obtained exactly by accurate, in-place measurements.

(a) What conditions must the load-strain diagram of the member satisfy in order that the engineer can consider using a Markov-chain model of this process? In particular, does the diagram have to be linear and elastic, or nonlinear and elastic, or can it be generally nonlinear and nonelastic?

(b) Assume that the material is linearly elastic. What conditions must the load process satisfy in order that the deformation process can be assumed to be a Markov process? (See also Prob. 3.49.)

3.49. The transition matrix below represents a simplified (homogeneous) version of a Markov model of the beam-deformation problem discussed in Prob. 3.48.

	1	2	3	4
1	0.4	0.3	0.2	0.1
2	0.3	0.3	0.2	0.2
3	0.1	0.4	0.3	0.2
4	0.1	0.4	0.4	0.1

(a) Find the long-run fractions of the peak strains which will be of levels 1, 2, 3, and 4.

(b) A simple cumulative damage law for fatigue states that the "total damage" at any time is the sum of the "individual damages" contributed by each peak and these individual damages grow with the strain level of the peak. For the four states above the individual damages are 5×10^{-7}, 5×10^{-6}, 5×10^{-5}, and 5×10^{-4}, respectively. What is the expected total fatigue damage 30 years and 60 years after construction (if, of course, the use of the room remains such that the assumption of homogeneity holds)?

4

Probabilistic Models and Observed Data

A probabilistic model remains an abstraction until it has been related to observations of the physical phenomenon. These data yield numerical estimates of the model's parameters, and provide an opportunity to verify the model by comparing the observations against model predictions. The former process is called *estimation;* it is the subject of Sec. 4.1. The latter, comparative, process includes *verification* of the entire model (Sec. 4.4), but more broadly it includes the search for *significance* in a batch of statistical data. Conventional methods of significance testing are discussed in Sec. 4.2.

Consider the engineer who has developed a model of some physical phenomenon leading to a proposed functional form (e.g., lognormal) of the governing probability distribution. He must next estimate its parameters and then judge the validity of the model. Both these processes, estimation and verification, require data for the resolution. Repeated experiments will yield a sample of observed values of the random variable, say n in number. The resulting set of n numbers might be treated by the data-reduction techniques discussed in Chap. 1, yielding sample moments and a histogram. If we were to repeat the sequence of n tests, it is the nature of the probabilistic phenomenon that we would get (except coincidentally) a *different* set of values for the n numbers and therefore for the sample moments and histograms. Any reasonable rule for extracting parameter estimates from such sequences of numbers (e.g., a rule which says "estimate the parameter m by the average of the sample") would consequently yield *different* estimates of the values of the model's parameters. No single sequence of observations finite in number can be expected to give us exact model parameter values, because the data itself is a product of the "randomness" which characterizes the phenomenon. We must be prepared then to accept a data-derived parameter value as no more than an *estimate* of the true value, subject to an error of uncertain but, as we shall see, not unquantifiable magnitude.

We can also use data to evaluate the validity of an hypothesized model by comparing the model's predictions with observed occurrences. The relative frequency of particular events, for example, can be measured in the experiment and compared with probabilities computed from the model. Consistent and large discrepancies between such numbers would cast doubt on the ability of the hypothesized model to describe the physical phenomenon, while close correspondence would tend to lend credence to the proposed model. A moment's reflection, however, will reveal that, again owing to the probabilistic nature of the situation, one can seldom state that certain observed data are sufficient to reject or accept categorically a proposed model. An apparent discrepancy between model and reality may simply be the result of the occurrence in that particular data of a relatively rare but not impossible set of events. On the other hand, there may be a number of other models capable of yielding (with small or large likelihood) a set of observed data judged to be in accordance with our model's predictions. Consequently such evaluations will have to be qualified and phrased in a way that reflects the possibilities of such errors in the conclusions.

In short, if parameter estimation and model verification are based on practical sample sizes, there will remain uncertainty in the parameter values and perhaps uncertainty in the validity of the model itself. We recognize, then, that the engineer faces at least *three* distinct types of uncertainty. The first is the *natural, intrinsic, irreducible,* or *fundamental* uncertainty of the probabilistic phenomenon itself. This was emphasized in previous chapters. The second type of uncertainty can be called *statistical.* It is associated with failure to estimate the model parameters with precision. It is not fundamental, in that it can be virtually eliminated at the expense of a very large sample size. The third kind of uncertainty, that in the *form* of the model itself (the *model uncertainty*), can, in principle, also be eliminated with sufficient data.

For example, a Bernoulli trial model says the number of heads in k flips of an unbalanced coin is a binomial random variable $B(k,p)$. The outcome of any experiment (of k flips) is *fundamentally uncertain,* even given the validity of the binomial distribution and given a known value of p. With only a limited number of experiments upon which to base an estimate, however, the true value of p will remain in some doubt, i.e., subject to *statistical uncertainty.* A limited amount of data will not provide enough comparisons between histograms and the binomial PMF to guarantee that the model holds (e.g., there may not be independence among the flips); *model uncertainty* remains. More data will tend to reduce the magnitude of statistical and model uncertainty. The fundamental uncertainty remains unchanged by the number of past experiments.

This chapter represents a small but appropriate sampling of *applied statistics,*† a widely studied and broadly applied discipline dealing with the relationships among abstract probabilistic models and real data gathered from situations in engineering, industry, and the sciences. We shall see that convenient conventions have been established for reporting and discussing the very difficult questions surrounding the second and third kinds of uncertainty, those associated with estimation and model verification.

4.1 ESTIMATION OF MODEL PARAMETERS

Having hypothesized the type or functional form of his model, the engineer's first problem upon being confronted with numerical data is the estimation of the values of the parameters of these functions. In this operation he chooses the parameters' estimates so that the model, in some sense, best fits the observed behavior. We seek estimators (certain functions of the random variables in the sample) which promise this fit, and then we study the nature of these estimators as random variables.

4.1.1 The Method of Moments

We choose to introduce this topic through illustrations. A structural engineer concerned with strength of reinforced-concrete structures needs information concerning the force at which the tensile reinforcement will yield. A yield-strength theory for ductile steel argues that a bar yields only when its many individual "fibers" have all yielded, and, since

---

† Recommended first references in this field include Hald [1952] and Bowker and Lieberman [1959], two applied books with industrial engineering flavor, and Lindgren and McElrath [1966], Freeman [1963], and Brunk [1960], three books of increasing difficulty dealing in an introductory way with the mathematical theory of statistics.

each such fiber contributes in an additive manner to the total force, it should be expected that X, the yield force of a bar, would be normally distributed[†]:

$$f_X(x) = \frac{1}{\sigma \sqrt{2\pi}} e^{-\frac{1}{2}[(x-m)/\sigma]^2} \qquad -\infty \leq x \leq \infty \qquad (4.1.1)$$

The engineer tests a number of specimens of no. 4 (nominally $\frac{1}{2}$ in. in diameter) mild steel reinforcing bars. The data from 27 specimens are shown in Table 4.1.1. The sample mean $\bar{x}$ and sample standard devia-

[†] There are, in fact, several weaknesses in this argument. See Johnson [1953] and Vorlicek [1963] for presentation and discussion of such theories.

Table 4.1.1 Tests of no. 4 reinforcing bars from a single lot

Test No.	Yield point, lb
1	8560
2	8700
3	8420
4	8580
5	8460
6	8440
7	8460
8	8440
9	8550
10	8430
11	8620
12	8580
13	8650
14	8460
15	8500
16	8420
17	8500
18	8620
19	8440
20	8420
21	8420
22	8410
23	8700
24	8360
25	8310
26	8330
27	8510

$\bar{x} = 8490$ lb, $s = 103$ lb
$s^2 = 10,522$ lb^2, $v = 0.0120$

tion s (Sec. 1.2), are 8490 and 103 lb, respectively.† Accepting the conjecture that the yield force of such a bar has a normal distribution, what values of the parameters m and σ should be used when describing the behavior of a no. 4 reinforcing bar in this lot?

It would be contrary to the engineer's common sense to consider estimates of m and σ which differ much, if any, from the sample statistics $\bar{x}$ and s. The moments of the numerical data provide obvious and reasonable estimates of the corresponding moments of the proposed mathematical model.

This intuitively satisfactory rule for calculating model parameter estimates from data is given the name *the method of moments*. It says that *the estimator* $\hat{m}_X$ of m_X should be the sample mean, and the estimator $\widehat{\sigma^2}$ of the variance σ^2 is the sample variance. Similar rules hold for higher moments of the variable and the data, although these two estimators are sufficient for two-parameter distributions.

In the example, the *estimators* (rules) lead to the following *estimates* (values)‡ of the mean and variance

$$\hat{m}_X = \bar{x} = \frac{1}{n} \sum_{i=1}^{n} x_i = 8490 \text{ lb} \tag{4.1.2}$$

$$\widehat{\sigma^2} = s^2 = \frac{1}{n} \sum_{i=1}^{n} (x_i - \bar{x})^2 = (103)^2 \text{ lb}^2 \tag{4.1.3}$$

Given these parameter values the engineer can define completely his model [Eq. (4.1.1)] of the bar-force random variable and he is prepared to make such statements as "The probability is 5 percent that the yield force of any particular bar will be less than $m - 1.65\sigma = 8490 - 1.65(103) = 8490 - 173 = 8317$ lb."

Illustration: Exponential distribution   It is not always true, of course, that the parameters of a distribution are equivalent to moments. The parameters n and p of the binomial distribution (Sec. 3.1.3) are examples. The moments are, however, functions of the parameters, and these equations can normally be solved to yield the parameters in terms of the moments.§ Consider the following simple example. A traffic engineer who had adopted the common assumption of Poisson traffic (Sec. 3.2) measured the gap or time-between-

† The sample coefficient of variation, $v = {}^{103}\!/_{8490} = 1.2$ percent, is typical of steel strength variation within bars from a common lot. When many lots are sampled, a value of 7 to 8 percent is more common. See Granholm [1965].

‡ In general, both the estimator and the estimate of an unknown parameter b will be denoted by a "hat," thus, $\hat{b}$.

§ A case where this is not possible except by trial and error occurs in reliability testing with the Weibull or type III extreme-value distribution (Sec. 3.3.3).

Table 4.1.2   Observed gaps in traffic: Arroyo Seco Freeway (Los Angeles)*

Gap length, sec	No. of gaps observed	Gap length, sec	No. of gaps observed
0–1	18	14–15	3
1–2	25	15–16	3
2–3	21	16–17	6
3–4	13	17–18	4
4–5	11	18–19	3
5–6	15	19–20	3
6–7	16	20–21	1
7–8	12	21–22	1
8–9	11	22–23	1
9–10	11	23–24	0
10–11	8	24–25	1
11–12	12	25–26	0
12–13	6	26–27	1
13–14	3	27–28	1
		28–29	1
		29–30	2
		30–31	1

Interval midpoint is used as value of t_i; $\Sigma x_i = 1640$; total no. of observations $= 214$; $\bar{t} = 7.66$ sec; $\Sigma t_i^2 = 21{,}396$ sec². * *Source:* Gerlough [1955].

successive-cars data shown in Table 4.1.2. The sample mean of this data is 7.66 sec. The mathematical model states that this distribution is exponential with parameter λ:

$$f_T(t) = \lambda e^{-\lambda t} \qquad t \geq 0 \tag{4.1.4}$$

The mean of this distribution is (Sec. 3.2.2)

$$m_T = \frac{1}{\lambda} \tag{4.1.5}$$

Hence the parameter, as a function of the moments, is simply

$$\lambda = \frac{1}{m_T} \tag{4.1.6}$$

By the method of moments [Eq. (4.1.2)] the estimator of m_T is the sample mean. The corresponding estimator of the parameter is the reciprocal of the sample mean. For this data the estimate of λ is

$$\hat{\lambda} = \frac{1}{\bar{t}} = \frac{1}{7.66} \text{ sec} = 0.130 \text{ sec}^{-1} \tag{4.1.7}$$

Adopting this parameter value, the engineer now can state, for example, that the probability of a gap length of 8 to 10 sec is

$$F_T(10) - F_T(8) = 1 - e^{-0.130(10)} - [1 - e^{-0.130(8)}]$$

$$= e^{-1.04} - e^{-1.30} = 0.081$$

4.1.2   The Properties of Estimators: Their First- and Second-order Moments

The sample moments have been proposed as reasonable estimators of the moments of the mathematical model. The sample moments, it should be remembered, were proposed in Chap. 1 only as logical numbers to summarize a batch of numbers observed in some physical experiment. They were given definitions [Eqs. (4.1.2) and (4.1.3)] in terms of those observed numbers, $x_1, x_2, \ldots, x_n$. The moments m and σ^2, on the other hand, are constants. They are parameters of some distribution defining the behavior of a random variable X, which is a purely mathematical model. Once we have adopted the assumption, however, that this mathematical model is controlling the physical experiment, we can interpret the x_i's as observed values of a random variable X. In this new light, the sample average $\bar{x}$ is the numerical average of n observations of the random variable X.

Sample statistics as random variables   With the ultimate goal of gaining some knowledge of how good an estimator of the mean m the sample average $\bar{x}$ is, let us back up one step to a point *before* the physical experiments have been carried out, i.e., to a point *before* the observed values $x_1, x_2, \ldots, x_n$ are known, and then let us see what can be said about this sample mean. Define the random variables $X_1, X_2, \ldots, X_n$ as the first, second, . . . , nth outcomes of the sequence of n experiments involved in the process† of sampling. Each X_i has the same distribution as X, sometimes called in this context the *population* to be sampled. In particular, each variable has mean and variance equal to the population mean m_X and population variation σ_X^2. Then the sample mean which we intend to compute can be written

$$\bar{X} = \frac{1}{n} \sum_{i=1}^{n} X_i \tag{4.1.8}$$

and the sample variance is

$$S^2 = \frac{1}{n} \sum_{i=1}^{n} (X_i - \bar{X})^2 \tag{4.1.9}$$

† Note that in principle when considering the sampling or repeated observing of a random variable X, we are dealing with a simple random process, $X_1, X_2, \ldots,$ with "time" parameter $i = 1, 2, \ldots$. This process interpretation of sampling will be fully discussed in Chap. 6. At any "time" i, the distribution of the value of the process is the same, $f_X(x)$.

Capital letters are now used for the sample mean and the sample variance to emphasize the fact that, *as functions of the random variables X_1, X_2, . . . , X_n, these sample averages are themselves random variables.* After the experiments, the outcomes $X_1 = x_1$, $X_2 = x_2$, . . . , $X_n = x_n$ will be recorded, and the observed values $\bar{x}$ and s^2 of the random variables $\bar{X}$ and S^2 will be computed. Functions $g(X_1, X_2, . . . , X_n)$ of a sample X_1, X_2, . . . , X_n are called *sample statistics.* The sample mean and sample variance are just two of many such sample statistics of potential interest. Others include $X^{(n)} = \max [X_1, X_2, . . . , X_n]$, the largest observed value, and $X^{(i)}$, the ith smallest value in the sample. (The latter is called an "order statistic." See Prob. 3.46.) Our present interest in sample statistics is due to their role as estimators of unknown parameters.

Moments of the sample mean $\bar{X}$ As a random variable, the sample mean, $\bar{X}$ is open to study of such questions as, "What is its average value?" "What is its variance?" and "What is its distribution?" In particular, since we use $\bar{X}$ as its estimator, what is the relationship between answers to these questions and the mean m_X of the random variable X? In this section we concentrate only on first- and second-order moments of this estimator. The moments of $\bar{X}$ are independent of the form of the distribution of X and they are sufficient to answer many questions about estimators. To go further we will have to specify the PDF of X. Do not forget that we study $\bar{X}$ only because we ultimately intend to use observations of $\bar{X}$ to estimate an unknown m_X.

Consider first, then, the mean of the sample mean [Eq. (4.1.8)]:

$$E[\bar{X}] = E\left[\frac{1}{n}\sum_{i=1}^{n} X_i\right] = \frac{1}{n}\sum_{i=1}^{n} E[X_i]$$

$$= \frac{1}{n}\sum_{i=1}^{n} m_{X_i} = \frac{1}{n}\sum_{i=1}^{n} m_X = \frac{nm_X}{n} = m_X$$

In short,

$$E[\bar{X}] = m_X \tag{4.1.10}$$

In words, the expected value of the sample mean is equal to the mean of the underlying random variable. Any other answer would undoubtedly leave us somewhat uneasy as to the merit of $\bar{X}$ as an estimator of m_X.

As a random variable it cannot be expected, however, that $\bar{X}$ will often, if ever, take on values exactly equal to the mean and hence give a perfect estimate of m_X. The *estimation error*, $\bar{X} - m_X$, is a random variable with a mean of zero. A measure of the average magnitude of the random error associated with an estimator can be gained through the expected value of its square (called the *mean square error*). This second-order moment reduces in this case to the variance of $\bar{X}$:

$$E[(\bar{X} - m_X)^2] = E[(\bar{X} - E[\bar{X}])^2]$$
$$= \text{Var}\,[\bar{X}] \qquad\qquad (4.1.11)$$

The next question is then how $\text{Var}\,[\bar{X}]$ depends on $\text{Var}\,[X]$:

$$\text{Var}\,[\bar{X}] = \text{Var}\left[\frac{1}{n}\sum_{i=1}^{n} X_i\right]$$
$$= \frac{1}{n^2}\text{Var}\left[\sum_{i=1}^{n} X_i\right]$$

Using Eq. (2.4.81b) for the variance of a sum of random variables,

$$\text{Var}\,[\bar{X}] = \frac{1}{n^2}\sum_{i=1}^{n}\text{Var}\,[X_i] + \frac{2}{n^2}\sum_{i=1}^{n}\sum_{j=i+1}^{n}\text{Cov}\,[X_i,X_j]$$
$$= \frac{\sigma_X^2}{n} + \frac{2}{n^2}\sum_{i=1}^{n}\sum_{j=i+1}^{n}\rho_{X_i,X_j}\sigma_{X_i}\sigma_{X_j} \qquad (4.1.12)$$

We shall restrict our attention to those cases where the sampling of X is carried on in such a manner that it may be assumed that the individual observations $X_1, X_2, \ldots, X_n$ are mutually independent random variables. Such sampling is called *random sampling*. The main reason for this restriction is simply that, even if the stochastic dependence involved lent itself to description, nonindependent types of sampling are usually more difficult to analyze.†

It should be pointed out that in practice it is virtually impossible to ascertain that true random sampling is being achieved, but, if the engineer wishes to exploit the theory of statistics to give him a quantifiable measure of the errors in estimation, it remains necessary that he take precautions to insure that there is little systematic variation or "persistence" from experiment to experiment. In sampling a population of reinforcing bars, for example, one would avoid taking more than one specimen from any one bar because correlation among specimens within a bar is to be expected.

An immediate consequence of assumption of random sampling is the great simplification of Eq. (4.1.12). For independent samples, the

---

† This is not to say that nonrandom sampling is unimportant. Note, for example, that if the sampling can be carried out so that there is negative correlation between the pairs of outcomes, the variance of $\bar{X}$ can be reduced from σ^2/n, the value associated with random sampling.

correlations are zero and

$$\text{Var}\,[\bar{X}] = \frac{1}{n}\sigma_X{}^2 \qquad (4.1.13)$$

$$\sigma_{\bar{X}} = \frac{1}{\sqrt{n}}\sigma_X \qquad (4.1.14)$$

The simplicity of these formulas belies their great practical significance. We now have a measure of the expected error associated with the rule that states that m_X should be estimated by $\bar{X}$. No matter what the distributions of X and $\bar{X}$, through Chebyshev's inequality [Eq. (2.4.11)] we are now in a position to make probability statements regarding the likelihood that the estimator $\bar{X}$ will be within certain limits about m_X, say $\pm k\sigma_{\bar{X}}$. Equation (4.1.14) also shows how, for fixed probability (or fixed k), the width of these limits decreases as the sample size increases. Notice, for example, that to halve the interval width from that associated a single observation ($n = 1$) of X, one must increase the sample size to $n = 4$. To halve it again, n must equal 16, and so forth. The distributions of $\bar{X}$ for these cases are sketched in Fig. 4.1.1. It becomes increasingly expensive to continue to reduce the magnitude of $\sigma_{\bar{X}}$, the measure of error in the estimator $\bar{X}$. It should be remembered, however, that $\sigma_{\bar{X}}$ is a measure of *expected* error. In a state of ignorance regarding the true mean we can only say that the sample averages

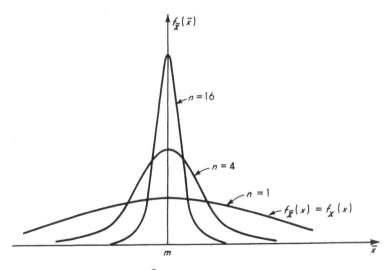

Fig. 4.1.1 Distribution of $\bar{X}$ for $n = 1, 4, 16$.

based on larger n are *more likely* to fall closer to m_X than those derived from less information.

Moments of the sample variance; bias   A brief, similar study of the first- and second-order moment behavior of the (random variable) sample variance S^2 will lead us to some additional conclusions.   We are interested in S^2 because we have proposed using S^2 as an estimator of the parameter σ^2.   By definition,

$$S^2 = \frac{1}{n} \sum_{i=1}^{n} (X_i - \bar{X})^2 \tag{4.1.15}$$

Expanding the square and taking expectations leads directly to the conclusion that (for random samples)

$$E[S^2] = \frac{n-1}{n} \sigma_X{}^2 \tag{4.1.16}$$

The striking and somewhat disturbing conclusion is that the expected value of the second central moment of the sample $E[S^2]$ is *not* equal to the second central moment of the probability distribution σ^2. On the average over many different samples, S^2 will not equal the parameter $\sigma_X{}^2$ for which it has been proposed to serve as an estimator, but will equal something smaller.†

Estimators, like S^2, whose expected values are not equal to the parameter they estimate are said to be biased estimators.   On the assumption that bias is an undesirable property, one can easily determine an unbiased estimator of the variance.

The estimator S^{2*}, which we define as

$$S^{2*} = \frac{n}{n-1} S^2 \tag{4.1.17}$$

or

$$S^{2*} = \frac{1}{n-1} \sum_{i=1}^{n} (X_i - \bar{X})^2 \tag{4.1.18}$$

is a sample statistic whose expected value is $n/(n-1)$ times the expected value of S^2 or

$$E[S^{2*}] = \sigma^2 \tag{4.1.19}$$

† Notice, however, that if n is reasonably large, there is little numerical difference between σ^2 and $[(n-1)/n]\sigma^2$.

Hence S^{2*} is an *unbiased estimator* of σ_X^2. For this reason the formulas

$$s^{2*} = \frac{1}{n-1} \sum_{i=1}^{n} (x_i - \bar{x})^2 \tag{4.1.20}$$

$$= \frac{1}{n-1} \left(\sum_{i=1}^{n} x_i^2 - n\bar{x}^2 \right) \tag{4.1.21}$$

are found in many texts discussing data reduction in place of Eqs. (1.2.2a) and (1.2.2b) for s^2. However, without the formal mechanism of probability and statistics and without the number of assumptions (e.g., independence and desirability of an unbiased estimator) which led us to Eq. (4.1.19), there is no theoretical reason to favor s^{2*} over the more natural s^2 as a numerical summary of a batch of numbers.

In the bar-yield-force illustration, the estimator S^2 led to an estimate of $(103)^2$ for σ^2, and the unbiased estimator S^{2*} yields an estimate of $(105)^2$.

Even if S^{2*} is unbiased, however, it is not necessarily true that S^{2*} is a "better" estimator of σ^2 than S^2. In general, the error between any (biased or unbiased) estimator $\hat{\theta}$ and the parameter it estimates θ is $\hat{\theta} - \theta$. (The former value is random; the second is a constant.) The mean square error $E[(\hat{\theta} - \theta)^2]$ is found easily upon expanding the square to equal

$$E[(\hat{\theta} - \theta)^2] = \text{Var } [\hat{\theta}] + (m_{\hat{\theta}} - \theta)^2 \tag{4.1.22}$$

which depends on *both* the variance and the bias, $m_{\hat{\theta}} - \theta$, of the estimator. Therefore, the mean square errors of the two estimators S^2 and S^{2*} are

$$E[(S^2 - \sigma^2)^2] = \text{Var } [S^2] + \left(\frac{n-1}{n} \sigma^2 - \sigma^2 \right)^2$$

$$= \text{Var } [S^2] + \frac{1}{n^2} \sigma^4 \tag{4.1.23}$$

and

$$E[(S^{2*} - \sigma^2)^2] = \text{Var } [S^{2*}] = \frac{n^2}{(n-1)^2} \text{Var } [S^2] \tag{4.1.24}$$

It is not immediately clear which estimator has the smaller mean square error. For a special case, that where X is normally distributed, some simple algebra will show that†

$$\text{Var } [S^2] = \frac{2(n-1)}{n^2} \sigma^4 \tag{4.1.25}$$

---

† Notice the dependence of the mean square error of an estimator on n. Characteristically, it is approximately inversely proportional to n.

Substituting into the previous two equations for the mean square errors, we can calculate the ratio $q(n)$ of the expected squared error of S^2 to that of S^{2*}:

$$q(n) = 1 - \frac{3n - 1}{2n^2} \tag{4.1.26}$$

The ratio is less than 1 for all values of n. For example, $q(2) = 1 - \frac{5}{8} = \frac{3}{8}$; $q(4) = \frac{21}{32}$; $q(10) = 0.855$; and $q(100) = 0.985$. In other words, although S^2 has an expected value different from σ^2, the expected value of its squared error is smaller than that of the unbiased estimator S^{2*}. The latter estimator leads more frequently to larger errors. This condition is sketched in Fig. 4.1.2 for small n where it is most extreme. In customary statistical terminology, although *unbiased*, S^{2*} is not a *minimum-mean-square-error estimator*† of σ^2.

Choice of estimator One can conclude generally that the choice of the sample statistic (the function of the observed values) which he should use to estimate a parameter is not necessarily obvious. Any reasonable statistic can be considered as a contending estimator. The sample median, sample mode, or even the average of the maximum and minimum value in the sample are not unreasonable estimators of the mean of many distributions. The method of moments is one general rule for selecting estimators. The method of maximum likelihood (Sec. 4.1.4) is yet another, very common general rule. The question of choosing the

† In the face of this dilemma the authors recommend the use of S^{2*} as an estimator of σ^2 if only for the sake of promoting uniformity among reports of experimental data.

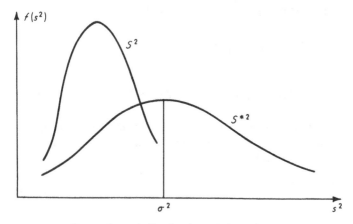

Fig. 4.1.2 Comparison of distributions of S^2 and S^{2*}, an extreme case.

best estimator is ambiguous until "best" is defined. Such intuitively desirable properties† of an estimator as its being unbiased and as its being the estimator which promises the minimum mean square error are unfortunately not necessarily combined in the same sample statistic, as was seen in comparing S^2 and S^{2*} as contending estimators of σ^2. The search for and study of "best" or just "good" estimators is one of the tasks of mathematical statisticians. We refrain from venturing further into the questions of defining other desirable properties, or of choosing among contending estimators. In Sec. 4.1.4 we present an alternative general method, the method of maximum likelihood. It is an intuitively appealing method of choosing estimators which are, on the whole, equal to or better (in the mean-square-error sense) than those estimators suggested by the method of moments.

We have seen that a sample statistic's mean and variance, which we were able to find *without reference to the distribution of X*, are sufficient to describe the salient characteristics of the statistic as an estimator of a model parameter. They define its central tendency relative to the parameter and its dispersion (mean square error) about the true parameter; in particular they indicate the dependence of this latter factor on sample size. More precise, quantitative statements of the uncertainty in parameter values are possible if the entire distribution of the sample statistic is utilized. (This distribution depends upon the distribution of X.) We shall find that one vehicle for such statements is the *confidence interval*.

4.1.3 The Distributions of Estimators and Confidence–interval Estimation

The distribution of a sample statistic, which is a function of the random variables $X_1, X_2, \ldots, X_n$ in the sample, depends upon their (common) distribution. The methods of derived distributions (Sec. 2.3) are in general necessary to determine the distribution of the statistic.‡ We shall focus our attention here on the distribution of the sample mean, which is the distribution of a constant times a sum of random variables.

Distribution of the sample mean of Bernoulli variables For example, suppose the sampled random variable X has a simple Bernoulli distribution with

$$p_X(x) = \begin{cases} p & x = 1 \\ 1 - p & x = 0 \end{cases} \qquad (4.1.27)$$

† A number of such properties will be briefly defined in Sec. 4.1.4.
‡ The study of the distributions of sample statistics is central to the field of *mathematical statistics*. Clearly these problems are ones of deriving distributions of functions of random variables and hence are problems of pure probability theory. The studies retain the name of statistics, however, because their motivation is statistical applications.

with

$$m_X = p \tag{4.1.28}$$

and

$$\sigma_X^2 = p(1 - p) \tag{4.1.29}$$

Then the distribution of the average of a sample of size n,

$$\bar{X} = \frac{1}{n} \sum_{i=1}^{n} X_i$$

where each X_i has the distribution above, can be determined by inspection. The sum

$$Y = \sum_{i=1}^{n} X_i$$

representing the number of successes (observations of "1's") in n Bernoulli trials, is binomial, $B(n,p)$ [Sec. 3.1.2, Eq. (3.1.4a)]. Hence the distribution of $\bar{X} = Y/n$ is [Sec. 2.3.1]

$$p_{\bar{X}}(\bar{x}) = p_Y(\bar{x}n) = \binom{n}{n\bar{x}} p^{n\bar{x}}(1 - p)^{n-n\bar{x}}$$

$$\bar{x} = \frac{0}{n}, \frac{1}{n}, \frac{2}{n}, \cdots, \frac{n}{n} \tag{4.1.30}$$

$$m_{\bar{X}} = m_X = p \tag{4.1.31}$$

$$\sigma_{\bar{X}}^2 = \frac{\sigma_X^2}{n} = \frac{p(1 - p)}{n} \tag{4.1.32}$$

which is a binomial-like distribution defined on noninteger values.

Illustration: Quality indicator In the process of setting up proposed statistical methods of quality control in highway construction, engineers have investigated data from old tests of a typical material quality indicator, such as the density of the subbase. The data were selected only from those sections of roads which had subsequently proved with use to show *good* performance. The purpose of this effort is to determine the reliability of the material tests as indicators of final road performances. It is desired to estimate what proportion p of these tests indicated poor material and consequently incorrectly suggested that the road would be of unsatisfactory quality. If X is a $(0,1)$ Bernoulli random variable, indicating by a value "1" that the test would have rejected the (satisfactory) road, then in a sample of four sections of a road, the estimator $\bar{X}$ of m_X and hence of p has the distribution

$$p_{\bar{X}}(\bar{x}) = \begin{cases} (1 - p)^4 & \bar{x} = 0 \\ 4p(1 - p)^3 & \bar{x} = \frac{1}{4} \\ 6p^2(1 - p)^2 & \bar{x} = \frac{2}{4} \\ 4p^3(1 - p) & \bar{x} = \frac{3}{4} \\ p^4 & \bar{x} = \frac{4}{4} \end{cases}$$

Clearly the estimate of p is restricted to one of these five values, since only an integer number, 0, 1, 2, 3, or 4, of tests can be observed to be "bad." The mean and variance of the estimator are

$$m_{\bar{X}} = p$$

$$\sigma_{\bar{X}}{}^2 = \frac{1}{4}p(1 - p)$$

For example, if the true value of p were 10 percent, then $\bar{X}$, the estimator of p, would have this distribution:

$$p_{\bar{X}}(\bar{x}) = \begin{cases} 0.6561 & \bar{x} = 0 \\ 0.2916 & \bar{x} = 0.25 \\ 0.0486 & \bar{x} = 0.50 \\ 0.0036 & \bar{x} = 0.75 \\ 0.0001 & \bar{x} = 1.00 \end{cases}$$

Thus, for example, if the engineers look at only four observations of a test of this kind, i.e., one which is actually incorrect only 10 percent of the time, there is a 5 percent chance that they will *estimate* that the test "rejects" potentially good roads half the time. If many independent samples of size four were taken (say, four sections from many different roads) and p estimated from each sample, the average (mean) of the estimate would be

$$E[\bar{X}] = 0.1$$

and the square root of the mean square error (*rms*) of these many estimates would be

$$\sigma_{\bar{X}} = \sqrt{\sigma_{\bar{X}}{}^2} = \sqrt{\frac{0.1(0.9)}{4}} = 0.15$$

Normally distributed sample mean $\bar{X}$ From Chap. 3 we know that the distribution of the sample means in the two illustrations in Sec. 4.1.1 are, respectively, normal, $N(m_X, \sigma_X/\sqrt{n})$, and gamma,† $G(n, n\lambda)$, for the normally distributed bar-yield-force variable and the exponentially distributed traffic-interval variable. Assuming that the true distribution of the bar-force data is normal, with mean 8500 lb and standard deviation 100 lb, this distribution and those of the sample means for two sample sizes, $n = 9$ and $n = 27$, are shown in Fig. 4.1.3. The distribution of $\bar{T}$, the mean of the exponentially distributed traffic intervals is skewed right for small sample sizes. As n increases, this gamma distribution will be almost perfectly symmetrical. In a few such cases the distribution of the sample mean is one of the common distributions discussed in Chap. 3. But in general when dealing with the distribution of the sample mean we rely upon another observation.

† The reader can convince himself of this by recognizing that $\sum_{i=1}^{n} T_i$ is $G(n, \lambda)$ (see Sec. 3.2.3) and then finding the moments and parameters of $\bar{T} = 1/n\Sigma T_i$.

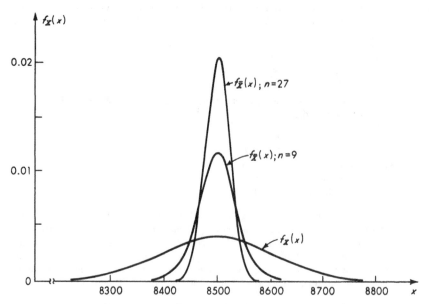

Fig. 4.1.3 Bar-yield-force illustration—distributions of X and of two sample means $\bar{X}$; $n = 9$, $n = 27$.

If n is reasonably large, we can invoke the central limit theorem (Sec. 3.3.1) to state that, no matter what the distribution of X, the sample mean will be approximately normal. Its mean is m_X and its standard deviation is $\sigma_X/\sqrt{n}$. As the most important single case, we choose the normally distributed sample mean to illustrate the use to which the distribution of the sample statistic is put in estimation situations, namely, in the determination of *interval estimates.*

Confidence intervals   Consider again the problem of the yield force in reinforcing bars. Let us assume that through a history of tests on similar material and other bar sizes the engineer knows with relative certainty that the standard deviation of the yield force is 100 lb.† His problem is to estimate the mean, and he focuses attention quite logically upon the sample mean $\bar{X}$. This statistic is, prior to making the sample, a random variable normally distributed‡ with unknown mean m and known standard deviation $100\sqrt{1/n}$. Thus the variable $(\bar{X} - m)/100\sqrt{1/n}$ is $N(0,1)$. Anticipating an experiment involving 27 speci-

---

† Treatment of the situation in which σ^2 cannot be assumed to be known will follow.
‡ Given the engineer's hypothesis that X is normally distributed, this statement is exactly correct. Under other conditions it remains approximately true (for n large).

mens, the engineer can say, for example,

$$P\left[-1.96 \leq \frac{\bar{X} - m}{100\sqrt{\frac{1}{27}}} \leq 1.96\right] = 0.95 \qquad (4.1.33)$$

$$P\left[-1.96\frac{(100)}{\sqrt{27}} \leq \bar{X} - m \leq 1.96\frac{(100)}{\sqrt{27}}\right] = 0.95$$

or

$$P[-37.7 \leq \bar{X} - m \leq 37.7] = 0.95 \qquad (4.1.34)$$

The statement, *before* experimentation, is that the sample average will with probability 0.95 lie within 37.7 lb of the true mean.

But what can be said after the experiment when the results in Table 4.1.1 are observed and $\bar{X}$ is found to take on the value $\bar{x} = 8490$ lb? It is tempting to substitute $\bar{x}$ for $\bar{X}$ in the formula above, yielding

$$P[-37.7 \leq 8490 - m \leq 37.7] = 0.95$$

$$P[8452.3 \leq m \leq 8527.7] = 0.95 \qquad (4.1.35)$$

As inviting as it is to use the statement "the probability that the true mean lies in the interval 8452.3 and 8527.7 is 0.95" as a concise way of incorporating our uncertainty about the true value of the mean, the statement is, strictly speaking, inadmissible. The true mean of X has been considered, to this point,[†] to be a constant, not a random variable. As such, m_X has a specific (although unknown) value, and it is not free to vary from experiment to experiment. It either lies in the stated interval or it does not.

Although the probability statement is not strictly correct in the present context, intervals formed in the manner above do convey a measure of the uncertainty associated with incomplete knowledge of parameter values. Such intervals are termed *confidence intervals* or *interval estimates*. The number which is the measure of this uncertainty is called the *confidence coefficient*. In this case the "95 percent confidence interval" that was observed in this particular experiment is 8452.3 to 8527.7.

The confidence interval provides an *interval estimate* on a parameter as opposed to the *point estimates* we considered earlier in the chapter. When the distribution whose parameter is being estimated is to be embedded in a still larger probabilistic model, the engineer can seldom afford the luxury of carrying along anything more than a point estimate,

---

† The more natural consideration of the parameter as a random variable will be discussed in Chap. 6. This intuitively more satisfactory treatment will remove the misgivings that many engineers feel when they are told by a large class of statisticians that they "must" not make a probability statement about an unknown parameter such as m.

i.e., a single number, to represent the parameter. But the statement of an interval estimate does serve to emphasize that the value of a parameter is not known precisely and to quantify the magnitude of that uncertainty.

Confidence intervals on the mean Consider a more general statement concerning the confidence interval of the mean when σ is known. The statistic $U = (\bar{X} - m)/\sigma \sqrt{1/n}$ is assumed to be approximately (if not exactly) distributed, $N(0,1)$. Hence

$$P\left[a \leq \frac{\bar{X} - m}{\sigma \sqrt{1/n}} \leq b\right] = F_U(b) - F_U(a) \qquad (4.1.36)$$

U being the standardized normal random variable. The limits a and b are usually chosen symmetrically,† and we write

$$P\left[-k_{\alpha/2} \leq \frac{\bar{X} - m}{\sigma \sqrt{1/n}} \leq k_{\alpha/2}\right] = 1 - \alpha \qquad (4.1.37)$$

in which $k_{\alpha/2}$ is defined to be that value such that $1 - F_U(k_{\alpha/2}) = \alpha/2$. It is found in tables of the standardized normal distribution. Then

$$P\left[m - \frac{\sigma k_{\alpha/2}}{\sqrt{n}} \leq \bar{X} \leq m + \frac{\sigma k_{\alpha/2}}{\sqrt{n}}\right] = 1 - \alpha \qquad (4.1.38)$$

or

$$P\left[\bar{X} - \frac{\sigma k_{\alpha/2}}{\sqrt{n}} < m < \bar{X} + \frac{\sigma k_{\alpha/2}}{\sqrt{n}}\right] = 1 - \alpha \qquad (4.1.39)$$

which is of the form

$$P[g_1(\bar{X}) \leq m \leq g_2(\bar{X})] = 1 - \alpha \qquad (4.1.40)$$

In this form it is clear that the two limits of the interval, $g_1(\bar{X})$ and $g_2(\bar{X})$, are sample statistics and random variables.‡ Equation (4.1.40) states that the probability that these two random variables will bracket the true mean is $1 - \alpha$. The probability that *either* the lesser random variable $g_1(\bar{X})$ will exceed m *or* the greater, $g_2(\bar{X})$, will be less than m is α. Figure 4.1.4 demonstrates these points.

The confidence interval on m with confidence coefficient $(1 - \alpha)100$ percent is formed after the experiment by substituting the observed value $\bar{x}$ for $\bar{X}$. Thus the $(1 - \alpha)100$ percent confidence interval on m is $\{g_1(\bar{x}), g_2(\bar{x})\}$ or $\{\bar{x} - \sigma k_{\alpha/2}/\sqrt{n}, \bar{x} + \sigma k_{\alpha/2}/\sqrt{n}\}$.

† This choice gives the shortest possible interval in this case.
‡ These two statistics are perfectly correlated random variables, since $g_2(\bar{X})$ equals simply $g_1(\bar{X})$ plus a constant $(2\sigma k_{\alpha/2}/\sqrt{n})$.

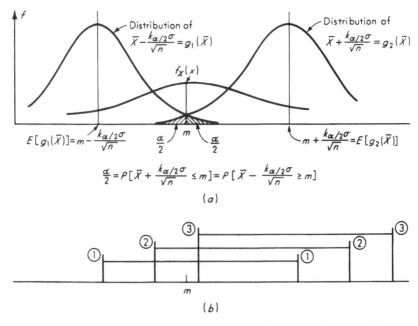

Fig. 4.1.4 (a) Distributions of confidence-limit statistics, $g_1(\bar{X})$ and $g_2(\bar{X})$; (b) three possible observed values of confidence limits $g_1(\bar{X})$ to $g_2(\bar{X})$ (two contain m, the third does not).

Factors influencing confidence intervals The engineer would prefer to report very tight intervals, that is, well-defined estimates of m. But notice that the width of the interval between these limits depends on three factors: σ, α, and n. The larger the standard deviation of X, the wider these intervals will be, and the intervals will be narrower if $1 - \alpha$ is permitted to be smaller. The value of $1 - \alpha$ can be interpreted as the long-run proportion of many $(1 - \alpha)100$ percent confidence intervals which will include the true mean. In the long run $\alpha100$ percent will fail to bracket the true mean. The choice of the value of α is the engineer's, and depends in principle upon the importance associated with making the mistake of stating that an observed interval contains m when in fact it does not. In practice, the 90, 95, or 99 percent confidence intervals are reported. Decreasing α requires increasing the interval width, i.e., making a less precise statement about the value of m.

It is most important to observe that these confidence limits depend upon the sample size n. For a fixed confidence level, $1 - \alpha$, the interval can be made as short as the engineer pleases at the expense of additional sampling. If, on the other hand, the sample size has been fixed (and α is prescribed), the engineer must be content with the interval width associated with this amount of information. Precise statements with high confidence come only at the price of a sufficiently large sample.

Confidence intervals: typical procedure Confidence limits on a parameter are obtained, in general, in the following way:

1. Writing a probability statement about an appropriate sample statistic
2. Rearranging the statement to isolate the parameter
3. Substituting in the inequality the observed value of the sample statistic

Illustrations will make this procedure clear and bring out some additional points.

Illustration: The exponential case: one-sided intervals and confidence intervals on functions Recall the exponential illustration in Sec. 4.1.1. There the sample size was large, so we may assume that the estimator $\bar{T}$ of the mean $1/\lambda$ is approximately normally distributed with mean $m_{\bar{T}} = m_T = 1/\lambda$ and variance $\sigma_{\bar{T}}^2 = \sigma_T^2/n = 1/n\lambda^2$. The large sample size implies that the coefficient of variation of $\bar{T}$, $(1/\sqrt{n}\,\lambda)/(1/\lambda) = 1/\sqrt{n}$, is small and that consequently the method-of-moments estimator of λ, $\hat{\lambda} = 1/\bar{T}$, is also approximately normally distributed† with approximate mean $m_{\hat{\lambda}} = \lambda$ and variance $\sigma_{\hat{\lambda}}^2 = \lambda^2/n$. The implication is that the sample statistic $(\hat{\lambda} - m_{\hat{\lambda}})/\sigma_{\hat{\lambda}} = (\hat{\lambda} - \lambda)/\lambda\sqrt{1/n}$ is approximately $N(0,1)$.

In the traffic-interval study suppose that interest lies in the possibility of the average arrival rate λ being so great that the gaps between cars are not long enough on the average to permit proper merging of traffic from a side road. The engineer would like a "conservative" lower bound estimate on the value of λ.

A *one-sided* probability statement on the sample statistic is called for. It should be

$$P\left[\frac{\hat{\lambda} - \lambda}{\lambda\sqrt{1/n}} \leq k_\alpha\right] = 1 - \alpha \tag{4.1.41}$$

in which k_α is that value of an $N(0,1)$ variable which will be exceeded with probability α. Isolating the parameter of interest within the statement,

$$P\left[\hat{\lambda} \leq \lambda\left(1 + k_\alpha\sqrt{\frac{1}{n}}\right)\right] = 1 - \alpha$$

$$P\left[\lambda \geq \frac{\hat{\lambda}}{1 + k_\alpha\sqrt{1/n}}\right] = 1 - \alpha \tag{4.1.42}$$

The random variable in this statement is, recall, $\hat{\lambda}$. After the sample is taken, a *one-sided* $(1 - \alpha)100$ percent confidence interval on the parameter λ is found by substituting for the observed value of the random variable $\hat{\lambda} = 1/\bar{T}$. In this example, suppose a 95 percent level of confidence is desired; then

† See Sec. 2.4.4. This approximation will also be justified in Sec. 4.1.4.

$\alpha = 5$ percent and $k_{0.05} = 1.64$ (Table A.1). A sample (Table 4.1.2) of size $n = 214$ yielded an observed value of $\hat{\lambda} = 0.130 \text{ sec}^{-1}$. Thus "with 95 percent confidence" λ is greater than

$$\frac{\hat{\lambda}}{1 + k_\alpha \sqrt{1/n}} = \frac{0.130}{1 + 1.64 \sqrt{\frac{1}{214}}} = 0.116 \text{ sec}^{-1}$$

in words, 0.166 is a *lower* one-sided 95 percent confidence limit on λ.

Exact Distribution   It is worth outlining how such a confidence interval would be obtained if the sample size were not sufficiently large to justify the assumption of normality. In principle the exact distribution of $\hat{\lambda} = 1/\bar{T}$ is needed, when, as pointed out above, $\bar{T}$ has a gamma distribution. In this case, however, we can use the distribution of $\bar{T}$ directly, since

$$P[\hat{\lambda} \le a] = P\left[\frac{1}{\bar{T}} \le a\right] = P\left[\bar{T} \ge \frac{1}{a}\right] = 1 - \alpha \tag{4.1.43}$$

For given α and n, tables (Sec. 3.2.3) of the gamma distribution will give $1/a$ as a function of the parameter λ. A rearrangement will isolate λ in the probability statement permitting the substitution of the observed value of $\hat{\lambda}$ and yielding an exact confidence interval on λ.

Confidence Interval on a Function of a Parameter   Engineering interest in estimated quantities usually goes beyond the values of the parameters themselves. If the model is incorporated into the framework of a larger problem, confidence intervals on functions of the parameters may be required. Consider the following elementary example. In the traffic situation above, the engineer might be interested in an estimate of the proportion of gaps which are larger than t_0, the value required to permit another car to enter from a side road.† This proportion is

$$P[T \ge t_0] = 1 - F_T(t_0) = e^{-\lambda t_0} \tag{4.1.44}$$

which is a function of the unknown parameter λ.

A point estimate of $e^{-\lambda t_0}$ is simply $e^{-\hat{\lambda} t_0}$. If an *upper*, one-sided confidence interval on $e^{-\lambda t_0}$ is desired, the starting point is again an appropriate, rearranged probability statement on the sample statistic [Eq. (4.1.42)]:

$$P\left[\lambda \ge \frac{\hat{\lambda}}{1 + k_\alpha \sqrt{1/n}}\right] = 1 - \alpha$$

Manipulating the inequality to isolate, not λ, but the function of λ of interest, namely, $e^{-\lambda t_0}$, we obtain

$$P\left[-\lambda t_0 \le \frac{-\hat{\lambda} t_0}{1 + k_\alpha \sqrt{1/n}}\right] = 1 - \alpha$$

† A second similar example arises in studying the reliability of systems where one wants to have a lower-bound estimate on the *reliability*, i.e., on the probability that the system will perform satisfactorily for a time t_0 when the system's time to failure T, is a random variable.

and finally

$$P \left[e^{-\lambda t_0} \leq \exp \left\{ \frac{-\lambda t_0}{1 + k_\alpha \sqrt{1/n}} \right\} \right] = 1 - \alpha \qquad (4.1.45)$$

In words, a $(1 - \alpha)100$ percent upper confidence interval on the proportion of gaps greater than t_0 is $\exp[-\lambda t_0/(1 + k_\alpha \sqrt{1/n})]$ with the observed value of λ substituted.

In the numerical example above, with $t_0 = 10$ sec, the 95 percent upper confidence interval on the proportion is

$$\exp \left[\frac{-(0.133)(10)}{1 + 1.64 \, (1/\sqrt{214})} \right] = e^{(-1.16)} = 0.313$$

In words, the engineer would report that "with 95 percent confidence the proportion of gaps greater than 10 sec is less than 0.313."

Notice that in this case the result coincides with the value obtained by substituting the *lower* bound of the 95 percent confidence interval on λ into the function of λ, $e^{-\lambda t_0}$. When a monotonic function is involved, such a simple substitution is always possible provided that thought is given to the type of bound on the parameter, upper or lower, which corresponds to the desired bound on the function. More difficult operations, analogous to those used in Sec. 2.3, are needed when confidence intervals on nonmontonic functions are encountered.

Illustration: Variance of the normal case Confidence intervals on the variance of a random variable can be formed once the distribution of S^2 is known. We consider here a special case—that in which the underlying variable X is normally distributed.

The determination of the distribution of S^2 follows from available results. Consider the statistic S^2 in Eq. (4.1.15). Adding and subtracting m and expanding the square gives

$$S^2 = \frac{1}{n} \sum_{i=1}^{n} (X_i - m)^2 - (\bar{X} - m)^2 \qquad (4.1.46)$$

Multiplying by n/σ^2, we find

$$\frac{nS^2}{\sigma^2} = \sum_{i=1}^{n} \left(\frac{X_i - m}{\sigma} \right)^2 - \left(\frac{\bar{X} - m}{\sigma \sqrt{1/n}} \right)^2 \qquad (4.1.47)$$

The first term on the right-hand side is the sum of the squares of n independent normally distributed random variables each with zero mean and standard deviation 1. According to the results in Sec. 3.4.2, this sum has the χ^2 distribution with n degrees of freedom. But the second term is also the square of a $N(0,1)$ random variable since $\bar{X}$ has been shown to be $N(m,\sigma \sqrt{1/n})$. Hence the second term has a χ^2 distribution with 1 degree of freedom. Rewriting Eq. (4.1.47), we obtain

$$\sum_{i=1}^{n} \left(\frac{X_i - m}{\sigma} \right)^2 = \frac{nS^2}{\sigma^2} + \left(\frac{\bar{X} - m}{\sigma \sqrt{1/n}} \right)^2 \qquad (4.1.48)$$

Equation (4.1.48) states that a χ_n^2 random variable is the sum of nS^2/σ^2 and a χ_1^2 random variable. Knowing, as shown in Sec. 3.4.2, that the sum of two *independent* χ^2 random variables, one with r degrees of freedom and the other with p, is χ^2-distributed with $r + p$ degrees of freedom, it is not unexpected (what is shown here is not a proof†) that nS^2/σ^2 *is* χ^2-*distributed with* $n - 1$ *degrees of freedom.* The same is true of $(n - 1)S^{2*}/\sigma^2$.

With this knowledge confidence intervals follow easily. For example, begin with the probability statement

$$P\left[\frac{nS^2}{\sigma^2} \geq \chi_{1-\alpha,n-1}^2\right] = 1 - \alpha \tag{4.1.49}$$

in which $\chi_{1-\alpha,n-1}^2$ is that (tabulated) value‡ such that, if Y is $\mathrm{CH}(n - 1)$, then $F_Y(\chi_{1-\alpha,n-1}^2) = \alpha$. We are led to a $(1 - \alpha)100$ percent one-sided confidence interval on σ^2 by rearrangement:

$$P\left[\sigma^2 \leq \frac{nS^2}{\chi_{1-\alpha,n-1}^2}\right] = 1 - \alpha \tag{4.1.50}$$

For example, in the bar-yield-force illustration, the 99 percent upper confidence limit on σ^2 is (Table A.2)

$$\frac{ns^2}{\chi_{1-\alpha,n-1}^2} = \frac{27(103)^2}{\chi_{0.99,26}^2} = \frac{(27)(103)^2}{12.2} = 2.2(103)^2 = 23{,}500 \text{ lb}^2$$

Owing to the monotonic relationship between the variance and standard deviation, the square roots of variance confidence interval bounds are the bounds of the corresponding confidence limits on the standard deviation. In this example a 99 percent one-sided upper confidence interval on σ is $\sqrt{2.2(103)^2}$ or $\sqrt{2.2}\,103 = 158$ lb.

Illustration: Mean of the normal case, variance unknown; experimental design Interval estimates on the mean are usually based on the assumption that the sample statistic $\bar{X}$ is normally distributed with mean m and variance σ^2/n. This case has been considered in detail above under the assumption that σ^2 was known. If, as is more common, the variance is not known, other approaches are necessary. In some cases the unknown variance may be simply related to the unknown mean. This is the case if the distribution of X has only a single unknown parameter. (Examples are: the exponential distribution, the binomial distribution, or the normal distribution with known coefficient of variation.) In this case relatively simple intervals are possible, as before. See, for example, the first paragraph of the preceding illustration of the exponential parameter. In other cases, if the sample size is sufficiently large, it is reasonable (see Brunk [1960]) to set up the confidence interval assuming that σ^2 is known and then simply replace σ^2 by its estimate S^2 after the test results are obtained. In other words, the statistic $(\bar{X} - m)/\sigma\sqrt{1/n}$ used in Eq. (4.1.36) is replaced by the statistic $(\bar{X} - m)/S\sqrt{1/n}$ or $(\bar{X} - m)/S^*\sqrt{1/n}$, but the uncertainty associated with not knowing σ^2 is ignored.

† The needed and somewhat unexpected independence of nS^2/σ^2 and $(\bar{X} - m)/\sigma\sqrt{1/n}$ is discussed further in the next illustration.

‡ $\chi_{1-\alpha,n-1}^2$ is the rather unfortunate common notation. Caution: it does not denote a random variable, but a *value* which a random variable with a χ^2 distribution with $n - 1$ degrees of freedom will exceed with probability $1 - \alpha$.

In smaller samples this contribution cannot be ignored, however, and, if one of the latter statistics is to be used, its true distribution must be found. This has been done for the case when X is normal. Multiplying numerator and denominator by $\sqrt{n-1}/\sigma$, the second statistic[†] above, i.e., the so-called *t statistic*, takes on a recognizable form:

$$T = \frac{\left(\dfrac{\bar{X}-m}{\sigma\sqrt{1/n}}\right)\sqrt{n-1}}{\sqrt{n-1}\,S^*/\sigma} = \frac{(\bar{X}-m)}{S^*\sqrt{1/n}} \tag{4.1.51}$$

The numerator is $\sqrt{n-1}$ times a $N(0,1)$ random variable. The denominator being the square root of the χ^2-distributed variable, $(n-1)S^{2*}/\sigma^2$, is χ-distributed, CHI$(n-1)$, as defined in Sec. 3.4.3. In that same section a variable of the form of T was said to have the "t distribution" with $n-1$ degrees of freedom, if numerator and denominator are independent. That S^{2*} and $\bar{X}$, both functions of the same n normally distributed random variables X_1, $X_2, \ldots, X_n$, are independent is hardly to be expected, but is in fact true. The verification is lengthy and will not be shown here.[‡]

The t statistic, Eq. (4.1.51), has a somewhat broader distribution than the corresponding normally distributed statistic, $(\bar{X}-m)/\sigma\sqrt{1/n}$, associated with a known standard deviation σ. See Fig. 3.4.4. This fact merely reflects the greater uncertainty which is introduced if there is a lack of knowledge of the true value of σ. For the same value of $1-\alpha$, we should expect somewhat wider confidence intervals on m in this case. This will be demonstrated.

Having at hand the distribution of the statistic $(\bar{X}-m)/S^*\sqrt{1/n}$ (see Table A.3), we can find a confidence interval on m by following the usual pattern.

Write an appropriate probability statement:

$$P\left[-t_{\alpha/2,n-1} \le \frac{\bar{X}-m}{S^*\sqrt{1/n}} \le +t_{\alpha/2,n-1}\right] = 1-\alpha \tag{4.1.52}$$

in which $t_{\alpha/2,n-1}$ is that (tabulated) value which a random variable with a t distribution with parameter $n-1$ exceeds with probability $\alpha/2$.

Isolate the parameter of interest:

$$P\left[\bar{X} - \frac{S^*}{\sqrt{n}}t_{\alpha/2,n-1} \le m \le \bar{X} + \frac{S^*}{\sqrt{n}}t_{\alpha/2,n-1}\right] = 1-\alpha \tag{4.1.53}$$

Substitute in the inequality the observed values of the sample statistics. The $(1-\alpha)100$ percent two-sided confidence interval on m is

$$\left\{\bar{x} - \frac{s^*}{\sqrt{n}}t_{\alpha/2,n-1}\right\} \qquad \text{to} \qquad \left\{\bar{x} + \frac{s^*}{\sqrt{n}}t_{\alpha/2,n-1}\right\} \tag{4.1.54}$$

[†] We choose here, for historical reasons, S^{2*} over S^2. The results using S^2 are parallel. $(\bar{X}-m)\sqrt{n-1}/S$ is $T(n-1)$ distributed. S^* is the square root of S^{2*}.
[‡] Proof is available in Brunk [1960], Freeman [1963], or Hald [1953], to name but a few locations.

In the bar-yield force example with $n = 27$, $\bar{x} = 8490$, and $s^* = 105$, the 95 percent confidence intervals on m are

$$8490 \pm \frac{105}{\sqrt{27}} t_{0.025,26}$$

which, upon substitution from Table A.3 of $t_{0.025,26} = 2.056$, yields

$$\{8490 \pm 43\} \quad \text{or} \quad \{8447, 8533\}$$

These limits are wider by a factor $2.056/1.963 = 1.05$ than those which would be obtained if σ had been assumed to be known with certainty to equal the observed value of S^*, namely, 105 lb. For sample sizes of 5, 10, and 15 this factor is 1.31, 1.16, and 1.08. For sample sizes larger than about 25, it is clearly usually sufficient to use the test based on the normal distribution assuming $\sigma = s^*$. In other words, for large n, T is very nearly $N(0,1)$ (see Fig. 3.4.4).

Choice of Sample Size The engineer is occasionally in a situation where he can "design" his sampling procedure to yield confidence limits of a desired width with a prescribed confidence. In other words he can determine n, the sample size needed to yield an estimate of m, say, with specified accuracy and confidence.

Consider the following example. It is typical of practice recommended by the American Concrete Institute for determining the required number of times to sample a batch of concrete in order to estimate its mean strength for quality control purposes. The strength is assumed to be normally distributed. Assume that we desire a one-sided, lower, 90 percent confidence interval on the mean which has an interval width of no more than about 10 percent of the mean. Considering a one-sided version of Eq. (4.1.54), we want to pick n such that

$$\frac{s^*}{\sqrt{n}} t_{0.10,n-1} \approx 0.10\bar{x}$$

It has been observed that the coefficient of variation of concrete strength under fixed conditions of control remains about constant when the mean is altered. (See Prob. 2.45.) Here, say it is about 15 percent. Thus we expect s^* to be about $0.15\bar{x}$, and we seek n such that

$$\frac{t_{0.10,n-1}}{\sqrt{n}} = \frac{0.10\bar{x}}{0.15\bar{x}} = 0.67$$

Using Table A.3, we find that a value of n equal to 5 gives

$$\frac{t_{0.10,4}}{\sqrt{5}} = \frac{1.53}{\sqrt{5}} = 0.68$$

A sample of size five should be used to meet the stated specifications.

After taking a sample of size 5 and observing, for example, $\bar{x} = 4000$ psi and $s^* = 700$ psi, the engineer could say, with 90 percent confidence, that the true mean is greater than

$$4000 - \frac{700}{\sqrt{5}} (1.53) = 4000 - 476 = 3524 \text{ psi}$$

Notice that the required number of samples depends upon the desired confidence, the desired width of the interval, and the estimated coefficient of variation of the concrete strength (the last is considered a measure of the quality of the control practiced by the contractor).

Comments For the reasons stated in this section the engineer who has, for example, observed $\bar{x}$ and calculated confidence limits should not say "The probability that m lies between $\bar{x} \pm \sigma k_{\alpha/2}/\sqrt{n}$ is $1 - \alpha$," but rather just "the $(1 - \alpha)100$ percent confidence limits on m are $\bar{x} \pm \sigma k_{\alpha/2}/\sqrt{n}$." Superficially, the distinction may seem to be one of semantics only, but it is of fundamental importance in the theory of statistics. The reader will, however, hear statements like the former one made by engineers who use statistical methods frequently. Such "probability statements" remain the most natural way of describing the situation and of conveying the second kind of uncertainty, that surrounding the value of a parameter. Consequently, such statements should probably not be discouraged, as they seem to express the way engineers operate with such limits. Fortunately, the more liberal interpretation of probability, namely, that associated with degree of belief (Sec. 2.1 and Chap. 5) permits and encourages this pragmatic, operational use of confidence limits, as we shall see in Chap. 6.

The use of confidence limits is most appropriate in reporting of data. It provides a convenient and concise method of reporting which is understood by a wide audience of readers, and which encourages communication of measures of uncertainty as well as simply "best" (or point) estimates of parameters. In spite of the somewhat arbitrary nature of the conventionally used confidence levels, and in spite of the philosophical difficulties surrounding their interpretation, confidence limits remain, therefore, useful conventions.

4.1.4 The Method of Maximum Likelihood†

A widely used rule for choosing the sample statistic to be used in estimation often replaces the method of moments described in Sec. 4.1.1. This is called the method of maximum likelihood. In addition to prescribing the statistic which should be used, it provides an approximation of the distribution of that statistic, so that approximate confidence intervals may be constructed.

The likelihood function The method of maximum likelihood makes use of the *sample-likelihood function*. Suppose for the moment that we know the parameter(s) θ of the distribution of X. Then we can write down the joint probability distribution of a (random) sample $X_1, X_2, \ldots, X_n$,

† This section may be omitted on first reading.

by inspection, as:

$$f_{X_1,X_2,\ldots,X_n}(x_1,x_2,\ldots,x_n \mid \theta) = f_{X_1}(x_1)f_{X_2}(x_2) \cdots f_{X_n}(x_n)$$
$$= \Pi f_X(x_i \mid \theta) \quad (4.1.54)$$

The expression has been written in a manner, $f(\cdot \mid \theta)$, which emphasizes that the parameter is known. If, for example, the underlying random variable is T and it has an exponential distribution with parameter λ,

$$f_{T_1,T_2,\ldots,T_n}(t_1,t_2,\ldots,t_n \mid \lambda) = \lambda e^{-\lambda t_1}\lambda e^{-\lambda t_2} \cdots \lambda e^{-\lambda t_n}$$
$$= \lambda^n e^{-\lambda \Sigma t_i} \quad (4.1.55)$$

Placing ourselves now in a position of having sampled and observed specific values $X_1 = x_1$, $X_2 = x_2$, . . . , $X_n = x_n$, but *not* knowing the parameter θ, we can look upon the right-hand side of the function above [Eq. (4.1.54)] as a function of θ only. It gives the *relative likelihood* of having observed this particular sample, $x_1, x_2, \ldots, x_n$, as a function of θ. To emphasize this, it is written

$$L(\theta \mid x_1,x_2,\ldots,x_n) = \prod_{i=1}^{n} f_X(x_i \mid \theta) \quad (4.1.56)$$

and is called the *likelihood function* of the sample. For the exponential example, the observed values $t_1, t_2, \ldots, t_n$ determine a number $\sum_{i=1}^{n} t_i$, and the likelihood function of the parameter λ is

$$L(\lambda \mid t_1,t_2,\ldots,t_n) = \lambda^n e^{-\lambda \Sigma t_i} \quad \lambda \geq 0 \quad (4.1.57)$$

The likelihood function is sketched in Fig. 4.1.5 for the data reported in Table 4.1.2. The vertical scale is arbitrary, since only relative values will be of interest.

The likelihood function is subject also to a looser, inverse interpretation. It can be said to give, in addition to the relative likelihood of

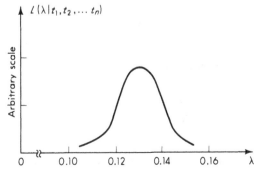

Fig. 4.1.5 Traffic-interval illustration: likelihood function on λ.

the particular sample for a given θ, the relative likelihood of each possible value of θ given the sample. This intuitively appealing interpretation suggests that values of θ for which the observed sample was relatively speaking more likely, are, relatively speaking, the more likely values of the unknown parameter θ. In Fig. 4.1.5, for example, values of λ between 0.12 and 0.14 are, in light of the observed sample, "more likely" to be the true value of λ than a value near 0.10 or 0.16. Likelihood functions will be used extensively in Chaps. 5 and 6 to represent the information contained in the observed sample about the relative likelihoods of the various possible parameter values.

Maximum-likelihood criterion for estimation The likelihood function is used to determine parameter estimators. Formally, the rule for obtaining a maximum-likelihood estimator is:

The maximum-likelihood estimator of θ is the value $\hat{\theta}$ which causes the likelihood function $L(\theta)$ to be a maximum.

For example, in the exponentially distributed traffic-interval illustration, it is clear from Fig. 4.1.5 that the most likely value of λ, i.e., the maximum-likelihood estimate of λ, is about 0.13. More generally we seek $\hat{\lambda}$ to maximize Eq. (4.1.57):

$$L(\lambda \mid t_1,t_2, \ldots ,t_n) = \lambda^n \exp \left(-\lambda \sum_{i=1}^{n} t_i \right) \tag{4.1.58}$$

Any of the many techniques for seeking the maximum of a function are acceptable. One which is frequently appropriate in maximum-likelihood estimation problems is to find the maximum of the logarithm of the function† by setting its derivative equal to zero. Here:

$$\ln L = \ln [L(\lambda \mid t_1,t_2, \ldots ,t_n)] = n \ln \lambda - \lambda \sum_{i=1}^{n} t_i \tag{4.1.59}$$

The derivative (with respect to λ), when set equal to zero, yields

$$\frac{n}{\lambda} - \sum_{i=1}^{n} t_i = 0 \tag{4.1.60}$$

Solving for $\hat{\lambda}$, the value which maximizes $\ln L$ (and L),

$$\hat{\lambda} = \frac{n}{\sum_{i=1}^{n} t_i} \tag{4.1.61}$$

---

† Owing to the monotonic, one-to-one relationship between the likelihood function and its logarithm, they have a maximum at the same value $\hat{\theta}$.

which equals $1/\bar{t}$ and hence (only coincidentally) is equivalent to the estimator suggested by the method of moments, Eq. (4.1.7).

The two estimators, given by the method of moments and the method of maximum likelihood, may not be identical. The reader is asked to show in Prob. 4.20 that the method-of-moments estimator of the parameter b of a random variable known to have distribution uniform on the interval 0 to b is $2\bar{X}$, and the maximum-likelihood estimator is $X^{(n)} = \max_{i=1}^{n} [X_i]$, the largest observed value in the sample. Note that the method of moments might lead to an estimated value of the theoretical upper value b that is less than the largest value observed!

Two or more parameters It should be clear that these notions can be easily generalized to those situations where two or more (say r) parameters must be estimated to define a distribution. Call this vector of parameters $\theta = \{\theta_1, \theta_2, \ldots, \theta_r\}$. The method of maximum likelihood states: choose the estimators $\hat{\theta} = \{\hat{\theta}_1, \hat{\theta}_2, \ldots, \hat{\theta}_r\}$ to maximize the likelihood

$$L(\theta \mid x_1, x_2, \ldots, x_n) = \prod_{i=1}^{n} f(x_i \mid \theta) \qquad (4.1.62)$$

or, equivalently, the log likelihood (if appropriate to the particular problem)

$$\log [L(\theta \mid x_1, x_2, \ldots, x_n)] = \log \left[\prod_{i=1}^{n} f(x_i \mid \theta) \right] = \sum_{i=1}^{n} \log [f(x_i \mid \theta)]$$
$$(4.1.63)$$

This maximization will typically require solving a set of r equations

$$\sum_{i=1}^{n} \frac{\partial}{\partial \theta_j} \log [f(x_i \mid \theta)] = 0 \qquad j = 1, 2, \ldots, r \qquad (4.1.64)$$

The solution may also be subject to certain constraints, for example, $\hat{\lambda} > 0$ or $\hat{b} \geq \max [x_i]$. The solution of the system is not always possible in closed form; in complex problems, computer-automated trial-and-error search techniques are often employed to find the $\hat{\theta}$ values which maximize the likelihood function.

Illustration: Maximum-likelihood estimators in the normal case As an illustration, we seek the maximum-likelihood estimators of the parameters m and σ of the normal distribution when both are unknown. The likelihood function is

$$L(m, \sigma \mid x_1, x_2, \ldots, x_n) = \left(\frac{1}{\sqrt{2\pi}\sigma} \right)^n \exp \left[-\frac{1}{2\sigma^2} \sum_{i=1}^{n} (x_i - m)^2 \right] \qquad (4.1.65)$$

The presence of exponentials suggests in this case an attack on the log-likelihood function

$$\ln [L(m,\sigma \mid x_1,x_2, \ldots ,x_n)] = -n \ln \sqrt{2\pi} - n \ln \sigma$$

$$-\frac{1}{2\sigma^2} \sum_{i=1}^{n} (x_i - m)^2 \quad (4.1.66)$$

To choose $\hat{m}$ and $\hat{\sigma}$ to maximize this function, we set the partial derivatives equal to zero:

$$\frac{\partial \ln L}{\partial m} = \frac{1}{\hat{\sigma}^2} \sum_{i=1}^{n} (x_i - \hat{m}) = 0 \quad (4.1.67)$$

$$\frac{\partial \ln L}{\partial \sigma} = -\frac{n}{\hat{\sigma}} + \frac{1}{\hat{\sigma}^3} \sum_{i=1}^{n} (x_i - \hat{m})^2 = 0 \quad (4.1.68)$$

and solve these simultaneous equations for $\hat{m}$ and $\hat{\sigma}$. Since $0 < \sigma < \infty$, the former equation is satisfied only if

$$\hat{m} = \frac{1}{n} \sum x_i = \bar{x} \quad (4.1.69)$$

Substituting this expression into the second equation and solving for $\hat{\sigma}$ gives

$$\hat{\sigma} = \sqrt{\frac{1}{n} \sum_{i=1}^{n} (x_i - \bar{x})^2} \quad (4.1.70)$$

Properties of maximum-likelihood estimators As sample statistics or functions of the observations, the estimators are, before the experiment, random variables and can be studied as such. In fact, the major advantage of maximum-likelihood estimators is that their properties have been thoroughly studied and are well known, virtually independently of the particular density function under consideration. Noting the summation form of Eq. (4.1.64), one should not be surprised to learn that the estimators $\hat{\theta}$ are approximately (for large n), normally distributed.[†] Their means are (asymptotically, as $n \to \infty$) equal to the true parameter values θ; that is, the estimators are *asymptotically unbiased*. Their variances (or mean square errors) are (asymptotically) equal to

$$\text{Var } [\hat{\theta}_i] = -\frac{1}{nE\left[\dfrac{\partial^2 \log f(x \mid \theta)}{\partial \theta_i^2} \right]} \quad (4.1.71)$$

For example, the approximate variance of the estimator of the parameter

† See for example, Hald [1952] or Brunk [1960].

λ in the exponential illustration is found as follows:

$$f(t \mid \lambda) = \lambda e^{-\lambda t}$$
$$\ln [f(t \mid \lambda)] = \ln (\lambda) - \lambda t$$
$$\frac{\partial^2 \ln [f(t \mid \lambda)]}{\partial \lambda^2} = -\frac{1}{\lambda^2}$$

In this case the result is simply a constant independent of the random variable T. Therefore, the expectation is simply that of a constant or the constant itself:

$$E \left[\frac{\partial^2 \ln [f(t \mid \lambda)]}{\partial \lambda^2} \right] = -\frac{1}{\lambda^2}$$

and [Eq. (4.1.71)]

$$\text{Var} [\hat{\lambda}] \approx \frac{\lambda^2}{n}$$

An estimate of this approximate variance is found by substituting $\hat{\lambda}$ for λ, which gives $\hat{\lambda}^2/n$.

In the car-following data the standard deviation of $\hat{\lambda}$, the estimator of λ, is approximately

$$\sqrt{\frac{\hat{\lambda}^2}{n}} = \frac{\hat{\lambda}}{\sqrt{n}} = \frac{0.130}{\sqrt{214}} = 0.0088 \text{ sec}^{-1}$$

Note once again the characteristic decrease in the expected square error of an estimator by a factor $1/n$.

In addition to being asymptotically unbiased (that is, $\lim_{n \to \infty} E[\hat{\theta}] = \theta$), the maximum-likelihood estimators can be shown (e.g., Freeman [1963]) to have, at least asymptotically, the additional desirable properties that they have minimum expected squared error among all possible unbiased estimators (hence they are said to be *efficient*). They also generally give the same estimator for a function of a parameter $g(\theta)$ as that found by simply substituting the estimator† of θ, that is, $\widehat{g(\theta)} = g(\hat{\theta})$ (i.e., they are said to be *invariant*); they will with increasingly high probability be arbitrarily close to the true values (i.e., they are said to be *consistent*); and they make maximum use of the information contained in the data (i.e., they are said to be *sufficient*). If the available sample sizes are large, there seems little doubt that the maximum-likelihood estimator is a good choice. It should be emphasized, however, that the properties above

† For example, the mean of the exponential distribution is a function of λ, $g(\lambda) = 1/\lambda$. The maximum-likelihood estimator of the mean of the exponential distribution can be found to be $\bar{X}$, while also $g(\hat{\lambda}) = 1/\hat{\lambda} = 1/(1/\bar{X}) = \bar{X}$.

are asymptotic (large n), and better (in some sense) estimators† may be available when sample sizes are small.

4.1.5 Summary

In this section we have investigated the problem of obtaining from data *estimates* of the parameters of a probabilistic model, such as a random variable X. We worked with *estimators* (i.e., certain functions of the sample, $X_1, X_2, \ldots, X_n$); the estimators considered were those suggested by general rules, either the method of moments (Sec. 4.1.1) or the method of maximum likelihood (Sec. 4.1.4).

The customary estimator of the population mean m_X is the sample mean $\bar{X}$:

$$\bar{X} = \frac{1}{n} \sum X_i$$

while the conventional estimate of the variance $\sigma_X{}^2$ is the sample variance, defined either as

$$S^2 = \frac{1}{n} \sum (X_i - \bar{X})^2$$

or as

$$S^{2*} = \frac{1}{n-1} \sum (X_i - \bar{X})^2$$

The choice between these estimators is not clear-cut, but the latter is used somewhat more frequently in practice.

An estimator $\hat{\theta}$ is studied by familiar methods once it is recognized that it is a random variable and a function of $X_1, X_2, \ldots, X_n$. The estimator's mean $E[\hat{\theta}]$ is compared with the true value of the parameter θ. If they are equal, the estimator is said to be unbiased. The mean square error of any estimator is the sum of its variance and its squared bias, $E[\hat{\theta}] - \theta$,

$$E[(\hat{\theta} - \theta)^2] = \text{Var}\,[\hat{\theta}] + (E[\hat{\theta}] - \theta)^2$$

Generally we seek unbiased estimators with small mean square errors.

The distribution of an estimator can be used to obtain confidence-interval estimates on a parameter at prescribed *confidence levels*. These are obtained by "turning inside out" probability statements on the estimator $\hat{\theta}$ to obtain random limits $g_1(\hat{\theta})$ and $g_2(\hat{\theta})$ which are $(1 - \alpha)100$

† For instance, the maximum-likelihood estimator of the variance of a normal distribution is S^2 [Eq. (4.1.70)]. The properties of this estimator have been compared in Sec. 4.1.2 with S^{2*}. S^2 was, recall, a biased estimator for small n.

percent likely to contain the true parameter.   After a sample has been observed, these limits are evaluated and called the $(1 - \alpha)100$ percent confidence interval on the true parameter.

4.2 SIGNIFICANCE TESTING

A very common and most vexing problem that engineers face is the assessment of *significance* in scattered statistical data.   For example, given a set of observed concrete cylinder strengths, say 4010, 3880, 3970, 3780, and 3820 psi, should a materials engineer report that the true mean strength of the batch of concrete is less than the design value of 4000 psi?   Or to put the question differently, in light of the inherent variation in such observations, is the observed sample average

$$(\tfrac{1}{5})(4010 + 3880 + 3970 + 3780 + 3820) = 3892 \text{ psi}$$

significantly less than the presumed value of 4000 psi?   If the engineer had twice as many samples, but their average happened to be the same, 3892 psi, would this not be a "more significant" deviation from the presumed mean?   Suppose an engineer has installed a new traffic-control device in an attempt to enhance the safety of an intersection. In the 3 months prior to the installation there were four major accidents there; in the 3 months following the installation there were two.   Can the engineer conclude with confidence that the device has been effective? That is, has it reduced the mean accident rate?   Is this limited data sufficient to justify such a conclusion?   In different words, is the difference between 4 and 2 significant when one recognizes the inherently probabilistic nature of accident occurrences?

Generally speaking we state such problems as follows.   The engineer considers a particular assumption about his model (e.g., "the true mean strength is 4000 psi" or "the mean accident rate is unchanged"). Given data, he wants to evaluate this tentative hypothesis.   If the hypothesis is assumed to be correct, the engineer can calculate the expected results of an experiment.   If the data are observed to be significantly different from the expected results, then the engineer must consider the assumption to be incorrect.   The technical question to be answered is, "What represents a *significant* difference?"

Although many such problems in engineering are most effectively treated in the context of the modern decision theory to be presented in Chaps. 5 and 6, a simpler procedure has also been developed.   Like confidence limits, this method of assessing the significance of data has the merit of being a widely used convention.   Its use is therefore a convenient way to communicate to others the method by which the data were analyzed and the criterion by which the assumption or model was judged.

In this section we shall concentrate on relatively simple model assumptions; in particular, we shall test hypotheses about model parameter values. In Sec. 4.4 we shall treat the verification of the model or distribution as a whole, including its shape as well as its parameters.

4.2.1 Hypothesis Testing

This conventional procedure for drawing simple conclusions from observed statistical data is called *hypothesis testing*. Typical examples include: a product-development engineer who wants to decide whether or not a new concrete additive has a significant influence on curing time; a soils engineer who wants to report whether a simple, in-the-field "vane test" provides an unbiased estimate of the lab-measured unconfined compressive strength of soil; and a water-resources engineer who wants to verify an assumption used in his system model, namely, that the correlation coefficient between the monthly rainfalls on two particular watersheds is zero.

We shall develop the conventional treatment of such problems through the following simple example. Even under properly controlled production, asphalt content varies somewhat from batch to batch. A materials-testing engineer must report, on the basis of three batches, whether the mean asphalt content of all batches of material leaving a particular plant is equal to the desired value of 5 percent. The concern is that unknown changes in the mixing procedure or raw materials supply might have increased or decreased this mean. Based on previous experience the engineer is certain that, no matter what the mean, the asphalt content in different batches can be assumed to be normally distributed with a known standard deviation of 0.75 percent.

It is important to appreciate the basic dilemma. The engineer must use information in the sample to make his decision. This information is usually summarized in some sample statistic, such as the sample average $\bar{X}$. The finite size of the sample implies that any sample statistic calculated from the observations is a random variable with some nonzero variance. For example, $\bar{X}$ is a normally distributed random variable with a variance of $(0.75)^2/3 = (0.43)^2 = 0.184$ and a mean m equal to the (unknown) true mean of the plant's present production. Therefore the observed value of this sample statistic may lie some distance from the expected value of 5 even if this is the true value of the mean. On the other hand, if the true value of the mean asphalt content is no longer 5 but has slipped to 4.5 or increased to 5.6, the observed sample average may nonetheless lie close to the hypothesized value of 5. Thus, with limited data, the engineer cannot hope to be correct every time he makes a report. In some cases he will report that the mean is not 5 when it is; and in other cases he will report that it is 5 when it is not. At best he can ask for an *operating rule* which recognizes the inherent variability of the data and the size of his sample, and which promises some consistency in approach to this and similar questions he must report on. Preferably he would like to have some measure, too, of the percentage of times he will make one mistake or the other.

Hypothesis testing is a method which is commonly used in such routine decisions. It provides most of these desired consistency properties, and, being a widely used convention, it simplifies communication among engineers. If the materials engineer reports that he "accepted the hypothesis that the true mean is 5 at the 10 percent significance level," trained readers of the report will know first, and most importantly, that the engineer has studied and considered the scatter in the data, and, second, precisely the way by which he reached his conclusion, complete with the implied risks and limitations. The method is objective in that, under the convention, two engineers faced with the same data will report the same conclusion.

Application of the method In this example, the engineer using an hypothesis test would tentatively propose the *null hypothesis H_0* that the true mean is 5. He seeks a rule by which he can judge (i.e., *accept* or *reject*) this hypothesis on the basis of the observed asphalt content in the three specimens. A reasonable choice for this *operating rule*† is:

> Accept the null hypothesis if the sample average $\bar{X}$ lies within $\pm c$ of 5. Otherwise reject this hypothesis.

Rejection of the null hypothesis implies (under this method) acceptance of the *alternate hypothesis H_1*. In this case, the alternate hypothesis is simply that the mean is *not* 5 percent.

Once the form of this rule is chosen, the method of hypothesis testing provides a convention by which to determine the value of c. The method suggests:

> Choose c such that there is only a prescribed (small) probability α, say 10 percent, that the sample mean of any three specimens will lie outside $5 \pm c$, *given* that the null hypothesis holds.

We know the probability distribution of $\bar{X}$ *given* that the null

---

† In modern literature this is called a *decision rule*, but we prefer to restrict the word decision to analyses which explicitly account for the consequences of actions and outcomes, as will be discussed in Chaps. 5 and 6.

hypothesis H_0 holds.   That distribution is $N(5, 0.184)$.   Therefore, we can easily find c from the statement (Fig. 4.2.1a)

$$P[5 - c \le \bar{X} \le 5 + c \mid H_0] = 1 - \alpha = 90 \text{ percent} \qquad (4.2.1)$$

In terms of the standardized normal distribution we want c such that

$$F_U\left(\frac{(5 - c) - 5}{\sqrt{0.184}}\right) = \frac{\alpha}{2} = 5 \text{ percent}$$

From tables,

$$-\frac{c}{\sqrt{0.184}} = -1.65$$

$$c = 1.65\sigma_{\bar{X}} = (1.65)(\sqrt{0.184}) = 0.71 \qquad (4.2.2)$$

Thus the operating rule becomes:

Accept the null hypothesis that the true mean is 5 if the observed value of the sample average lies between 5 ± 0.71 (that is, 4.29 and 5.71).   Otherwise, reject the hypothesis.

Discussion   The intuitive justification for this rule of operation is that if the observed sample mean is a "significant distance" from the assumed value, the assumption is probably wrong.   Under the method, significant distance can be interpreted quantitatively as being a discrepancy c which would only be exceeded with a small probability if the hypothesis were true.   If a discrepancy of that size or greater is observed, the method suggests that the engineer act as if his assumption were not true rather than accept the observation as a "rare event."   Although the conclusion will clearly depend upon the value of α chosen, the method does demand a rational analysis of uncertainty, namely, the study of the probabilistic behavior of the sample statistic used in the operating rule.   If, for example, the standard deviation between individual batches (σ_X or 0.75) is different, or if the sample size is different, the value of c in the operating rule will also change.

Type I errors   The values used for α, which is called the *significance level of the test*, are virtually always chosen *by convention* to be 10, 5, or 1 percent.   In principle, but seldom in practice, one chooses α to reflect the consequences of making the mistake of *rejecting the hypothesis when in fact it is true*.   If one always uses $\alpha = 5$ percent, he will make this kind of mistake (the so-called *type I error*) 5 percent of the time in the long run. It is the authors' experience and recommendation that the significance

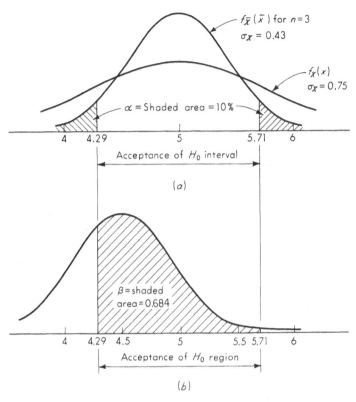

Fig. 4.2.1 Type I and type II error probabilities for the asphalt-content operating rule. (a) Distribution of $\bar{X}$ if $m = 5$, showing α, the probability of making type I error; (b) distribution of $\bar{X}$ if $m = 4.5$ showing the probability of type II error β.

level of a hypothesis test continue to be chosen simply by convention, say at the most commonly used 5 percent level. A thorough decision analysis requires explicit inclusion of costs or consequences and is best done using statistical decision theory (Chaps. 5 and 6).

Once the operating rule has been determined, the engineer need only obtain the data, calculate the statistic used in the rule, compare it with the acceptance interval, and accept or reject the hypothesis according to prescription. For example, if the materials engineer measured the asphalt content in three batches and found $x_1 = 4.2$, $x_2 = 4.7$, and $x_3 = 3.7$, the observed sample mean would be $\bar{x} = 4.2$. Since this value lies outside the interval 4.29 to 5.71, he should reject the null hypothesis that the true mean is 5.0. In this case, the observed statistic would be said to be *statistically significant* at the 10 percent level, and the engineer would report that† he rejected the hypothesis that $m = 5.0$ at the 10 percent significance level.

Type II errors   It must be recognized that one cannot afford to make the probability α of making a type I error arbitrarily small.   This probability of rejecting H_0 when it is true can only be made small at the expense of increasing the probability of a *type II error*, which is one of *accepting H_0 when it is not true.*

Consider again the asphalt tests.   The engineer might be concerned if either the true mean asphalt content were low, say 4.5 percent, or if it were high, say 5.5 percent.   In either situation there is a possibility that the sample average will lie in the rule's acceptance interval, 4.29 to 5.71. If this happens, the operating rule which the hypothesis-testing method has proposed will mistakenly suggest that the engineer act as if the true mean were 5.   The probability of this possibility is called β.   It can be evaluated for any particular value of the true mean.   For example, if the true mean is 4.5, then for this operating rule (recall that $\alpha = 10$ percent) the probability of incorrectly accepting the assumption that the mean m is 5 is

$$\beta = P[4.29 \leq \bar{X} \leq 5.71 \mid m = 4.5] \tag{4.2.4}$$

If the true mean is 4.5, the sample average is distributed $N(4.5, 0.184)$, implying (Fig. 4.2.1b)

$$\beta = F_U\left(\frac{5.71 - 4.5}{\sqrt{0.184}}\right) - F_U\left(\frac{4.29 - 4.5}{\sqrt{0.184}}\right)$$
$$= 0.998 - 0.314 = 0.634 \tag{4.2.5}$$

Clearly, when the true mean is 4.5, the engineer using this method will, more often than not, mistakenly assume that the true mean is 5.   If the true mean is 4, the type II error probability is reduced to about 25 percent, and if the true mean is 6.5, the β probability is only 3 percent. One can plot, as in Fig. 4.2.2, the error probability β as a function of the true mean m.   Such a curve is called the *operating characteristic* (OC) *curve* of the rule.   (Note that it equals $1 - \alpha$ at $m = 5$.)

If the type II error probabilities implied by the operating rule are not tolerable, the engineer can, with some expense, reduce them.   Either he may reduce the width of the acceptance interval $\pm c$ at the expense of increasing α, the type I error probability (see Fig. 4.2.1), or he may increase the sample size from three to some larger value.   At the same

---

† It should *not* be said that the hypothesis is false with probability 90 percent.   Like confidence levels, significance levels cannot be treated as probabilities, despite engineers' common desire to do so.   This restriction is lifted in Sec. 6.2.2.

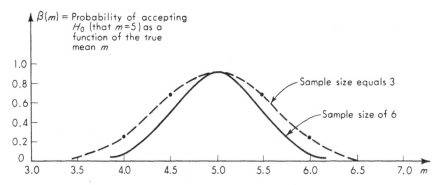

Fig. 4.2.2 The operating characteristic curve for the asphalt-content operating rule, $\alpha = 10$ percent, with a sample size of three or six.

significance level, $\alpha = 10$ percent, the increased sample size will reduce $\sigma_{\bar{X}}$ and hence the interval width $\pm c$ [Eq. (4.2.2)]; at the same time it will change the shape of the distribution of $\bar{X}$ given that $m = 4.5$, say. The net effect will be a set of β values everywhere less than those for a sample size of three. For example, for a sample size of six, and $\alpha = 10$ percent, $c = 0.50$ and the OC curve of β probabilities appears as indicated in Fig. 4.2.2.

Choice of sample size   If the engineer would like to obtain some pre-scribed level of "reliability" in his reporting method, he can calculate the sample size necessary to achieve it. For example, to keep α at 10 percent and also to reduce β to at most 20 percent when the true mean is 4.5, one can easily calculate that the sample size must be at least 14. If this amount of testing is unduly expensive, some tradeoff between sampling cost, type I error, and type II errors must be made. Such decisions are probably best made in a totally economic context such as that to be discussed in Chaps. 5 and 6. We will restrict attention in this chapter to analysis of data from a fixed sample size and under a conventional level of significance (say, $\alpha = 5$ percent), accepting the implied levels of β.

4.2.2 Some Common Hypothesis Tests

In this section we will summarize the hypothesis-testing method and then present briefly a number of common hypothesis tests, using them to bring out some additional points about the method.

Procedure summary   Generally speaking, the engineer has an assump-tion about his model and certain data. The object is to determine if the observations display characteristics significantly different from those expected under the assumption. The steps are:

1. Identify the *null hypothesis* H_0 and the *alternate hypothesis* H_1.
2. *Select* an appropriate *sample statistic* to be used in the test.
3. *Select* an appropriate *form* for the *operating rule*, usually up to a constant c. The rule must divide the sample space of the sample statistic into two mutually exclusive, exhaustive events, one to be associated with acceptance of H_0, the other with rejection.
4. *Select a significance level* α at which to run the test.
5. Use the distribution of the sample statistic to *determine the value of the constant c* so that when H_0 is true, there is a probability α of rejecting that hypothesis.
6. If appropriate and feasible, *investigate the type II error probabilities* associated with the chosen rule, assuming that H_1 rather than H_0 holds.
7. *Carry out the test* by calculating the observed value of the sample statistic and comparing it with the acceptance and rejection regions defined by the rule.
8. If the observed sample statistic falls in the rejection region (or *critical region*), announce that it is "significant at the α percent level" and reject the hypothesis. If the statistic falls in the acceptance region, accept H_0.

We saw in the last section an extended example of a case which can be summarized as follows:

H_0: $m = 5.0$
H_1: $m \neq 5.0$
Statistic: $\bar{X}$
Rule form: Accept H_0 if $5.0 - c \leq \bar{X} \leq 5.0 + c$
                Otherwise, reject H_0.
Significance level: $\alpha = 10$ percent
Acceptance region: Solution of

$$P[5.0 - c \leq \bar{X} \leq 5.0 + c \mid m = 5.0] = 1 - \alpha \qquad \text{for } c$$

Result: $c = 0.71$

In this example, the variance of the underlying random variable X was assumed to be known to be $(0.75)^2$; the sample size was three; $\bar{X}$ was assumed to be normally distributed with variance $(0.75)^2/3$. Notice that this would be the case no matter what the distribution of X as long as the engineer is prepared to assume that $\bar{X}$ is normally distributed.   In general, of course, the exact distribution of $\bar{X}$ should be used. As we have seen in Sec. 4.1, this will depend upon the distribution of X. For example, if X is Bernoulli, $\bar{X}$ is binomial (Sec. 4.1.3).   In such discrete cases one cannot usually find a value of c (since nc must be integer) to exactly meet the desired significance level α.

One-sided versus two-sided tests; simple versus composite hypotheses
Other forms of hypotheses might be more appropriate in the asphalt illustration.   For example, the engineer may know that if the mean is not equal to m_0 (say, 5) then it is surely equal to m_1 (say, 4.0).   This situation might result when a particular kind of change in the mixing procedure or raw materials is known to occur at times.   Then the hypothesis test becomes.

H_0: $m = 5.0$
H_1: $m = 4.0$
Rule form: Accept H_0 if $\bar{X} \geq c$
Acceptance region: Solution of

$P[\bar{X} \geq c \mid m = 5.0] = 1 - \alpha$ for c

Result: $c = m_0 - 1.28\sigma_{\bar{X}} = 5.0 - 0.55 = 4.45$

Notice that the logical form of the operating rule has changed to reflect the engineer's belief that m is either 5.0 or a smaller value (namely, 4.0). Such a test is called *one-sided*, as compared with the previous example, which was a *two-sided test*.   Notice, too, that the alternate hypothesis H_1 is a *simple hypothesis*, meaning only a single value of m is associated with the hypothesis.   In contrast, in the previous example, H_1 was $m \neq 5.0$, which is called a *composite hypothesis* in that it is composed of a collection of values of m, in this case infinitely many.   One implication of having a simple alternate hypothesis is that there is a unique value for β, the type II error probability.   In this case the probability of incorrectly accepting H_0 when H_1 is true is simply

$$\beta = P[\bar{X} \geq 4.45 \mid m = 4.0] = 1 - F_U \left(\frac{4.45 - 4.0}{\sqrt{0.184}} \right) = 0.145 \quad (4.2.6)$$

Consider another case where the engineer must report only that the mean is *above* a certain value m_0 or below it.   For example, if the materials engineer represents the state highway department, he might guard only against too small quantities of asphalt content; he is not responsible for identifying excessive asphalt content, which is uneconomical to the supplier.   The hypothesis test in this case would logically be:

H_0: $m \geq 5.0$
H_1: $m < 5.0$
Statistic: $\bar{X}$
Rule form: Accept if $\bar{X} \geq c$
Significance level: $\alpha = 10$ percent

Notice that there is no unique solution for c since H_0 is now a composite hypothesis. There is a different value of c for each value of m equal to or greater than 5.0. In practice c would be chosen so that α is *no more than* 10 percent. In this case the implication is solving for c when m is set equal to the smallest value permitted by H_0, namely, 5.0. The result is: $c = 4.45$ just as in the previous example where H_0 and H_1 were both simple hypotheses.

The use of other statistics in the test   The sample mean $\bar{X}$ has been used above in hypothesis tests concerning the true mean m. It should be emphasized that other sample statistics may, for a variety of reasons, be more appropriate. For example, if the underlying variable X is normally distributed, $N(m,\sigma^2)$, but neither m nor σ is known, it is desirable to work with the "t statistic" if the sample size is small. This statistic was also used in similar circumstances to obtain a confidence-interval estimate† of m in Sec. 4.1.3. Consider again the original asphalt-content example. If the materials engineer were not confident of the value of σ^2, owing, say, to a limited experience with the plant setup, he should use the following test, involving an estimator of σ:

H_0: $m = 5.0$
H_1: $m \neq 5.0$

Statistic: $T = \dfrac{(\bar{X} - m)}{S^*/\sqrt{1/n}}$

Rule form: Accept H_0 if $-c \leq T \leq +c$
Significance level: $\alpha = 10$ percent
Acceptance region: Solution of

$$P[-c \leq T \leq c \mid H_0] = 1 - \alpha \qquad \text{for } c$$

Result: $c = t_{\alpha/2,n-1} = 2.90$

---

† The reader will have noted by now many parallel features in confidence-interval estimates and hypothesis tests on parameters. In fact, one can easily show, for example, that if the $(1 - \alpha)$ percent confidence-interval estimate of the mean contains the value of the mean associated with the null hypothesis H_0, then H_0 should be accepted at the α percent significance level. The relationship between confidence-interval and hypothesis testing is more computational than fundamental, but similar attitudes with respect to not treating unknown parameters as random variables exist in both.

Recall (Sec. 4.1.3) that $(S^*)^2 = S^{2*}$ is the unbiased estimate of σ^2 and $t_{\alpha/2,n-1}$ is the value from a table of the t statistic (Table A.3) which a t-distributed random variable with $n - 1$ degrees of freedom (in this case, $n - 1 = 3 - 1 = 2$) will exceed with probability $\alpha/2$.

With the same observations of asphalt content obtained in the previous example, namely, $x_1 = 4.2$, $x_2 = 4.7$, and $x_3 = 3.7$, the sample mean is $\bar{x} = 4.2$, and

$$s^* = \sqrt{\frac{1}{3-1}\,[(4.2 - 4.2)^2 + (4.7 - 4.2)^2 + (3.7 - 4.2)^2]} = 0.5$$

Thus, given H_0 or $m = 5$, the observed t statistic is

$$t = \frac{(4.2 - 5.0)\,\sqrt{3}}{0.5} = -2.79$$

Since this value is between ± 2.90, the null hypothesis should be accepted at the 10 percent level. Notice that this is in contrast to the conclusion of rejection suggested by the original test. The change in conclusion results (despite the fact that the observed value of S^* is considerably smaller than the previously given value of $\sigma = 0.75$) from the fact that the "allowance for" uncertainty in σ implied by the use of the t statistic leads to wider acceptance regions when the sample size is small. (Compare $t_{\alpha/2,n-1} = 2.90$ for $\alpha/2 = 5$ percent and $n - 1 = 2$ with the value from a normal table of $k_{\alpha/2} = 1.64$ for $\alpha/2 = 5$ percent.)

In quality-control practice it is common to use other, more easily calculated sample statistics in routine hypothesis testing. Simply to avoid calculation in the field of the sample average, the engineer might have set up the following test in one of the examples above:

H_0: $m \geq 5.0$
H_1: $m < 5.0$

Statistic: Sample minimum $X^{(1)} = \min_{i=1}^{n} [X_i]$

Rule form: Accept H_0 if $X^{(1)} \geq c$
Significance level: $\alpha = 10$ percent
Acceptance region: Solution of

$$P[X^{(1)} \geq c \mid H_0] = 1 - \alpha \qquad \text{for } c$$

Result: $c = 3.54$

Again assuming that σ is known to equal 0.75 and using the smallest value of m permitted by the composite hypothesis H_0 (namely, $m = 5$), c is found simply by

$$P[X^{(1)} \geq c \mid m = 5] = P[\text{all three } X_i \geq c \mid m = 5]$$
$$= (P[\text{any } X \geq c \mid m = 5])^3 = 1 - \alpha$$

or

$$P[X \geq c \mid m = 5] = (1 - \alpha)^{\frac{1}{3}} = (0.90)^{\frac{1}{3}} = 0.975$$

$$= P\left[\frac{X - 5.0}{0.75} \geq \frac{c - 5.0}{0.75}\right] = 0.975$$

$$c = 5.0 - 0.75(1.96) = 3.54$$

With the data obtained, 4.2, 4.7, and 3.7, the null hypothesis should be accepted since the minimum value, 3.7, is greater than 3.54.

In principle, when considering two different test forms, such as those above involving either $\bar{X}$ or $X^{(1)}$, the sample mean or the sample minimum, one should study the behavior of the type II errors in detail. For a fixed value of α, one would like a test whose β probabilities are smaller than those for all other tests (at the same significance level); that is, a test with an OC curve below the OC curves of other tests under consideration. Such a test, if it can be found, is said to be the *most powerful* test.† Such comparative studies are usually left to mathematical statisticians, who recommend the most appropriate tests. In routine testing, the civil engineer must, however, be prepared to balance a loss in test power versus a simplification of the application of the test in the field. For this reason, many *quality-control* specifications are written in terms such as "accept the material if no more than 2 of 10 samples fail to meet the prescribed level (of strength, say)," rather than in terms of sample averages and variances.

Tests on dispersion Suppose that a development engineer is interested in the influence of a new earth-moving device on construction projects. Dispersion in haul times is known to cause idle equipment and long queues. On a particular job, the present equipment has a variance in haul times of σ_0^2. The new equipment is to be tested for a short time. The engineer must report whether there is a reduction in σ^2. He must attempt to abstract significance from the limited trial data as to whether there has been a change. His test might be:

H_0: $\sigma^2 \geq \sigma_0^2$
H_1: $\sigma^2 < \sigma_0^2$
Test statistic: S^2, the sample variance
Rule form: Accept H_0 if $S^2 \geq c$
Significance level: $\alpha = 5$ percent
Acceptance interval: Solution of

$$P[S^2 \geq c \mid H_0] = 1 - \alpha \qquad \text{for } c$$

† The complement of the OC curve, i.e., the plot of $1 - \beta$, is called the *power function*. We would like a test whose power function is highest everywhere.

If the underlying individual times are normally distributed, then the distribution of S^2 is related to the χ^2 distribution (Sec. 4.1.3), and c can be easily found for any particular sample size, with σ^2 set equal to σ_0^2, the lowest value permitted by H_0.

Notice that there is freedom for the engineer to choose whether H_0 be $\sigma^2 \geq \sigma_0^2$ or $\sigma^2 \leq \sigma_0^2$. The form of the test will be different, the type II errors will be different, and even the conclusions from the same data may be different. In the first case the engineer knows that the (maximum) probability of saying that $\sigma^2 < \sigma_0^2$ when it is not is 5 percent, whereas in the second case, the engineer knows that the (maximum) probability of concluding that $\sigma^2 > \sigma_0^2$ when it is not is 5 percent. Which choice is appropriate depends in principle upon the consequences implicit with each kind of error. In practice, a conventional value of α is used and the engineer chooses H_0 to be $\sigma^2 \geq \sigma_0^2$ or $\sigma^2 \leq \sigma_0^2$, depending upon the position he wants to bias himself toward. For example, if the equipment-development engineer felt that he should be conservative and not suggest that the device was better (i.e., that it had σ^2 smaller than σ_0^2) unless he were quite sure, then he should set H_0 as $\sigma^2 \geq \sigma_0^2$; that is, he should assume that the new device is worse unless the data is significantly contradictory and "forces" him to reject the hypothesis. On the other hand, if the engineer is convinced that the new device is better and he seeks only supporting evidence, he might set H_0 as $\sigma^2 \leq \sigma_0^2$; that is, he might assume that it is better unless the data deviates significantly from the expectation and forces him to reject that assumption. A similar duality exists in statistical quality control. The organization which must *accept the material* will usually assume that the mean material property (say, strength) is *less* than some fixed value unless the data indicate that it is "significantly" above. This encourages the supplier to produce material with low variance and to provide large samples. In routinely *controlling the output* of his own plant, however, the producer may assume that the mean is *above* some fixed value unless the samples are significantly low to cause him to take expensive corrective action. The difference in attitude will appear in the choice of the null hypothesis.

In routine testing, the *sample range* (maximum observation minus minimum observation) is sometimes used as the test statistic when considering dispersion. It is trivial to calculate in the field, and its distribution has been tabulated, at least for normal variables (e.g., Burlington and May [1953]).

Tests on two sets of data In certain situations the engineer may want to look at two sets of data, $x_1, x_2, \ldots, x_n$, and $y_1, y_2, \ldots, y_r$, and decide if one set comes from the same population as the other or if one set represents data from a sample with a higher mean than the other. The data from one set may represent durability tests on concrete with a new additive designed to increase workability. The other data represent the *control group*, one which is treated in all ways in a similar manner, but which lacks the additive.

If the variances of X and Y are known, the test statistic usually used is the difference in the sample means, $\bar{X} - \bar{Y}$. If X and Y are normal, then $\bar{X} - \bar{Y}$ is normal with mean $m_X - m_Y$ and variance $\sigma_X^2/n + \sigma_Y^2/r$. Therefore, one might, for example, set up a one-sided test:

H_0: $m_X \leq m_Y$

H_1: $m_X > m_Y$

Rule form: Accept H_0 if $\bar{X} - \bar{Y} \leq c$

Significance level: $\alpha = 5$ percent

Acceptance region: Solution of

$$P[\bar{X} - \bar{Y} \leq c \mid H_0] = 1 - \alpha \qquad \text{for } c$$

Result: $c = 1.64 \sqrt{\dfrac{\sigma_X^2}{n} + \dfrac{\sigma_Y^2}{r}}$

There exist many kinds and forms for tests involving the means of two variables. Some of these are "nonparametric," meaning that the distribution of X and Y need not be assumed. The reader is referred to standard statistics texts for their enumeration (e.g., Hald [1952], Bowker and Lieberman [1959], or Freeman [1963]).

Another test involving two populations is that which tests for equality of the two variances (or standard deviations). The test statistic used if the variables are both normal is the *ratio* of the unbiased sample variances:

$$F = \frac{S_X^{*2}}{S_Y^{*2}} \qquad\qquad (4.2.7)$$

The statistic has a widely tabulated distribution, the so-called F distribution (Sec. 3.4.3), if the null hypothesis, $\sigma_X^2 = \sigma_Y^2$, holds. In that case

$$F = \frac{S_X^{*2}/\sigma_X^2}{S_Y^{*2}/\sigma_Y^2} = \frac{S_X^{*2}}{S_Y^{*2}}$$

The former ratio is the ratio of two χ^2 random variables, $(S_X^{*2}/\sigma_X^2)(n - 1)$ and $(S_Y^{*2}/\sigma_X^2)(r - 1)$, divided by their degrees of freedom, $n - 1$ and $r - 1$. As discussed in Sec. 3.4.3, this random variable has two parameters, $n - 1$ and $r - 1$. A typical one-sided test would be of the form:

H_0: $\sigma_X = \sigma_Y$
H_1: $\sigma_X > \sigma_Y$
Rule form: Accept H_0 if $F \leq c$
Significance level: $\alpha = 5$ percent
Acceptance region: Solution of

$$P[F \leq c \mid H_0] = 1 - \alpha \qquad \text{for } c$$

Result: $c = F_{\alpha; n-1, r-1}$

in which $F_{\alpha; n-1, r-1}$ is that value (found in Table A.6) which an F-distributed random variable with $n - 1$ and $r - 1$ degrees of freedom will exceed only with probability α.

A final test we will mention involving two random variables is a test for lack of correlation, that is $\rho_{X,Y} = 0$. If X and Y are jointly normally distributed, it can be shown that the sample correlation coefficient (Sec. 1.3)

$$R_{X,Y} = \frac{1}{n} \frac{\sum\limits_{i=1}^{n} (X_i - \bar{X})(Y_i - \bar{Y})}{S_X S_Y}$$

has a distribution related to the t distribution with $n - 2$ degrees of freedom *under* the null hypothesis H_0 that $\rho = 0$. If $\rho \neq 0$, the distribution of R is complicated even for X and Y jointly normal.†

In the joint normal case this becomes a test for independence of the following form:

H_0: $\rho_{X,Y} = 0$
H_1: $\rho_{X,Y} \neq 0$

Statistic: $T = \dfrac{R \sqrt{n - 2}}{\sqrt{1 - R^2}}$

Rule form: Accept H_0 if $-c \leq t \leq +c$
Significance level: $\alpha = 5$ percent
Acceptance region: Solution of

$$P[-c \leq t \leq c \mid H_0] = 1 - \alpha \qquad \text{for } c$$

Result: $c = t_{\alpha/2, n-2}$

For example, using the beam-crack and ultimate-load data of Sec. 1.3, Table 1.2.2, the sample correlation coefficient of 15 specimens was found to be 0.12. If the loads are assumed to be normally distributed, the

† This distribution is tabulated, however, in various places (see Owen [1962]).

research engineer could report that at the 5 percent significance level the crack load and ultimate load were found to be independent, since

$$t = \frac{r\sqrt{n-2}}{\sqrt{1-r^2}} = \frac{0.12\sqrt{13}}{\sqrt{1-(0.12)^2}} = 0.43$$

This value is between ± 2.160, the range from Table A.3 beyond which a t variable with 13 degrees of freedom will be 5 percent of the time.

Discussion   The method of hypothesis testing provides the service of extracting significance from data, but it suffers from several limitations with respect to arbitrariness in selecting significance levels and forms of hypotheses.   In practice, engineers will sometimes beg the question of significance level by reporting only that significance level at which the hypothesis would have to be rejected given the observed value of the statistic.   For example, in the original example in Sec. 4.2.1, the observed value of $\bar{X}$ was $\bar{x} = 4.2$.   This value lies $(5.0 - 4.2)/\sqrt{0.184} = 1.86$ standard deviations, $\sigma_{\bar{x}}$, below the expected value of $\bar{X}$ (given the null hypothesis).   Therefore the hypothesis would have to be rejected at all significance levels *greater* than 6.2 percent.   This value is sometimes called the *p level* of the test.

In any case the method of hypothesis testing provides a convenient, conventional method for reporting the basis on which a conclusion was reached in the light of scatter in the data.   The reader of such a report must retain, however, his critical attitude, in the broad sense of the word.   On the basis of the report, he knows how the conclusion was reached, and he must judge the conclusion in the face of known limitations in the method.   For the engineer, the necessity for choosing somewhat arbitrarily among conventional values of significance levels and among forms of hypotheses will often prove unsatisfactory.   In many such situations, more powerful tools for decision making in the face of uncertainty (Chaps. 5 and 6) should be considered instead.

4.2.3 Summary

A conventional method for "extracting significance" from disperse data has been presented in brief.   Hypothesis testing is an aid in verifying certain assumptions that the engineer might make about his model. Described here are various tests on assumptions about model parameters.

The procedure requires that the engineer state two exclusive hypotheses—H_0, the null hypothesis, and H_1, the alternate hypothesis.   Based on a chosen sample statistic to be used in the test, the appropriate form of the operating rule is selected.   Given a prescribed value for α, which is the probability of the type I error (rejecting H_0 when it is true), an acceptance region (or, its complement, *the critical region*) is calculated in the sample space of the sample statistic such that there is a probability of $1 - \alpha$ that the sample statistic will lie in the acceptance region, given that H_0 holds.

The hypothesis test is run by obtaining the data, calculating the observed statistic, and comparing it with acceptance region. The engineer will state that "H_0 is accepted at the α significance level" if the observed statistic lies in the range. The observed statistic is said to be *significant* if it lies in the critical region, suggesting that the null hypothesis is not true.

In principle, the user must also consider for any test the probability β of a type II error (accepting H_0 when it is not true). If H_1 is a composite hypothesis, it is necessary to plot β versus all possible values of the true parameter permitted by this hypothesis. Such a plot is called an *operating characteristic* (OC) curve for the test.

A number of examples showed one-sided and two-sided tests, simple and composite hypotheses, and cases involving means, variances, and correlation coefficients as parameters of one or two sets of data.

4.3 STATISTICAL ANALYSIS OF LINEAR MODELS†

As in mechanics, linear models of physical phenomena are widely used in applied probability and statistics. In this case, by a linear model it is meant that a random variable Y is affected in its mean in a linear manner by another variable x or variables, $x_1, x_2, \ldots, x_r$. The other variables may or may not be random. The underlying model says that $E[Y \mid x_1, x_2, \ldots, x_r] = \alpha + \beta_1 x_1 + \beta_2 x_2 + \cdots + \beta_r x_r$. In this section we will provide a brief, introductory look at the formulation and the statistical analysis of such linear models. By statistical analysis we mean *parameter estimation* and *significance testing*. We shall ask such questions as, "What are estimates of the β_i's?" and "Is the observed dependence of the mean of Y on x_j statistically significant?" Much of this statistical analysis falls under the customary title of *regression analysis*. In one sense we will simply be applying the tools of Secs. 4.1 and 4.2 to a very useful new class of models.

4.3.1 Linear Models

A wide variety of practical problems can be represented by linear models. In this section we shall discuss the characteristics of some of these probability models, concentrating on the simpler cases but indicating the extensions to other problems. The section ends with a discussion

---

† The new material introduced in this section is not necessary for subsequent sections of the text. Sec. 4.3 can be bypassed at first reading.

of the statistical interests we should have in such models.　In Sec. 4.3.2, we shall demonstrate the statistical analysis of the simpler cases.

Model 1　In the simplest case a linear model states that the mean of the random variable Y is linearly functionally dependent on x, which is a *nonrandom* quantity but one which may change from experiment to experiment.　Let us denote this by saying

$$E[Y]_x = \alpha + \beta x \tag{4.3.1}$$

In any single experiment x will have some value x_i, and the *expected* value of Y_i will be $\alpha + \beta x_i$.　Y_i itself can be written

$$Y_i = \alpha + \beta x_i + E_i \tag{4.3.2}$$

in which E_i is *mean zero* random variable, sometimes called the "*error term.*"

Consider some examples.　In a particular deep layer of soil, the soil will be sampled at a number of depths, $x_1, x_2, \ldots, x_n$.　The sample will be recovered and tested.　It may be reasonable to assume that the expected value of Y_i, the unconfined compressive strength at depth x_i, is $\alpha + \beta x_i$.　The basic assumption of the model is that within this geological layer there is a general linear trend in the strength with depth. Owing to small-scale local soil variations and due to sampling and testing variability,[†] the observed unconfined compressive strengths will show scatter or dispersion about this trend; that is, Y_i is a *random variable* with mean $\alpha + \beta x_i$, with some variance σ_i^2, and some distribution $f_{Y_i}(y)$.　(In much of the statistical analysis to follow we will make the specific assumptions that σ^2 does not depend on x_i and that $f_{Y_i}(y)$ is normal.)　Figure 4.3.1 shows observed data that may follow this model.　Note that if the slope β were zero, the model would simply be that soil strength throughout the layer is a random variable Y.　In this case, the problem analysis would simply follow that discussed in previous sections of the text.

In studying the growth of strength of concrete with age x at low curing temperatures, it is reasonable to assume that the mean compressive strength Y grows linearly with age during the initial curing period, when form removal might be contemplated.　Individual cylinders strengths are random variables with some distribution about this mean.

In this so-called *model 1*, it is important to recognize that the x's are not *random* variables, but values selected by the engineer prior to carrying out the experiment.

---

† The sampling and testing procedure may also introduce a *bias*, say by weakening the soil while (unavoidably) disturbing it.　If known, the bias can be subtracted out. In this case we avoid the issue by *defining* Y to be the *laboratory* test quantity: unconfined compressive strength of a soil specimen.

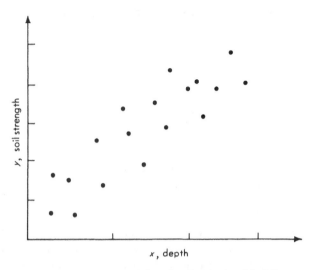

Fig. 4.3.1 Scattergram of data for simple (model 1) linear model analysis. Unconfined compressive strength versus depth.

Model 2 In this class of linear models interest focuses on *prediction*—prediction of one random variable Y given an observation of another random variable X (or others X_1, X_2, . . . , X_r). The *linear* model states that the *conditional* expected value of Y given $X = x$ (see Sec. 2.4.3) is a linear function of x:

$$E[Y \mid X = x] = \alpha + \beta x \tag{4.3.3}$$

or, more simply, in this chapter we use the notation

$$E[Y]_x = \alpha + \beta x \tag{4.3.4}$$

Conditional on the value x of X, the random variable Y has some (conditional) distribution about this mean and some (conditional) variance σ^2. In much of the statistical analysis to follow, it is assumed that this conditional distribution is normal with a variance not dependent on the specific value x. We have already seen one case in which all of these assumptions hold, namely, that in which X and Y are jointly normally distributed, Sec. 3.6.2. If X and Y are jointly normal, recall that

$$E[Y \mid X = x] = m_Y + \rho \frac{\sigma_Y}{\sigma_X} (x - m_X) \tag{4.3.5}$$

which is of the linear form assumed above with

$$\beta = \rho \frac{\sigma_Y}{\sigma_X} \tag{4.3.6}$$

and

$$\alpha = m_Y - \rho \frac{\sigma_Y}{\sigma_X} m_X = m_Y - \beta m_X \tag{4.3.7}$$

m_Y, m_X, σ_Y, and σ_X are the marginal means and standard deviations of Y and X.

Again it is sometimes convenient to write that, conditional on $X = x_i$, Y can be written

$$Y_i = \alpha + \beta x_i + E_i \tag{4.3.8}$$

in which E_i is a random variable distributed exactly like Y_i except that it has zero mean. Notice that if it is assumed (as we shall later) that for *any* x_i, Y_i is normal with mean $\alpha + \beta x_i$ and with the same variance σ^2, then we are assuming that the E_i are *identically* distributed, $N(0, \sigma^2)$.

Examples of these prediction problems include the following. For a system of soil-embankment flood levees, engineers in charge of design, construction, and maintenance want to be able to judge in-place shear strengths quickly and economically over wide areas. To do so they will use a simple field-shear-vane test. Because most design methods and experience are related to shear strengths determined from unconfined compressive tests, it is necessary to predict this latter strength index Y given an observed value of the field-vane test X. Both these quantities are random owing to spatial material variation and to testing variability. The analytical model adopted says that, given a field-vane value x at a particular point, the expected value of the conventional shear strength measure Y is $\alpha + \beta x$. In a sense the model represents a probabilistic calibration of the field test.

Similarly, an engineer might assume that the total rainfall on a watershed in a particular month Y is a random variable whose conditional mean is $\alpha + \beta x$, given that a rainfall of $X = x$ was observed in the preceding month. In developing a highway user's-cost model, an engineer has assumed that the user's total maintenance cost Y can be predicted from his total fuel cost X in the sense that the *expected* value of Y, given $X = x$, is $\alpha + \beta x$. Both X and Y, of course, vary from user to user.

The fact that both X and Y are random variables implies that, if appropriate, the engineer could consider the alternate but analogous prediction problem, $E[X \mid Y = y] = \alpha' + \beta' Y$. Whether this is appropriate depends upon how the model will be used in practice. That is, will the engineer predict X given $Y = y$ or Y given $X = x$, (the rainfall in month k given rainfall in month $k + 1$ or vice versa)? It also depends upon whether the linearity assumption remains valid. One case in which it does is when X and Y are jointly normally distributed, when

$$E[X \mid Y = y] = m_X + \rho \frac{\sigma_X}{\sigma_Y} (y - m_Y) \tag{4.3.9}$$

Notice that in general this equation would *not* represent the same line on a plot of y versus x as $E[Y \mid X = x]$, Eq. (4.3.5). The slope of y on x in the latter case, for instance, is $\rho\sigma_Y/\sigma_X$, while in the former case it is $(1/\rho)(\sigma_Y/\sigma_X)$. The two slopes are the same only if $\rho = 1$, which holds only if there is no scatter (of probability mass or of observed data) about the line. In general the two (regression) lines do pass through m_X and m_Y, however.

The distinction between model 1 and model 2 lies in the interpretation of x, whether it represents an essentially preselected value (measured without error) or whether it represents the observed value of a random variable X. In either case the fundamental assumption of a linear model is, for *a given number* x, that the expected value of Y is linear in x, $\alpha + \beta x$. Virtually none of the statistical analysis which follows distinguishes between the two models, but the distinction in interpretation is fundamental to the engineering user.

Statistical questions From a set of n experiments the engineer will have n pairs of observations $\{y_i, x_i\}$, $i = 1, 2, \ldots, n$, such as those represented by points in Fig. 4.3.1. As with the simple models discussed in previous sections, the statistical analysis of a linear model involves two kinds of questions, *estimation* of parameters and *verification* of model assumptions. In the simplest model we generally have three parameters to estimate: the intercept α, the slope β, and the variance term σ^2. We shall be interested in finding estimators for these parameters. As before, we shall study the probabilistic behavior of these estimators in order to quantify the uncertainty we have in the parameter estimates. In addition we shall want to express the uncertainty we have in an estimate of the mean of Y, $E[Y]_x = \alpha + \beta x$ for any value of x. For example, consider the engineer dealing with the soil strength versus depth problem mentioned above. He has been asked for, as input to a stability-analysis computation, the spatial average of the soil strength over a horizontal strip of finite thickness. For the linear assumption adopted here, this spatial average is simply the value of the mean of Y at x_g, the mid-depth of the strip, $\alpha + \beta x_g$. The engineer might choose to report both his best estimate of $\alpha + \beta x_g$ and, as a "conservative" estimate of $\alpha + \beta x_g$, the 95 percent lower confidence limit on $\alpha + \beta x_g$.

A number of important questions about the model can be asked. Among the most important is whether the data indicate a *significant* dependence of the mean of Y on x. Formally, we consider the null hypothesis H_0 that $\beta = 0$. If there is no dependence,† the model can be simplified by ignoring x and treating Y as a simple random variable. The implication in the soil strength versus depth example is that the strength is independent of depth. In a model 2 situation, such as the example of rainfall in two successive months, the implication of no (linear) dependence of the conditional means of Y on x is that Y and X are uncorrelated [e.g., see Eq. (4.3.5)]. In a calibration situation like that of vane test versus lab test of shear strength mentioned above, the engineer might want to determine whether the estimated value of the intercept α is significantly different from zero. If not, again the model could be simplified, now to $E[Y]_x = \beta x$. The engineer in this situation might want to know, too, if β is unity. If $E[Y]_x = x$, then the vane test is a direct, unbiased predictor of the lab index of strength.

We shall analyze these estimation and statistical-significance questions in Sec. 4.3.2.

It should be pointed out that in the case of model 2, with both X and Y random, it would be equally appropriate to estimate from the data the five customary parameters, m_X, m_Y, σ_X, σ_Y, and ρ. This is called statistical *correlation analysis*, which is distinct from the *regression* or *prediction analysis* we have been considering. Correlation analysis has the property that it is "symmetrical" in the two variables X and Y. Regression, on the other hand, is directly concerned with what may be the ultimate application, namely, one or the other of the *prediction* equations, $E[Y \mid X = x]$ or $E[X \mid Y = y]$. These equations are *not* symmetrical, as discussed above. We shall find in Sec. 4.3.2, however, that there is a close relationship between the two kinds of statistical analyses.

Extensions   Although their statistical analysis is beyond the scope of this text, it is worthwhile to point out extensions of the simple linear probability models mentioned above. The reader is referred to any of many texts that discuss the problems of estimation and hypothesis testing surrounding these models (e.g., Freeman [1963], Hald [1952], Brownlee [1965], or Graybill [1961]).

Transformations   The easiest extension is to note that x can represent some transformation or function of the variable of interest. For example, in the concrete–strength-versus-age example above the relationship

† It should be emphasized that presence of a value of β significantly different from 0 *does* not imply any causal relationship between the variables. It only suggests that one is *useful* in predicting the other.

$E[Y]_x = \alpha + \beta \ln$ (age) might be more appropriate at higher curing temperatures. If we define $x = \ln$ (age), then the model reverts to the linear form. Also, suppose that the relationship between the dependent variable Z and the independent† variable w is

$$Z = \alpha' w^\beta Q \tag{4.3.10}$$

in which Q is a random factor. Then, by taking logarithms, we obtain

$$\ln Z = \ln \alpha' + \beta \ln w + \ln Q$$

Defining $Y = \ln Z$, $\alpha = \ln \alpha'$, $x = \ln w$, and $E = \ln Q$, we again obtain a linear model like that in Eq. (4.3.2). It must be cautioned that any assumptions about normality or about independence of σ with respect to x adopted in the statistical analysis of the model must be made on E or $\ln Q$, not on Q itself. With this warning the analysis methods in Sec. 4.3.2 can also be used for these transformed problems. The analysis requires, in fact, only linearity in the relationship between $E[Y]_x$ and the *parameters* α and β.

Multivariate extensions More interesting extensions involve the introduction of an arbitrary number of independent variables $x_1, x_2, \ldots, x_r$, so that the model becomes

$$E[Y]_x = \alpha + \beta_1 x_1 + \beta_2 x_2 + \cdots + \beta_r x_r \tag{4.3.11}$$

in which the symbol $\mathbf{x}$ implies that the values $x_1, x_2, \ldots, x_r$, a vector of variables, are specified. With this model (for $r = 3$) the soils engineer could study trends in the soil strength in a particular geological layer under an entire dam site, in two plan dimensions as well as in depth. Similarly, the engineer investigating "persistence" in monthly rainfall could assume that rainfall in the next month depends, in the mean, on the past r months.

In both cases (the first a model 1 and the second a model 2), the engineer has observed data $\{y_i, x_{i1}, x_{i2}, \ldots, x_{ir}\}$, for $i = 1$ to n, the sample size. He is interested in estimating α, the β_j's $(j = 1, 2, \ldots, r)$, and σ^2, the (usually assumed constant) variance of Y given the x_j values. Questions of major interest would include whether the value of any particular β_j or the values of any particular subset of the β_j's were zero. For example, do the data indicate a significant dependence of the mean soil strength on the direction x_2, perpendicular to the dam's axis? And is the expected rainfall in the next month significantly dependent on months more than 1 month in the past (i.e., are the $\beta_j = 0$ for all $j \geq 2$)?

All fields in civil engineering have made use of such multivariate

---

† Note that the use of dependent and independent here is in the functional sense, not the stochastic sense.

models† and their statistical analysis. The method is usually referred to as *multiple-regression analysis*. It should be pointed out that with multiple independent variables it is possible to construct models which are best interpreted as a mixture of model 1 and model 2. For example, the study of lab shear-strength prediction of soil might be enhanced if the engineers asked for the expected value of this strength given the field-vane-test observation x_1 *and* the sample depth x_2. The first of these independent variables is the observed value of a random variable X_1, and the second is of the preselected, nonrandom kind.

If all the independent variables are of the model 2 type, then Y and the X_j's are, in fact, $r + 1$ jointly distributed random variables.‡ It is appropriate in such cases to carry out a *correlation analysis* to estimate means, variances, and an array of pairwise correlation coefficients. In addition, correlation analysis may include computation of *partial correlation coefficients*, which in the multivariate normal case represent the correlation coefficient of the *conditional* distribution of Y and X_k, given that some partial set of the X_j's have taken on fixed (but unspecified) values.§ These coefficients and correlation analysis can be of major value in the study of such multivariate problems, but we shall not pursue them further. See, for example, Graybill [1961].

Extensions: Analysis of variance   We shall consider one further extension or interpretation of linear models. It will serve to indicate another class of problems open to this kind of statistical analysis. The original simple model assumed that the mean of Y was a continuous linear function of a single continuous variable x. Suppose now that the independent quantity upon which the mean of Y depends is discrete and nonquantitative, perhaps not even amenable to any kind of ordering. For example, suppose an engineer would like to study the expected traffic demand between center city and suburbs as a function of the socioeconomic class (high, middle, or low) of suburbs.¶ The linear model is usually written

$$E[Y]_k = \alpha + \beta_k \qquad\qquad (4.3.12)$$

in which β_k is factor which depends on suburb class ($k = 1$, high; $k = 2$,

---

† Note that transformations permit treatment of *multiplicative* models ($Y = \alpha' x_1{}^{\beta_1} x_2{}^{\beta_2} \cdots x_r{}^{\beta_r} Q$) and *polynomial* models ($E[Y]_x = \alpha + \beta_1 z + \beta_2 z^2$, the latter becoming $E[Y]_x = \alpha + \beta_1 x_1 + \beta_2 x_2$ with $x_2 = z^2$).

‡ The implication is that there are $r + 1$ different linear prediction models that the engineer could construct. Which of these is appropriate depends in part on the future use, i.e., upon which factor the engineer will want to predict, given observations of the others.

§ See for example, Prob. 2.67.

¶ Other factors such as population and distance may also be important of course. We shall see that they can be considered too.

middle; $k = 3$, low), and in which α is selected such that $\beta_1 + \beta_2 + \beta_3 = 0$. We note with interest that the model can be considered as familiar linear regression model if we consider three variables x_1, x_2, and x_3,

$$E[Y]_x = \alpha + \beta_1 x_1 + \beta_2 x_2 + \beta_3 x_3 \qquad (4.3.13)$$

with x_1, x_2, and x_3 defined such that for, say, a middle-class suburb, $x_1 = 0$, $x_2 = 1$, and $x_3 = 0$. Only one of the three x_i's is nonzero at a time; if nonzero, its value is 1.

Again the statistical analysis begins with sets of data and seeks parameter estimates and answers to model verification or statistical-significance questions. In the example above, the engineer would like to know if the parameter estimates β_1, β_2, and β_3 are significantly different from zero. That is, does socioeconomic class have any influence on traffic demand?

Notice that *two* (or more) factors may be involved in a model of this kind. For example, in a study of beach erosion prevention, a research engineer may select four different types of protective material and three different shapes of the beach profile for test in a small-scale physical model with a wave generator. If Y represents a measure of erosion rate, the analytical model might be

$$E[Y]_{k,j} = \alpha + \beta_k + \tau_j \qquad k = 1, 2, 3, 4; j = 1, 2, 3 \qquad (4.3.13)$$

The β's are associated here with material types and the τ's with shape types. In regression form, the model is

$$E[Y]_x = \alpha + \sum_{k=1}^{4} \beta_k x_k + \sum_{j=5}^{7} \beta_j x_j \qquad (4.3.14)$$

in which for any test only one of x_1, x_2, x_3, or x_4 will be unity, depending on the protective material, *and* only one of x_5, x_6, and x_7 will be unity, depending on the beach shape. It is important to recognize that, contrary to common belief, one can make statements about, say, the significance of the beach shape (i.e., are β_5, β_6, and β_7 all zero?) *without* testing all 12 possible combinations of protective material and shape. Recognition of this fact can reduce the cost of physical experiments.

In this example, the engineer might believe, in addition, that there exists the possibility of "interaction" between the material and the beach shape, that is, that material 1 is more effective on a convex shape than on a concave one, whereas material 2 may be just the reverse. A model including interaction is normally written

$$E[Y]_{k,j} = \alpha + \beta_k + \tau_j + \gamma_{kj} \qquad k = 1, 2, 3, 4; j = 1, 2, 3 \qquad (4.3.15)$$

in which the γ_{kj} are the 12 so-called "interaction terms," one of which acts when the material is the kth type and the shape the jth type. In regression form, the model becomes

$$E[Y]_x = \alpha + \sum_{k=1}^{4} \beta_k x_k + \sum_{j=5}^{7} \beta_j x_j + \sum_{k=1}^{4} \sum_{j=5}^{7} \beta_l x_j x_k \qquad (4.3.16)$$

in which there are $(3 \times 4 =)$ 12 additional parameters β_l, $l = 8, 9,$ $\ldots$, 19, one for each combination of material and shape, x_k and x_j. Note that x_j and x_k being restricted to (0,1) variables implies that their product is unity if and only if they are both acting. Important hypothesis tests would now include questions with respect to the values of the interaction parameters β_l. In particular, "Are they zero?"

The kind of model we have just been discussing, that with discrete levels for the independent variables, is customarily given its hypothesis-testing analysis under the title *analysis of variance* or *design of experiments*.† The interpretation given there is that various "treatments" (e.g., the beach shape or the beach protection material) contribute various components to the total variance of Y. Analysis of the components leads to tests (F tests) on the significance of any treatment, or interaction between treatments. The regression formulation has the advantage of uniformity‡ in appearance and treatment, and there exist widely available computer programs for performing the numerical analysis.

4.3.2 Statistical Analysis of Simple Linear Models

Having discussed the formulation of a wide variety of linear models, we turn now to demonstrate the statistical analysis of the simplest such model:

$$E[Y]_x = \alpha + \beta x \qquad (4.3.17)$$

We consider first parameter estimation and then significance testing. Additional assumptions about the properties of the linear model will be introduced as necessary.

Point estimates of parameters Any of a variety of parameter estimators, $\hat{\alpha}$ and $\hat{\beta}$, could be considered; these estimators are usually obtained by the *least squares method*.§ This method states that $\hat{\alpha}$ and $\hat{\beta}$ be chosen to minimize the sum of the squared differences between the observed values

---

† See, for example Freeman [1963], Hald [1952], Brownlee [1965], or Graybill [1961].
‡ Note that continuous variables and discrete "treatments" can be mixed; e.g., the traffic-demand model might include the continuous variables suburb population and distance, as well as socioeconomic class.
§ Under more restrictive assumptions, the maximum-likelihood method, Sec. 4.1.4, will suggest the same estimators. See, for example, Brownlee [1965] or Graybill [1961].

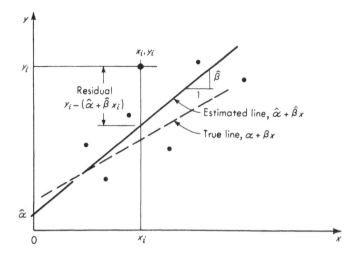

Fig. 4.3.2 The residual, or the "vertical" distance between y_i and the estimated line, $\hat{\alpha} + \hat{\beta}x_i$.

y_i and the estimated expected values $\hat{\alpha} + \hat{\beta}x_i$; that is, one should choose $\hat{\alpha}$ and $\hat{\beta}$ to minimize

$$\sum_{i=1}^{n} [y_i - (\hat{\alpha} + \hat{\beta}x_i)]^2 \tag{4.3.18}$$

As shown in Fig. 4.3.2, these differences, called the *residuals*, are the "vertical" distances between the y_i and an estimate $\hat{\alpha} + \hat{\beta}x$ of the true line $\alpha + \beta x$.

To find the estimates of $\hat{\alpha}$ and $\hat{\beta}$, we simply take partial derivatives of the sum of squares expression with respect to $\hat{\alpha}$ and $\hat{\beta}$, set these two derivatives equal to zero, and solve for the $\hat{\alpha}$ and $\hat{\beta}$ which minimize the sum. Thus

$$\frac{\partial}{\partial \hat{\alpha}} \left\{ \sum_{i=1}^{n} [y_i - (\hat{\alpha} + \hat{\beta}x_i)]^2 \right\} = \sum \frac{\partial}{\partial \hat{\alpha}} [y_i - (\hat{\alpha} + \hat{\beta}x_i)]^2$$

$$= \sum 2[y_i - (\hat{\alpha} + \hat{\beta}x_i)](-1) = 0 \tag{4.3.19}$$

and

$$\frac{\partial}{\partial \hat{\beta}} \left\{ \sum_{i=1}^{n} [y_i - (\hat{\alpha} + \hat{\beta}x_i)]^2 \right\} = \sum \frac{\partial}{\partial \hat{\beta}} [y_i - (\hat{\alpha} + \hat{\beta}x_i)]^2$$

$$= \sum 2[y_i - (\hat{\alpha} + \hat{\beta}x_i)](-x_i) = 0 \tag{4.3.20}$$

Carrying out the summations and simplifying yields two simultaneous linear equations in $\hat{\alpha}$ and $\hat{\beta}$:

$$n\bar{y} - n\hat{\alpha} - \hat{\beta}n\bar{x} = 0 \tag{4.3.21}$$

$$\Sigma x_i y_i - n\hat{\alpha}\bar{x} - \hat{\beta}\Sigma x_i^2 = 0 \tag{4.3.22}$$

in which $\bar{y}$ is the average of the y_i's, $(1/n)\Sigma y_i$, and $\bar{x}$ is the average of the x_i's, $(1/n)\Sigma x_i$.

Solution of these, the so-called *normal equations*, is trivial.

$$\hat{\alpha} = \bar{y} - \hat{\beta}\bar{x} \tag{4.3.23}$$

$$\hat{\beta} = \frac{\Sigma x_i y_i - n\bar{x}\bar{y}}{\Sigma x_i^2 - n(\bar{x})^2} \tag{4.3.24}$$

Using Eqs. (1.2.2b) and (1.3.2) (also see Prob. 1.19), it is clear that we can also write†

$$\hat{\beta} = \frac{s_{X,Y}}{s_X^2} = \frac{r_{X,Y}s_Y}{s_X} \tag{4.3.25}$$

in which $s_{X,Y}$ is the sample covariance, $r_{X,Y}$ is the sample correlation coefficient, and s_X and s_Y are the sample standard deviations, all defined in Chap. 1.

Example   Consider the data in Fig. 4.3.3 of y_i's, the true, lab-measured water content in randomly selected field soil specimens from a particular site, and the corresponding x_i's ,the water content estimated by a fast, inexpensive method which measures the gas pressure created when the soil is mixed with a chemical which reacts with water. If it is sufficiently accurate, the second method will provide an inexpensive way to obtain more frequent water-content samples during the quality control of soil compaction on a highway project. The purpose is to predict true water content given an observed value of the fast test. This is a model 2 example.

The sample size is $n = 67$. The calculated values needed are

$$\bar{y} = 13.8$$

$$\bar{x} = 13.8$$

$$\Sigma x_i^2 = 13,260$$

$$\Sigma x_i y_i = 13,145$$

---

† Compare this form with Eq. (4.3.6) for jointly distributed random variables. Recall from Chap. 1 that these sample averages can be computed whether the x's and y's represent observations of random variables or not. In particular, in a model 1 case (Sec. 4.3.1), the x's are preselected, *nonrandom* variables, but $\bar{x}$ and s_X can still be computed.

(The equality of $\bar{y}$ and $\bar{x}$ is coincidence.)   Therefore the estimates of $\hat{\alpha}$ and $\hat{\beta}$ are

$$\hat{\beta} = \frac{13{,}145 - 67(13.8)(13.8)}{13{,}260 - 67(13.8)^2} = \frac{13{,}145 - 12{,}750}{13{,}260 - 12{,}750} = 0.776$$

$$\hat{\alpha} = 13.8 - (0.776)(13.8) = 3.04$$

This estimated regression line is shown with the data in Fig. 4.3.3.

Also shown in the figure is an estimate, $\sqrt{s^2}$, of σ, the assumed constant standard deviation of Y given x.   The obvious estimator of σ^2, $\widehat{\sigma^2}$, is the average of the squared residuals.   That is,

$$s^2 = \widehat{\sigma^2} = \frac{1}{n-2} \sum_{i=1}^{n} [y_i - (\hat{\alpha} + \hat{\beta}x_i)]^2 \qquad (4.3.26)$$

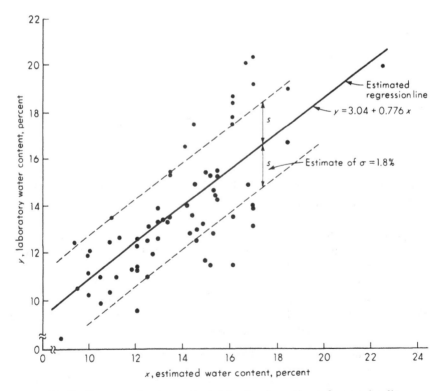

Fig. 4.3.3   Scattergram of water-content data showing estimated regression line.

which upon substitution for $\hat{\alpha}$ and $\hat{\beta}$ can be reduced for ease in computation to†

$$s^2 = \widehat{\sigma^2} = \frac{n}{n-2}\left(1 - \frac{s_{X,Y}^2}{s_X^2 s_Y^2}\right)s_Y^2 = \frac{n}{n-2}(1 - r_{X,Y}^2)s_Y^2 \quad (4.3.27)$$

Note the use of a divisor of $n - 2$ rather than n in Eq. (4.3.26). This is customary, since it can be shown to produce an unbiased estimator of σ^2.

In our example (with $\Sigma y_i^2 = 13{,}291$),

$$s_{X,Y} = \frac{1}{n}(\Sigma x_i y_i - n\bar{x}\bar{y}) = \frac{1}{67}[13{,}145 - 67(13.8)(13.8)]$$

$$= 5.31$$

$$s_X^2 = \frac{1}{n}[\Sigma x_i^2 - n(\bar{x})^2] = \frac{1}{67}[13{,}260 - 67(13.8)^2]$$

$$= 6.85$$

$$s_Y^2 = \frac{1}{n}[\Sigma y_i^2 - n(\bar{y})^2] = \frac{1}{67}[13{,}291 - 67(13.8)^2]$$

$$= 7.32$$

implying

$$r_{X,Y}^2 = \frac{s_{X,Y}^2}{s_X^2 s_Y^2} = \frac{(5.31)^2}{(6.85)(7.32)} = 0.562$$

Thus

$$s^2 = \widehat{\sigma^2} = \frac{n}{n-2}(1 - r_{X,Y}^2)s_Y^2 = \frac{67}{65}(1 - 0.562)(7.32) = 3.29$$

and

$$s = \sqrt{3.29} = 1.8$$

This completes the process of obtaining point estimates of the model parameters, α, β, and σ^2.

Properties of estimators In order to determine confidence intervals or to carry out significance tests we must, as in Sec. 4.1, study the properties of the estimators as random variables. This requires adopting a position prior‡ to the experiment, when the Y_i's are random variables, and determining the mean, variance, and distribution of these estimators as functions of the Y_i's. We shall demonstrate one case only, and simply state readily verifiable conclusions for others.

† Compare this form with Eq. (2.4.98b) for the conditional variance of Y given X when X and Y are jointly normal.

‡ In the case of model 2 we assume that the X_i's have been observed, but the Y_i's have not been, as yet.

Consider the estimator of the slope, $\hat{\beta}$. Emphasizing its character prior to the experiment as a random variable, we write it as B. Then, from Eq. (4.3.24),

$$B = \frac{\Sigma x_i Y_i - n\bar{x}\bar{Y}}{\Sigma x_i{}^2 - n(\bar{x})^2} \tag{4.3.28}$$

The x_i's are all known constants. The Y_i's and $\bar{Y}$ are random variables. The numerator of B can be expanded to

$$B = \frac{\Sigma x_i Y_i - \bar{x}\Sigma Y_i}{\Sigma x_i{}^2 - n\bar{x}^2} = \frac{\Sigma(x_i - \bar{x})Y_i}{\Sigma x_i{}^2 - n\bar{x}^2} = \Sigma c_i Y_i \tag{4.3.29}$$

in which $c_i = (x_i - \bar{x})/(\Sigma x_i{}^2 - n\bar{x}^2)$. The purpose of the expansion is to recognize the important fact that B, the least squares estimator of β, is a *linear combination* of the Y_i's. As we have seen, this simplifies analysis greatly. For example, the expected value of B is quickly found to be

$$E[B] = E[\Sigma c_i Y_i] = \Sigma c_i E[Y_i] = \Sigma c_i(\alpha + \beta x_i)$$
$$= \alpha \Sigma c_i + \beta \Sigma c_i x_i = \beta \tag{4.3.30}$$

since $\Sigma c_i = 0$ and $\Sigma c_i x_i = 1$. The implication is that B is an unbiased estimator of β.

Similarly, *if we now make the assumption that the Y_i's are uncorrelated,* e.g., if we have a random sample,† then

$$\text{Var}[B] = \text{Var}[\Sigma c_i Y_i] = \Sigma c_i{}^2 \text{Var}[Y_i] \tag{4.3.31}$$

If we also assume now that the Y_i's have the same variance σ^2, no matter what x_i is, then

$$\sigma_B{}^2 = \text{Var}[B] = \sigma^2 \Sigma c_i{}^2 = \sigma^2 \frac{\Sigma(x_i - \bar{x})^2}{(\Sigma x_i{}^2 - n\bar{x}^2)^2} = \frac{\sigma^2}{\Sigma(x_i - \bar{x})^2} = \frac{\sigma^2}{n s_X{}^2} \tag{4.3.32}$$

Note the characteristic decay in an estimator's variance (like $1/n$) and note that, if, as in a model 1 situation, the engineer is able to preselect the x_i's, he can reduce the uncertainty in the slope by spreading the x_i's as much as possible to increase $s_X{}^2$.

With a large enough sample size, we could safely assume (by the central limit theorem) that B is approximately normally distributed, no matter how the Y_i are distributed. With this information, and with an

† There are important practical cases when this condition does not hold. For example, if the Y_i's are creep strains of compressed concrete cylinders and the x_i's are times since load was applied, the Y_i's taken from the *same* cylinder will *not* be uncorrelated. Such problems should be treated by time-series analysis (e.g., see Bendat and Piersol [1966]), which is beyond the scope of this book.

estimate s^2 for σ^2, we could prepare confidence limits on β. For example, $1 - \alpha$ two-sided confidence interval would be

$$\hat{\beta} \pm k_{\alpha/2}\sigma_B$$

in which $k_{\alpha/2}$ is taken from a normal table (Sec. 4.1). For smaller sample sizes, we must adopt a distribution for the Y_i's. *If we assume that the Y_i's are normal and independent,* then B is also normal, since it is a linear combination of independent normal variables. But for small n and unknown σ, it should be clear from experience in Secs. 4.1 and 4.2 that such confidence intervals must be based on a value from a table of the t distribution, $t_{\alpha/2,n-2}$, and s_B, the estimate of σ_B based on replacing σ by its estimate s in Eq. (4.3.32). The confidence interval is then

$$\hat{\beta} \pm t_{\alpha/2,n-2}s_B \tag{4.3.33}$$

Note that the $n - 2$ degrees of freedom are used in entering the t table (A.3).

In the example just considered

$$s_B{}^2 = \frac{s^2}{ns_X{}^2} = \frac{3.29}{67(6.85)} = 0.0072$$

implying that a 90 percent confidence interval on the true slope β is

$$\hat{\beta} \pm t_{0.05,65}s_B$$

or

$$0.776 \pm 1.67(0.085) \qquad \{0.634, 0.918\}$$

(Assuming B normal would have replaced 1.67 by 1.64, i.e., $t_{0.05,65}$ by $k_{0.05}$.)

Without proof, it is easily shown that A is also a linear combination of the Y_i's, that it is an unbiased estimator of α, and that its variance is

$$\sigma_A{}^2 = \frac{\sigma^2}{n}\left(1 + \frac{\bar{x}^2}{s_X{}^2}\right) \tag{4.3.34}$$

In our example, the estimate of σ_A is

$$s_A = \sqrt{\frac{s^2}{n}\left(1 + \frac{\bar{x}^2}{s_X{}^2}\right)} = \sqrt{\frac{3.29}{67}\left[1 + \frac{(13.8)^2}{6.85}\right]} = \sqrt{1.42} = 1.20$$

The 90 percent confidence intervals are also based on a t distribution:

$$\hat{\alpha} \pm t_{0.05,65}s_A \tag{4.3.35}$$

or

$$3.04 \pm 1.67(1.20) \qquad \{1.50, 4.58\}$$

Under the same assumptions the estimator S^2 of σ^2 is unbiased with a distribution related to the χ^2 distribution. Specifically,

$$\frac{S^2}{\sigma^2/(n-2)} \tag{4.3.36}$$

is χ^2-distributed with $n - 2$ degrees of freedom. Therefore, for example, a 90 percent one-sided lower confidence interval on σ^2 is found from the statement (Sec. 4.1)

$$P\left[\frac{S^2(n-2)}{\sigma^2} \le \chi^2_{0.9,n-2}\right] = 0.9$$

which implies that

$$P\left[\sigma^2 \ge \frac{S^2(n-2)}{\chi^2_{0.9,n-2}}\right] = 0.9$$

The lower limit in our example is

$$\sigma^2 \ge \frac{s^2(n-2)}{\chi^2_{0.9,n-2}} = \frac{(3.29)(65)}{81} = 0.80(3.29) = 2.64$$

As a final example we can find the properties of the estimator of $E[Y]_x = \alpha + \beta x$ for any particular x, say x_0. Calling this estimator $\bar{Y}_{x_0}$, we can find that it, too, is a linear combination of the Y_i's with mean

$$E[\bar{Y}_{x_0}] = E[A + Bx] = \alpha + \beta x$$

and variance

$$\sigma^2_{\bar{Y}_{x_0}} = \frac{\sigma^2}{n}\left[1 + \frac{(x_0 - \bar{x})^2}{s_X^2}\right] \tag{4.3.37}$$

Note that the variance depends on x_0. At x_0 equal to $\bar{x}$, that is, near the middle of the data, $\sigma^2_{\bar{Y}_{x_0}}$ is equal to simply σ^2/n, while for values of x_0 removed from $\bar{x}$, for example, near the extremes of the data, the variance in the estimator of $E[Y]_{x_0}$ grows. Note that for $x_0 = 0$, $\sigma^2_{\bar{Y}_{x_0}}$ equals the variance of A, the estimator of the intercept, as it should.

In our example, for a rapid water-content test value of $x_0 = 15$, the best estimate of the expected value of Y, the true water content, is

$$\widehat{E[Y]}_{x_0} = \bar{y}_{x_0} = \hat{\alpha} + \hat{\beta}x_0 = 3.04 + 0.776(15) = 14.68$$

The 90 percent confidence limits on this conditional mean are

$$\hat{\alpha} + \hat{\beta}x_0 \pm t_{\alpha/2,n-2}s_{\bar{Y}_{x_0}}$$

$$14.68 \pm 1.67\sqrt{\frac{3.29}{67}\left[1 + \frac{(15-13.8)^2}{6.85}\right]}$$

or

$$14.68 \pm 0.408 \qquad \{14.27, 15.09\}$$

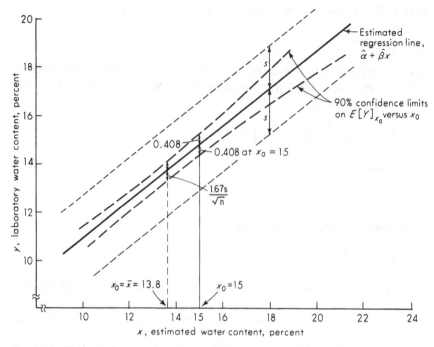

Fig. 4.3.4 Estimated regression line and 90 percent confidence limits on $E[Y]_{x_0}$. Water-content illustration.

This value and the locus of such confidence limits for a range of values of x_0 are shown in Fig. 4.3.4.

It should be emphasized that this confidence interval is on the *mean* of Y given x_0, not on Y itself; that is, in our example, the confidence interval is on the *average* true water content of many soil specimens whose estimated water contents are all 15. If the engineer wants to make some kind of statement about the true water content Y of an *individual* specimen with an estimated water content of $x = 15$, he must *also* allow for the inherent variation σ^2 about the mean value. In most texts, it is suggested that one also consider this wider interval as a confidence interval, now on Y, a single future value. It is found by simply replacing $s_{\bar{Y}x_0}^2$ by $(s_{\bar{Y}x_0}^2 + s^2)$ in the expressions above for confidence intervals on $E[Y]_{x_0}$.

Significance tests   Now that we have the distributions of the estimators, it is very easy to conduct hypothesis tests on the parameters. We demonstrate three such cases.

Suppose the engineer wants to report whether the data indicate an intercept significantly different from zero. Then we have, following the pattern in Sec. 4.2,

H_0: $\alpha = 0$
H_1: $\alpha \neq 0$
Test statistic: A, the estimator of α
Rule form: Accept if $-c \leq A \leq +c$
Significance level:† α
Acceptance interval: Solution of

$$P[-c \leq A \leq +c \mid H_0] = 1 - \alpha \quad \text{for } c$$

In this case we work with a t statistic and find, as expected,

$$c = t_{\alpha/2, n-2} s_A \tag{4.3.38}$$

For example, in the soil-water-content illustration, with $\alpha = 10$ percent, we find

$$c = 1.67(1.20) = 2.00$$

Since the observed value of A, $\hat{\alpha} = 3.04$, lies outside‡ the interval ± 2.00, we reject the hypothesis and state that the intercept estimate is significantly different from zero at the 10 percent significance level.

Similarly, we could test if the slope β is, say, unity by comparing the observed value of B with the acceptance interval

$$1 \pm t_{\alpha/2, n-2} s_B$$

(In our example, $\hat{\beta}$ *is* significantly different from zero at the 10 percent level.)

The most common test, however, is whether there is significant dependence of $E[Y]_x$ on x. This is tested as follows:

H_0: $\beta = 0$

H_1: $\beta \neq 0$

As above, the acceptance interval is

$$0 \pm t_{\alpha/2, n-2} s_B$$

Note that for model 2, in which X and Y are both random, asking whether β is zero is equivalent to asking if the correlation coefficient ρ is zero. Compare the test on ρ here with that discussed in Sec. 4.2. Note that both are based on a t distribution.

For example, in the reinforced-concrete-beam data in Chap. 1, the

† Do not confuse α the intercept and α the significance level. They are both common notation.

‡ Recall from Sec. 4.2 that we can in fact determine the outcome of the significance test by simply observing that the hypothesized value 0 does not fall in the $1 - \alpha$ confidence limit {1.50,4.58} found above.

estimate of the slope of a regression of ultimate load Y, on cracking load X, is

$$\hat{\beta} = \frac{s_{X,Y}}{s_X{}^2} = \frac{94,200}{(1690)^2} = 0.033$$

while

$$s_B{}^2 = \frac{s^2}{n s_X{}^2} = \frac{1}{n-2} \left(1 - \frac{s_{X,Y}{}^2}{s_X{}^2 s_Y{}^2} \right) \frac{s_Y{}^2}{s_X{}^2}$$

$$= \frac{1}{13} \left[1 - \frac{(94,200)^2}{(1690)^2(440)} \right] \frac{(440)^2}{(1690)^2} = 0.0051$$

At the 10 percent significance level

$$\pm t_{\alpha/2, n-2} s_B = 1.77 \sqrt{0.0051} = 0.13$$

The conclusion is that the slope (and the correlation coefficient) are not indicated by the data to be different from zero.

Note that all calculations involving the estimate and variance of the estimator of the slope involve only the sample variances, never the averages $\bar{x}$ or $\bar{y}$. As one could have predicted, a shift in the origin will *not* change $\hat{\beta}$ or s_B, but it will alter $\hat{\alpha}$ and s_A. There is certain computational convenience if one shifts the origin to $\{\bar{x},\bar{y}\}$, working then with $x_i - \bar{x}$ and $y_i - \bar{y}$. Two resulting simplifications,† for example, are that $\hat{\alpha} = 0$ and that $s_A{}^2 = s^2/n$.

An important class of significance tests arises when *simultaneous hypotheses* are tested. For example, if the engineer were considering the feasibility of reducing a model from $E[Y]_x = \alpha + \beta x$ to simply $E[Y]_x = x$ (as would be appropriate, for instance, with the water-content "calibration" example above), then in principle he should not first test $\{H_0: \alpha = 0\}$ and then $\{H_0: \beta = 1\}$, but rather the simultaneous hypothesis

$$H_0: \quad \alpha = 0 \quad and \quad \beta = 1$$

Such tests involve *both* estimators, A and B, and their *joint* distribution. In regression analysis the customary test statistic is an F-distributed variable. We shall not pursue these tests further, but we do point out their particular importance in analysis of variance problems (Sec. 4.3.1) where a multiple hypothesis such as $\{H_0: \beta_1 = 0, \beta_2 = 0,$ and $\beta_3 = 0\}$ is needed to test whether a particular "treatment" (e.g., socioeconomic class) has a significant influence on the dependent variable (e.g., travel demand). The reader is referred to Rao [1965], page 196, for a general treatment of simultaneous hypothesis tests in multivariate regression.

---

† A fundamentally more important simplification is that with this origin, the estimators A and B are independent random variables. In general, they are not.

We chose not to delve into the statistical analysis of multivariate linear models. But the reader with a moderate amount of additional preparation from other texts (e.g., Hald [1952]) and with the help of the widely available computer programs for multivariate regression can find himself quickly in command of a powerful set of analytical models and data-analysis capability.

4.3.3 Summary

Linear models are of the general form

$$E[Y]_{\mathbf{x}} = \alpha + \beta_1 x_1 + \beta_2 x_2 + \cdots + \beta_r x_r$$

in which the independent variables, $x_1, x_2, \ldots, x_r$, are either pre-selected nonrandom values (model 1) or observed values of random variables $X_1, X_2, \ldots, X_r$ (model 2). In general, interest is in predicting Y for a specified set of the x_i. As such it is closely related (in the model 2 case) to linear prediction (Sec. 2.4.3) and jointly normally distributed random variables (Sec. 3.6.2) and correlation analysis.

The model can be used to represent nonlinear relationships and situations in which an independent "variable" or treatment occurs at discrete levels, possibly not even "orderable or quantifiable" (analysis of variance).

The statistical analysis of the model includes parameter estimation and significance testing. For the simplest case,

$$E[Y]_x = \alpha + \beta x$$

the parameter estimates are

$$\hat{\alpha} = \bar{y} - \hat{\beta}\bar{x}$$

$$\hat{\beta} = \frac{s_{X,Y}}{s_X^2}$$

$$\hat{\sigma}_2 = \frac{n}{n-2}(1 - r_{X,Y}^2)s_Y^2$$

in which $\bar{x}, \bar{y}, s_X, s_Y, s_{X,Y}$, and $r_{X,Y}$ are sample averages defined in Chap. 1, and σ^2 is the (assumed constant) variance of Y about its regression line (i.e., the variance of Y given x). The estimators are linear functions of the Y_i's, and they are unbiased. Their variances are

$$\sigma_B^2 = \frac{\sigma^2}{n s_X^2}$$

$$\sigma_A^2 = \frac{\sigma^2}{n}\left(1 + \frac{\bar{x}^2}{s_X^2}\right)$$

in which A and B are, respectively, $\hat{\alpha}$ and $\hat{\beta}$ treated as random variables.

Confidence intervals and hypothesis tests on the parameters or functions of the parameters (for example, $E[Y]_{x_0} = \alpha + \beta x_0$) follow from study of the estimators' distributions. Assuming that the Y_i's are normally distributed, t statistics are typically involved when parameters α and β are central, while the χ^2 distribution governs the estimator of σ^2.

4.4 MODEL VERIFICATION

Throughout the text to this point we have always assumed that the form of the model was known—e.g., normal, Poisson, etc. In Sec. 4.1, for example, we discussed the estimation of the parameters of a given model or distribution. At the outset of the chapter, however, we recognized that a third source of uncertainty always exists, namely, that surrounding the validity of the mathematical model being proposed or hypothesized. Is it or is it not an accurate representation of the physical phenomenon? We take the position in this section that the engineer has for one reason or another adopted a mathematical model of a physical phenomenon and seeks, after gathering data, to *test the validity of that particular model.*

We introduce in Secs. 4.4.1 graphical ways of comparing the form of the model and that of the observed data. An understanding of the behavior of these observed shapes vis-à-vis the underlying distribution is critical to their proper interpretation. In Sec. 4.4.2, we will use this understanding to develop quantitative rules that lead to acceptance or rejection of a particular hypothesized model. In Sec. 4.5, this same understanding will underlie the discussion of the different, but related, problem of *choosing from among several* convenient mathematical forms the one which best fits observed data.

4.4.1 Comparing Shapes: Histograms and Probability Paper

Observed physical data provide the engineer an opportunity to verify the model he has developed. He seeks to support or reject such statements as, "Annual floods at station A have a type II extreme value distribution," or "Contractor B produces concrete whose compressive strength is normally distributed," or "The traffic on highway C follows a Poisson process."

In principle he must compare some observation with some model prediction in order to verify the validity of the latter. He could, for example, compare higher† sample moments with the corresponding moments of the mathematical model. Although occasionally such moment comparisons are useful checks, the engineer usually finds more

† Higher than those used to estimate the model's parameters, Sec. 4.1.

benefit in looking at a graphical comparison of the "shapes" of the data and the model.   These may deal either with a histogram of the data and a PDF (or PMF) of the model or with a cumulative histogram and a CDF.   In either case the visual comparison permits an immediate assessment of the proximity of observed and predicted results and, additionally, provides information about the regions of discrepancy, if any.‡   Intuitively, such a shape comparison seems preferable to a simple higher-moment comparison in that it brings more of the information contained in the sample into the comparison and in that it tends to counter the criticism that many models may have nearly identical second- or third-order moments, but still have markedly different shapes.

Histogram versus PDF   Consider the two comparisons made in Figs. 4.4.1 and 4.4.2 between the histograms and the PDF's of the proposed models in the traffic and yield-force illustrations we considered in Sec. 4.1. Notice that the probability density functions of the models have been

‡ Depending on his purposes, the engineer may be prepared to accept such discrepancies in regions of less interest, for example, upper tails.   An illustration appears in Sec. 4.5.2.

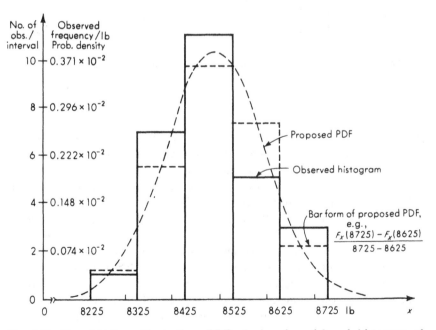

Fig. 4.4.1 Bar-yield-force illustration: PDF of normal model and histogram of observed data.

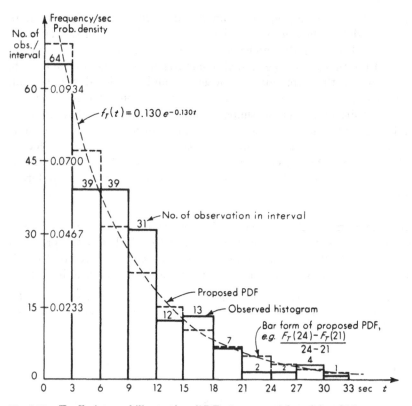

Fig. 4.4.2 Traffic-interval illustration: PDF of exponential model and histogram of observed data.

shown in a *bar form*, coinciding with the intervals of the histograms. Owing to the grouping of data into intervals necessary to construct a histogram, it is clearly the corresponding bar form of the PDF which must be used in the comparison, and *not* the complete, continuous form. This requirement is unfortunate.   The fine features, including the tails, of the proposed model are "smeared" and lost in the process.   Narrower interval choices reduce this smearing, but at the expense of fewer observations per interval and subsequently larger fluctuations in the observed bar heights, as discussed in Chap. 1 and later in this section.

Within this restriction, however, the data seem to suggest that the traffic and yield-force models are not unreasonable.   There is no apparent systematic discrepancy in the shapes, but some of the individual observed frequencies differ considerably from the predicted values.   Are some of these values unduly large? Is the apparent argeement good enough?

Does the engineer dare wish for a closer fit? In short, is the observed histogram shape *significantly* different from the predicted shape?

In order to ask such questions more precisely and to attempt to answer them, we must try to understand the histogram as well as we do the sample mean and other simple sample statistics. To achieve this end, we must, as always, assume that the model is known and then describe the probabilistic behavior of the statistics to be formed from samples taken from that model. Consider for illustration a simple triangular distribution on the interval 0 to 1, as shown in Fig. 4.4.3f. The results of several samples of size nine drawn from this distribution (using random numbers as described in Sec. 2.3.3) are shown in Fig. 4.4.3a to e. These histograms were constructed such that there was an equal probability, ⅓, of observing a value in each of the three intervals, 0 to 0.58, 0.58 to 0.82, and 0.82 to 1. The random nature of these histograms is obvious. They vary in shape from one which exaggerates the true triangular PDF (e) through one which coincides "perfectly" with the expected frequencies (b), on to bell-shaped (c), and uniform shapes (a), and to a reversed triangular shape (d). Most important for our purposes here is to consider the reaction to any one of these observed histograms that an engineer who had correctly hypothesized the underlying triangular PDF might have had. In several cases he undoubtedly would have considered the data as substantiating his proposal, but at least one (d) might have caused him (erroneously) to doubt the model.

Probability law of a histogram Given the true distribution and a set of intervals, the observed histogram interval frequencies are clearly random variables that are directly proportional to the random numbers of sample observations which fall into those intervals. These numbers are jointly distributed random variables possessing a multinomial distribution (Sec. 3.6.1). If the interval probabilities are $p_1, p_2, \ldots, p_k$ and the sample of size is n, the numbers in each interval, $N_1, N_2, \ldots, N_k$, have the joint PMF [Eq. (3.6.5)]:

$$p_{N_1,N_2,\ldots,N_k}(n_1,n_2, \ldots ,n_k) = \frac{n!}{n_1!n_2! \cdots n_k!} p_1^{n_1}p_2^{n_2} \cdots p_k^{n_k}$$

$$(4.4.1)$$

If the interval probabilities are equal, $p_1 = p_2 = \cdots = p_k = 1/k$, a simpler result is obtained:

$$p_{N_1,N_2,\ldots,N_k}(n_1,n_2 \ldots ,n_k) = \frac{n!}{n_1!n_2! \cdots n_k!} \left(\frac{1}{k}\right)^n$$

$$(4.4.2)$$

since $n_1 + n_2 + \cdots + n_k$ must equal n.

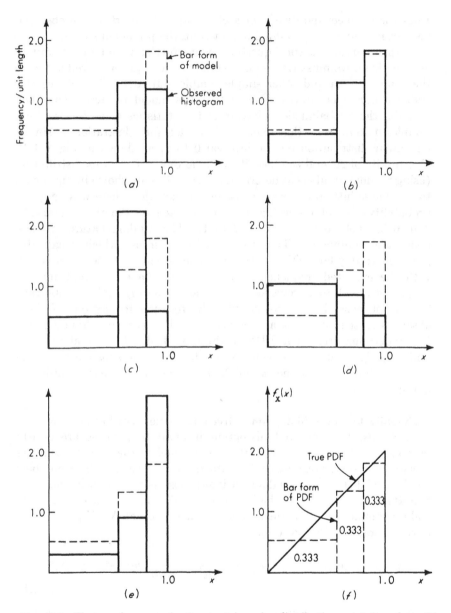

Fig. 4.4.3 Five random samples from a triangular distribution. (a) Sample a; (b) sample b; (c) sample c; (d) sample d; (e) sample e; (f) true PDF and bar-form frequencies in three equally likely intervals.

With this formula we can compute the probability of obtaining any particular histogram. For example, consider the triangular distribution discussed above. For simplicity, we take $k = n = 3$, that is, three intervals as above, but a sample size of only three. Then the likelihood of a histogram with two observations falling in the first interval and one in the last is

$$p_{N_1,N_2,N_3}(2,0,1) = \frac{3!}{2!0!1!} \left(\frac{1}{3}\right)^3 = \left(\frac{1}{9}\right)$$

Since the interval probabilities, were chosen to be equal, this probability is the same for any one of the six possible histograms with two observations in one interval and one in another. A histogram with all three observations in the same interval, say the last, has probability

$$p_{N_1,N_2,N_3}(0,0,3) = \frac{3!}{0!0!3!} \left(\frac{1}{3}\right)^3 = \left(\frac{1}{27}\right)$$

The "perfect fit" is obtained only if there is a single observation in each and every interval:

$$p_{N_1,N_2,N_3}(1,1,1) = \frac{3!}{1!1!1!} \left(\frac{1}{3}\right)^3 = \frac{2}{9}$$

The "perfect fit" is in fact always the *most* likely single histogram, but, as here, it is *not* a *highly* likely outcome. The probability of not finding a perfect fit is $7/9$. This number rapidly approaches 1 as the sample size increases. Therefore the engineer should not be discouraged if the data do not compare perfectly with his model. This is the major distinction between verifying the more familiar deterministic models and verifying probabilistic models. In the former case the model is correct if and only if data and predictions fit perfectly. In the latter case this perfect fit is a rare event.

Close fits Recognizing that the perfect fit is not likely even if the model holds, we might consider circumstances under which "close" fits are likely. We can investigate the degree of variability of the histogram frequency about the model predicted value. The expected number of observations in the ith interval is

$$E[N_i] = np_i = n\left(\frac{1}{k}\right) \tag{4.4.3}$$

assuming a sample size n and histogram intervals designed to yield k equally likely intervals. The variance of N_i is

$$\text{Var }[N_i] = np_i(1 - p_i) = \frac{n}{k}\left(1 - \frac{1}{k}\right) \tag{4.4.4}$$

It is, of course, the *proportion* N_i/n of the observations in any interval that will be compared with the model prediction of $1/k$. This proportion has the following mean, variance, and coefficient of variation:

$$E\left[\frac{N_i}{n}\right] = \frac{1}{n}\frac{n}{k} = \frac{1}{k} \tag{4.4.5}$$

$$\text{Var}\left[\frac{N_i}{n}\right] = \frac{1}{n^2}\frac{n}{k}(1-k) = \frac{1}{nk}\left(1-\frac{1}{k}\right) \tag{4.4.6}$$

$$V_{N_i/n} = \frac{\sqrt{(1-1/k)/nk}}{1/k} = \frac{\sqrt{k-1}}{n} \tag{4.4.7}$$

Let us use this coefficient of variation of the interval proportion† as a measure of how close a fit we can "expect." For a fixed number of intervals k, an increase in the sample size n, can make V arbitrarily small, i.e., can make close fits almost certain. On the other hand, *reducing* the number of intervals k will have the same effect. One should expect, then, close fits if he has adopted the crude bar-form representation of his model that results from using only a small number of intervals. Under these circumstances, however, a large number of different models will have nearly identical bar-form representations. Therefore an apparent close fit supports these models as strongly as it does the engineer's model. This smearing can be reduced and the underlying model shape more finely resolved only by using more intervals k. But this comes at the expense of higher variability in the observed values of individual histogram interval frequencies. Close fits are not likely in this case, and it is therefore not likely that the engineer will find strong visual support for his model in the data.

In short, when comparing the shape of the data with the PDF of the model, the engineer must make a tradeoff between larger smearing and larger variability. He does this at the time he chooses the histogram intervals. When he judges his model on the basis of this visual comparison, he must keep in mind that if he takes a small number of intervals, many other models will have nearly identical bar forms and that if he chooses a large number of intervals, similarity in shapes is not really likely. Again, any of many laws could potentially have generated the histogram.

Comparison of cumulatives Rather than comparing the observed histogram of the data to the PDF or PMF of the hypothesized model, one may also compare the cumulative frequency polygon (Sec. 1.1) with the CDF of the model. As before, one seeks to answer the question, "Is the

† There is degree of correlation among the interval proportions, Sec. 3.6.1. It is small if k is moderately large.

'shape' of the data predicted by the model?" Alternatively, we might ask, "Is the shape of the data so different from the prediction that it is unreasonable to retain the proposed model?" The now-familiar problems remain. We must expect variation about the model CDF even if it holds, and another probability law may in fact be governing the generation of the data even if it appears to agree with the proposed law.

Although these fundamental difficulties remain, comparison of cumulative shapes has a marked advantage over comparing the histogram and PDF. With cumulative shapes there is no need for grouping both data and mathematical model into selected intervals. Lumping of the data into intervals, as is necessary with histograms, implies ignoring some of the information in the sample, since the exact values of the observations are lost. At the same time the necessary smearing of the PDF into a bar form leads to a loss of the details in the model. Cumulative shapes, on the other hand, are compared by plotting each observation as a specific point side by side with the complete, continuous CDF of the model.

Probability paper In practice, both the plotting and comparison of cumulative curves can be simplified by scale changes, that is, by special plotting paper. Such graph paper is called *probability paper*. This probability paper provides properly scaled ordinates such that the cumulative distribution function of the probability law plots as a straight line.† With such paper, comparison between the model and data is reduced to a comparison between the cumulative frequency polygon of the data (plotted on the paper) and a straight line.

We illustrate first using the simple triangular PDF considered above. The density function is

$$f_X(x) = 2x \qquad 0 \le x \le 1 \tag{4.4.8}$$

while

$$F_X(x) = x^2 \qquad 0 \le x \le 1 \tag{4.4.9}$$

The CDF is plotted in Fig. 4.4.4. If we take the square root of both sides, we obtain a linear relationship between the square root of $F_X(x)$ and x:

$$\sqrt{F_X(x)} = x$$

Consequently, the square root of $F_X(x)$ plots as a straight line versus x, Fig. 4.4.4b. The simple change of the scale of the ordinate, also shown

---

† The reader is undoubtedly familiar with the use of semilog paper which provides a scale such that a relationship like $y = ab^{cx}$ will plot linearly and hence facilitate the comparison between experimental data and a proposed formula of such a form.

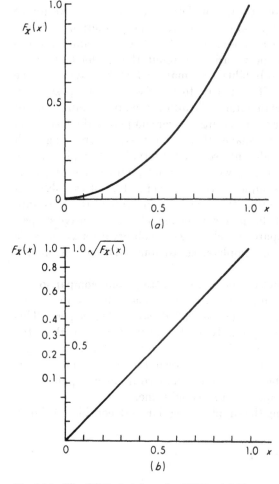

Fig. 4.4.4   The CDF of a triangular PDF.   (a) Numerical scale; (b) probability paper scale.

in that figure, permits reading $F_X(x)$ directly. This is probability paper for the triangular distribution $f_X(x) = 2x$, $0 \leq x \leq 1$.

If a set of observations had been generated by such a law, we would expect its cumulative frequency polygon to approximate the shape in Fig. 4.4.4a or b depending on the scale used to plot cumulative frequency. Using the second scale, we can compare the model and the data by the simpler task of judging the cumulative histogram with respect to a straight line.

In plotting the cumulative frequency polygon, we have learned in Chap. 1 to order the set of observations $x_1, x_2, \ldots, x_n$ in increasing value. Calling these ordered values $x^{(1)}$, $x^{(2)}$, . . . , $x^{(n)}$, we suggested plotting $x^{(i)}$ versus i/n. In terms of comparing predicted and "observed" or estimated probabilities, i/n represents an obvious estimate of $F_X(x^{(i)})$. On the other hand, a somewhat preferable estimator† or "plotting point" is $i/(n + 1)$, since it can be shown to be an unbiased estimator. It also avoids "losing" a data point, if, owing to an unlimited distribution, the probability paper scale cannot show probabilities of 0 or 1. Additionally, as will be seen, it is sometimes more convenient to plot $1 - F_X(x)$ rather than $F_X(x)$. In this case the original data, x_1, x_2, . . . , x_n, might be arranged in *decreasing* order. If plotted versus $i/(n + 1)$, the same estimate of $F_X(x)$ will result by either ordering technique; otherwise, it will not.

Suppose we had a set of nine observations: 0.10, 0.29, 0.30, 0.41, 0.59, 0.61, 0.74, 0.83, 0.90, suspected of coming from the triangular law. Using plotting points $i/(n + 1)$, we would plot 0.10 versus 1/10, 0.29 versus 2/10, 0.30 versus 3/10, etc. This is done in Fig. 4.4.5a on the appropriate probability paper. All five sets of data whose histograms appear in Fig. 4.4.3 are plotted on this probability paper in Fig. 4.4.5. Clearly, even though the data represent random samples from this tri-angular distribution, variation about the expected line is the rule rather than the exception. It is worthwhile to study the two figures to under-stand the manner in which a feature in a histogram is manifested on a particular probability paper. Histograms of samples a and e have essentially triangular shapes, one flatter and one steeper than the PDF of the model; their cumulative histograms appear as straight lines, flatter and steeper, than the true line. Sample c, more "bell-shaped" as a histogram, becomes S-shaped, while sample d, of "opposite" tri-angular shape as a histogram, shows a marked curvature on probability paper. Analogous kinds of behavior will be exhibited on other types of probability paper.

Exponential probability paper For other laws, probability papers differ only in the scale needed to "straighten out" the CDF of the model. Consider the exponential distribution. Here it is simpler to work with the complementary cumulative of the distribution.

$$G_T(t) = 1 - F_T(t) = e^{-\lambda t} \qquad (4.4.10)$$

Taking natural logarithms of both sides, we find

$$\ln G_T(t) = -\lambda t \qquad (4.4.11)$$

---

† Notice the unusual nature of this estimating situation. Instead of using a sample statistic, say $\bar{X}$, to estimate a specified parameter, say m, we are using a specified number [i/n or $i/(n + 1)$] to estimate $F_X(X^{(i)})$, where $X^{(i)}$, the ith largest observa-tion, is a sample statistic (called an order statistic).

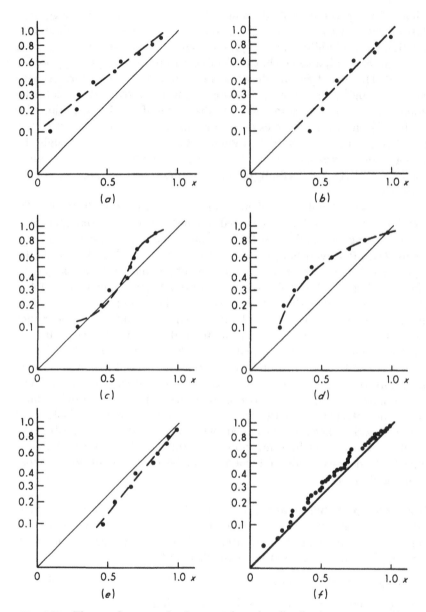

Fig. 4.4.5 Five random samples from a triangular distribution on true probability paper. (a) Sample a; (b) sample b; (c) sample c; (d) sample d; (e) sample e; (f) true CDF and pooled sample.

which is a linear relationship in t. The complementary CDF of the exponential distribution will plot as a straight line on semilog paper. $G(x)$ for $\lambda = 1$ is shown in Fig. 4.4.6a and b to a linear scale and on exponential probability paper, with a graphical indication of the construction of the latter. If the exponential model is valid, we should expect a graph of the complementary cumulative frequency polygon of the data also to plot approximately as a straight line on such paper.

In Fig. 4.4.7 we see the traffic-interval data plotted on ordinary three-cycle semilog paper. We have plotted an estimate of $G_T(t^{(i)}) = 1 - i/(n + 1)$ versus $t^{(i)}$, where $t^{(i)}$ is the ith in an increasing ordered list of the observed values. In this example, there were in most circumstances a number of values observed at each value of t. In an ordered list, for example, the 212th and 213th values were both 29.5 sec. The exponential model proposed seems to be confirmed by this data as the points appear to lie roughly upon a straight line passing through $G_T(0) = 1$.

The model based on the method-of-moments point estimate, $\lambda = 0.130 \text{ sec}^{-1}$, has a complementary CDF given by the straight line indicated in Fig. 4.4.7. It should be realized, however, that the verification of the exponential *family* of shapes was possible even *before* a specific value of λ was specified. A very nearly straight cumulative histogram suggests that *some* exponential distribution generated it. Different slopes imply different values of the parameter. In fact, one can first fit (by eye or by more elaborate methods, such as least squares) a straight line on the probability paper and then, by determining the slope, calculate an

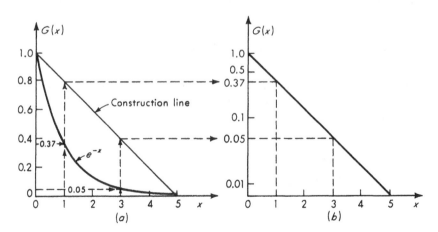

Fig. 4.4.6 Exponential distribution; complementary CDF. (a) Linear scale; (b) probability paper.

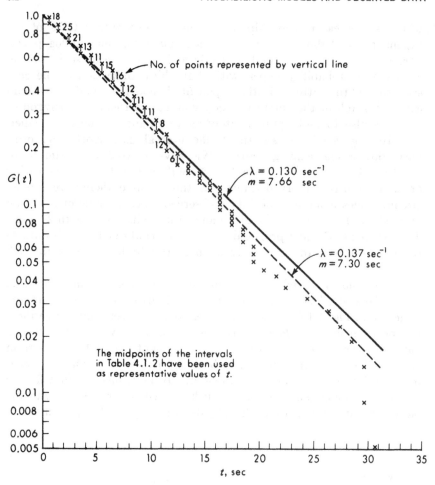

Fig. 4.4.7 Traffic-gap data on exponential probability paper.

estimate of λ. In this example, a straight-line fit by eye is shown dashed.
At $t = 20$ sec it passes through $G_T(20) = 0.064$. This implies that the
corresponding model has

$$\lambda = \frac{-\ln 0.064}{20}$$

$$\lambda = \frac{-(-2.75)}{20} = 0.137 \text{ sec}^{-1}$$

and mean

$$\frac{1}{\lambda} = 7.30 \text{ sec}$$

Thus the probability paper can be used conveniently both to judge the validity of the proposed type of distribution and to estimate the particular parameters that seem to fit the data best.   Unless a systematic method† is used to fit the line, however, it is not possible to determine the statistical properties of parameter estimates found in this way.

Normal probability paper   Normal probability paper and its graphical construction from the CDF are illustrated in Fig. 4.4.8.   So that the bar-yield data can be compared to a normal model (Fig. 4.4.9) a convenient horizontal scale is determined and the values of $i/(n + 1)$ are plotted versus x^i.   As with the exponential case, comparison of the data with a normal model can be made independently of specific parameter values by asking if the data points fall (*within reasonable sampling variation*) upon a straight line.   In this case a positive answer again seems justified.   Comparison with a specific normal model, say $N(8490,103^2)$, is possible once the straight line corresponding to these parameters is plotted (see Fig. 4.4.9).   As before, one can also fit by any reasonable

† See Gumbel [1958].

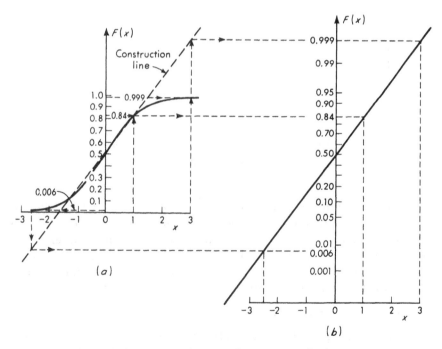

Fig. 4.4.8 Construction of normal probability paper.   (*a*) Linear scale; (*b*) probability paper.

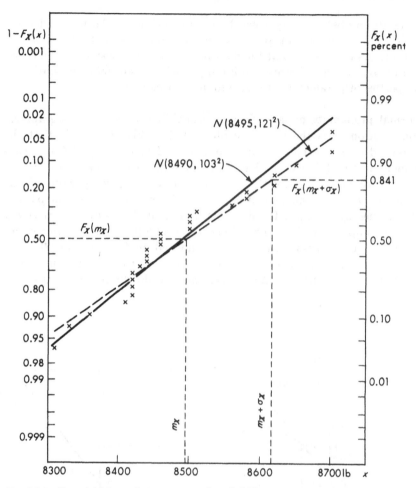

Fig. 4.4.9 Bar-yield-force data on normal probability paper.

method (by eye in the case of the dashed line shown) a straight line to
the data and use it to estimate the parameters. Given such a line, the
corresponding mean is found at the abscissa x of the line's intersection
with $F_X(x) = 0.50$, and the abscissa x of $F_X(x) = 0.841$ is approximately
the mean plus 1 standard deviation. The dashed line, then, corresponds to the normal distribution with $m = 8495$ lb and

$$\sigma = 8616 - 8495 = 121 \text{ lb}$$

In general, changing the mean shifts the straight line laterally, while
increasing the standard deviation flattens it. Thus both the location
and scale parameters of the distribution can be estimated by the paper.

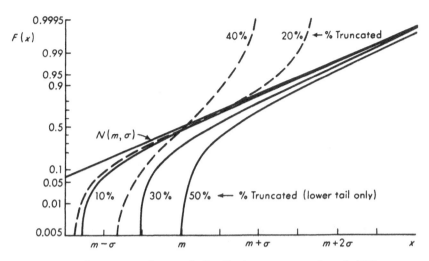

Fig. 4.4.10 Five truncated normal distributions on normal probability paper. (Adapted from Hald [1952].)

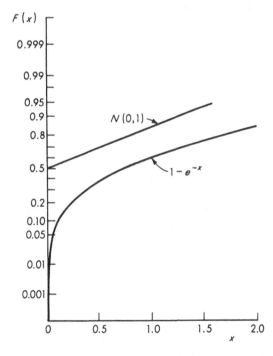

Fig. 4.4.11 The exponential distribution on normal probability paper.

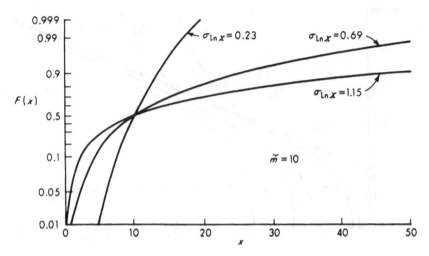

Fig. 4.4.12 Three log-normal distributions on normal probability paper. (Adapted from Hald [1952].)

Different models on the same paper It is important to be able to recognize how the CDF of one distribution appears on the probability paper of another. A number of these cases are shown in Figs. 4.4.10 to 4.4.15.

Notice (Figs. 4.4.10 and 4.4.11) that distributions with abruptly

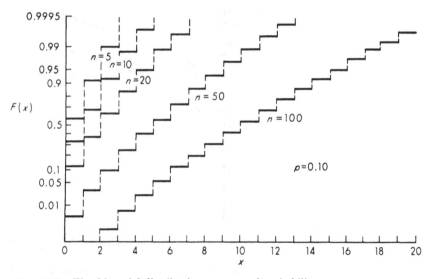

Fig. 4.4.13 Five binomial distributions on normal probability paper.

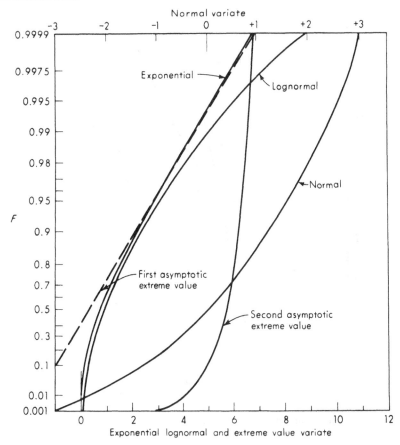

Fig. 4.4.14 First asymptotic extreme-value probability paper. (Adapted from Gumbel [1958].)

falling PDF's (e.g., exponential or truncated normal), lead to sharply falling (or rising) curves on normal paper (depending on which tail of the curve the fall occurs in). The effects of skew are also observable. Notice that a normal distribution fit in the region of the mean would overestimate $F(x)$ in the upper tails of positively skewed distributions (e.g., Fig. 4.4.12). The CDF of a discrete random variable shows jumps at certain values (Fig. 4.4.13).

Other probability papers have also been constructed and found useful. Figure 4.4.14 is extreme-value paper for the Type I asymptotic distribution. Lognormal probability paper is most easily obtained by simply transforming the horizontal, numerical scale of normal paper to a logarithmic one. Figure 4.4.15 shows several normal distributions plotted on lognormal paper of this kind. By a similar transformation

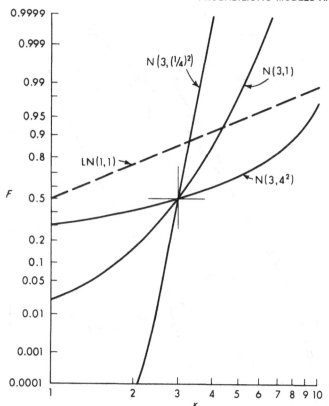

Fig. 4.4.15 Lognormal probability paper.

(Sec. 3.3.3) Type II and Type III asymptotic extreme-value papers can be constructed from Type I paper (if the parameters ω or ϵ of the Type III distribution are known). The papers do not provide estimates for these third parameters of three parameter distributions.

Discussion In order to interpret properly the qualification that the model is acceptable if the data points fall "within reasonable sampling variation" about a straight line, we should study the cumulative frequency polygon as a (complicated, vector-valued) sample statistic. Various approaches are possible,† but it is sufficient for our immediate purposes to state that, by analogy with the histogram, the conclusions of

---

† See, for example, Gumbel [1958] or Pieruschka [1963] on "control curves," the locus of confidence limits on the value of $F_X(x)$.

such a study would be:

1. The cumulative frequency polygon is a sample statistic, exhibiting random variation about the true CDF.
2. Perfect agreement between the observed cumulative frequency polygon and the model CDF may be the single most likely outcome, but it is a very unlikely event.
3. If another model is in fact controlling the observations, the observed cumulative frequency polygon may or may not seem to support (incorrectly) the engineer's hypothesized model. The likelihood that it will depends on how nearly similar in shape the true law is to the hypothesized one.

In Sec. 4.4.2 a simpler statistic will be introduced which will permit quantification of many of these conclusions. In addition, plots on probability paper in that section (Figs. 4.4.16 and 4.4.17) will demonstrate graphically another most important fact. When compared on probability paper, wide excursions of the cumulative frequency polygon are very likely to occur in regions of distributions where the paper's scale causes major distortions. These include the upper tail of an exponential distribution and both tails of a normal distribution. Such excursions appear in the data plotted in Figs. 4.4.7 and 4.4.9. The important implication is that what seem to be large discrepancies between the last few data points in these tails and the straight-line fit through points in the central portion of the data can seldom be interpreted as a sufficient reason to reject the proposed model. Such occurrences are simply not unlikely, even if the model holds.

4.4.2 "Goodness-of-fit" Significance Tests

In this section we shall combine some of those ideas in the previous section about the nature of observed shapes of data with the technique of the hypothesis test developed in Sec. 4.2. In this case the null hypothesis will be the model proposed by the engineer. Thus one might encounter such a hypothesis as

H_0: X is normal with $m = 100$ and $\sigma = 15$

or

H_0: Y is Poisson with parameter 2

or, lacking a specific parameter value,

H_0: Z is exponentially distributed

The alternate hypothesis, namely, that the variable is not as H_0 claims, is grossly composite. For example, if X is not $N(100,15^2)$ it might be $N(110,15^2)$ or $N(100,20^2)$ or $N(120,12^2)$ or it might not even be normally distributed at all, but rather lognormal or gamma or any of an infinite variety of named and unnamed continuous, discrete, or mixed distributions. While the computation of the type I error probability will turn out to be possible, it is clear that β, the type II error probability, is a function of not only other parameter values, but also other distribution types, and it is hardly amenable to complete description.

We shall consider two different hypothesis tests with similar forms, forms that differ only in the specific test statistic. We require some sample statistic D, which, roughly speaking, measures the deviation of the observed data shape from the model predicted shape. Such a statistic might, for example, be related to the differences between the histogram proportions N_i/n and the predicted fraction $1/k$, discussed in Sec. 4.4.1. The form of the operating rule is:

Accept H_0 if $D \leq c$
Accept H_1 (reject H_0) if $D > c$

If we know that under H_0, the statistic D has some particular distribution, we can calculate the value of c associated with any desired significance level α. In principle we might also be able to compute the value of β for any particular alternate model, but clearly there is no end to such alternate distributions and no meaningful way to summarize such results.

Having computed c, one would observe the underlying random variable n times, find the value of D, compare this with c, and report the conclusion to accept or reject the proposed model.

Observe that the purpose of the test is solely to answer the question of whether the data deviate a statistically *significant* amount from the model prediction.

χ^2 **Test**   The first such hypothesis test we shall consider is by far the more popular, having been proposed by an early statistician, Karl Pearson, in 1900. It is called the χ^2 *goodness-of-fit test*. The statistic used is related to the histogram's deviations from the predicted values. Suppose we have a model for a discrete random variable X. If we (randomly) sample that random variable n times, the number N_i of observations of the ith possible value x_i of X is a random variable with the binomial distribution, since the probability of observing the ith possible value in each test is constant, with value $p_i = P[X = x_i] = p_X(x_i)$. The expected value of N_i is np_i, and its variance is $np_i(1 - p_i)$ (see Sec. 3.1.2). If np_i is not too small, the standardized random deviation from the expected value

$$\frac{N_i - np_i}{\sqrt{np_i(1 - p_i)}} \tag{4.4.12}$$

is approximately $N(0,1)$, as shown in Sec. 3.3.1. It should not be unexpected then that D_1', the sum of the squares of these standardized deviations,

$$D_1' = \sum_{i=1}^{k} \frac{(N_i - np_i)^2}{np_i(1 - p_i)} \tag{4.4.13}$$

has approximately the χ^2 distribution (Sec. 3.4.3). The sum might be over the k possible values of the random variable X if each np_i is sufficiently large that the approximation is valid. If not, we shall resort to lumping, which will be demonstrated shortly.

Owing to the lack of independence among the N_i (notice that a linear functional relationship $N_k = n - \sum_{i=1}^{k-1} N_i$ exists), the number of degrees of freedom of the χ^2 variable is not k but $k - 1$ and each term is decreased by a factor $(1 - p_i)$. These corrections can be hueristically justified by noting that if the square of one deviation is larger than average owing to the chance occurrence of an "extra" observation at that value, then the square of some other deviation (that associated with the value which was "shorted") is also necessarily larger than average. More formal derivations are available in Hald [1952] and Cramer [1946]. Finally, then, the statistic

$$D_1 = \sum_{i=1}^{k} \frac{(N_i - np_i)^2}{np_i} \tag{4.4.14}$$

is approximately χ^2-distributed with $k - 1$ degrees of freedom. From tables of this distribution we can determine critical values $\chi^2_{\alpha, k-1}$ for any value of $k - 1$ such that

$$P[D_1 \geq \chi^2_{\alpha, k-1}] = \alpha \tag{4.4.15}$$

After observing a sample of the random variable X, we can compute the observed value of the D_1 statistic and compare it to the value $\chi^2_{\alpha, k-1}$.

Illustration: Discrete random variable with specified parameter   Consider, for example, the rainstorm data reported in Table 4.4.1. Under the null hypothesis that rainstorms occur at these (independent) stations with an average rate of occurrence of one per year and that the occurrences are Poisson arrivals

(Sec. 3.2), the distribution of the number of arrivals X each year at any station is Poisson-distributed with mean (and parameter) 1.

$$p_X(x) = \frac{(1)^x e^{-1}}{x!} = \frac{0.368}{x!} \qquad x = 0, 1, 2, \ldots$$

If the model holds, the expected numbers of observations of each value of x in a sample of size 360 are as shown in column 1 of Table 4.4.1. Subtracting each expected value from its observed value, squaring, and dividing by the expected number yields the normalized squared deviations (column 2, Table 4.4.1). Notice that the last two categories have been lumped together to give a single category with a larger expected number of observations. This is done to insure that the approximation of the statistic's distribution by the χ^2 distribution is adequate. A conservative value for the smallest expected number of observations permissible in any category is 5 to 10 (Cramer [1946] and Hald [1952]), but more recent experiments† have shown that tests based on expected values as low as two or even one give satisfactory results (Lindley [1965], part II, page 167; Fisz [1963], page 439).

The observed statistic is $d_1 = 16.68$. Under the null hypothesis, the distribution of D_1 is approximately χ^2 with 5 degrees of freedom, since there are, after lumping, $k = 6$ categories considered here. The probability of such a χ^2 random variable exceeding 16.68 is less than 0.005.‡ In particular, for a significance level of $\alpha = 5$ percent, the critical value is

$$\chi^2_{0.05,5} = 11.1$$

Since $d_1 = 16.68 \geq \chi^2_{0.05,5} = 11.1$, the hypothesis should be rejected at the 5 percent level. In words, the engineer should report that there is a significant difference between the shape of the data and the model, a Poisson distribution with parameter 1.0 storms per year.

χ^2 Test: estimated parameters

It is rarely the case that the engineer's experience is sufficient to suggest a hypothesis which includes a value of the distribution's parameter. Normally he hypothesizes the *form* of the distribution, but estimates the parameter or parameters from the data itself. Is it reasonable to use this same data to help judge the goodness of fit of the model? It is if we account for the fact that the data have also been used to estimate the model parameters. The more parameters a distribution has, the closer it can be fit to given data. We account for this effect by simply reducing the number of degrees of freedom of the distribution of the statistic by one for *each* parameter estimated from the data (Hald [1952], Cramer [1946]). This reduction of the degrees of freedom has the effect of *reducing* the value c which the statistic must be less than in order that the null hypothesis be accepted. In other

---

† How might such experiments be conducted? (See Lindley [1965], part II, page 167, Vessereau [1958], and Williams [1950].)

‡ This number, unlike α, is computable only after the data are observed, and is sometimes called the *p level* (Sec. 4.2). It is the highest significance level which would lead to acceptance of H_0.

Table 4.4.1 The annual number of rainstorms recorded at a number of stations

No. of rainstorms per station per year	Observed no. of occurrences	Model: $\lambda = 1$ specified		Model: $\lambda = \bar{x} = 1.18$ estimated	
		(1) Expected no. of occurrences	(2) Normalized squared deviations	(3) Expected no. of occurrences	(4) Normalized squared deviations
0	102	$360(0.368) = 132.48$	$\dfrac{(102 - 132.48)^2}{132.48} = 7.01$	110.52	0.66
1	144	$360(0.368) = 132.48$	$\dfrac{(144 - 132.48)^2}{132.48} = 1.00$	130.32	1.44
2	74	$360(0.184) = 66.24$	$\dfrac{(74 - 66.24)^2}{66.24} = 0.91$	76.68	0.09
3	28	$360(0.061) = 21.96$	$\dfrac{(28 - 21.96)^2}{21.96} = 1.66$	30.24	0.17
4	10	$360(0.015) = 5.40$	$\dfrac{(10 - 5.40)^2}{5.40} = 3.92$	9.00	0.11
5	2	$360(0.003) = 1.08 \Big\}\,1.44$	$\dfrac{(2 - 1.44)^2}{1.44} = 2.18$	$2.16 \Big\}\,3.24$	0.48
6	0	$360(0.001) = 0.36 \Big\}$		$1.08 \Big\}$	
$\bar{x} = 1.18$	$\overline{360}$	$\overline{360.00}$	$\overline{16.68}$ $\chi^2_{0.05,5} = 11.1$	$\overline{360.00}$	$\overline{2.95}$ $\chi^2_{0.05,4} = 9.49$

Source: E. L. Grant [1938].

words, because we have used the same data to partially fit the model to the data, the statistic must be smaller to be acceptable at the same significance level.   It can be said that in such cases we are testing the proposal that the distribution is of a particular type or shape, parameter values being *unspecified* (at least until after the observations).

Formally, the number of degrees of freedom of the statistic D_1 is

$$k - r - 1 \tag{4.4.16}$$

where k is the number of categories considered and r is the number of parameters estimated from the data.   Clearly if $r = k - 1$, there could be perfect agreement between observed and expected numbers.   This occurs, for example, if one attempts to apply a goodness-of-fit test to a sample of a zero or one Bernoulli ($k = 2$) random variable whose single ($r = 1$) parameter p has been estimated by the ratio of the number of successes to the number of trials.   Recall, by analogy, that we can always fit a polynomial of degree $k - 1$ through any k points.

Illustration: Discrete random variable, estimated parameter It is more likely, for example, that the engineer studying rainstorms would have hypothesized simply a Poisson model, only later adopting a parameter (mean) equal to the estimate provided by the observed sample mean, $\bar{x} = 1.18$ storms per year. With this parameter the expected number of observations are given in column 3 and the normalized squared deviations in column 4 of Table 4.4.1. The observed statistic is now $d_1 = 2.95$, suggesting a considerably better fit between model and data.   In this case we must compare, however, with a χ^2 random variable with $6 - 1 - 1 = 4$ degrees of freedom.   The critical value at the 5 percent level of significance is $\chi^2_{0.05,4} = 9.49$.   Since $2.95 \leq 9.49$, this hypothesized model is not to be rejected.   Given that the model holds, the probability (the p level) was about 60 percent that a value of D_1 as large as or larger than 2.95 would be observed.

χ^2 Test: Continuous random variables The χ^2 goodness-of-fit test can also be applied to continuous random variables.   It is only necessary to divide the region of definition of the variable into a finite number of intervals and compute (through integration over each of the intervals) the probability p_i of the random variable's being in each of the intervals. Such intervals are often chosen in much the same manner as those for histograms (Chap. 1.1).   In fact, as discussed in Sec. 4.4.1, what is being compared by the test is the degree of fit between an observed histogram and a density function lumped into a corresponding bar-form PDF or discrete mass function.

If we consider, as an example, the yield-force data and the intervals used in the histogram in Fig. 4.4.1, we are led to Table 4.4.2.   The expected number of occurrences in an interval is computed by multiplying the sample size times the probability of the random variable's

Table 4.4.2 Histogram intervals for bar-yield-force model and data

Interval	No. of observed values	Model: normal $(8490, 103^2)$ expected no. of observations		Normalized squared deviations
$-\infty$–8325	1	$27[F(8325) - F(-\infty)] = 27[Fv(-2.57) - 0] = 27(0.0548) =$	1.48	0.16
8325–8425	7	$27[F(8425) - F(8325)] = 27[Fv(-1.60) - Fv(-2.57)]$		0.13
		$= 27(.2265)$	$= 6.12$	
8425–8525	11	$\cdots\cdots\cdots\cdots\cdots\cdots\cdots\cdots\cdots\cdots\cdots\cdots\cdots\cdots\cdots\cdots\cdots\cdots\cdots\cdots = 9.50$		0.23
8525–8625	5	$= 7.34$		0.75
8625–∞	3	$= 2.56$		0.07
	27		27.00	1.34

taking a value in the interval; this probability is determined as the difference between the values of the cumulative distribution function at each end of the interval. The hypothesized model is normal with the parameters estimated from the data: $\hat{m} = 8490$; $\hat{\sigma} = 103$. Hence the number of degrees of freedom of the statistic D_1 is $5 - 2 - 1 = 2$. At the 5 percent significance level the critical value is $\chi^2_{0.05,2} = 5.99$, a value much greater than the observed value of 1.34. (The probability that a χ^2 random variable with 2 degrees of freedom will exceed 1.34 is about 50 percent.) This data are not in significant contradiction to the model.

It is desirable,[†] particularly when dealing with continuous random variables, to seek intervals with equal p_i (as in Sec. 4.2.1). Once a number of categories k has been decided upon, the interval division points can be determined from a table as those values of x where, for the hypothesized model, $F(x) = 0, 1/k, 2/k$, etc. Notice that this approach removes the arbitrary process of interval selection associated with histograms.[‡]

That the choice of intervals may be important can be observed in Table 4.4.3, where 10 intervals were used for the yield-force data, giving an expected number of observations of 2.7 in each category. This number is somewhat smaller than the commonly recommended smallest expected number, 5. The observed value of D_1 is $d_1 = 17.07$, and the appropriate number of degrees of freedom is $10 - 2 - 1 = 7$, yielding $\chi^2_{0.05,7} = 14.1$. Since $d_1 > \chi^2_{0.05,7}$ the test suggests that the model be rejected at the 5 percent level of significance, a decision counter to that arrived at previously. The hypothesis would, however, be accepted at the 1 percent level, where $\chi^2_{0.01,7} = 18.5$. The reader may wish to show that if a test based on five intervals each of probability 20 percent and expected number of observations 5.4 is adopted, the hypothesis would be

† See Mann and Wald [1942].
‡ It also simplifies computations, since the p_i are all equal. The statistic in Eq. (4.4.14) now becomes $(1/n)[k\Sigma(N_i^2) - n^2]$.

Table 4.4.3†   Uniform probability intervals for bar-yield-force model and data

Interval	Observed no.	np_i	$(N_i - np_i)^2$
$-\infty$–8357	2	2.7	$(0.7)^2$
8357–8403	1	2.7	$(1.7)^2$
8403–8437	6	2.7	$(3.3)^2$
8437–8464	6	2.7	$(3.3)^2$
8464–8490	0	2.7	$(2.7)^2$
8490–8516	4	2.7	$(1.3)^2$
8516–8543	0	2.7	$(2.7)^2$
8543–8577	1	2.7	$(1.7)^2$
8577–8622	4	2.7	$(1.3)^2$
8622–∞	3	2.7	$(0.3)^2$
	27	27.0	46.1

$$\dagger \frac{1}{np} \sum (N_i - np_i)^2 = \frac{46.1}{2.7} = 17.07$$

rejected at both of these levels of significance.   It is not clear how the engineer should react when two tests on the same data lead to contradictory conclusions.   What is important is to realize that the popular χ^2 test is open to such contradictions.   This limitation should be kept in mind when reviewing the reported results of χ^2 tests.

Kolmogorov–Smirnov test   A second quantitative goodness-of-fit test is based on a second test statistic.   It concentrates on the deviations between the hypothesized *cumulative* distribution function $F_X(x)$ and the observed *cumulative* histogram†

$$F^*(X^{(i)}) = \frac{i}{n} \tag{4.4.17}$$

in which $X^{(i)}$ is the ith largest observed value in the random sample of size n.   Consider the statistic

$$D_2 = \max_{i=1}^{n} \left[|F^*(X^{(i)}) - F_X(X^{(i)})| \right] = \max_{i=1}^{n} \left[\left| \frac{i}{n} - F_X(X^{(i)}) \right| \right] \tag{4.4.18}$$

In words, D_2 is the largest of the absolute values of the n differences between the hypothesized CDF and the observed cumulative histogram (sometimes called the *empirical CDF*) evaluated at the observed values

---

† The cumulative histogram differs little from the cumulative frequency polygon (Sec. 1.1).   The former is a stair-step function, while the latter has continuous straight-line segments connecting the "toes" of the steps.

in the sample. It has been found (Massey [1951], Birnbaum [1952], Fisz [1963]) that the distribution of this sample statistic is independent of the hypothesized distribution of X. Its only parameter is n, the sample size. Knowing this statistic's distribution, we can compute the critical value c and perform an hypotheses test at a prescribed significance level. Such a test is called the *Kolmogorov-Smirnov goodness-of-fit test*.

As before

H_0: X has a specified distribution

H_1:   the distribution of X is other than that specified

The form of the operating rule is

Accept H_0 if $D_2 \leq c$

The test statistic is now D_2, given in Eq. (4.4.18).

The determination of d_2, the observed value of the statistic, is facilitated if a graphical comparison of the cumulative histogram and the CDF of the hypothesized model is available. In the case of the bar-yield-force example the plot on probability paper, Fig. 4.4.9, can be searched quickly to find likely locations of the near maxima of $|i/n - F(x^{(i)})|$. [Notice that if probability paper plotting points i/n rather than $i/(n+1)$ had been used, the points on this figure would define exactly rather than approximately the cumulative histogram $F^*(x^{(i)})$, Eq. (4.4.17)]. See the footnote on page 468. By inspection, possible locations for the maximum include $x^{(12)} = 8440$, $x^{(15)} = 8460$, and $x^{(18)} = 8500$. For these values:

$x^{(i)}$	$F^*(x^{(i)}) = i/n$	$F_X(x^{(i)})$	$\lvert F^* - F_X\rvert$
$x^{(12)}$	$1\tfrac{2}{2}7 = 0.444$	$F_U\left(\dfrac{8440 - 8490}{103}\right) = 0.313$	0.131
$x^{(15)}$	$1\tfrac{5}{2}7 = 0.555$	$F_U\left(\dfrac{8460 - 8490}{103}\right) = 0.386$	0.169
$x^{(18)}$	$1\tfrac{8}{2}7 = 0.667$	$F_U\left(\dfrac{8500 - 8490}{103}\right) = 0.539$	0.128

Consequently, $d_2 = 0.169$. The critical value for $n = 27$ and $\alpha = 0.05$ is, from Table A.7, about 0.256. Since d_2 is less than the critical value, we conclude that the model should not be rejected. (Recall, incidentally, that two of three χ^2 tests led to the model's rejection.)

Discussion of Kolmogorov-Smirnov test   The Kolmogorov-Smirnov test
has an advantage over the χ^2 test in that it does not lump data and compare discrete categories, but rather compares all the data in an unaltered
form.   The value of the statistic D_2 is also usually computed more easily
than D_1, especially if the data has been plotted on probability paper.
Unlike the χ^2 test, the Kolmogorov-Smirnov test is an exact test for all
sample sizes.   On the other hand, the Kolmogorov-Smirnov goodness-
of-fit test is strictly valid only for continuous distributions and only when
the model is hypothesized wholly independently of the data.   Nevertheless, the test is often used[†] for discrete distribution tests, when it is conservative, and for the case in which parameters have been estimated from
the same data, when it is not known quantitatively how to adjust the
distribution in a manner analogous to subtracting degrees of freedom in
the χ^2 test.   If the parameters have been estimated from the data, it can
be said only that the critical value should be reduced in magnitude.

Graphical execution of Kolmogorov-Smirnov test   It is also possible to
carry out quantitative goodness-of-fit evaluations in a graphical manner.
It is worthwhile to consider this topic if for no other reason than its
ability to help us visualize the effects of statistical uncertainty on the
distribution as a whole.   This graphical technique can best be carried
out directly on probability paper as we shall do here.

Notice first that the Kolmogorov-Smirnov test just discussed can
be conducted by plotting curved lines above and below the hypothesized
CDF offset by an amount equal to the critical value determined from the
table of the distribution of the statistic D_2.   In the bar-yield-force example just considered this value was 0.256.   In Fig. 4.4.16, the hypothesized distribution $N(8490,103^2)$ is plotted, along with curves offset by an
amount 0.256.   That none of the observed plotting points falls outside
this line implies that the hypothesized model should not be rejected at
the 5 percent significance level.[‡]   Curves for the 20 percent significance
level (with critical value 0.20) are also shown in order to indicate the
effect of the level of significance.   At this level also the hypothesized
model should not be rejected.   The tighter curves indicate a larger likelihood of rejecting the model when it is correct, but there is less chance that
the model will not be rejected when another model holds (type II error).

The most striking feature of the curves representing the acceptance

---

† See, for example, Fisz [1963] page 447.
‡ These points were plotted at $i/(n + 1)$ to agree with the the procedure outlined
in Sec. 4.4.1.   The Kolmogorov-Smirnov test is designed to compare values i/n.
Strictly, each point should be moved up by a factor $(n + 1)/n$ before carrying out
this test.   Some of these moved points are shown circled in Fig. 4.4.16.   The conclusion is unaltered in this case.

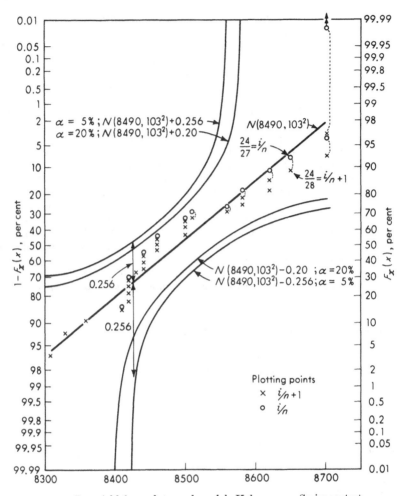

Fig. 4.4.16 Bar-yield-force data and model: Kolmogorov-Smirnov test.

region is their wide flare in the tails of the distribution. This is a graphical representation of the uncertainty surrounding tail areas of distributions, owing to the small likelihood associated with values of the random variable in these regions. Very large deviations of the larger and smaller values of the sample must occur before they can be considered "practically impossible" (i.e., significantly unlikely, <5 percent, say) to have occurred given this model.

It should be stated again that with this Kolmogorov-Smirnov test no correction is available to account for the fact that the hypothesized model parameters were estimated from the data itself. Such a correction would take the form of tighter curves about the hypothesized line.

Illustration: Traffic-gap data Let us conduct the quantitative goodness-of-fit tests on the traffic-gap data presented in Table 4.1.2 and plotted in Fig. 4.4.2. The hypothesized model (based recall, on the assumption that the traffic is Poisson) is that a gap time is an exponential random variable. The parameter λ has been estimated from the data to be 0.130 sec^{-1}.

To perform a χ^2 goodness-of-fit test we shall first consider using 20 categories of equal expected number of observations, $(1/20)(214) = 10.7$. The corresponding interval divisions can be determined sufficiently accurately by using the solid straight line in Fig. 4.4.7, which corresponds to the hypothesized model. These intervals are indicated in Table 4.4.4. Owing to the rounding off to the nearest second which occurred in the reporting of the data, several of the first intervals do not contain abscissas (even half-seconds) at which plotting took place. One might consider lumping the first 10 categories to form 5 each containing a plotting abscissa and each with expected number $(2/20)(214) = 21.4$, as shown in Table 4.4.4. Such lumping, first of the observed data and then of the categories for the χ^2 test, is undesirable in regions of large expected numbers, since it introduces artificial errors in the numbers actually observed in each interval used in the test. If the reported data have been lumped (owing, for instance, to lack of measuring instrument sensitivity or, as here, to ease reporting), it is preferable to pick goodness-of-fit test intervals to coincide with reporting intervals and to determine (nonequal)

Table 4.4.4 Goodness-of-fit test: equal-probability intervals for traffic-gap-data illustrations

Interval, sec	Observed no. of occurrences	Expected no. of occurrences
0.0–0.38 0.38–0.75	18	21.4
0.75–1.25 1.25–1.75	25	21.4
1.75–2.10 2.10–2.75	21	21.4
2.75–3.25 3.25–3.80	13	21.4
3.80–4.60 4.60–5.25	11	21.4
5.25–6.10	15	10.7
6.10–7.00	16	10.7
7.00–8.00	12	10.7
8.00–9.25	11	10.7
9.25–10.60	19	10.7
10.60–12.10	12	10.7
12.10–14.60	12	10.7
14.60–17.75	13	10.7
17.75–23.00	9	10.7
23.00–∞	6	10.7
	214	214.0

Table 4.4.5   Goodness-of-fit test for traffic-gap-data illustration

Interval	Observed no.	Expected no.	Normalized squared difference
0.0–1.0	18	26.1	2.39
1.0–2.0	25	23.1	0.16
2.0–3.0	21	20.0	0.05
3.0–4.0	13	18.0	1.38
4.0–5.0	11	15.4	1.26
5.0–6.0	15	13.6	0.14
6.0–7.0	16	11.8	1.49
7.0–8.0	12	10.5	0.21
8.0–9.0	11	9.2	9.35
9.0–10.0	11	8.1	1.04
10.0–12.0	20	13.2	3.50
12.0–13.0	6	5.6	0.03
13.0–16.0	9	12.6	1.03
16.0–19.0	13	8.6	2.25
19.0–24.0	6	8.6	0.79
24.0–∞	7	9.6	0.70
	214	214.0	16.77

expected numbers for each. Such 1-sec intervals and their corresponding expected numbers are shown in Table 4.4.5. When the expected number dropped below 10, eight intervals with expected values averaging approximately 10 were determined, still with boundaries coinciding with those used in reporting. The observed value of the test statistic is found to be 16.77. The statistic has $16 - 1 - 1 = 14$ degrees of freedom. The critical value at the 5 percent level of significance is 23.7 (Table A.2), implying that the model should not be rejected at this level.

The Kolmogorov-Smirnov test is carried out graphically in Fig. 4.4.17. The critical value at the 5 percent significance level is 0.093 ($= 1.36/\sqrt{214}$) (Table A.7). Lines of this value either side of the hypothesized model are drawn as shown. The model should not be rejected, since all points lie within the band. The modifications to the plotting points associated with using i/n in place of $i/(n + 1)$ clearly would not affect this decision.

Illustration: Highway-construction quality-control data   Existence of a physical model of the underlying mechanism may not be the reason that an engineer proposes a particular distribution for a variable. Computational convenience or other professional circumstances may suggest that a particular model be used if possible.

A first step in setting up statistical quality-control methods in the highway-construction field has been the determination of the distributions of the commonly measured and controlled variables, such as embankment compaction, asphalt content, depth of subbase, aggregate gradation indices, etc. Statistical

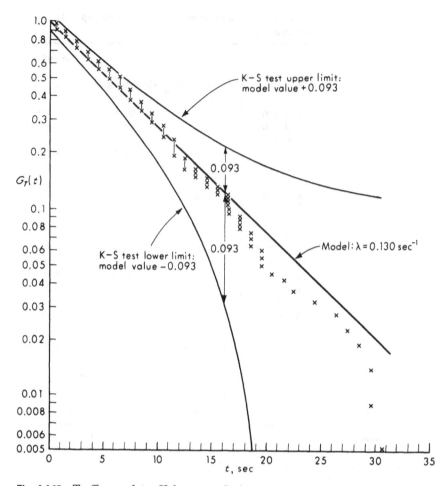

Fig. 4.4.17 Traffic-gap data: Kolmogorov-Smirnov test.

methods of quality control are best developed and simplest to use if the under-lying variable is normally distributed. Successful field implementation requires simple and uniform procedures.

Suppose that the normal distribution has been judged to be a satisfactory representation of the other variables of interest and, now, based on data, a decision must be made about whether "percent relative compaction" can also be treated as a normally distributed random variable when the new methods are introduced. To be forced to reject the hypothesis that this variable was normal would require setting up a special procedure for it (tables different from those used for the other variables, etc.), which would make the whole new program less palatable and possibly delay its acceptance by the industry. Consequently, the engineer seems justified in adopting a very small significance level, say $\alpha = 1$ percent, when conducting a goodness-of-fit test on this data.

The data are given in Table 4.4.6 and, as a histogram, in Fig. 4.4.18 (Beaton

Table 4.4.6 Kolmogorov-Smirnov test for illustration involving percent relative compaction data

Order i	Observed value $x^{(i)}$	$F^*(x^{(i)})$	$F_X(x^{(i)})$	$F^*(x^{(i)}) - F_X(x^{(i)})$
1, 2	80	0.006, 0.011	0.007	0.004
3, 4	82	0.017, 0.023	0.018	0.005
5–7	83	0.028–0.039	0.027	0.012
3–12	84	0.045–0.068	0.041	0.027
13–13	85	0.074–0.102	0.059	0.043
19–26	86	0.108–0.148	0.082	0.066
27–32	87	0.153–0.182	0.113	0.069
33–38	88	0.188–0.215	0.151	0.064
39–42	89	0.221–0.238	0.198	0.040
43–47	90	0.242–0.267	0.255	0.013
48–57	91	0.273–0.324	0.316	0.043
58–66	92	0.330–0.375	0.382	0.052
67–80	93	0.381–0.455	0.452	0.071
81–85	94	0.461–0.483	0.524	0.063
86–100	95	0.489–0.569	0.594	0.105 (maximum)
101–115	96	0.575–0.653	0.666	0.091
116–127	97	0.659–0.699	0.729	0.088
128–142	98	0.705–0.808	0.785	0.080
143–146	99	0.813–0.830	0.834	0.021
147–160	100	0.836–0.910	0.875	0.039
161–166	101	0.916–0.942	0.908	0.034
167–173	102	0.948–0.983	0.934	0.048
174–176	103	0.989–1.000	0.955	0.045

Source: Beaton [1967].

[1967]). Should the null hypothesis that this variable is normally distributed be accepted at the 1 percent significance level?

The parameter estimates are:

$\hat{m} = \bar{x} = 93.64$ percent

$\hat{\sigma} = s = 5.52$ percent

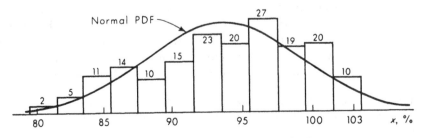

Fig. 4.4.18 Percent relative compaction of roadway embankment.

The values of $F^*(x^{(i)}) = i/n$ and $F_X(x^{(i)})$ are tabulated in the same table. The largest discrepancy is:

$$|F^*(95) - F_X(95)| = 0.105$$

For an α level of 1 percent and a sample size of 176, the critical value of the Kolmogorov-Smirnov statistic is $1.63/\sqrt{176} = 0.122$ (Table A.7). Therefore, the hypothesis need not be rejected at this significance level. The variable may be treated as if it were normally distributed.

It is important to realize that had the engineer, for other reasons, set up a null hypothesis that this variable was gamma-distributed or lognormally distributed or any one of many other choices, the conclusion might very well have been the same: that he could accept that hypothesis. It cannot be emphasized too strongly that the goodness-of-fit test is *not* designed to discriminate among two or more distributions or to help the engineer choose from among several contending distributions. Rather its purpose is to help answer the question, "Can I use the distribution toward which I have a certain predisposition or is the data in such apparent discord with this model that I must reject it?"

Discussion Two general comments on goodness-of-fit tests are in order before closing this section. The first is on the one-sided form of the rule, "Accept H_0 if $D \le c$." Recall from the previous section that perfect fits ($D = 0$) are very unlikely. Also, since many different observed histograms will lead to the same (nonzero) value of D, it is no longer even true that $D = 0$ is necessarily the *most likely* outcome. Consider the χ^2 statistic. It has a gammalike distribution (Sec. 3.4.3) with a mean equal to the number of degrees of freedom, $k - r - 1$. For $k - r - 1$ greater than 1, the mode, or most likely† value, of D is $k - r - 3$. For larger numbers of degrees of freedom at least, very small values of D are as unlikely as very large. There have been proposals, in fact, that two-sided goodness-of-fit tests be used, although it is not clear what alternate hypothesis is considered consistent with very small values of D. Certainly alternative distributions can only tend to yield large values of D since for these models $E[N_i - np_i]$ does not equal zero (p_i being based on the null hypothesis).

Secondly, goodness-of-fit hypothesis tests should be open to the same critical review given any hypothesis tests (Sec. 4.2). The selection of the significance level remains solely a matter of convention. The balancing of type I error versus type II error probabilities, which should be carried out in principle, becomes impossible in light of the form of the alternative hypothesis. As mentioned, it contains not only the same model with other parameter values, but also an infinite variety of other forms of

---

† We will make further use of this observation in Sec. 4.5 where we use this statistic in a maximum-likelihood manner to help choose among several contending distributions.

distributions.   In addition the goodness-of-fit tests have certain limitations of their own, as discussed above (dependence on interval choice,
for example).   Nonetheless, if the reader of a report which includes a
goodness-of-fit test to substantiate the model choice uses proper judgment in its interpretation, there is value and convenience in the conventional application of these tests.

4.4.3 Summary

A proposed model can be verified only by comparing its predictions with
observed data.   In the case of probability distribution models, the comparison is usually between the shape of the distribution (PDF or CDF)
and the shape of the data (histogram or cumulative histogram).

The usual comparison is made difficult by the fact that the data
"shape" is a sample statistic subject to random fluctuations.   In the case
of histograms the comparison is further complicated by the dependence
of the variability of this statistic on the number of intervals used.   This
number also dictates the degree of resolution or lack of resolution ("smearing") associated with the bar-form PDF vis-à-vis the complete PDF.

Analytical hypothesis tests are also available to compare data and
model.   The χ^2 goodness-of-fit test is based on a statistic

$$D_1 = \sum_{i=1}^{k} \frac{(N_i - np_i)^2}{np_i}$$

which measures the discrepancy between histogram and bar-form PDF.
The Kolmogorov-Smirnov test uses the deviations between the observed
cumulative histogram and the CDF.

4.5 EMPIRICAL SELECTION OF MODELS

In the previous section we dealt with the situation in which a particular
distribution had been suggested and the validity of that assumption was
in question.   In other situations the problem may be one of selecting a
distribution from among a number of contending mathematical forms
when no single one is preferred on the basis of the physical characteristics
of the phenomena.   The situation may also arise when any convenient,
"reasonable" distribution is adequate for the purposes at hand.   The
engineer may simply be looking for a tractable, smooth mathematical
function to summarize the available data because subsequent dealings
with the variable will be simplified by this representation.

The treatment of such selection problems will be considered briefly
in this section.   These selections are often strongly influenced by professional considerations, so generalizations are difficult.   Statistical theory

offers little quantitative help. Our format here, therefore, will be to present illustrative examples and to raise our major points within the context of these problems. A brief summary of these points will then suffice for guidance in other situations.

4.5.1 Model Selection: Illustration I, Loading Times

In heavy construction operations, variations in the time required for various steps, for example, in the loading, the hauling, the dumping, and the return of individual dump trucks, affect the interaction between pieces of equipment. For example, owing to fluctuations in truck-arrival times, a power shovel loading the trucks may have to wait for the next truck to return, or several trucks may have to wait in a queue before being filled. An engineer is developing a large computer program to simulate such operations. It will be used by project engineers to plan better future construction operations. In particular, for example, it could be used to predict the best number of trucks to use with a shovel in order to minimize the expected losses due to waiting equipment.

As just one part of this program the engineer must decide what distribution to use for the time required to load a truck. Of particular concern to him is the *type* of distribution to be used. The specific values of the distribution's parameters for a particular job will be input by future users of the computer program.

Although he recognizes that the total time of each loading operation is made up of the sum of the times required for each of a (small) number of buckets of material, there is no strong prior physical justification for a particular mathematical model. On the other hand, facility in model description and manipulation favors one of the common distributions. The engineer intends to choose from among several contenders on the basis of a small amount of data available to him from a previous job.

The data, a total of 18 values, are presented in Table 4.5.1 and in Fig. 4.5.1. The predominant characteristics are the shift away from the origin and the apparent skew toward higher values. The sample mean, standard deviation, and skewness coefficient are (Table 4.5.1):

$$\bar{x} = 1.93 \text{ min}$$

$$s = 0.24 \text{ min}$$

$$g_1 = 0.97$$

Uniform model The simplest probability density function to use is the uniform distribution

$$f_T(t) = \frac{1}{b-a} \qquad a \leq t \leq b \qquad (4.5.1)$$

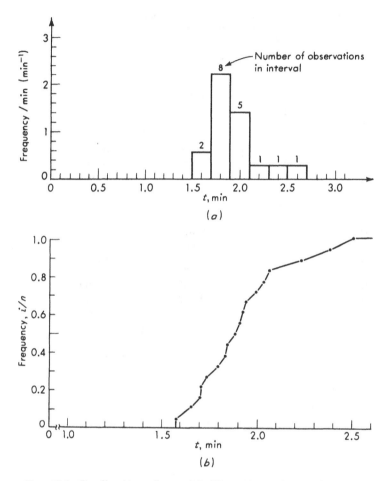

Fig. 4.5.1 Loading-time data. (*a*) Histogram; (*b*) cumulative frequency polygon.

To compare this distribution with data, one must obtain estimates of the parameters a and b. These could be based on the method of moments (Sec. 4.1.1) or, more simply in this case, on the maximum-likelihood estimators (Prob. 4.20)

$$\hat{b} = \max [t_i] = 2.52$$
$$\hat{a} = \min [t_i] = 1.58$$

This density function and the corresponding CDF are plotted in Fig. 4.5.2 along with the observed histogram and cumulative frequency polygon. This simple distribution does not appear to be a reasonable repre-

Table 4.5.1   Loading-time data and sample moments

i	Ordered observations $t^{(i)}$	For use with frequency polygons i/n	For use with probability papers $i/(n+1)$
1	1.58	0.055	0.053
2	1.67	0.111	0.105
3	1.71	0.167	0.158
4	1.72	0.222	0.210
5	1.74	0.278	0.263
6	1.80	0.333	0.316
7	1.84	0.388	0.369
8	1.85	0.445	0.421
9	1.88	0.500	0.474
10	1.92	0.556	0.527
11	1.93	0.611	0.580
12	1.94	0.667	0.632
13	2.00	0.723	0.685
14	2.03	0.778	0.738
15	2.06	0.833	0.790
16	2.23	0.889	0.842
17	2.38	0.945	0.895
18	2.52	1.000	0.941
	$\Sigma 34.80$		

$$\bar{t} = \frac{34.80}{18} = 1.93$$

$$s^2 = \tfrac{1}{18}\Sigma(t_i - 1.93)^2 = 0.0568; \quad s = 0.24$$

$$g_1 = \tfrac{1}{18}\Sigma(t_i - 1.93)^3/s^3 = 0.2352/18(0.0135) = 0.97$$

sentation of the phenomenon. It does not reproduce either the peak around 1.8 min or the relatively infrequent nature of the longer times.

Notice that for the uniform distribution, simple arithmetic graph paper is the appropriate probability paper. A straight line, marked "alternate CDF" in Fig. 4.5.2b, can be fit by eye on this probability paper to represent very well all but the relatively large observations, which it predicts to be impossible. If these tail values were unimportant, the uniform model might be justified solely on the basis of its simplicity. These occasional longer loading times may, however, have a significant influence on the performance of the operation. The presence or absence of these larger times may influence an engineer's equipment-allocation decision, and consequently some reasonable attempt should be made to reflect them in the model chosen.

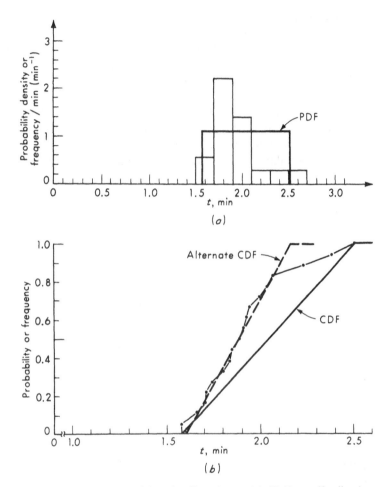

Fig. 4.5.2 Uniform model of loading time. (*a*) Uniform distribution, PDF; (*b*) Uniform distribution, CDF.

Normal model A gaussian distribution can also be considered. It is a very tractable model analytically and a very convenient distribution for hand calculations or computer simulations. Parameter estimates are:

$$\hat{m} = \bar{t} = 1.93$$

$$\hat{\sigma} = s = 0.24$$

The corresponding density function and CDF are shown in Figs. 4.5.3 and 4.5.4. The peak in the data is reproduced by the normal PDF,

although it is shifted to the right somewhat. The values away from the
mean are represented perhaps less well. Once again, on probability paper
(Fig. 4.5.4, alternate model), it is clear that if the larger values could be
ignored, a very good representation could be produced by a normal model
with a smaller mean and variance. In other words, except for these
larger observations, either a uniform distribution or a normal distribution
would appear to represent the data very well. This is true despite the
marked differences in the shapes of these two distributions. When the
amount of data available is small, it may seem to be fit very well by
several, nonsimilar distributions. This is an inherent difficulty in empiri-

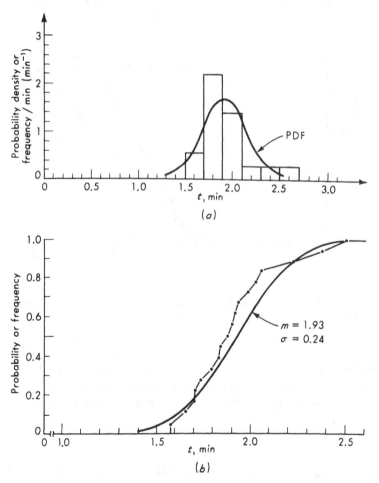

Fig. 4.5.3 Normal model of loading time. (*a*) Normal distribution,
PDF; (*b*) normal distribution, CDF.

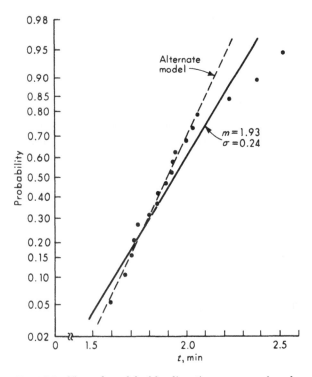

Fig. 4.5.4 Normal model of loading time on normal probability paper.

cal mathematical modeling (whether the models are probabilistic or not). Small samples simply do not contain sufficient information to permit fine discrimination among models.

The intended use of this model again dictates that a more determined attempt be made to represent the skew of data. Several convenient distributions capable of representing positive skew will be considered: the gamma distribution, the lognormal distribution, and the beta distribution.

Gamma model The gamma distribution, Sec. 3.2.3, has positive skewness and might be considered as a possible way to represent the data here. The PDF is [Eq. (3.2.18)]:

$$f_T(t) = \frac{\lambda(\lambda t)^{k-1}e^{-\lambda t}}{\Gamma(k)} \qquad t \geq 0 \tag{4.5.2}$$

with

$$m_T = \frac{k}{\lambda} \tag{4.5.3}$$

$$\sigma_T = \frac{\sqrt{k}}{\lambda} \tag{4.5.4}$$

and

$$\gamma_1 = \frac{2}{\sqrt{k}} \tag{4.5.5}$$

If the method of moments is used to estimate parameters k and λ, we must solve two equations for the estimates $\hat{k}$ and $\hat{\lambda}$:

$$\frac{\hat{k}}{\hat{\lambda}} = \bar{t} = 1.93$$

$$\frac{\sqrt{\hat{k}}}{\hat{\lambda}} = s = 0.24$$

The solution is

$$\hat{k} = 67.8 \qquad \text{and} \qquad \hat{\lambda} = 35.0$$

The implication is that the skewness is

$$\hat{\gamma}_1 = \frac{2}{\sqrt{\hat{k}}} = \frac{2}{\sqrt{67.8}} = 0.24$$

which is significantly less than the observed sample value of 0.97. In fact, with k so large, this gamma distribution is not very different from a normal distribution (Sec. 3.3.1); hence the same weaknesses will be encountered in the use of the gamma distribution as found with the normal distribution, but without the relative convenience of the latter.

Shifted gamma model The major characteristics of the observed shape can be reproduced, however, if a *shifted* gamma distribution is used. Such a distribution will have the following form (Sec. 3.5.1):

$$f_T(t) = \frac{\lambda[\lambda(t-a)]^{k-1}e^{-\lambda(t-a)}}{\Gamma(k)} \qquad t \geq a \tag{4.5.6}$$

in which a is an added, shifting parameter. Since $T - a$ has the same distribution as T in Eq. (4.5.2), it is clear that the mean will become

$$m_T = a + \frac{k}{\lambda} \tag{4.5.7}$$

while the variance, standard deviation, and skewness coefficient, all being related to central moments, will remain unchanged in their relationship to the parameters.

Again using the method of moments,† the parameter estimates become the solution of the three equations

$$a + \frac{\hat{k}}{\hat{\lambda}} = \bar{t} = 1.93$$

$$\frac{\sqrt{\hat{k}}}{\hat{\lambda}} = s = 0.24$$

$$\frac{2}{\sqrt{\hat{k}}} = g_1 = 0.97$$

The solutions are

$$\hat{k} = \frac{4}{(0.97)}\,2 = 4.25$$

$$\hat{\lambda} = \frac{\sqrt{4.25}}{0.24} = 8.60 \text{ min}^{-1}$$

$$\hat{a} = 1.93 - \frac{4.25}{8.60} = 1.43 \text{ min}$$

This shifted gamma distribution is plotted in Fig. 4.5.5.‡ It is clear that distribution is capable of reproducing all the pertinent properties observed in the data. This ability, combined with mathematical tractability and widely available tables and computer routines to evaluate gamma functions, make the shifted gamma distribution a strong contender to represent loading time, at least with the data available.

Shifted lognormal model   A second distribution that can successfully be fitted to data of this shape is a shifted lognormal distribution:

$$f_T(t) = \frac{1}{\sqrt{2\pi}\,\sigma_{\ln Y}(t - a)} \exp\left\{-\frac{1}{2}\left[\frac{\ln(t - a) - m_{\ln Y}}{\sigma_{\ln Y}}\right]^2\right\} \qquad t \geq a$$

$$(4.5.8)$$

in which $\sigma_{\ln Y}$ is the standard deviation of the natural logarithm of $Y = T - a$ and $m_{\ln Y}$ is the mean of $\ln Y$.

Method-of-moments estimation† of the parameters is somewhat

---

† Maximum-likelihood estimators (Sec. 4.1.4) can be found, but they require an iterative simultaneous solution of three coupled, nonlinear equations (Markovic [1965]).
‡ It is not possible to construct a general probability paper for gamma distributions.

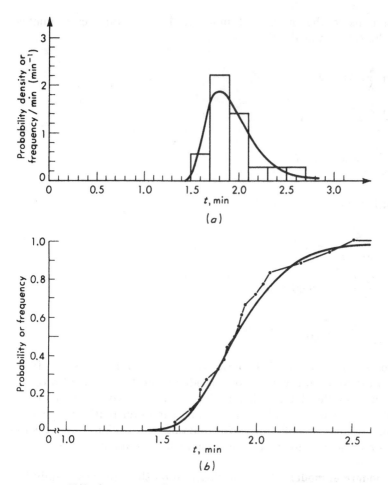

Fig. 4.5.5 Shifted gamma model of loading time. (*a*) Shifted gamma distribution, PDF; (*b*) shifted gamma distribution, CDF.

more awkward, but it is straightforward. Since the standard deviation and skewness coefficient of Y equal those of T, which are in turn estimated by the corresponding sample moments, we can solve Eq. (3.3.38) for the mean of Y. That equation is

$$\gamma_1 = 3 \frac{\sigma_Y}{m_Y} + \left(\frac{\sigma_Y}{m_Y}\right)^3 \tag{4.5.9}$$

implying

$$m_Y{}^3 - \frac{\sigma_Y{}^3}{\gamma_1} - \frac{3\sigma_Y m_Y{}^2}{\gamma_1} = 0 \tag{4.5.10}$$

With estimates of σ_Y and γ_1 this cubic equation can be easily solved by trial and error for an estimate of m_Y. Then the coefficients $\sigma_{\ln Y}$ and $m_{\ln Y}$ can be estimated as in Sec. 3.3.2 [Eqs. (3.3.35) and (3.3.36)], and the shift parameter is simply

$$a = m_T - m_Y \qquad\qquad (4.5.11)$$

The results are

$$\hat{m}_{\ln Y} = -0.338$$

$$\hat{\sigma}_{\ln Y} = 0.316$$

$$\hat{a} = 1.18 \text{ min}$$

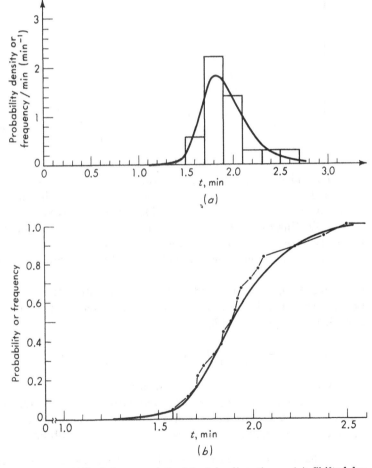

Fig. 4.5.6 Shifted lognormal model of loading time. (*a*) Shifted lognormal distribution, PDF; (*b*) shifted lognormal distribution, CDF.

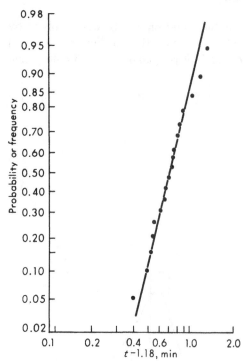

Fig. 4.5.7 Shifted lognormal model of loading-time on probability paper.

The shifted lognormal distribution with these parameters is plotted in Figs. 4.5.6 and 4.5.7.† Everything said above for the shifted gamma distribution also holds for this shifted lognormal distribution. It is capable of reflecting the observed shift and skew, and it is a reasonably tractable distribution. Without further information it must be considered equally as good an empirical model as the shifted gamma distribution.

Beta model As a final consideration, a beta distribution (Sec. 3.4.2) will be fit. In general this distribution has four parameters [Eq. (3.4.17)], here denoted p, r, a, and b.

$$f_T(t) = \frac{1}{(b-a)^{p-1}B}(t-a)^{r-1}(b-t)^{p-r-1} \qquad a \leq t \leq b \qquad (4.5.12)$$

With the wide variety of shapes possible (Fig. 3.4.1) and with a total of four parameters (one more than either of the previous two distributions) free to be used in the fitting process, the beta distribution can be made to reproduce very closely virtually any observed histogram.

---

† Notice that the plot on lognormal probability paper is possible only for a fixed value of the shift parameter; in other words, $Y = T - 1.18$ is actually plotted on standard lognormal paper.

If the method of moments is used for estimation, it is clear that at least the first four moments of the data would be matched identically. In practice, however, other factors may influence the values of the parameters used. The four equations which must be solved to obtain the estimates can only be solved by arduous trial and error in several variables simultaneously.† Tables are not complete for a wide range of the parameters. The estimates of the limits a and b may turn out to be such that an observed value lies outside of them.‡ This failure in the model may or may not be important, depending on its intended use.

In this specific case it would be desirable to use a value of r greater than unity in order to have the density function zero at the lower limit (Sec. 3.4.2). In fact, the reader can verify that a beta distribution with $a = 1.58$, $b = 3.44$, $r = 1.5$, and $p = 8.0$ has mean and standard deviation equal to the observed values, a skewness coefficient of 0.96 (which is very close to the observed value of 0.97), and a shape very much like the shifted gamma distribution. Lack of convenient tables for noninteger r suggests, however, the use of $r = 1.0$ and $p = 5.0$, which yield a skewness coefficient [Eq. (3.4.14)] of 1.05. Solving Eqs. (3.4.18) and (3.4.19) for a and b estimates gives $\hat{a} = 1.63$ and $\hat{b} = 3.13$.

This beta distribution is plotted in Fig. 4.5.8. The cumulative seems to fit well, but one might object to the fact that the PDF does not start from a zero value at the lower limit. In addition, the estimated value of a is larger than the smallest observed value, 1.58. These facts, combined with the rather unsatisfactory parameter-estimation procedure, lead to the judgment that the beta distribution may not be as useful a model for the simulation program as either the shifted gamma or the shifted lognormal.

In the light of the available data there seems little evidence to dictate the choice of the shifted gamma over the shifted lognormal or vice versa. The final choice should be made on the basis of ease of use in the intended program. Consequently no general recommendations can be made here.

Closeness-of-fit statistics   If the engineer needs for some reason to make a final selection between the remaining contending models in an "objective" way, influenced only by the data, a closeness-of-fit statistic such as the χ^2 statistic (Sec. 4.4.2) can be used in a somewhat heuristic manner to accomplish this choice. It should be emphasized once again that good-

---

† The same is true of the maximum-likelihood estimates.

‡ The same event can also occur, of course, with the method-of-moments estimates of the shift parameter in the shifted gamma or lognormal distributions. Maximum-likelihood estimates can, however, be constrained to avoid this problem, for example, $\hat{a} \leq \min [X_i]$ and $\hat{b} \geq \max [X_i]$, although this may complicate their evaluation.

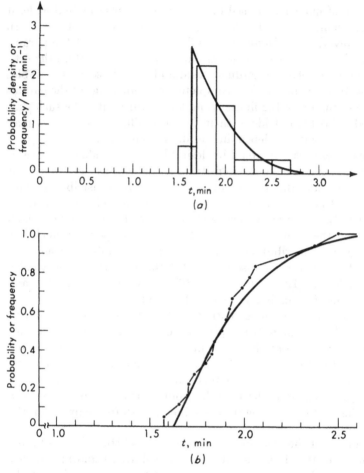

Fig. 4.5.8 A beta model of loading time. (a) Beta distribution, PDF;
(b) beta distribution, CDF.

ness-of-fit hypothesis tests (Sec. 4.4.2) are *not* designed to choose among
contending models but rather to suggest that a given proposed model
should or should not be retained.

    To use closeness-of-fit statistics to aid in the model-selection pro-
cess we adopt an attitude analogous to that used in maximum-likelihood
parameter estimation: We shall select that model for which the likelihood
of the observed value of the corresponding closeness-of-fit statistic is the
largest. (We do *not*, notice, select the model with the smallest value of
the statistic.)

Consider the use of the χ^2 statistic for choosing between the shifted gamma and shifted lognormal models. In this illustration we chose to lump the data into six intervals (of equal likelihood if the assumed model holds†). The expected and observed numbers of observations are listed in Table 4.5.2. The observed numbers can be read directly from cumulative plots (Fig. 4.5.9).

In terms of this small number of intervals the shifted gamma distribution happened to yield a "perfect fit." The statistic is zero. This may not, however, be as likely an outcome as the value $\frac{2}{3}$ observed with the shifted lognormal model. To determine the more likely of the two observed values of the statistics, however, their distributions must be known. In this example, both models contain three data-estimated parameters. In general, of course, the contending models may have different numbers of data-estimated parameters.‡ In both cases, then, the statistics have a χ^2 distribution with $\nu = 6 - 1 - 3 = 2$ degrees of freedom (Sec. 4.4.2). The expected value of a χ^2 random variable is ν or 2 in this case. The lognormal value of $\frac{2}{3}$ is in fact closer to the expected value than is the gamma value of 0. But a maximum-likelihood approach is preferred here. The χ^2 distribution with $\nu = 2$ is sketched in Fig. 3.4.3 and happens to coincide with an exponential distribution with parameter $\frac{1}{2}$. Clearly with a distribution of this shape, smaller values

† This particular division is unnecessary, of course, but it avoids arbitrary divisions and is easier to apply. The expected number of observations per interval, three, is undesirably small (Sec. 4.4.2).
‡ In order to compare likelihoods directly, one should choose the number of histogram intervals such that all statistics have the same number of degrees of freedom.

Table 4.5.2 χ^2 statistic evaluation

Interval i	Expected no. np_i	Shifted gamma model		Shifted lognormal model	
		Obs. no. N_i	$(N_i - np_i)^2$	Obs. no. N_i	$(N_i - np_i)^2$
1	3	3	0	2	1
2	3	3	0	4	1
3	3	3	0	3	0
4	3	3	0	3	0
5	3	3	0	3	0
6	3	3	0	3	0
		18	0	18	2
$\sum \dfrac{(N_i - np_i)^2}{np_i} =$			0		$\frac{2}{3}$

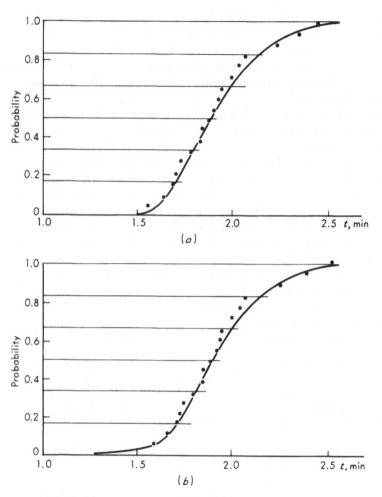

Fig. 4.5.9 Evaluation of χ^2 statistic, loading time. (a) Shifted gamma; (b) shifted lognormal.

are more likely than larger ones.† The observed value of the shifted gamma's statistic, 0, is more likely than that of the shifted lognormal's statistic, $\frac{2}{3}$. The former distribution is therefore to be preferred on this basis.

† It should be pointed out that the χ^2 statistic is often used incorrectly in model comparisons, the model with the smallest observed statistic being selected no matter how many degrees of freedom are involved. For ν greater than 2, zero is *not* the most likely value of the statistic. In this case the PDF of the χ^2 distribution (Sec. 3.4.3) must be evaluated to determine which model has yielded the most likely value of the statistic and inversely, therefore, which is the "most likely" model.

Discussion It is important to distinguish between the more familiar curve fitting associated with deterministic functional relationships between variables and the process here of selecting a probabilistic model on the basis of a comparison between mathematical functions (PDF's and CDF's) and observed values or shapes. In the latter case the closest fit cannot be said to be the best fit. To a point, as in the earlier part of this illustration, closeness of fit of a shape of distribution to the observed shape of data can be used to select from among models.† This is reasonable for the gross characteristics (skewness, shift, etc.) which are more reliably represented by the data. At a finer scale, however, between distributions which reproduce the grosser characteristics equally well, the expected variation must be recognized and the closeness-of-fit criterion dropped. In its place either one can state that the data are insufficient to discriminate when professional factors (convenience, etc.) will govern, or he may elect to choose on the basis of a maximum-likelihood criterion using a convenient statistic such as the χ^2 statistic.

Without further evidence, no extrapolations should be made or physical significance placed on the distribution chosen to represent the observed data. In particular, in this illustration, the estimated lower limits of the shifted distributions should not be interpreted as values which will always be exceeded. One should place no great confidence in the predictions based on the tails of empirically adopted models. In the same sense that we expect the observed sample mean and sample standard deviation to be close but not perfect estimates of the variable's moments, so the shape in the central portion and other gross characteristics of the fitted model, including its mean and standard deviation (which, of course, are usually chosen to equal the corresponding sample values), are probably close but not perfect representations of the same characteristics of the physical variable. As with the parameters, the larger the sample upon which it is based the more confidence one can have that these characteristics are close to the true ones.

If for some reason loading time proves to be a critical component in the total simulation of construction operations, the engineer would undoubtedly review his choice of distribution. More data from the same or other jobs would help, although combining the data from situations where the parameters are different is not a simple task. More data must be expected to alter the estimated moments, including the coefficients of variation and skewness; further, the additional data may even indicate that some of the gross shape characteristics of the adopted model are incorrect. We cannot stress frequently enough the variation inherent in

---

† Presumably it is just a fast way of excluding models whose χ^2 statistics would be large and unlikely.

samples of small size and hence the small confidence one can have in conclusions based on an observation of such a sample.

On the other hand, the engineer may prefer to study the loading operation more carefully in order to justify the adoption of a particular distribution as a reasonable model of the physical mechanism generating the variation. He might be guided in this investigation by knowledge of the distributions which he found to fit the data well and the mechanisms known to generate them (Chap. 3).

For example, in this case the fact that a gamma variable is the sum of other independent gamma variables (including the exponential variable) might suggest that the engineer investigate the validity of a model which recognizes that the total time is the sum of the times required to load a (small) number of power-shovel buckets into a truck. Each bucket load requires a fixed time plus a random time. If, for example, the number of buckets required was k, if the fixed time was a/k, and if the random times were exponential with common parameter λ, then the total loading time would have a shifted gamma distribution with parameters a, k, and λ, where k is integer. This physical model could be verified by collecting appropriate data and testing goodness-of-fit hypotheses (Sec. 4.4.2) on the shifted gamma distribution and on the individual exponential distributions. Questions as to the validity of the assumptions of independence of these latter times, of a fixed number of buckets per load, etc., could be raised, but more elaborate models† without these simplifying assumptions could always be constructed (with correspondingly more complicated distributions for T). The point is simply that the distribution suggested empirically by the data may be a starting point for a more detailed, more physically based investigation, if necessary.

4.5.2 Model Selection: Illustration II, Maximum Annual Flows

In this illustration it is desired to select a model to describe the observations of annual maximum flood flows on the Feather River at Oroville, California. The data for 1902 to 1960 are presented in Table 4.5.3. A histogram of this data appears in Fig. 4.5.10. The model will be used for economic studies in the design of small flood-control projects (dikes, channels, etc.) in the area.

Physical models Some engineers have given arguments that such floods, as results of a chain of influencing factors, should be governed by the lognormal distribution (Sec. 3.3.2); others have argued that, as the maximum of daily or weekly flows, such flows should be governed by an extreme-value distribution (Sec. 3.3.3). There are reasons for question-

---

† For example, a compound model (Sec. 3.5.3) in which the number of buckets k was treated as a discrete random variable.

Table 4.5.3 Maximum annual flows,
Feather River, Oroville, California

Year	Flood, cfs	Year	Flood, cfs
1907	230,000	1935	58,600
1956	203,000	1926	55,700
1928	185,000	1954	54,800
1938	185,000	1946	54,400
1940	152,000	1950	46,400
1909	140,000	1947	45,600
1960	135,000	1916	42,400
1906	128,000	1924	42,400
1914	122,000	1902	42,000
1904	118,000	1948	36,700
1953	113,000	1922	36,400
1942	110,000	1959	34,500
1943	108,000	1910	31,000
1958	102,000	1918	28,200
1903	102,000	1944	24,900
1927	94,000	1920	23,400
1951	92,100	1932	22,600
1936	85,400	1923	22,400
1941	84,200	1934	20,300
1957	83,100	1937	19,200
1915	81,400	1913	16,800
1905	81,000	1949	16,800
1917	80,400	1912	16,400
1930	80,100	1908	16,300
1911	75,400	1929	14,000
1919	65,900	1952	13,000
1925	64,300	1931	11,600
1921	62,300	1933	8,860
1945	60,100	1939	8,080
1952	59,200		

ing either of these physical models.   Both are used in practice, however, and in this case the engineer wishes to choose between these two models after looking at the data.

Qualitative comparisons   Consider first the gross characteristics of the two models as compared to the data.   The lognormal distribution is limited to positive values, as the data are, while the Type I extreme-value distribution is not.   To overcome this limitation one might quite reasonably consider a Type II extreme-value distribution limited at zero; there is no evidence as to the form of the distribution of the underlying random variables whose maximum this flow represents, and hence there is no

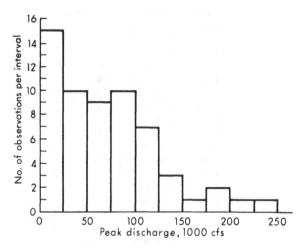

Fig. 4.5.10 Annual maximum flood discharge, Oroville, California, 1902–1960.

fundamental reason to prefer Type I to Type II. We chose here to continue to consider the Type I distribution, however, for two reasons. First, civil engineers who have used extreme-value distributions to describe floods have commonly used Type I only, probably simply for computational convenience. Our example here will retain this tradition because it will permit those familiar with the problem to compare the conclusions given here with similar studies. Second, and more important, it permits us to reinforce the idea that the use to which a model will be put strongly influences its choice. In this case, since the interest in flood prevention dictates a focus on the upper tail of the distribution, we need not be overly concerned with agreement in the lower tail. In this case the future engineering use permits a wider freedom in model selection than would have been the case had we been seeking a "scientific" description of all the observed data. In short we shall *not* reject the Type I extreme-value distribution in favor of the lognormal simply because one predicts no negative values while the other does.

   The histogram of data indicates a decided skew to the right. Such skew can have a critical influence on upper-tail probabilities. Observing Figs. 3.3.3 and 3.2.3, it is clear that both models demonstrate skew in this same direction.

Probability paper plots The data are plotted on the appropriate probability paper for each of the models in Figs. 4.5.11 and 4.5.12. Straight lines have been fit by eye in both cases. The indicated bands, defined by the value that the Kolomogorov-Smirnov statistic would exceed with probability 20 percent, suggest that in neither case can the sample be considered unlikely, if in fact the model holds.   Such bands provide a rough, but sample-size-dependent initial screening.   Smaller sample sizes would lead to even more widely spread curves.

Except for the anticipated discrepancies between the extreme-value model and the very small observed values, both models seem to be capable of representing the data adequately in the sense that systematic, major deviations of the data away from straight lines are not evident in either case.   (This is true despite the fact that if either cumulative distribution were plotted on the probability paper of the other, a gentle curve would result, as the reader can easily verify.)

Use of a goodness-of-fit statistic   The χ^2 statistic can also be used to compare the two distributions in the manner suggested in the last section. In each case, a histogram with intervals chosen to be equally likely (if the

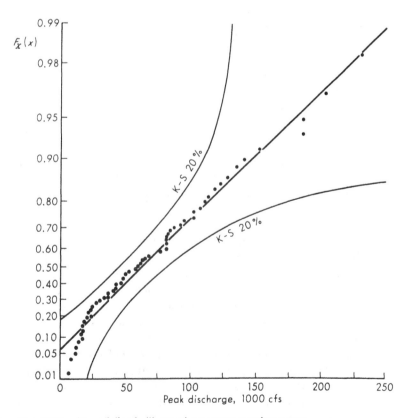

Fig. 4.5.11   Annual-floods illustration: extreme-value paper.

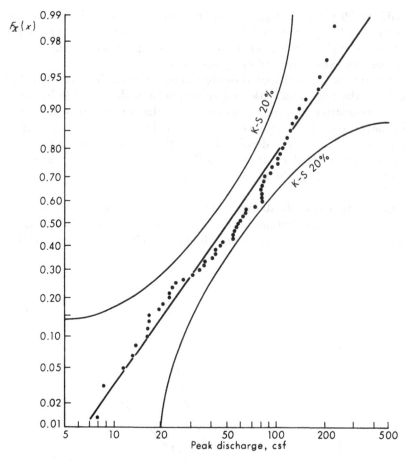

Fig. 4.5.12   Annual-floods illustration: lognormal paper.

particular model were true) was constructed. The results are tabulated in Table 4.5.4. A χ^2 random variable with $\nu = 10 - 1 - 2 = 7$ degrees of freedom has a mean of 7 and a mode, or most likely value, of $\nu - 2$ or $7 - 2 = 5$. Thus, the observed value of 6.6 associated with the lognormal assumption is quite close to the most likely value. On the other hand, the standard deviation of the χ^2 variable is $\sqrt{2\nu} = \sqrt{14} = 3.75$, implying that even the large value of 14.6 observed under the extreme-value assumption is only 2 standard deviations from the mean. At the same time the major contribution to this large value comes from a single interval in the lower tail, where, as discussed above, the failure to fit was anticipated and is not considered important. In short this statistic cannot be used successfully in this example to help in choosing between the

Table 4.5.4

		Lognormal model		Extreme-value model	
Interval i	Expected no. np_i	Obs. no. N_i	$(N_i - np_i)^2$	Obs. no. N_i	$(N_i - np_i)^2$
1	5.9	7	1.2	2	15.2
2	5.9	7	1.2	13	50.2
3	5.9	3	8.4	5	0.8
4	5.9	3	8.4	6	0
5	5.9	5	0.8	5	0.8
6	5.9	7	1.2	3	8.4
7	5.9	6	0	9	9.6
8	5.9	9	9.6	6	0
9	5.9	8	4.4	5	0.8
10	5.9	4	3.6	5	0.8
		59	38.8	59	86.6
	$\sum \dfrac{(N_i - np_i)^2}{np_i} =$		6.6		14.6

distributions, as it cannot properly account for the engineering judgment that these smaller values are not of concern.

Discussion In summary, the available tools provide no strong reason to favor one model over the other as providing a "better" description of the observed data. The data are simply insufficient to permit a clear-cut choice between them.

The final choice must necessarily remain a professional one. The engineer must consider the validity of the physical justifications for the models, the implications of the two models, and the consequences associated with an improper choice. In this case, the authors favor the extreme-value physical argument[†] over the multiplicative one. The implications in terms of relative computational convenience depend upon such factors as the availability of electronic computational facilities and the characteristics of a possible subsequent, larger model, of which this distribution will become a part. (For example, if the flood level is going to be the devisor in a ratio of capacity to demand, representing a "safety factor," computations with the lognormal distribution may very well be simpler since X^{-1} is also lognormal if X is.)

The consequences of an improper model choice will lie, in this case, in a nonoptimal design for the flood-control facilities. Notice that larger floods are predicted to be more frequent by the lognormal model than by

† But, as mentioned, the question of which type of extreme-value distribution is left open.

the extreme-value model. For example, the 50-year flood in the former case is 300,000 cfs, while in the latter it is only 220,000 cfs.† If, for example, the lognormal model is used when the extreme-value model is in fact more accurate, overconservative capabilities will be designed. The extra money spent on the initial investment in these structures would no longer be available to construct similar flood facilities in another location. On the other hand, if the extreme-value model were used as a basis for design when the lognormal model was a better representation, smaller than optimal capacities might result with a subsequently higher risk of flood damage than appropriate. If time and design costs permit it, the engineer might develop preliminary designs on the basis of both models to assess the sensitivity of the final decision, namely the facility's capacity and cost, to his choice of model. In many, diverse problems this sensitivity has been found to be significantly smaller than one might expect. Finally, of course, the engineer must simply make what appears to him at the time, with the evidence available, to be the better choice.

4.5.3 Summary

In summary, the selection of a convenient distribution to represent a random variable when observed empirical data are the only source of information is a two-part process:

1. The first step is an initial qualitative screening of contending distributions to eliminate those that are not capable of reproducing the important, gross characteristics of the data. In this gross sense closeness of fit is the criterion governing the choices.
2. The final selection step is one in which, if professional reasons have not already dictated the choice, a quantitative "model estimation" procedure may be used. Here maximum likelihood rather than closeness of fit is a more reasonable criterion.

In both phases professional judgment will influence the choices. Such judgment is necessary to weigh the relative importance of computational tractability, accuracy of representation, and failure of a model to reproduce certain aspects displayed in data (limits, skew, etc.). In all such considerations the ultimate use of the model will be the single criti-

† This comparison emphasizes the earlier statement that tail probabilities of the adopted empirical model should not be given great confidence. Unfortunately it is all too often the civil engineer's plight that he must make design decisions related to events with small probabilities when insufficient data is available to provide reliable estimates of these probabilities. Since the engineer must make some decision, he often is forced to base it upon unreliable estimates for lack of acceptable substitutes. Lacking reasons to the contrary, extrapolation is probably a reasonable approach, since it is systematic and consistent.

cal factor. If, for example, the model will be used in part of an engineering decision-making process, two models are equally accurate if the same decision is reached in either case. If they are equally accurate in this sense, the distribution which is easier to use is to be preferred.

Caution must be urged in any form of extrapolation of empirically fit distributions. The data are seldom sufficient to discriminate among at least two or three acceptable distributions, yet each may predict quite different probabilities for particular events. The discrepancies may be particularly large in the tails of these distributions; here there may exist order-of-magnitude differences in the small probabilities based on extrapolations of different distributions, even though these distributions appear to fit central observed data equally well.

With small samples, two distributions of quite dissimilar density functions may both seem to fit the data, particularly if viewed on graphs as cumulative distributions. Histograms of the same small set of data may take on different shapes by merely changing interval lengths and starting points. Density functions of one shape may, simply by the nature of random variables, yield a histogram of quite dissimilar shape (Sec. 4.4.1). Such an event is more likely with smaller samples.

In short, there may be need for and convenience in looking at data, choosing some mathematically defined distribution, and attributing to the physical variable the properties of this model, but the engineer must at all times be aware of the limitations and potential inaccuracies of the process. He must keep these difficulties in mind when assessing engineering predictions and decisions based on the adopted model.

4.6 SUMMARY OF CHAPTER 4

This chapter presents the conventional statistical methods available for dealing with the uncertainty in the parameters of a probabilistic model and with the uncertainty in the form of the model itself. In both cases real data, representing observations of a random phenomenon, must be available. The degree of uncertainty in the parameters and in the model is inversely related to the amount of data available.

Section 4.1 considers parameter estimation. Two methods, the method of moments and the method of maximum likelihood, are introduced for generating rules for extracting estimates from data. These rules or estimators are functions of the random variables X_1, X_2, . . . , X_n in the sample, and therefore are themselves *random variables*. Study of the moments and distributions of estimators permits comparison among contending estimators or rules, and permits quantitative statements to be made about the relationship between the estimator and the parameter it estimates (e.g., bias, mean square error, and confidence limits). Confidence limits, although not strictly probabilities, represent quantitative statements about the degree of uncertainty in a parameter.

A significance (or hypothesis) test is a scheme for extracting significance from dispersed statistical data. It provides a quantitative comparison between observed deviations and the deviations expected in light of the probabilistic nature of the phenomenon. If the deviation of the estimate (or sample statistic) from the hypothesized value (or values) is larger than considered "likely," it is said to be statistically significant and the hypothesis is rejected. The measure of what is "likely" or not is the significance level α.

The methods of estimation and hypothesis testing are applied in Sec. 4.3 to a simple multivariate linear model of the form

$$Y_i = \alpha + \beta x_i + E_i$$

More general models of the linear form are also discussed. They form a widely used class of models of which the reader should be aware.

The form of the model itself is evaluated in Sec. 4.4. The "shape" of the data is compared with that of the model. Proper evaluation requires an appreciation of the inherent variability of the histogram or cumulative histogram of the data and the factors upon which that variability depends. Again, significant differences between data and model can be evaluated by conventional hypothesis tests, namely by the χ^2 (chi-square) or Kolmogorov-Smirnov goodness-of-fit tests. These tests are designed to evaluate a particular model, not to compare several models.

Finally, the problem of choosing from among several empirical models given statistical data is discussed (Sec. 4.5). Professional as well as closeness-of-fit criteria are important. Quantitative statistics can be used in a maximum-likelihood manner, but it must be recognized that the smallest value of, say, a χ^2 statistic is not necessarily the most likely value.

REFERENCES

General

Bowker, A. H. and G. J. Lieberman [1959]: "Engineering Statistics," Prentice-Hall, Inc., Englewood Cliffs, N.J.

Brownlee, K. A. [1960]: "Statistical Theory and Methodology in Science and Engineering," John Wiley & Sons, Inc., New York.

Brunk, H. D. [1960]: "An Introduction to Mathematical Statistics," Ginn and Company, Boston.

Cramer, Harald [1946]: "Mathematical Methods of Statistics," Princeton University Press, Princeton, N.J.

Freeman, H. [1963]: "Introduction to Statistical Inference," Addison-Wesley Publishing Company, Inc., Reading, Mass.

Fisz, M. [1963]: "Probability Theory and Mathematical Statistics," 3d ed., John Wiley & Sons, Inc., New York.

Graybill, F. A. [1961]: "An Introduction to Linear Statistical Models," vol. I, McGraw-Hill Book Company, New York.

Hald, A. [1952]: "Statistical Theory with Engineering Applications," John Wiley & Sons, Inc., New York.

Lindgren, B. W. and McElrath, G. W. [1966]: "Introduction to Probability and Statistics," 2d ed., The MacMillan Company, New York.

Lindley, D. V. [1965]: "Introduction to Probability and Statistics," part 1, "Probability," part 2, "Inference," Cambridge University Press, Cambridge, England.

Neville, A. M. and J. B. Kennedy [1964]: "Basic Statistical Methods for Engineers and Scientists," International Textbook Company, Scranton, Pa.

Rao, C. R. [1965]: "Linear Statistical Inference and Its Applications," John Wiley & Sons, Inc., New York.

Specific text references

Alexander, B. A. [1965]: Use of Micro-Concrete Models to Predict Flexural Behavior of Reinforced Concrete Structures Under Static Loads, *M. I. T. Dept. Civil Engr. Res. Rept.* R65-04, Cambridge, March.

Allen, C. R., P. St. Amand, C. F. Richter, and J. M. Nordquist [1965]: Relationship between Seismicity and Geologic Structure in the Southern California Region, *Bull. Seismol. Soc. Am.*, vol. 55, pp. 753–797.

Beard, L. R. [1962]: "Statistical Methods in Hydrology," U.S. Engineer District, Army Corps of Engineers, Sacramento, Calif., January.

Beaton, J. L. [1967]: Statistical Quality Control in Highway Construction, *ASCE Conf. Preprint* 513, Seattle, Washington, May.

Bendat, J. S. and A. G. Piersol [1966]: "Measurement and Analysis of Random Data," John Wiley & Sons, Inc., New York.

Birnbaum, Z. E. [1952]: Numerical Tabulation of the Distribution of Kolmogorov's Statistic for Finite Sample Size, *J. Am. Statist. Assoc.*, vol. 47, pp. 425–441.

Chow, V. T. and S. Ramaseshan [1965]: Sequential Generation of Rainfall and Runoff Data, *J. Hydraulics Div., Proc. ASCE*, vol. 91, HY4, pp. 205–223, July.

Gerlough, D. L. [1955]: "The Use of Poisson Distribution in Highway Traffic," The Eno Foundation for Highway Traffic Control, Saugatuck, Conn.

Granholm, H. [1965]: "A General Flexural Theory of Reinforced Concrete," John Wiley & Sons, Inc., New York.

Grant, E. L. [1938], Rainfall Intensities and Frequencies., *ASCE Trans.*, vol. 103, pp. 384-388.

Greenshields, B. D. and F. M. Weida [1952]: "Statistics with Applications to Highway Traffic Analysis," The Eno Foundation for Highway Control, Saugatuck, Conn.

Gumbel, E. J. [1958]: "Statistics of Extremes," Columbia University Press, New York.

Haight, F. A. [1963]: "Mathematical Theory of Traffic Flow," Academic Press, Inc., New York.

Johnson, A. I. [1953]: Strength, Safety, and Economical Dimensions of Structures, *Roy. Inst. Technol. Div. Bldg. Statics Structural Eng. (Stockholm), Bull.* 12.

Langejan, A. [1965]: Some Aspects of the Safety Factor in Soil Mechanics, Considered as a Problem of Probability, *Proc. 6th Intern. Conf. Soil Mech. Foundation Des.*, Montreal.

Liu, T. K. and T. H. Thornburn [1965]: Statistically Controlled Engineering Soil Survey, *Univ. Illinois Civil Eng. Studies Soil Mech. Ser.*, no. 9, January.

Mann, H. B. and A. Wald [1942]: On the Choice of the Number of Class Intervals in the Application of the Chi-square Tests, *Ann. Math. Statist.*, vol. 13, p. 306.

Markovic, R. D. [1965]: Probability Function of Best Fit to Distributions of Annual Precipitation and Rainfall, *Colorado State Univ. Hydrology Paper No.* 8, August.

Massey, F. J. [1951]: The Kolmogorov Test for Goodness of Fit, *J. Am. Statist. Assoc.*, vol. 46, pp. 68–78.

Mills, W. H. [1965]: Development of Procedures for Using Statistical Methods for Process Control and Acceptability of Hot Mix Asphalt Surface, Binder, and Base Course, *Proc. Highway Conf. Res. Develop. Quality Control Acceptance Specifications Using Advan. Technol.*, Washington, D.C., pp. 304–378, April.

Natrella, M. G. [1963]: "Experimental Statistics," National Bureau of Standards Handbook 91.

Pattison, A. [1964]: Synthesis of Rainfall Data, *Stanford Univ. Dept. Civil Eng. Tech. Rept. No.* 40, July.

Pieruschka, E. [1963]: "Principles of Reliability," Prentice-Hall, Inc., Englewood Cliffs, N.J.

Shook, J. F. [1963]: Problems Related to Specification Compliance on Asphalt Concrete Construction, *Proc. Highway Conf. Res. Develop. Quality Control Acceptance Specifications Using Advan. Technol.*, vol. 1, Washington, D.C., April.

Venuti, W. J. [1965]: A Statistical Approach to the Analysis of Fatigue Failure of Prestressed Concrete Beams, *J. Am. Concrete Inst., ACI Proc.*, vol. 62, pp. 1375–1394, November.

Vesserau, A. [1958]: Sur les conditions d'application de criterion χ^2 de Pearson, *Rev. Statist. Appl.*, vol. 6, p. 83.

Vorlicek, M. [1963]: The Effect of the Extent of Stressed Zone upon the Strength of Material, *Acta Tech. Csav.*, no. 2, pp. 149–176.

Wald, A. and J. Wolfowitz [1943]: An Exact Test for Randomness in the Nonparametric Case Based on Serial Correlation, *Ann. Math. Statist.*, vol. 14, pp. 378–388.

Williams, C. A. [1950]: On the Choice of the Number and Width of Classes for the Chi-square Test of Goodness of Fit, *J. Am. Statist. Assoc.*, vols. 45, 77.

Tables

Burlington, R. S. and D. C. May, Jr. [1953]: "Handbook of Probability and Statistics with Tables," McGraw-Hill Book Company, New York.

Hald, A. [1952]: "Statistical Tables and Formulas," John Wiley & Sons, Inc., New York.

Owen, D. B. [1962]: "Handbook of Statistical Tables," Addison-Wesley Publishing Company, Inc., Reading, Mass.

Pearson, E. S. and H. O. Hartley [1966]: "Biometrika Tables for Statisticians," Cambridge University Press, Cambridge, England.

PROBLEMS

4.1. Use a table of random numbers to take five independent samples, each of size 10 observations, of a random variable uniformly distributed on the interval 0 to 100. From each of the five samples estimate the mean and standard deviation of the distribution.

All the estimates produced by the class will be plotted in a histogram and their mean and standard deviation will be estimated. What are the true values of these moments of the estimator of the mean?

4.2. The results of testing samples from 15 very large lots of concrete pipe for in-place porosity were as follows. Each sample included 100 sections of pipe:

Lot no. i	No. failing to meet standard, n_i
1	1
2	7
3	3
4	5
5	4
6	2
7	6
8	9
9	1
10	3
11	4
12	8
13	0
14	1
15	6

(a) Using *all* 1500 tests, estimate the fraction of failures, p by the observed fraction of failures.

(b) Using a normal approximation for the distribution of the estimator in part (a), form and compute 95 percent confidence limits on p. You will need an estimate of the variance of the estimator. Determine this variance as a function of p and estimate the variance by substituting the estimate of p.

(c) Of what form is the *true* distribution of $N_i/100$, the estimator, for a sample size of 100? *Hint:* What is the distribution of N_i?

(d) Plot a histogram of these 15 observed values of the estimator $\hat{p}_i = N_i/100$. Compare this to a bar chart of the true distribution of the estimator given that $p = 0.04$. Compare, too, the variance of this estimator with the observed sample variance of the 15 values of $\hat{p}$.

4.3.† The following 31 values of annual maximum runoff (in cfs) were observed in the Weldon River at Mill Grove, Missouri, from 1930 to 1960: 108.0, 53.6, 585.0, 98.1, 40.6, 472.0, 96.5, 217.0, 42.7, 208.0, 143.0, 93.7, 398.0, 298.0, 248.0, 441.0, 386.0, 567.0, 122.0, 151.0, 244.0, 400.0, 245.0, 114.0, 659.0, 132.0, 44.0, 72.5, 135.0, 635.0, 508.0.

(a) Using these data, estimate the parameters of a normal distribution model.

$$\Sigma x_i = 7970 \text{ cfs} \quad \text{and} \quad \Sigma (x_i - \bar{x})^2 = 1{,}120{,}000$$

(b) What are the "95 percent two-sided confidence limits" on m_X assuming that σ is known with certainty to equal 180 cfs?

(c) Estimate the parameters of a lognormal distribution model.

$$\sum_i \ln x_i = 181.8 \quad \text{and} \quad \sum_i \left(\ln x_i - \frac{1}{n} \sum_j \ln x_j \right)^2 = 21.837$$

4.4. Simulation sample-size determination. When designing a simulation study, it is important to determine how many times to sample the system behavior. Suppose it is desired to estimate the probability that some event p_i occurs (e.g., maximum system response exceeds some critical value). The obvious estimator of p_i is

$$\hat{p}_i = \frac{N_i}{n}$$

† *Source:* Markovic [1965].

in which N_i is the observed number of occurrences of the event in a sample of size n. N_i has a binomial distribution, but for large enough values of np_i, N_i is approximately normal. Use this approximation to

(a) Find $(1 - \alpha)100$ percent two-sided confidence limits $\hat{p}_i \pm c_\alpha$ on the true p_i given n and an observed value of the estimator $\hat{p}_i$. Replace p_i by $\hat{p}_i$ where necessary.

(b) Show that the necessary sample size to insure that the $(1 - \alpha)100$ percent confidence limits are within $\gamma100$ percent of the true value of p_i (that is, that $c_\alpha \leq \gamma p_i$) is

$$n \geq \frac{k_{\alpha/2}^2(1 - p_i)}{\gamma^2 p_i}$$

where $k_{\alpha/2}$ is that value which a standardized normal random variable exceeds with probability $\alpha/2$. The answer is a function of p_i which is unknown before the experiment is performed. This implies that the engineer must estimate the value of p_i before the experiment.

(c) In a rainfall generation study (Chow and Ramaseshan [1965]) it was desired to estimate p_i (which was thought to be about 0.15) to within $\gamma100 = 20$ percent of its true value with confidence $(1 - \alpha)100$ percent $= 90$ percent; find the required n.

4.5.[†] Data from an AASHO road test revealed that the deviations from the planned thickness of pavement layers were very nearly normally distributed with

$$\bar{x} = -0.02 \text{ in.} \qquad s = 0.195 \text{ in.}$$

The sample size was 4217.

(a) What are the 95 percent confidence limits on m assuming
(i) $\sigma = 0.195$ in.
(ii) σ is not known with certainty.

(b) Adopting the normal model using $\bar{x}$ and s as point estimates of m and σ, what is the probability that a particular specimen will exceed the tolerances -0.02 ft and $+0.04$ ft?

4.6. Control charts. To aid in controlling manufacturing processes, control charts on the sample mean are maintained by engineers responsible for quality control. If the mean m and standard deviation σ of the strength, say, of the material being produced are known, one can compute limits of a *control* band, $m \pm k\sigma/\sqrt{n}$, outside of which the average of a sample of size n should fall only very rarely (k usually equals 3) unless something has gone "wrong" with the process. By periodically (say, daily) taking a sample of size n and observing the average, the engineer can ascertain whether the process is remaining stable, i.e., within the limits of natural variation. The values of m and σ are presumably known from experience or are estimated during a stable period.

One control chart for a plant producing binder mix for a highway surface[‡] appeared as shown on page 505. The characteristic involved was the percent of the aggregate passing a no. 4 sieve. The parameters, estimated over a long stable period, are $m = 40.0$ and $\sigma = 4.76$. Each day a sample of size five was taken. The control band is $40.0 \pm (3)(4.76)/\sqrt{5} = 40 \pm 6.35 = \{33.65, 46.35\}$. Corrective action to the process is indicated by the observed value of $\bar{x}$ in the fourteenth lot. [Conventional (nonstatistical) specifications tolerances were, incidentally, 40 ± 4.0 percent.] Some control schemes call for additional testing if an observation lies outside, say, $\pm 2\sigma/\sqrt{n}$.

[†] *Source:* Shook [1963].
[‡] Adapted from Mills [1965].

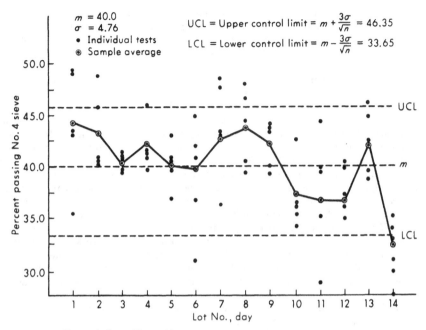

Fig. P4.6   Control-chart illustration.

Construct a control chart with $k = 3$ for the following reported† concrete test data.   Averages are of samples of size 10.

Lot no.	Average ultimate compressive stress
1	4004
2	3845
3	4195
4	4170
5	4043
6	3695
7	3900
8	3730
9	3670
10	3667

The mean has been estimated to be 3868 psi, and σ is 318 psi.   (The specified strength was 3500 psi.)   Is process correction indicated at any point?   Under the assumption that the sample mean is normally distributed, what is the likelihood that for any single lot process-control correction will be initiated unnecessarily, i.e., that the excursion of $\bar{X}$ beyond the control band is simply a rare random deviation?   What is the probability of *at least one* unnecessary correction during a job of 100 days or lots?

4.7. Specifications have been written to control concrete pavement thickness during construction which state that the proportion of slab less than 6 in. thick must not exceed 10 percent.

† Data provided by the Master Builders Co., Cleveland, Ohio, for an office building constructed in New York.

(a) To control his production the contractor periodically measures the thickness just as it is poured. The contractor sets up a null hypothesis that states that by his present operating procedure the true proportion p, of such thicknesses is just 0.1. He must go to the expense of changing his method of operation if forced to reject this hypothesis, and so he wants to set α at least as low as 10 percent. With a sample size of five, what should the value of c be for a test of the form: accept H_0 if R, the number of specimens less than 6 in. thick, is less than c.

(b) The state inspector, on the other hand, must take the position that the major loss will occur if he mistakenly accepts a poor-quality job. He has cores cut from the slab after it is completed. He sets up the null hypothesis that p, the proportion of specimens less than 6 in. thick is 0.2, that is, that the job is inadequate. He will only reject this hypothesis for the alternative that the proportion is only 0.1 if more than c specimens out of n are greater than 6 in. thick. To keep the risk of accepting a poor job low, he wishes to make the significance level less than 5 percent. He recognizes, however, that if he rejects the job (accepts the null hypothesis), the contractor will make many additional tests to attempt to demonstrate a good job. If the contractor proves that p is in fact 0.1, the state must bear the expense of the added testing and delay. Therefore the inspector wants to keep the type II error at least as low as 20 percent. How many specimens must he take to satisfy both criteria? (Assume in this part that the number is large enough that the binomial variable R is approximately normally distributed with the same mean and variance.)

Remark. The problem here is illustrative of the common statistical decision problems called "quality control" and "acceptance sampling." Note that one is carried out by the manufacturer and the other by the purchaser, not necessarily with the same α and β probabilities. More commonly the problem is set up with composite rather than simple hypotheses, for example, $p \geq 0.1$ versus $p < 0.1$. Many interesting problems can be explored. The contractor will usually carry out a two-sided test, for example, $p = 0.1$ versus $p \neq 0.1$, to avoid wasting material and profit. (See Prob. 4.6 on quality control charts.) To arrive at his proper sample size, the contractor must ask himself what is his acceptable β probability for failing to catch an out-of-adjustment operation. If p is too high and the job is rejected, he may have to redo it at a loss. Both parties should ask how expensive the observations are. (It is suggested in the problem statement that they are cheaper for the contractor than the inspector. This factor should in principle influence the error probabilities. The interested reader should reconsider this problem after studying Chap. 6.)

4.8.† Since, during a slip of the soil, small strengths in parts of a slope of generally homogeneous soil will be compensated for by larger strengths elsewhere, it has been suggested that the stability of the slope is related directly to the spatial *average* strength of the material. If the strength value (say, the shear strength) used in design is the *observed* mean $\bar{X}$ divided by a safety factor, find the probability that the design value used will be less than the true mean if a sample size of three is used and safety factor $= 1.5$. Assume that the underlying variable X is normally distributed with unknown mean and variance. *Hint:* $\bar{X}$ has, within some constants, the t distribution.

4.9.‡ In a test of the hypothesis that vehicle speeds are normally distributed, a χ^2 test was applied to a sample size of 100. The estimated mean and standard deviation were 40.8 and 9.31 mph, respectively. The number of intervals used in the test was seven; the observed statistic was $d_1 = 4.506$. Is the evidence such as to reject

† *Source:* Langejan [1965].
‡ *Source:* Greenshields and Weida [1952], page 172.

the hypothesis at the 10 percent level? If the hypothesis is true, what is the likelihood that a similar sample and test would lead to a test statistic greater than 4.506? (That is, what is the "p level" of the observed statistic?)

4.10. Triangular distribution probability paper. (a) Construct a probability paper for distributions with *symmetrical* triangular PDF's on the interval a to b. *Hint:* Use a graphical construction.

(b) A number of individuals measured the same item twice and each reported his average reading. If each of the two reading errors is uniformly distributed between -0.5 and $+0.5$ unit (the smallest division on a fine scale) and the two readings are independent, what is the distribution of each individual's average reading? (See Sec. 3.3.1.)

(c) The following readings were reported by eight individuals: 14.9, 15.0, 15.1, 15.15, 15.2, 15.25, 15.4, 15.6. Use the probability paper of part (a) to compare this data and the model of part (b). The true value is 15.2.

4.11.† Using Table A.5 to first construct Type I extreme-value paper, make Type III extreme-value paper for smallest values (with a lower bound w equal to zero). This is also known as the Weibull distribution and is commonly used in reliability studies. *Hint:* Recall (Sec. 3.3.3) the logarithmic relationship between Types I and III.

Based on a probability paper plot inspection, is a Weibull model appropriate for the following data on the modulus of rupture strength of concrete? The model is considered because the brittle nature of concrete in tension suggests that the weakest of many microscopic elements determines the strength of any specimen. Observed data: ($1 \times 1 \times 11$ in. specimens) (psi): 1047, 986, 922, 955, 940, 1005, 1093, 1074, 1074.

4.12.‡ Is the following set of data from Southern California consistent with the assumption that the occurrences of earthquakes (of magnitude greater than 3.0) are Poisson arrivals? That is, is the number per year in this region Poisson-distributed? (Consider using a normal approximation to the Poisson distribution.) Use probability paper, and a graphical (10 percent level) Kolmogorov-Smirnov test.

Year	Recorded number	Year	Recorded number
1934	171	1951	136
1935	317	1952	379
1936	209	1953	324
1937	179	1954	391
1938	178	1955	187
1939	209	1956	279
1940	269	1957	156
1941	215	1958	147
1942	235	1959	177
1943	173	1960	123
1944	158	1961	181
1945	97	1962	154
1946	302		
1947	290		
1948	207		
1949	310		
1950	212		

† *Data source:* Alexander [1965].
‡ *Data source:* C. R. Allen, P. St. Amand, C. F. Ricter, and J. M. Nordquist [1965].

4.13.† Among 29,531 drivers the following numbers of accidents per driver were observed (in Connecticut, 1931–1936):

Accidents/operator	Observed no. of operators
0	23,881
1	4,503
2	936
3	160
4	33
5	14
6	3
7	1
	29,531

The overall average is 0.24 accidents per operator. *If* the accident rate is the same for all operators, the number of accidents for any driver has a Poisson distribution. Test the hypothesis that there is no accident "proneness" in certain drivers by comparing these numbers with a Poisson distribution. Use $\alpha = 10$ percent. (Proneness includes here, of course, the factor that a driver simply may drive many more than the average number of miles and hence be exposed to more accident possibilities.)

4.14.‡ The following data on annual sediment load in the Colorado River at the Grand Canyon have been recorded (in millions of tons) 49, 50, 50, 66, 70, 75, 84, 85, 98, 118, 122, 135, 143, 146, 157, 172, 177, 190, 225, 235, 265, 270, 400, 480.

(a) Plot this data on lognormal probability paper and use a straight-line fit by eye to estimate the parameters.

(b) Does the model pass a Kolmogorov-Smirnov goodness-of-fit test at the 10 percent level of significance?

4.15. It is common in the study of fatigue failures to deal with the logarithm of the number of cycles to failure rather than with the number itself. One reason lies in the effect upon the distribution of observations (under a given level of cyclic loading). Compare the shapes of histograms of the following data. Choose between the use

N, millions	No. observed	log N, base 10	No. observed
0–10	13	5.5–5.75	1
10–20	17	6.25–6.5	3
20–30	12	6.5–6.75	3
30–40	4	6.75–7.0	6
40–50	7	7.0–7.25	15
50–60	1	7.25–7.50	16
60–70	1	7.50–7.75	10
70–80	1	7.75–8.0	2
110–120	1	8.0–8.25	1

† *Source:* Greenshields and Weida [1952], page 207.
‡ *Data source:* Beard [1962], page 42.

of two models. One says that the number N is normally distributed and the other that the log of the number is normally distributed. Use probability paper.

4.16. Two-population Kolmogorov-Smirnov test and simulation-model evaluation. When a stochastic simulation model has been constructed (Sec. 2.3.3) and must be evaluated, a valuable test of the model is to compare the distribution of some output generated by the model to the observed distribution of the corresponding historical data. A test is available based on the Kolmogorov-Smirnov statistic which permits such a comparison even though the true underlying distribution is not known. In other words the hypothesis is simply that the data are from the same (but unknown) distribution.

The test is accomplished by using the statistic D_2^*, the maximum value of the difference between $F_1^*(x)$ and $F_2^*(x)$, the observed cumulative histograms [Eq. (4.4.17)] of the two samples:

$$D_2^* = \max \left[|F_1^*(X) - F_2^*(X)| \right]$$

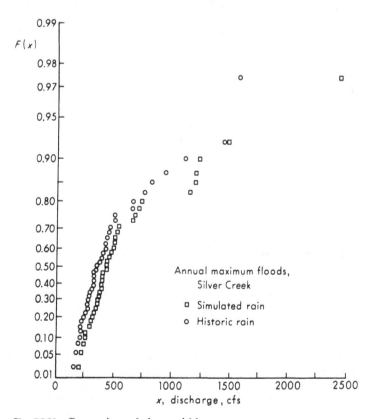

Fig. P4.16 Comparison of observed histograms.

The distribution of this statistic is the same as that of D_2 [Eq. (4.4.18)]. The value of n used should be (Fisz [1963])

$$n = \frac{n_1 n_2}{n_1 + n_2}$$

where n_1 and n_2 are the sizes of the two samples. The distribution of D_2 is given in Table A.7. The hypothesis that the two samples came from the same distribution (i.e., that the simulation model is a good one) should be rejected at the α percent level of significance if the observed value d_2^* of the statistic D_2^* is greater than the critical value, i.e., that value which the random variable D_2^* exceeds only with probability α percent.

Use this test to evaluate a model of rainfall proposed by Pattison [1964]. The cumulative histograms of historic annual floods† and the synthesized annual floods are shown in the figure on page 509 (plotted on extreme-value paper). In both cases the sample size is 40. Use $\alpha = 0.05$.

4.17. Using the data of Prob. 4.3, make plots on uniform, normal, and lognormal probability paper. Fit straight lines and compare the forecast runoffs with 10-, 30-, and 50-year return periods, according to each model. Discuss the variability of the forecasts. What values would you recommend for flood-control studies assuming that a major flood would involve (a) minor damage or (b) major damage?

4.18. Soil sampling sizes. Based on a pedologic soil survey map, an engineer has an initial idea as to types of soil which will be found along a proposed highway alignment. He wishes to distribute the total number of anticipated tests n among the r different soil types in a manner which accounts for their relative variability. Show that in order to achieve uniform "accuracy" (i.e., equal coefficients of variation) for the estimator $\bar{X}_i$ of m_i, the mean of the ith soil type, the number of independent tests in the region with this soil type should be‡

$$n_i = n \frac{V_i^2}{\sum_{j=1}^{r} V_j^2}$$

in which $V_i = \sigma_i / m_i$. It is assumed that previous tests and experience make initial estimates of the V_i possible.

In a road through Will County, Illinois, 11 soil types were encountered with the following preliminary estimates of coefficients of variation of the plasticity index. Use the values to determine how to proportion 100 samples among the various soil types:

---

† The stream flows in both cases were simulated by the Stanford watershed model. Rainfall, the input to this model, was either historic or simulated by Pattison's model.
‡ Adopted from Liu and Thornburn [1965], who make the additional suggestion that a factor proportional to the length of the region $L_i / \Sigma L_j$ also be included to account for the relative consequences should an error in estimation be made.

Soil type	Coefficient of variation of plasticity index, %
Lisbon silt loam	26.4
Harpster silty clay loam	62.0
Saybrook silt loam	27.9
Elliot silt loam	23.3
Drummer silty clay loam	64.6
Ashkum silty clay loam	32.4
Andres silt loam	32.4
Symerton silt loam	15.7
Lorenzo silt loam	69.4
Dresden silt loam	23.8
Homer silt loam	20.2

4.19. The contingency table: test for independence. A popular method of testing the independence of two events A and B is to form a *contingency table:*

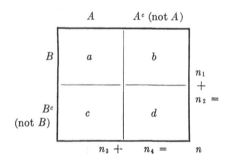

in which a, b, c, and d are the observed numbers of observations in which the corresponding attributes are observed; for example, b = number of tests in which B and A^c were observed.

Consider the logical estimates of $p_1 = P[A]$ and $p_2 = P[B]$; they are $(a + c)/n$ and $(a + b)/n$. Use them to compute the probability of all four possible events (such as $(B \cap A^c)$, $(A \cap B)$, etc.) *under the hypothesis* that A and B are independent, and finally apply the χ^2 goodness-of-fit test to show that the D_1 statistic reduces to

$$C = \frac{(ad - bc)^2 n}{n_1 n_2 n_3 n_4}$$

which is approximately χ^2-distributed with 1 degree of freedom, where $n = a + b + c + d$, $n_1 = a + b$, $n_2 = c + d$, $n_3 = a + c$, and $n_4 = b + d$.

The contingency table is often used to test whether there is relationship (dependence) between two factors. The two-way classification is often forced to simplify the test.† For example, to test whether the type of failure (tension T or compression C) is related to the fatique life (number of cycles to failure) of similar specimens of prestressed concrete beams, the accompanying data‡ can be simplified to a two-way

Type of failure	No. of cycles to failure
C	550
C	603
C	5,000
C	40,500
T	45.360
C	52,020
T	55.420
T	68,040
T	94,540
T	110,200
T	114,600
T	121,200
T	126,400
T	133,100
T	134,600
T	174,400
T	187,900
T	236,500

classification contingency table.

		Type of failure		
		T	C	
No. of	"low"	4	5	9
cycles to	"high"	9	0	9
failure		13	5	18

Test at the 5 percent significance level the hypothesis that the type of failure is unrelated to the number of cycles of failure.

4.20. For a particular type of heavy construction equipment, the breakdown time X is believed to be uniformly distributed on the interval 0 to θ, but θ is unknown. A sample of n values, $x_1, x_2, \ldots, x_n$, is observed $(n > 1)$. The largest is x_j.

(a) Find the likelihood function $L(\theta \mid x_1, x_2, \ldots, x_n)$. What are the limits over

† The test can be extended to more than two classifications. See, for example, Lindley [1965], sec. 7.6.

‡ *Source:* Venuti [1965].

which the function is nonzero? In particular, given the observations, what is the smallest value that θ can be?

(b) What is the maximum-likelihood estimator of the parameter? That is, for what value of θ is $L(\theta \mid x_1, x_2, \ldots, x_n)$ a maximum? A sketch may help.

(c) Show that the method-of-moments estimator is $\hat{\theta} = 2\bar{X}$.

(d) What are the two estimates of θ if the observed sample is 1, 1, 1.5, 2, 4? Is the method-of-moments estimate reasonable?

4.21. The timber-beam data of Prob. 1.1 is to be used to make probability statements about strength and deflection.

(a) Based on the histograms and professional judgment, select two possible models for each set of data. Check for fit with plots on probability paper and comparative χ^2 statistic likelihoods.

(b) Assuming that normal distributions provide satisfactory fits, state 90 percent confidence limits on the means and standard deviations.

4.22. The B-C Construction Company has found the following data for the last 20 jobs.

Job no.	$\dfrac{Actual\ labor\ cost}{Estimated\ labor\ cost}$ (100)
63-1	97.9
63-2	97.6
63-3	103.5
63-4	105.7
63-5	101.6
64-1	96.9
64-2	93.9
64-3	106.8
64-4	99.3
65-1	102.8
65-2	99.2
65-3	101.1
65-4	92.7
65-5	109.4
65-6	96.7
65-7	98.2
65-8	94.6
65-9	107.1
66-1	95.2
66-2	101.7

"Actual" costs were adjusted to eliminate the effects of change orders, etc. Make a study of the data with the objective of arriving at probability statements for use in bidding the next job. Where is the area of primary interest? Check for fit to a normal distribution using the χ^2 test. Is it reasonable to compare figures for jobs of different sizes?

4.23. The Neosho River originates in Kansas and flows into Oklahoma. Recorded mean-annual-flow Q data (in cfs) near the state border are:

141	522	1443	2582
146	589	1561	3050
166	696	1650	3186
209	833	1810	3222
228	888	2004	3660
234	1173	2013	3799
260	1200	2016	3824
278	1258	2080	4099
319	1340	2090	6635
351	1390	2143	
383	1420	2185	
500	1423	2316	

Make an analysis of the data. Try the normal and lognormal models using probability paper and compare the χ^2 statistic likelihoods. Which model seems the most reasonable on a physical basis?

A rule for allocating the use of the water between the two states is desired. Study the following rule using both probability models: *A fixed level x_0 for Kansas and an equal value on the average for Oklahoma* (that is, $2x_0$ = mean flow).

On the average, how often will Oklahoma have zero allocation under this rule?

4.24. The concrete pavement thickness for a series of county roads placed by one contractor was measured by taking cores. The numbers of cores at each thickness were as follows. (There is one core for each section for which a payment is made.)

No.	Thickness, in.	No.	Thickness, in.
1	5.25	10	6.05
2	5.30	11	6.10
2	5.35	14	6.15
3	5.40	9	6.20
4	5.45	15	6.25
7	5.50	5	6.30
6	5.55	4	6.35
4	5.60	5	6.40
3	5.65	3	6.45
4	5.70	4	6.50
11	5.75	3	6.55
8	5.80	2	6.60
11	5.85	2	6.65
5	5.90	2	6.70
8	5.95	1	6.75
8	6.00	1	7.90

(a) Make an analysis of the data to determine a reasonable model.

(b) The contractor assumes that the materials cost is 40 percent of the bid price, and the penalty clause in the contract is as follows.

Thickness, in.	Payment, %
≥ 5.85	100
$5.75 \leq$ to ≤ 5.85	94
$5.50 \leq$ to ≤ 5.75	89
$5.25 \leq$ to ≤ 5.50	81
$5.00 \leq$ to ≤ 5.25	74
≤ 5.00	10 (or remove and replace)

What percent of the total payment can the contractor expect to receive on similar jobs using the same construction practice?

(c) Is the penalty clause effective in making the thickness at least 6 in. (the design value)? Reduce the mean thickness and estimate the contractor's gain or loss.

4.26. Ratios of jack to anchor force for prestressed concrete cables were measured and reported in Prob. 1.3. The objectives of the study were to provide measures of mean and coefficient of variation and to obtain some idea as to the source of the variability.

(a) Make conventional two-sided 95 percent confidence interval statements on the mean and variance. Assume that the ratio is approximately normal.

(b) Fit the data to the normal probability distribution using the method of moments and probability paper. Check numerically for fit using χ^2 and Kolmogorov-Smirnov test of goodness of fit.

(c) Using the concepts of statistical inference or significance, what tests would be useful in making statements about the source of variability between the two sets of data?

4.27. A bridge is to be constructed at Waukell Creek (Prob. 1.4). The soils engineer must recommend allowable soil pressures for the footings based on the reported data.

(a) Is a normal distribution a satisfactory model considering both the histogram and professional considerations? Make such a fit.

(b) Fit a Type I extreme value distribution of smallest values to the data. Construct appropriate probability paper and plot.

(c) Compare the results of (a) and (b) as to fit and probability of finding a strength specimen equal to or less than 0.20 tons/ft^2.

4.28. A study of the relationship between floor (or column) loading in pounds per square foot (psf) and total area loaded can be made using the data of Table 1.2.1. Consider a particular bay. By working from the ninth floor down and assuming that all bay sizes are 400 ft^2, one can obtain total loads and then loads in psf for 400, 800, 1200, . . . , 3600 square feet of area.

(a) Make a plot of observed live load in psf against total area. Do this for each of three bays on one graph. Assuming that the loads on different floors are independent, discuss the applicability of the simplest linear regression model to this problem. In particular is σ independent of x (the area) and are the observations of total loads independent of each other? Is a Markov model appropriate?

(b) Compute the mean and variance of live load in psf, for 400, 800, 1200, . . . , 3600 ft^2 of floor area, using for each total area the data from 22 bays. Plot the mean and 95 percent probability limits on the graph, assuming that a normal distribution applies and that the sample moments are satisfactory population estimates. Compare the resulting plot with the code reduction of 0.08 percent/ft^2 of tributary area greater than 150 ft^2 (up to a maximum reduction of 60 percent). Assume a code-design live load of 100 psf.

(c) There is interest in the possible occurrence of extremely large loads, equal to or greater than 100, 150, and 200 psf, on small areas, say the "basic" floor area of 400 ft^2. Consider the data from all $(9)(22) = 198$ bays together. What are the estimated mean and variance? Compare probability statements for such magnitudes using normal, lognormal, and gamma models.

4.29. The discharge of the Ogden Valley artesian aquifer (Prob. 1.6) is to be allocated between two major users each of whom wants 5000 acre-ft.

(a) What probability models appear satisfactory considering the histogram and professional knowledge of such systems? Fit the normal model and one other model with a range from zero to infinity. For the two models compare probabilities of finding a discharge equal to or less than 10,000 acre-ft. Are the differences important? What is the probability of finding unsatisfactory discharges in both of the first two years? Assume that annual discharges are independent. (Excess water flows away.)

(b) If the two users share all the water equally, does the probability of each user obtaining less than 5000 acre-ft change?

(c) If one user has first chance at the available supply and takes all water up to 5000 acre-ft, leaving only the excess for the second user, what changes in the study must be made to yield probability statements on the water available to both users?

(d) Suppose a small storage dam is built to store any excess water for use in following drier years. Does the simple model of independence of annual water available remain valid?

4.30. A spillway is being designed for a dam on the Feather River at Oroville, California (Prob. 1.7). The problem is to make the most reliable forecasts of future maximum annual floods. Compare the fit of the data to the normal, lognormal, and Type I extreme-value model. Use probability paper.

(a) Compare forecasts of 10-, 50- and 100-year floods.

(b) What model appears to give the best fit?

(c) For each model, what is the most likely (i.e., the modal value of the) maximum annual flood to be found in any one year? What is the probability of finding a flood equal to or greater than double this magnitude? Are these magnitudes particularly sensitive to the choice of model?

4.31. Make a study of the design situation of Prob. 1.9 using probability models of Chap. 3. Is the decision sensitive to choice of model between those with similar degrees of fit? Consider at least two models consistent with the histogram.

4.32. Correlated samples. In an attempt to estimate the mean density of soil in a certain shallow layer under a building site, it has been proposed to do one of two things:

(i) Drill five holes at widely separated intervals and take single soil cores from each. Measure their densities and average them to estimate the mean at the site.
(ii) Drill a single hole, take five cores from it, and average their densities to estimate the mean at the site.

Experimental error in these tests is negligible. Soil density varies, however, from place to place within the layer owing to variations in the geological processes.

(a) Why, in words, is the latter method a less desirable way to estimate the mean density at the site?

(b) Why, in statistical terms, is the latter method less reliable? *Hint:* Interpret Eq. (4.1.12).

(c) If the core densities can be assumed uncorrelated by the former method, but by the second method there is a correlation coefficient of $\rho = 0.9$ between all pairs of variables, find the relative magnitudes of the mean square errors of the two testing techniques. How many independent samples are the five specimens in the latter method "equivalent" to? Assume that all cores have the same mean and standard deviation.

4.33. Reconsider Prob. 2.41. The project engineer is concerned with developing an operating rule by which the foreman can decide from his observation of a single roundtrip time whether to report that this particular process is functioning properly or to instigate a more careful inspection of the process. The significance test is of the form

H_0:   Process is operating properly
Operating rule: Accept H_0 if $X \leq$ minimum time $+ c$

(In Prob. 2.41, $c = 10$ min.) Find c such that the probability of making the inspection when it is unwarranted is only 10 percent.

4.34. Consider the water demands in Prob. 1.17. In previous years the air station was populated only by service personnel. The mean daily demand was 4,550,000 gal/day. In 1965 families replaced a portion of the service personnel. Is the daily mean demand in 1965 significantly greater than in the past? Assume that daily demands are normally distributed. Notice that the sample size is quite large.

4.35. Assuming normality of the beam deflections in Prob. 1.1, should the wood-products engineer report that the observed deflections indicate that the mean deflection is different from the calculated deflection? Use the actual dimensions (1.63 by 3.50 in.) when calculating the deflection. Note that the sample size is small.

4.36. Consider the aquifer recharge data in Prob. 1.6. There is some reason to believe that this may have changed in time. Assuming normality and equal variances, does there appear to be a significant difference between the mean recharge before 1943 and that after (and including) 1943? (Under the null hypothesis of equality, the best estimate of the common variance is simply the sample variance about the sample average of *all* 17 years. Use this estimate as an assumed *known* value of σ^2.)

4.37. In a traffic-counting study the number of cars arriving during n intervals, each of 1-min duration, were counted with observed values $x_1, x_2, \ldots, x_n$. Assume that the cars are Poisson arrivals with unknown average arrival rate ν per minute. Then X, the number in any minute, has a Poisson distribution with mean ν.

(a) Show that the likelihood function of ν given the observations is

$$L(\nu \mid x_1, x_2, \ldots, x_n) = e^{-\nu n} \nu^{\Sigma x_i} \qquad \nu \geq 0$$

(b) What is the maximum-likelihood estimator of ν? How does this compare with the method-of-moments estimator?

4.38. An engineer must report as to the kind of soil-pile behavior that exists at a site. The choice will be based on the indications from the blow counts from five preliminary piles.   The null hypothesis is that the behavior is type A which is known from experience to exhibit a mean blow count of 20/ft, while the alternate hypothesis is that the behavior is type B, which gives rise to a mean blow count of 30/ft.   In both cases the standard deviation of the number of blows per foot is 5.

(a) Propose a logical hypothesis test, i.e., a logical statistic and a reasonable form for the operating rule.

(b) Find the specific operating rule which gives a type I error probability of 25 percent.   Assume an appropriate, convenient distribution for the sample statistic adopted.

(c) What is the β probability for the test?

(d) What should the engineer report if the blow counts observed are 31, 22, 28, 25, 32?

4.39. The elevator in a high-rise apartment is presumed to behave like a simple Markov process in the following sense.   The elevator is observed each time it stops at or returns to the street floor.   Its state on arrival is either empty, 0; partially full, 1; or full, 2.

Consecutive observations of the elevator totaling 101 were made.   Therefore, 100 transitions were observed.   In 35 of them the car was initially empty; 5 of these were followed by full states, 15 by partially full states.   In 45 of the transition observations the car was initially partially full; 20 were followed by an empty car and 20 by a partially full car.   Of the 20 initially full cars, none was followed by an empty car, and 10 were followed by full cars.

Use the obvious fractions to estimate the nine transition probabilities for this (homogeneous) Markov chain.   Are these estimators independent random variables? Would estimators based on 100 nonconsecutive transition observations (e.g., observe one transition a day for 100 days) be "better" estimators (i.e., have smaller variance)? Would these nine estimators be independent?

4.40. A proposed estimator of the unknown median of a lognormally distributed random variable X is the "geometric mean" $\tilde{X}$ of the random sample, X_1, $X_2, \ldots, X_n,$

$$\tilde{X} = (X_1 X_2 X_3 \cdots X_n)^{1/n}$$

Show that $\tilde{X}$ is not an unbiased estimator of the median of X.   Does it become approximately unbiased as the sample size increases?

4.41. The depth to a bedrock layer under a site will be estimated by the sample average $\bar{X}$ of n independent readings of a sonic instrument.   Any reading is a random variable with a mean equal to the true (but unknown) depth m_X and with a known coefficient of variation V_X (i.e., the standard deviation increases with depth).

(a) What are the mean and variance of the estimator $\bar{X}$ in terms of m_X and V_X?

(b) Make a reasonable assumption for the distribution of $\bar{X}$ when $n = 16$, and write an expression for the 90 percent confidence interval of the true depth when $V_X = 0.2$.

(c) What is the 90 percent confidence interval estimate on true depth if the 16 tests are run and their average is 102 ft?

4.42. A building code's requirements are that a particular type of foundation can be used only if a particular type of soil exists below the site.   The soil is identified by a *mean* "index" value of 10.   At higher or lower values some other foundation type must be considered.

(a) Can the engineer accept at the 10 percent significance level the hypothesis that this mean value is 10 if he observes four soil specimens at different points around the site with index values of 6, 9, 10, and 7?   Assume that these values are observations of independent identically distributed normal random variables with standard deviation 2 and mean equal to the (unknown) mean index.

(b) Suppose, instead, that the soil specimens whose index values are listed above were taken quite close together on the site, so that they are not independent samples but rather have a common correlation coefficient, 0.5, between all pairs.   What moment of the sample statistic changes from the problem in part (a)?   Assuming that the sample mean is normally distributed, can the engineer accept the hypothesis that the (unknown) mean is 10 (at the 10 percent significance level)?

(c) Reconsider the situation in case (a) if the code states that the foundation can be used if the soil has an index of 10 *or more*.

4.43. A large school system is planning a renovation of a number of schools, all built prior to 1940.   While making estimates for the budget, it was necessary to forecast the likely cost of renovation based on accumulated experience in past similar jobs. The cost data on old jobs has been updated to current conditions using the *Engineering News-Record* index.   Data are as follows:

Cost	Rooms	Stories
$420,000	40	4
435,000	45	5
218,000	16	3
191,000	30	4
146,000	28	3
402,000	32	4
518,000	41	5
340,000	34	3
248,000	19	3
470,000	56	4
237,000	26	3

(a) Assume that the expected cost of renovation is linear with the number of rooms.   Estimate the regression line and show the relationship on a plot with the data.   Estimate σ and plot lines 1 standard deviation each side of the mean value function.   Estimate and plot 1-standard-deviation confidence limits on the mean value.

(b) If the first job to be studied contains 32 rooms, use the results of (a) to make statements about its expected cost and the standard deviation of this figure. Note that a single forecast of the random variable, i.e., cost, is desired.

(c) Data on the number of stories are given.   Repeat the analysis using number of stories to predict the cost.   The job contains three stories.   Which independent variable—number of rooms or number of stories—gives a "more reliable" prediction of cost?

4.44. Algae-fixed organic matter will be produced in a clear-water sample with zero chemical oxygen demand (COD) and biochemical oxygen demand (BOD) through the addition of controlled amounts of nitrogen. A laboratory study using 15-day specimens and sensibly constant phosphorus yields:

COD, mg/l	Nitrogen, mg/l
1780	27.5
1470	41.0
2080	33.5
1110	20.6
1050	20.4
745	15.1
1220	20.6
1020	16.0
830	31.0
183	0.18
1100	16.0

(a) Determine the regression line for COD versus nitrogen. Estimate α, β, and σ.

(b) Test if the slope estimate is significantly different from zero at the 5 percent significance level.

(c) Is the intercept zero? Test the hypothesis at the 5 percent level.

(d) What is a 90 percent upper confidence limit on the variance of COD for any value of nitrogen?

(e) Plot the residuals on normal probability paper. Is the normality assumption reasonable?

(f) Plot the residuals versus the independent variable. Does the data appear to satisfy the condition that σ is constant? Given a large amount of data, how might you test the hypothesis?

4.45. The tensile strength of concrete can be measured by the splitting test, in which a concrete cylinder is placed on its side in the testing machine and subjected to diametral compression (ASTM C496-66). The compressive and splitting tensile strengths of lightweight and ordinary concrete as a function of age of test for continuous moist cure specimens were reported by J. A. Hanson[†] for a particular mix design as follows:

† J. A. Hanson, Effects of Curing and Drying Environments on Splitting Tensile Strength of Concrete, *ACI J.*, Table 3, July, 1968.

Age, days	Lightweight		Ordinary	
	Compressive strength, psi	Splitting tensile strength, psi	Compressive strength, psi	Splitting tensile strength, psi
3	2620	276	2050	232
7	3300	342	3250	282
14	4320	407	3830	373
28	5120	420	4560	336
90	5850	458	5240	448
180	6620	494	5390	454
365	6840	513	5530	524
2 years	6760	492	5570	506

The compressive strengths are the average of two specimens and the splitting tensile strengths, the average of four specimens.

(a) Make a regression-analysis study of the growth of compressive strength with time. Use a model of the form:

$$\text{Strength} = \alpha' x^\beta = e^\alpha x^\beta$$

in which time is the controlled variable. Note that the data are incomplete; that is, only the average value of strength of two specimens is given for each time. Thus the information is incomplete for the purpose intended. This is a common situation in the reporting of data. An approximate adjustment is to use a value of n equal to two (or four) times the given value of 8. Is the estimate of β significantly less than 1? Find 95 percent confidence limits on α and β. Are the β's significantly different?

(b) Statements of tensile strength are desired based on observed compressive strength. How can such a relationship be studied. What significance tests would be of importance?

4.46. Regression analysis has been used in forecasting runoff from precipitation records. The 13-station (average) precipitation statistics from 1923 to 1966 for October, November, December, January, February, March, and April are given in the accompanying table for the Colorado River Basin along with the adjusted April–July runoff at the Grand Canyon. Assume that the expected April–July runoff is linear in the sum of the monthly precipitations and determine the regression equation. How would such an analysis be used? Is the estimate of the correlation coefficient between average monthly rainfall (October through April) and April–July runoff significantly greater than 0?

Colorado River Basin precipitation statistics

Water, year	Oct.	Nov.	Dec.	Jan.	Feb.	March	April	Adjusted April–July runoff
1923	.50	1.55	2.38	1.05	.80	1.70	1.58	11.84
1924	1.17	.79	1.32	.79	.45	1.94	.86	8.86
1925	1.61	.71	1.89	.86	.86	.95	.37	7.30
1926	1.83	.58	.68	.65	.81	1.27	2.08	9.94
1927	1.07	.89	1.50	1.05	2.09	1.58	1.00	11.34
1928	1.22	1.29	1.41	.42	.83	1.46	.59	10.58
1929	2.54	1.54	.72	1.16	1.04	2.00	1.25	12.51
1930	1.03	.54	.25	2.05	.95	.96	.94	7.90
1931	.57	1.15	.35	.24	.73	.81	1.15	3.66
1932	1.35	1.62	1.02	.82	1.46	1.32	1.55	11.38
1933	1.22	.25	1.22	.89	.64	.92	1.69	7.07
1934	.57	.73	.90	.58	1.57	.27	.67	2.25
1935	.12	1.09	.96	1.08	.59	1.13	2.04	7.60
1936	.84	.43	.47	1.26	2.79	.78	.57	8.65
1937	1.21	.31	1.37	1.04	1.70	1.50	.52	8.81
1938	.98	1.02	1.54	.96	1.09	2.16	1.00	11.31
1939	.93	.88	1.19	1.18	1.04	1.06	.37	5.79
1940	.75	.30	.41	1.66	1.42	.82	1.28	4.65
1941	1.25	.94	1.30	1.35	1.03	1.53	2.16	12.19
1942	3.53	.63	.96	.94	1.05	.94	2.12	11.53
1943	1.15	.64	1.11	1.37	.81	1.55	.50	7.87
1944	1.13	.48	.70	1.25	1.06	1.26	2.30	10.45
1945	.87	1.22	.96	.80	1.26	1.35	1.52	8.33
1946	1.13	.77	1.12	.45	.40	1.01	1.65	5.74
1947	1.91	1.44	.62	.76	1.04	.86	1.11	9.21
1948	2.21	.95	1.28	.70	1.50	1.73	1.22	9.62
1949	1.06	.97	1.78	2.10	.91	1.40	.97	10.74
1950	1.71	.39	1.05	1.85	.96	.89	1.04	7.68
1951	.21	.99	1.20	1.52	.95	.56	1.39	6.71
1952	1.54	.94	3.72	1.87	.71	1.75	1.11	14.53
1953	0	.82	1.00	1.05	.62	1.20	1.56	5.76
1954	1.61	1.23	.62	.71	.36	1.43	.50	3.51
1955	1.21	.94	.81	1.01	1.66	.59	.71	4.75
1956	.51	1.53	1.07	2.21	1.07	.47	1.05	6.70
1957	.92	.64	1.09	3.42	.97	1.26	2.23	13.68
1958	2.55	2.02	1.12	.65	1.32	1.72	.82	10.27
1959	.53	.83	.64	.71	1.62	.77	1.08	4.35
1960	1.94	.39	1.18	1.06	1.59	1.43	.83	6.58
1961	1.44	.93	.94	.27	.60	1.81	1.24	4.08
1962	1.55	.94	1.09	1.18	2.06	.60	1.19	11.45
1963	1.09	.84	.70	1.30	.76	1.03	1.08	3.66
1964	1.27	.75	.44	.61	.47	1.45	1.62	6.16
1965	.12	1.30	2.34	1.61	.84	1.58	1.47	12.64
1966	.73	1.87	1.69	.37	.72	.32	.71	5.14

4.47. The elastic limit and ultimate strength of reinforcing steel are often stated to be a function of bar size owing to differences in the rates of cooling when rolled. The following data were obtained by random sampling of the delivered reinforcing steel for a particular job.

Bar size number	Elastic limit, psi	Ultimate strength, psi
3	72,830	95,420
4	65,120	89,360
4	51,930	81,470
4	53,299	77,165
5	54,400	73,660
5	50,690	78,230
5	52,219	78,031
6	55,450	79,740
7	54,250	83,400
8	46,500	73,390
9	45,690	71,080
10	48,300	76,980
11	48,470	78,500

The steel was classed as "intermediate grade," and the allowable design working stress was 20,000 psi. Bar number equals the diameter in eighths of an inch.

(a) Make a regression analysis of the elastic limit data on bar size. The elastic limit may vary linearly with bar diameter or with bar area. Make a preliminary plot to determine which is the best measure for use in a regression analysis. Test the assumption that the elastic limit is a function of bar size.

(b) Repeat (a) for ultimate strength.

(c) Are elastic limit and ultimate strength correlated?

4.48. Plot the OC curve for the test using statistic $X^{(1)}$, page 413, and compare it with that for $\bar{X}$, Fig. 4.2.2. Which test is to be preferred on this basis? In field implementation, what other factors might influence the choice?

5

Elementary Bayesian
Decision Theory

The first four chapters of this text have been concerned with probability models and statistical analyses of these models. The emphasis has been on finding a realistic description of an unpredictable, natural phenomenon. Although scientific description or explanation may be an intermediate goal, in engineering practice the use of any analysis is ultimately in situations where *the engineer must make a decision*. This decision might be the choice of: (1) a flood-control channel's dimensions, when future flood magnitudes are unpredictable; or (2) the decision of whether to install a traffic control at an intersection where the number of long gaps between randomly passing vehicles determines the delays and safety risks to crossing vehicles on an intersecting street; or (3) the choice among alternate pavement designs when the amount of useful aggregate in the available borrow area is not well established. In this chapter and the next we shall deal with the analysis of decision making when unknown or unpredictable elements are involved in decisions.

In Chaps. 2 and 3 there was little discussion of the influence of the particular application and particular engineer on the description of the random variable. There was a tacit assumption that all engineers in all situations would arrive at substantially the same description of the random variable. In Chap. 4 it was recognized, however, that once we had left the world of purely mathematical models and were forced to relate these models to real data or to make elementary decisions about these models, there were no longer hard and fast rules of "correct" analysis. Two or more reasonable models may exist, different rules for choosing parameter estimators are available, and significance-test conclusions depend upon the convention for the choice of acceptable error probabilities. It is clear, then, that the particular engineer and hence the particular situation necessarily influence the results of applied probability and statistics. The discussions of the methods of previous chapters do not go far enough, however, in revealing the central role of engineering decisions and their influences.

In recent years a new philosophy of applied probability and statistics has been under development. This approach recognizes not only that the ultimate use of probabilistic methods is decision making but also that the individual, subjective elements of the analysis are inseparable from the more objective aspects. This new theory, Bayesian statistical decision theory, provides a mathematical model for making engineering decisions in the face of uncertainty. It derives its name from Thomas Bayes, an English mathematician who introduced the equation now used to relate certain probabilities in the decision model.

In this chapter we shall be concerned with situations where the consequences of a decision depend on some factor which is not known with certainty. We call this factor the "state of nature." The factor might be the total settlement of the soil below a proposed bridge footing, or the fraction of a suburban population which will use a proposed rapid-transit system, or the average annual maximum daily rainfall in a stream's watershed, or whether or not a potentially active geological fault exists in the bedrock several hundred feet below the site of a nuclear power plant. Recognizing that the uncertainty of what the true state of nature is can be expressed as probabilities, the engineer can analyze the alternative decisions facing him to determine which is the optimal choice. In this chapter, we present this method of analysis.

When uncertainty exists regarding the true state of nature, it is often feasible, but not necessarily economical, to obtain more information concerning the state. With the problems given above, for example, the engineer may use drill holes to learn the values of certain soil characteristics at a limited number of points below the footing; he may interview a representative group of the suburban population; he may install gauges and wait to collect several observations of the annual maximum daily rainfall; he may drill a pattern of holes hoping to intersect a fault plane and, if found, examine the recovered cores to assess the likelihood that it is potentially active. The data seldom permit a perfectly confident statement to be made about the true state of nature of interest, but they do provide new information.

Two major questions that we consider in this chapter are:

1. How to combine this new data with the previous probability assignments before making the decision analysis
2. Whether (and how) one should obtain more such information before making a final decision

Section 5.1 discusses the analysis of decisions with a given set of information. Sections 5.2 and 5.3 treat the processing of new or potential information. When this incorporation of new data has been completed, the method of analysis of the decision reverts to that given in Sec. 5.1.

In Chap. 6 we shall treat in some detail decision analysis when the consequences depend upon future outcomes of a process which generates a sequence of random variables, such as successive annual maximum floods or a sequence of concrete cylinders. The uncertain state of nature will be defined as the value of the parameters of the process. Information on the value of the parameters is made available through past observations of the sequence.

5.1 DECISIONS WITH GIVEN INFORMATION

In this section we treat those situations in which a decision must be made based on the available information about nature. In later sections we treat the questions of, first, deciding whether more information should be obtained and, second, how the new information should be processed.

5.1.1 The Decision Model

Components of the model Our concern is decision. What course of action should be taken when uncertainty exists regarding the "true state of nature," i.e., when some aspects of the problem have been treated by the engineer as probabilistic? The decision-making process will be formulated as the process of choosing an action a from among the available alternative actions $a_1, a_2, \ldots, a_n$, the members of an *action space A*. In practice this set or space will have already been reduced to a limited

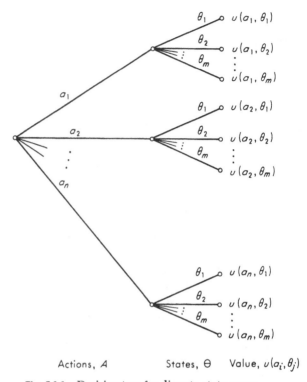

Fig. 5.1.1   Decision tree for discrete-state space.

number by the engineer's exclusion of all but the potentially optimal
courses of action.   Once the decision has been made, the engineer can
only wait to see which state of nature θ in the space of possible states Θ
is the true one.   As a result of having taken action a and having found
true state θ, the engineer will receive value $u(a,\theta)$, a numerical measure
(say, dollars) of the consequences of this action-state pair.

This formulation is displayed graphically in Fig. 5.1.1 as a *decision
tree*.   The process can be viewed as a game: the engineer chooses an
action, a state is chosen in a probabilistic manner "by chance," and the
engineer receives a payoff dependent upon his action and the outcome.

Examples   The formulation of decision problems will be demonstrated
by examples.   They include a construction engineer's problem of selecting a steel pile length when the depth to rock is uncertain.   The available
actions are driving a 40- or 50-ft pile, and the possible states of nature
are a 40- or 50-ft depth to bedrock.   The consequences of any action-

state pair can be given in a *payoff table:*

Table 5.1.1   A simple payoff table

States of nature	a_0 Drive 40-ft Pile	a_1 Drive 50-ft Pile
θ_0 Depth to rock is 40 ft	Correct decision given this state, *no loss*	10-ft piece of pile must be cut off and scrapped, 100-*unit loss*
θ_1 Depth to rock is 50 ft	Driving equipment and crew idle while pile is spliced and welded, 400-*unit loss*	Correct decision given this state, *no loss*

This situation is shown in decision-tree form in Fig. 5.1.2.   The losses have been entered as negative values.   (In this case they represent the "opportunity loss" associated with not making the best choice of action possible in light of the true state.   This is discussed more thoroughly in Prob. 5.7.)

Other examples of decisions under uncertainty include an engineer's choosing one of four cofferdam designs when the state of nature is the magnitude of next year's maximum flow.   Here the state space θ might logically be treated as continuous, 0 to ∞.   The magnitude of that future flow is uncertain to the designer.   Value to the engineer results from the initial cost of his design and its performance under the observed flood. The decision tree is shown in Fig. 5.1.3.

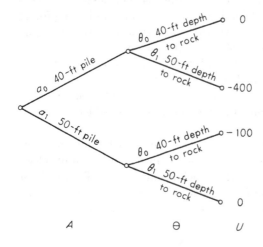

Fig. **5.1.2** Decision tree: pile-length choice.

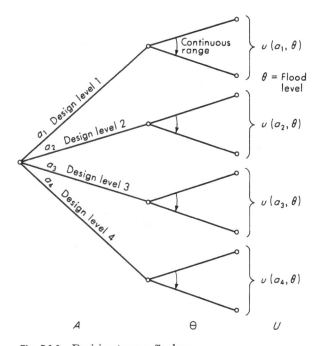

Fig. 5.1.3 Decision tree: cofferdam.

The performance and hence the value of a facility might depend both upon its capacity and upon the demand. If both are uncertain, the possible states are the possible pairs of capacity and demand levels. These pairs can be represented as a two-dimensional state θ_{ij}, representing the joint occurrence of demand level i and capacity level j. If these levels are discrete, the situation can be indicated by another level of branches in the decision tree, as in Fig. 5.1.4. The probabilities of the capacity levels, and possibly of the demand levels,† will depend upon the action or design chosen.

Another engineer might be responsible for designing a small flood-control channel for a 20-year lifetime. The state of nature could be taken as a *set* of values, the annual maximum flows in each of the next 20 years; that is, Θ is a highly multidimensional continuous-state space. The value received from taking an action and finding a point in this state space will depend on the initial cost and future performance associated with these 20 flows.

When choosing between two types of pavement, an engineer may recognize that the cost and performance of one type will depend on the

† For example, an unpredictable flow is induced on a new highway if shorter travel times are provided by increasing the number of lanes.

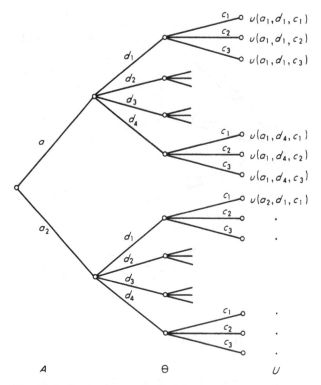

Fig. 5.1.4   Demand-capacity decision tree.

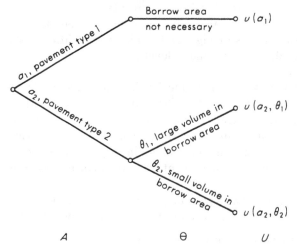

Fig. 5.1.5   Decision tree: pavement choice.

volume of a convenient aggregate borrow pit. A more expensive or lower-quality aggregate will have to be obtained if the volume in the available pit is inadequate. This quantity, the true state of nature, is uncertain (and, in fact, may never become known unless the designer chooses the pavement type requiring use of the pit). The decision tree might appear as in Fig. 5.1.5. Other examples will be found throughout this chapter.

Our problem now is to analyze such decision problems, that is, to develop a method which makes consistent use of all the engineer's information: his degrees of belief in the various possible states, his subsequent observed data, and his preferences among the various possible action-state pairs. We shall see that the method suggests the use of relative likelihoods or probabilities—no matter how subjectively defined and evaluated—to express the engineer's uncertainty, and the use of expected value to rank order the available actions. These concepts, probability and expectation, are not new to us. The validity of the expected-value criterion, however, relies on a properly chosen measure of preference among the outcomes. Hence we require a brief discussion of a new idea: *value theory*.

5.1.2 Expected-value Decisions

Our concern in this section is to demonstrate that if preferences or values of outcomes are properly expressed, then expected value is a logical basis for choosing among alternative actions. As early as Sec. 2.1 this criterion was suggested as reasonable. Here we shall explore its validity. Actually, this discussion is little more than suggestive of the more formal treatment that the subject requires in order that the criterion be firmly based. The reader is referred to other sources† for such discussions. Our purpose here is to suggest their content and necessity.

Consider a simple decision situation where the engineer must choose between actions a_1 and a_2. Action a_1 will lead with certainty to consequences B. Action a_2, on the other hand, involves uncertainty. The state of nature may be θ_1, in which case the consequences are A, or it may be θ_2, in which case C will be the consequences of this action and state. The decision tree is shown in Fig. 5.1.6. A simple example is the choice of pavement type mentioned in the previous section. Another example involves a construction engineer who must decide between a_1, paying for insurance against work stoppage due to rain during a critical 1-day erection operation, and a_2, buying no such insurance. In the latter case, if the true state of nature is θ_1, good weather on that day, the consequences are A, which includes no insurance costs. If, on the other hand, bad

† See Pratt, Raiffa, and Schlaifer [1965], and Hadley [1967], for example.

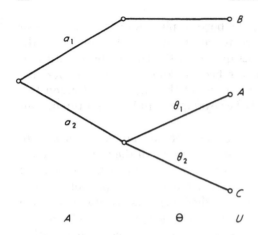

A   Θ   *U*   **Fig. 5.1.6** A basic decision process.

weather forces delay (that is, if state of nature proves to be θ_2), the consequences include monetary losses (and, possibly, professional embarrass-
ment for having failed to buy the insurance!). Before the true state is
known, however, the optimal decision depends upon the likelihood of θ_2
and the relative degrees of seriousness of these various consequences.

Utility function   In order to proceed with the decision-making analysis,
then, we require a numerical assessment by the designer of his preferences
among these outcomes.   Let us assume that he would prefer A to B and
B to C.   (Formally the theory also requires *transitivity*, i.e., that the
decision maker *also* prefers A to C. We shall not include in our present
discussion a number of similar but equally reasonable formal requirements.)   This simple preference statement can be expressed numerically
by any function u (that is, by any numerical values assigned to A, B,
and C) such that

$$u(A) > u(B) > u(C)$$

For example, $u(A) = 1.0$, $u(B) = 0.5$, and $u(C) = 0$; or, alternatively,
$u(A) = +100$, $u(B) = 93.6$, and $u(C) = -10^6$.   Our purpose is to find
a *particular* function u, which we will call a *utility function*, such that
it is logically consistent to make the decision between actions a_1 and a_2
by comparing $u(B)$ and *the expected value of the utility* of action a_2,
$pu(A) + (1 - p)u(C)$.   Here p is the probability that θ_1 is the true state.
Actually, there is no need for such a utility function in these simple
decision problems, but once we have established expected utility as a
valid criterion, we can continue to use it in very complicated problems.

We shall see that the choice of any *two* values of u may be arbitrary;
we may assign the two utilities $u(A)$ and $u(C)$, for example, the values

1 and 0, say, or $+10$ and -10. It is somewhat advantageous to assign the two arbitrary values to A and C, the most preferable and the least preferable outcomes. All other values of u will thus lie *between* $u(A)$ and $u(C)$ in numerical value. Assume then that $u(A)$ and $u(C)$ have been given convenient values. What value should be given $u(B)$ to make expected value a valid decision criterion? To answer this, we must return to our decision maker for a more precise preference statement. Notice that if p, the probability of θ_1 being the true state, were unity, the engineer would choose action a_2 over a_1 because he prefers A (now a sure consequence of a_2) to B. On the other hand, if p were zero, a_1 would be chosen over a_2 if the engineer were going to act consistently with his stated preferences. As for other values of p, it can be said that as p grows from zero, a_1 will be preferred less and less strongly over a_2 until some point p^*, after which a_2 will be preferred over a_1. We assume that by some means of interrogation we can find the crossover value p^* between 0 and 1, such that the engineer is *indifferent* between choosing a_1, with its certain consequence B, and choosing action a_2, the *lottery* with chance p^* of receiving A and $1 - p^*$ of receiving C.

Validity of expected-value criterion   Once we have that value p^*, we *choose to assign* $u(B)$ the value

$$u(B) = p^*u(A) + (1 - p^*)u(C)$$

(Note, in passing, that for any p^* between 0 and 1 the assigned value $u(B)$ will lie between $u(A)$ and $u(C)$ in value; this is consistent with the original preference statements.) By this assignment we have accomplished our intended purpose, that of establishing expected utility value as the consistent criterion by which to choose actions. *The decision maker should choose a_1 if and only if the expected utility given this action choice, $E[u \mid a_1]$, is greater than $E[u \mid a_2]$.* That this is true is verified by noting that:

1. *For all $p < p^*$:*

   (a) The engineer should, to be consistent with his preferences, as reflected in his p^* choice, choose a_1 over a_2, and it is also true that for these values of p (and only these values of p):

   (b) $E[u \mid a_1] > E[u \mid a_2]$

   as can be seen by evaluating the expectations:

   $$u(B) > pu(A) + (1 - p)u(C)$$

and substituting for $u(B)$ its assigned value

$$p^*u(A) + (1 - p^*)u(C) > pu(A) + (1 - p)u(C)$$
$$u(C) + p^*[u(A) - u(C)] > u(C) + p[u(A) - u(C)]$$

The inequality holds because $p^* > p$ and $u(A) - u(C)$ is positive.

2. *Similarly, for all $p > p^*$:*

(a) The engineer should choose a_2 over a_1, and also
(b) $E[u \mid a_1] < E[u \mid a_2]$, since

$$u(C) + p^*[u(A) - u(C)] < u(C) + p[u(A) - u(C)]$$

For example, assume that the engineer was given choices between action a_1 and a_2 for a selection of values of p until he showed by his choices to be indifferent at $p^* = 0.3$. We arbitrarily let $u(A) = 100$ and $u(C) = 0$. Then we should assign

$$u(B) = 0.3(100) + 0.7(0) = 30$$

If the available information suggests that the probability of θ_1 is $p = 0.4$, then the expected utilities of the actions are:

a_1: $E[u \mid a_1] = u(B) = 30$

a_2: $E[u \mid a_2] = pu(A) + (1 - p)u(C) = 0.4(100) + 0.6(0) = 40$

The engineer should choose action a_2, since $40 > 30$.

In other words, *if $u(B)$ is properly assigned to be consistent with the decision maker's stated preferences (here expressed in the statement A preferred to B preferred to C and in the indifference probability p^*), then ranking of the expected values of utility for the actions determines the ranking of the actions* which is consistent with those preference statements. The logic may appear circular and unnecessary. It is, however, precisely the circular reasoning which guarantees that the decision criterion remains valid when more complicated model structures make the decision choice far from obvious and the need for a consistent, formal analysis overwhelming.

Linear transformations   We note in passing, because it will be useful shortly, the validity of the arbitrary choice of any two values of u. To assign two other values, $\bar{u}(A)$ and $\bar{u}(C)$, to these two consequences is equivalent to saying $\bar{u} = c_0 + c_1 u$, that is, that a scaling and shifting (or linear transformation) is being made in the utility function. It is trivial to show that our conclusions remain true for a positive linear transformation of the utilities. Let us simply demonstrate this fact through the previous numerical example. Suppose that we assign a new utility function $\bar{u} = -1 + \frac{2}{100}u$. Then

$$\bar{u}(A) = -1 + \frac{2}{100} u(A) = -1 + \frac{2}{100} (100) = +1$$

$$\bar{u}(C) = -1 + \frac{2}{100} (0) = -1$$

and

$$\bar{u}(B) = -1 + \frac{2}{100} (30) = -0.4$$

This is the same value for the utility of B that we would have reached had we assigned, initially, A a utility of $+1$ and C a utility of -1, since then $\bar{u}(B) = p^*\bar{u}(A) + (1 - p^*)\bar{u}(C) = 0.3(1) + 0.7(-1) = -0.4$. The decision also remains unchanged, of course. Now we have, with $p = 0.4$,

a_1: $E[\bar{u} \mid a_1] = -0.4$

a_2: $E[\bar{u} \mid a_2] = 0.4(+1) + 0.6(-1) = -0.2$

The expected utility of action a_2 still exceeds that of action a_1.

The freedom in assigning the utility function is analogous to that in assigning a temperature scale. Note that, owing to this degree of arbitrariness, no significance can be attached to the relative magnitude of the expected utility values. In the example above, action a_1 cannot be said to be, say $|(40 - 30)/40|$ or "$\frac{1}{4}$ worse" than a_2, for with the $\bar{u}$ utility function it is $|[-0.2 - (-0.4)]/-0.2|$ or "100% worse." It can only be said that "a_1 ranks below a_2."

In simple decision situations such as this, nothing is gained by the steps of finding p^*, finding $u(B)$, and then finding the expected utilities in order to make a decision. It would be easier in this case simply to find p^* and ascertain whether p is greater or less than p^*. We shall become interested, however, in more complex decision problems where the latter approach is impossible, but where the determination of expected utilities will require nothing but familiar operations. The analysis of such complex decision situations will find us replacing lotteries (some embedded within others) by their expected utilities,[†] in order to reduce each action to an associated single number, the expected utility if that action is taken. By extension of the simple case above, we can say that *if* our utility assignment has been made in the manner outlined above, *maximization of expected utility is a criterion of action choice which is consistent with the decision maker's stated preferences.*

[†] See Prob. 5.17. An alternative approach is to replace outcomes by lotteries, and then to maximize the probability of a good outcome rather than expected utility.

Monetary value versus utility; risk aversion It is common in business and engineering practice to express the consequences of decisions and outcomes in monetary terms. It may or may not be appropriate to replace expected utility by expected dollar value as a decision criterion. Let us investigate this problem briefly.

If a decision tree has a number of terminal points, each the conse-quence of an action-state pair, and if dollar values are attached to these consequences, it is in general necessary to replace each of these dollar values by a proper utility in order to justify the use of the expected-value criterion. If the number of such terminal points is large, it is often easier to attempt to plot a smooth function of utility versus dollars, $u(d)$. Then one may simply read the utility value from this curve for each par-ticular dollar value of a consequence. The determination of this utility function follows the established pattern. Suppose the terminal-point dollar consequences are as shown in Fig. 5.1.7. The most preferable out-come is +\$10,000; assign this, arbitrarily, a utility of $u(\$10,000) = 100$. The least preferable value is a loss, −\$5,000; assign this a utility value of 0, $u(-\$5,000) = 0$. Then to find the utility of any other dollar value, say \$0, the engineer must be quizzed to determine at what value of p he would be indifferent between a sure gain of \$0 and a lottery in which he would win \$10,000 with probability p and lose \$5,000 with proba-bility $1 - p$. In personal financial matters most individuals would undoubtedly find themselves in the position where a gain of \$10,000 would be welcome but a loss of \$5,000 would be very serious. For any

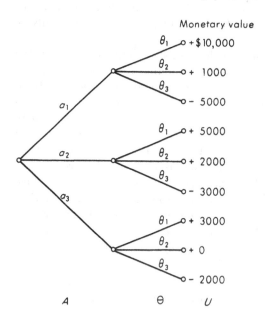

Fig. 5.1.7 Terminal-point dol-lar-value consequences.

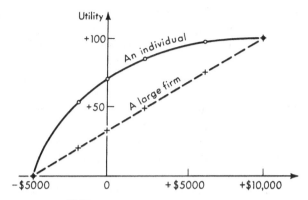

Fig. 5.1.8   Utility curves.

value of p less than, say, 0.7, they might prefer to take no risk at all, i.e., a sure \$0.   Above $p = 0.7$ they might feel that the potential large income was so likely that the small risk of loss was "worth the gamble." Arriving in this way at a p^* of 0.7, \$0 is assigned a utility of

$$u(\$0) = p^*u(\$10{,}000) + (1 - p^*)u(-\$5{,}000)$$
$$= (0.7)(100) + (0.3)(0)$$
$$= 70$$

A number of similar processes obtaining preferences for dollar values of, say, $-\$2{,}000$, $+\$2{,}000$, and $+\$6{,}000$ versus lotteries involving a gain of \$10,000 and a loss of \$5,000 will lead to corresponding values of p^*, and thence to the utilities of $-\$2{,}000$, $+\$2{,}000$, and $+\$6{,}000$.   For indifference p^*'s of 0.5, 0.85, and 0.95, respectively,

$$u(-\$2{,}000) = 50$$
$$u(+\$2{,}000) = 85$$
$$u(+\$6{,}000) = 95$$

A plot of these points and a smooth curve drawn through them is shown as a solid line in Fig. 5.1.8.   From this curve utilities can be assigned to all the dollar-valued consequences of the terminal points of the decision tree in Fig. 5.1.7.   The utility of $+\$1{,}000$, for example, is found to be about 75.   Given probabilities on θ_1, θ_2, and θ_3, say, 0.4, 0.4, and 0.2, the expected utility of taking action a_1 (Fig. 5.1.7) could be evaluated as

$$E[u \mid a_1] = 0.4u(+\$10{,}000) + 0.4u(+\$1{,}000) + 0.2u(-\$5{,}000)$$
$$= (0.4)(100) + (0.4)(75) + (0.2)(0)$$
$$= 40 + 30 + 0 = 70$$

Similar calculations for the other actions would lead to a choice of the action which is optimal in the light of the preferences reflected in the utility curve. We shall discuss these steps in detail in Sec. 5.1.4.

The shape of the solid-line utility curve in Fig. 5.1.8 has been commonly encountered (Grayson [1960]) among individuals making decisions involving dollar values which are large in relation to their working capital. This concave-downward curve reflects the facts that in such situations the utility of a gain of d dollars is less in magnitude than a loss of the same amount, and that, after a gain of d, a gain of d more has, incrementally, a smaller utility. Decision makers with such utility functions are said to be "risk averters."

In many practical circumstances the decisions may not involve dollar sums which are "large" for the individual or body (corporate or social) involved. For a large professional office, for a major business, or for a large political body the dollar values of +$10,000 and −$5,000 are not large. Many decisions may be made each day involving such numbers. For some individuals the comparable "not large" numbers might be +$10 and −$5. In these cases a strongly curved, concave utility assignment may be too conservative. As individuals we might assign an indifference probability of almost $\frac{1}{3}$ to a choice between a sure $0 (no gamble) and a lottery with a chance p of gaining $10 or $1 − p$ of losing $5. So too, a large firm might assign a breakeven probability of about $\frac{1}{3}$ to the earlier situation involving a sure $0 versus a lottery with payoffs +$10,000 or −$5,000. Assigning the largest and smallest dollar-value utilities of 100 and 0, as before, the proper utility assignment to $0 is 33 for the large firm.

This point and others for the same situation of a large firm or political body are shown as crosses in Fig. 5.1.8. The dashed line through them is linear. The implication is that utility can be expressed as $u = c_1 + c_2 d$ or, inversely, that dollars can be expressed as $d = c_3 + c_4 u$. Since, as we have seen, we retain a valid utility function after any linear transformation, it is clear that under these special circumstances dollars themselves are a valid utility measure. The important consequence is that *if utility is linear in dollars, expected monetary value is a valid criterion for choosing among alternative actions.* In most examples to follow, we shall assume that this linearity exists and hence that expected dollar values are adequate for decision-making purposes.

Extracting preferences Methods for interrogating individuals to determine their utility assignments can be found in other references (e.g., Schlaifer [1959] or Pratt, Raiffa, and Schlaifer [1965]). They usually involve choices among hypothetical lotteries in search of indifference points. The assessment may in practice prove difficult. Individuals may, for example, respond less favorably to hypothetical risky situations than to real ones. Some consequences (e.g., dam failures, formwork collapses, highway accidents) involve difficult, but fundamentally unavoidable problems in assessing utilities. When a highway planner puts the limited funds available to him into a new, time-saving highway link rather than into accident-reducing modifications at an existing intersection, he has made such an assessment implicitly. Explicit utility assessments could potentially bring more consistency to the decisions that must be made under such difficult circumstances.† Experience comparable to that being gained in major business decision applications (Howard [1966]) is needed in engineering situations.

Discussion A number of comments are in order about the validity of the expected utility decision criterion and about aspects of utility assignment which are peculiar to civil engineering.

It is important to note that for expected (or average) utility to become the valid criterion of choice, there has been no need for the situation to involve repeated trials. The decision may be, as most civil-engineering decisions are, a "one-shot" affair. Nor does the criterion say that one should always prefer a 50–50 lottery with yields $100,000 or $0 to a sure gain of, say, $40,000. For any number of reasons the decision maker may have an aversion to risk taking. The one-shot and risk-aversion elements will be reflected in the utility assignments which express the individual's preferences among *outcomes of the situation at hand*. For example, the engineer might feel that in this decision situation p^* is 0.7, in which case $u(\$40,000) = 0.7u(\$100,000) + 0.3u(\$0) = 0.7$, if we set $u(\$100,000) = 1$ and $u(\$0) = 0$. Then the preferred decision is the sure $40,000, since its expected utility is 0.7, whereas the lottery has an expected utility of only $(0.5)(1) + (0.5)(0) = 0.5$. Thus, the utility assignment has expressed the engineer's aversion to risking a sure $40,000 against a possible $100,000. He will accept the risk only if the likelihood of success is sufficiently high, for example, $p > 0.7$.

It should be clear that the utility assignments to consequences may be highly subjective. Depending on circumstances, such assignments may differ from individual to individual and even from time to time for one individual. They should represent the preferences of the moment of the individual responsible for the decision.

Civil engineers make many professional decisions under conditions

† A study aimed at making highway design decisions assessed the value to the French society as a whole of an individual French life as about $30,000 in 1957 (Abraham and Thedié [1960].)

of uncertainty about the future. Do these professional decisions bring them value? Can the value accruing from a particular professional decision be directly identified? Or is value only received over a long period of time from many decisions with few apparent direct relationships to the engineer, beyond the generalization that consistently wise decisions are associated with long-term economic gain and increase in professional stature?

In the field of business, where much of applied decision theory has been developed, an executive can often specifically identify value accruing from a specific decision. He is operating for himself or in the name of the concern, and an economic gain or loss is more apparent. The operations of a professional civil-engineering office do parallel ordinary business operations in that a profit or loss is found, but only a small part of the decisions made pertain to the management of the engineering business itself. Most of the decisions of interest are professional in nature. That is, the engineer is acting as an agent for his client, and he has a professional obligation to attempt to make a design with characteristics optimal to the owner. The situation is complicated since some aspects of decisions are predominantly client-related, whereas others are more closely engineer-related. For example, absolute minimum cost to the owner is rarely a goal for the engineer because of other professional factors that must be considered, such as public safety, appearance, etc. The values of these latter factors to the owner may be extremely difficult to ascertain. But the saving of $10,000 in the cost of construction of a building is of immediate importance to the client or owner. How important is this saving to the engineer who must make the critical design decision? Utility to the client is not equivalent to the utility to the engineer. The reverse may be in part true; the saving of $10,000 in construction costs may actually reduce the fee received by the engineer for the job. Another example will follow in a bridge-design illustration in Sec. 5.1.4.

How, in short, should the preferences of the client be reflected or transformed into the preference assessments of the engineer responsible for the decision? These are questions which have received no quantitative answers. But these are not new problems generated by the methodology of statistical decision theory. They are, in fact, only new ways of expressing some of the key issues underlying professionalism. By dealing with them in a more formal, mathematical framework, decision theory may provide a basis for further discussion of them and for a rational, quantitative resolution.

Despite the immediate practical difficulties in assessing utilities, the concept remains fundamental to the understanding of rational decision making under uncertainty.

Closure   It should be repeated that our development of the value theory underlying statistical decision theory has been nonrigorous and only suggestive of the basis for the expected utility criterion.   Briefly, the use of the criterion is justified by the method of making the utility assignments. The assignments are consistent both with the engineer's preferences and with the use of the expected utility criterion.   In the analysis of decision trees in Sec. 5.1.4, we shall use this fact over and over to replace an array of branches and associated outcomes by a single number—the expected utility.

　　We next must reconsider briefly the probability assignments on the decision tree.   We shall find that some of the earlier statements in Chap. 2 about the degree of belief or operational interpretation of probability can now be made more explicit.

5.1.3 Probability Assignments

The probability assignments on a decision tree are more familiar than utilities.   Their treatment has been the subject of all the preceding chapters.   The sources of these probabilities may include observed frequencies, deductions from mathematical models, and, in addition, measures of an engineer's subjective degree of belief regarding the possible states of nature.   The former two sources have been emphasized up to this point, but the broader, subjective interpretation can now be given a firmer basis.   As mentioned briefly in Sec. 2.1, such statements as "The odds are 2 to 1 that this job will be completed in 2 months," "I'll bet a quarter to your nickel that the depth to bedrock is less than 50 ft," or "The most likely value of the mean trip time is 8 minutes, but it may be as large as 10 or as small as 7" all represent expressions of the engineer's judgment about unknown quantities.   These quantities may represent repetitive experiments, one-of-a-kind situations, fixed but unknown factors, or even parameters of the distributions of repetitive random variables.   The subjective expressions above can all be related to probability distributions assigned to these quantities.   An excellent engineering argument for this position is presented by Tribus [1969].

　　Both observed frequencies and these "more subjective" probability assignments can now be interpreted as those numerical summaries of an individual's degrees of belief or relative likelihoods *upon which that individual is prepared to base his decision*.   The implication seems to be that another individual might make different probability assignments to the same decision tree.   This is true; it is consistent with the observation that different engineers have had different experiences and use different reasoning processes.

　　This operational interpretation of probability as an individual's decision-making aid is not, however, inconsistent in practice with the more restricted relative-frequency interpretation; if a large amount of relevant observed data is available, two reasonable individuals will undoubtedly use it in the same way, obtaining the same personal probability of the event.   The techniques for determining probability assignments when sufficient data and/or theoretical models are available have been discussed.

In those cases where the probabilities depend strongly upon judgment, however, the extraction of an engineer's degrees of belief in terms of personal or subjective probabilities may not be an easy task.   When accustomed to the idea, an engineer may be able to make direct statements.   For example, he might say:

> Based on the information available to me at the moment, the probability is 0.3 that the volume of useful aggregate in this borrow pit is about 4000 yd³ and 0.7 that it is about 5000 yd³.

Or, if a continuous-state space is desirable:

> There is a 50–50 chance that the volume is less than 4750 yd³. If it proves to be less than 4750 yd³, however, there is a 50–50 chance that it is less than 3800 yd³.   Given that it is greater than 4750 yd³, on the other hand, the probability is 50 percent that it is greater than 5000 yd³.   There is no significant chance that the volume is less than 3000 yd³ nor more than 5250 yd³.

The points implied by these latter statements and a smooth curve through them yield the CDF in Fig. 5.1.9.

Alternatively, these quantifications of degrees of belief can be extracted in a manner which more strongly reflects the decision situation at hand.   The engineer can be asked to state a preference between two lotteries, one artificial, the other involving the state of nature or event under consideration.   His preference will indicate a bound on the probability sought.   For example, assume that an engineer states a preference for an artificial "lottery," lottery I, which offers a sure $5, over lottery II, which offers him $0 if the amount of useful aggregate is found to be less than 4500 yd³, and $8 if it is found to be more than 4500 yd³.   The implication of this preference statement is that the expected utility of the first lottery is greater than that of the second, or

$$u(\$5) > pu(\$0) + (1 - p)u(\$8)$$

in which $p = P[\text{volume} \leq 4500]$.   Assuming, for simplicity, that utility and dollars have been found to be linearly related (Sec. 5.1.2) in this

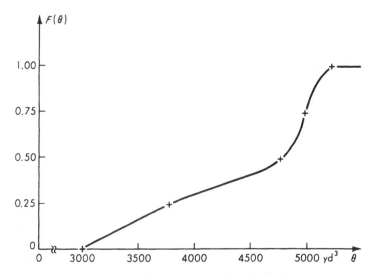

Fig. 5.1.9 Subjective cumulative probability distribution on usable aggregate volume.

range, and letting $u(\$d) = d$, the engineer's preference implies that

$$(1)(5) > (p)(0) + (1 - p)(8)$$

or

$$p > \tfrac{3}{8}$$

By changing the values of the rewards or the values of the probabilities (or both) in lotteries I and II, one can presumably find a situation where the engineer states that he is indifferent between the two lotteries. In this case the expected utilities can be set equal and the implied value of p solved for. This probability is clearly the one upon which the engineer is prepared to make a decision which rests in part, at least, upon the quantity of aggregate in the borrow pit.

Subjective probabilities have proved (Howard [1966]) a useful way to transmit judgments within an organization where numerous individuals have to contribute to the components of a larger decision. For example, they can help a consulting engineer communicate the uncertainty in his estimate of a factor in his field specialty. A soils consultant, who has pulled together various kinds of data and experience to estimate that the mean shear strength of a particular soil layer is 400 lb/ft², might go one step further and state that he is "50 percent sure" that the strength is with ± 10 percent of that estimate. The figure 10 percent might have been arrived at by varying that fraction until he was indifferent between lottery I, a coin flip offering \$0 or \$1, and lottery II, offering \$1 if the true mean strength is found to be within ± 10 percent of his estimate and $0 if it were outside of those limits. By transmitting two numbers—a best estimate 400 and a "probable error" ± 10 percent—rather than only a conservative estimate, say, 350 lb/ft², this engineering specialist will permit a balanced, economical design decision to be made by those supervising engineers who have gathered *all* such individual judgments together with the costs and the alternatives. This approach avoids uneconomical compounding of "safety factors" which can arise when each opinion is transmitted simply as a conservative estimate. This is true whether or not a formal decision analysis is made.

We can afford no further discussion on the subject of trying to get probability assessments from individuals, but we refer the reader to other sources (e.g., Schlaifer [1959], Pratt, Raiffa, and Schlaifer [1965], Grayson [1960], and Howard [1966]) for additional discussion. Interesting questions remain concerning the "policing" of inconsistent probability statements and lottery choices, the influence of the hypothetical nature of the lotteries, and the extension of the procedures to small probability values.

Like the procedure for utility assignments, the assessment of judgment-based probabilities utilizes the decision maker's preferences among simple, obvious lotteries to extract numbers which will become parts of the actual larger, more complicated decision problem. Many of the probabilities in a decision analysis will be determined by the methods developed in earlier chapters, that is, by manipulation of probabilities and random variables in a manner consistent with the rules of probability theory. Nonetheless, all applications rest fundamentally on an individual's probability assignment, whether it is extracted by preference statements or implied in the adoption of a particular distribution.

In the following section we will put together the pieces of the decision problem and analyze the problem to determine the best decision consistent with the probability and utility assignments.

5.1.4 Analysis of the Decision Tree with Given Information

We have discussed how, in principle, the elements of the decision problem are assembled. These elements are the available actions, the possible states with their assigned probabilities (representing the information available to the decision maker), and the utilities associated with the action-state pairs (representing the decision maker's preferences). The analysis of the problem determines the optimal action consistent with the individual's probability and preference assignments.

The criterion of choice among actions is maximum expected utility. That this is a valid criterion has been guaranteed by the method of utility assignment. The analysis of the decision situation reduces then to calculation of expected utilities and selection of the action which shows the largest expected value.

Finding expected values is a familiar process by this time, but we include several examples to illustrate its application in decision analysis. We shall adopt the short-hand notation $P[\theta_i]$ to mean "the probability that the true state of nature is θ_i." Notice in the examples that, if it is numerically valued, the state of nature θ will be treated as if it were a random variable. Since there is no opportunity for ambiguity in this chapter, we shall not, however, distinguish by upper and lower case between the random variable and its specific values.

Illustration: Pavement design, discrete-state space As a simple example we shall complete the analysis of the decision between two pavement types, Fig. 5.1.5. Uncertainty exists regarding the amount of aggregate in the available borrow pit. Assuming that it is sufficient to use just a pair of discrete states ($\theta_1 =$ "large" volume of aggregate, say about 5000 yd^3, and $\theta_2 =$ "small" volume, about 4000 yd^3), we shall use probability assignments extracted as in Sec. 5.1.3: 0.7 and 0.3, respectively. The engineer's utility assignments for the action-state pairs have been assessed to be 60 and 200 and 0 for $u(a_1)$, $u(a_2,\theta_1)$, and $u(a_2,\theta_2)$, respectively. The final tree with the numerical assignments indicated appears in Fig. 5.1.10. The expected value promised by choosing a particular action is indicated within a box at the right-hand end of the action branch.

For action a_2 it is 140. This expected value can be interpreted, recall, as the certain gain equivalent in value (in the decision maker's eyes) to the existing a_2 "lottery." In essence this is the "selling price" he would accept for his "ticket" in the lottery. Because 140 is greater than 60, the indication is that a_2 is the choice which is consistent with the engineer's assessment of the preferences and judgments (probabilities). It is important to realize that the value

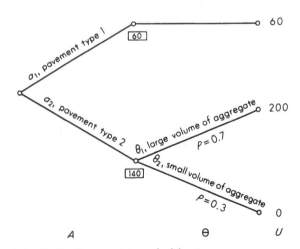

Fig. 5.1.10 Pavement-type decision tree.

which will be received if a_2 is taken will *not* be 140 but rather either 0 or 200.

It is also important to recognize the distinction between a *good decision* and a *good outcome*. The analysis suggests that a_2 is the better decision, but there is a possibility that the volume will turn out to be inadequate; that is, the outcome may be bad. Such an outcome does *not* imply that a bad decision was made. At the earlier time when the choice had to be made, the *available information* did not include the fact that the volume was inadequate. The available information at that time was summarized by the engineer as a probability of only 0.3 that volume was inadequate.

Notice, too, that even if the engineer had judged that the volume was "somewhat more likely than not" to be inadequate, say $P[\theta_1] = 0.4$ and $P[\theta_2] = 0.6$, action a_2 would have remained the better decision; the conclusion is that a decision should not necessarily be based on the "most likely" state, without consideration of the consequences.

Illustration: Pavement design, continuous-state space The engineer may have preferred a more detailed model of the state space in the pavement-choice example, a space which permitted the possibility of any of a range of values. Assume, then, that the state space of aggregate available in the borrow pit is the positive line, 0 to ∞. Assume, too, that utility is linearly related to construction cost dollars. We shall assume arbitrarily a scale of $u(-\$100,000) = -100,000$ and $u(\$0) = 0$. In other words, we shall use expected monetary value as a criterion. Pavement type II requires 4750 yd^3 of aggregate. If this amount is not available in the borrow pit, the additional material, $(4750 - \theta)$ yd^3, will have to be hauled a great distance at added cost. Assume, then, that the construction cost of type II is \$95,000 if the pit volume exceeds 4750 yd^3. But if the volume is less than 4750 yd^3, the cost will be \$95,000 plus \$10 per imported cubic yard, or $\$95,000 + \$10(4750 - \theta)$. Pavement type I has a cost of \$100,000. In short, the utility of choosing type I is $-\$100,000$ and the utility of the type II choice is

$$u(a_2,\theta) = \begin{cases} -95,000 - 10(4750 - \theta) & \theta \leq 4750 \\ -95,000 & \theta \geq 4750 \end{cases}$$

or

$$u(a_2\theta,) = \begin{cases} -142,500 + 10\theta & \theta \leq 4750 \\ -95,000 & \theta \geq 4750 \end{cases}$$

Assume that the engineer's probability assignment on the available volume θ, is the CDF indicated in Fig. 5.1.9. The decision tree can be indicated as shown in Fig. 5.1.11. Since the discrete-state "branches" are no longer appropriate, the value of the tree as a visual aid is weakened when continuous variables are involved. It remains useful conceptually however. The expected values associated with actions a_1 and a_2 are again shown in boxes.

The determination of the expected value of a_2 now requires integration:

$$E[u \mid a_2] = \int_0^\infty u(\theta,a_2)f(\theta)\, d\theta \tag{5.1.1}$$

This evaluation requires that a PDF be determined from the given CDF and that, analytically or numerically, the integration be accomplished. The approximate result is indicated in Fig. 5.1.11 to be $-97,900$. Action a_2 is preferable. The continuous-state model may be more detailed than the discrete-state representation, but any model which is accurate enough to give the proper action

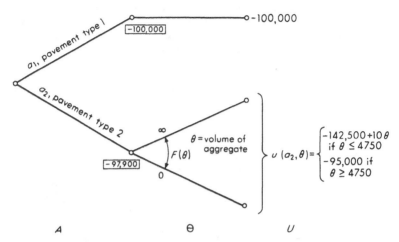

Fig. 5.1.11 Continuous-state-space model of pavement-type decision.

choice is sufficient. In all further examples in this chapter we shall restrict our attention to discrete models.

Recall that the expected utility criterion is sufficient only to obtain a ranking of actions. It should *not* be concluded, after inspecting the expected values −\$100,000 and −\$97,900, that a_2 is about 2 percent "better" than a_1, only that a_2 is better. Recall that assignment of a different, linearly transformed utility scale could yield a different proportion between the expected values, but never a different conclusion as to the better decision.

Illustration: John Day River bridge Two spans of a bridge over the John Day River near Rufus, Oregon, failed during the floods of December, 1964. The primary cause was "abnormal" flooding of the river. The foundation of the center pier was undermined by the unusually fast current, and two 200-ft spans collapsed. The bridge was 1 year old and had cost \$2,500,000.

The original plans for the bridge called for this pier to bear on bedrock, but after the contractor had excavation difficulties, he was allowed to found the pier on compacted sand and gravel. The outcome was bad, but the decision may not have been. The highway department explained that the change was allowed because the bridge would eventually be in the reservoir formed by the John Day dam, and by then it would not be exposed to fast currents. The change speeded construction and saved \$150,000 in construction cost. It is interesting to note that after the failure the decision was publicly called a "clear error in judgment."

The foundation was "designed" for 37,000 cfs. The river reached 40,000 cfs during the flood. Three older bridges on the same river were undamaged even though some of their piers also bear on gravel; performance under a given "load" cannot be precisely predicted when such a complex process as scour and erosion are involved. The cost of reconstruction was estimated to be \$765,000.†

† See *Engineering News-Record* of January 7, 1965, and March 11, 1965, for further details.

This type of decision situation is very often encountered in civil-engineering practice. It is common to consider design changes when new information is obtained during construction. Reanalysis of this foundation-design decision provides an excellent opportunity to illustrate the use of statistical decision theory. All probability statements and conclusions that follow are entirely fictitious and are designed only to illustrate the methods.

The decision tree is shown in Fig. 5.1.12. The engineer must choose

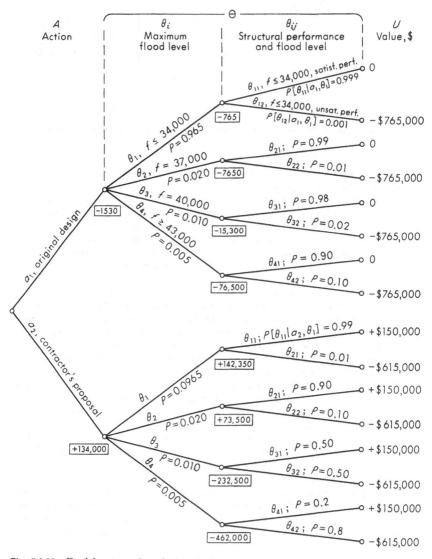

Fig. 5.1.12 Decision tree: foundation design.

between action a_1, requiring strict compliance with original plans, and action a_2, accepting the contractor's proposal to found the pier on sand and gravel at a saving of construction money and time.

Based on historical records, the engineer assigns probabilities to the four possible discrete flood states that he thinks are sufficient to describe this aspect of the problem. These must be the probabilities of the *largest* flow in the time before the foundations are protected by the dam. The states selected are:

$$\theta_1 = [\text{flow} \leq 34{,}000 \text{ cfs}]$$
$$\theta_2 = [\text{flow} \approx 37{,}000 \text{ cfs}]$$
$$\theta_3 = [\text{flow} \approx 40{,}000 \text{ cfs}] \qquad (5.1.2)$$
$$\theta_4 = [\text{flow} \geq 43{,}000 \text{ cfs}]$$

Their probabilities are independent of the action taken and are estimated semi-subjectively from available records of past flows (see, for example, Sec. 4.5.2):

$$P[\theta_1] = 0.965$$
$$P[\theta_2] = 0.020$$
$$P[\theta_3] = 0.010 \qquad (5.1.3)$$
$$P[\theta_4] = 0.005$$

The conditional probabilities of successful or unsuccessful performance of the original design and the proposed design, *given* the maximum flood level, appear in Table 5.1.2 and in the decision tree. Note, for example, that the

Table 5.1.2 Conditional probabilities of performance given the design and the maximum flood level

Original design a_1

Maximum flood level / *Performance*	$\leq 34{,}000$	$37{,}000$	$40{,}000$	$\geq 43{,}000$
Satisfactory	0.999	0.99	0.98	0.9
Unsatisfactory	0.001	0.01	0.02	0.1

Proposed design a_2

Maximum flood level / *Performance*	$\leq 34{,}000$	$37{,}000$	$40{,}000$	$\geq 43{,}000$
Satisfactory	0.99	0.9	0.5	0.2
Unsatisfactory	0.01	0.1	0.5	0.8

(fictitious) engineer has estimated that the probability of the original design's failing under a peak flow of 37,000 cfs is about 10 percent of that of the revised design. The states are labeled θ_{i1} and θ_{i2} for flood state i and satisfactory or unsatisfactory performance, respectively. Such performance probabilities can only be subjective assignments reflecting the engineer's judgment about the reliability of these particular designs under these particular conditions. Related experience and professional knowledge represent the only sources of information. Data cannot be obtained before the designs are completed.

The gains and losses in dollars are estimated as shown in the decision tree. The cost of poor performance, i.e., foundation and pier failure, will be taken as $765,000; the construction-cost savings associated with the proposed design are $150,000. Failure of a revised design has a net cost of $765,000 − $150,000 or $615,000. Assume that the engineer is making a decision to maximize the expected utility of the people of Oregon and hence that utility to the people and dollars can be assumed to be linearly related.

Analysis of the tree proceeds by finding the expected utilities or monetary values at each branching point. The right-hand-most expected values on the decision tree are the expected values *given* a particular action and flood level θ_i. For example, the top boxed value is:

$$E[u \mid a_1, \theta_1] = P[\theta_{11} \mid a_1, \theta_1]u(\theta_{11}, a_1) + P[\theta_{12} \mid a_1, \theta_1]u(\theta_{12}, a_1)$$
$$= (0.999)(0) + (0.001)(-765,000)$$
$$= -765 \tag{5.1.4}$$

The expected utilities for each action are found in turn by weighting the expected utilities given the flood level by the probabilities of these levels. For example:

$$E[u \mid a_1] = \sum_{i=1}^{4} E[u \mid a_1, \theta_i]P[\theta_i]$$
$$= (-765)(0.965) + (-7650)(0.020) + (-15,300)(0.010)$$
$$+ (-76,500)(0.005)$$
$$= -1530 \tag{5.1.5}$$

The better decision is to choose the proposed revised design because $E[u \mid a_2]$ is greater than $E[u \mid a_1]$.

Several observations are in order. First, the problem could also have been represented by a simple tree with only one set of state branches, as shown in Fig. 5.1.13. There are eight such states, the combinations of four flood levels and two performance levels. Call them $\bar{\theta}_k$, $k = 1, 2, 3, \ldots, 8$. Note that each $\bar{\theta}_k$ corresponds to an earlier θ_{i1} or θ_{i2}, $i = 1, 2, 3, 4$. For each action, there is associated with each state its probability of occurrence. Each probability is the product of a probability of flood level and a conditional probability of performance level given that flood level. The products of these eight probabilities and the eight dollar values give the expected monetary value of an action. In symbols, the calculation is

$$E[u \mid a_j] = \sum_{k=1}^{8} u(\bar{\theta}_k, a_j)P[\bar{\theta}_k] = \sum_{i=1}^{4}\sum_{n=1}^{2} u(\theta_{in}, a_j)P[\theta_{in} \mid a_j]$$
$$= \sum_{i=1}^{4}\sum_{n=1}^{2} u(\theta_{in}, a_j)P[\theta_{in} \mid \theta_i, a_j]P[\theta_j] \tag{5.1.6}$$

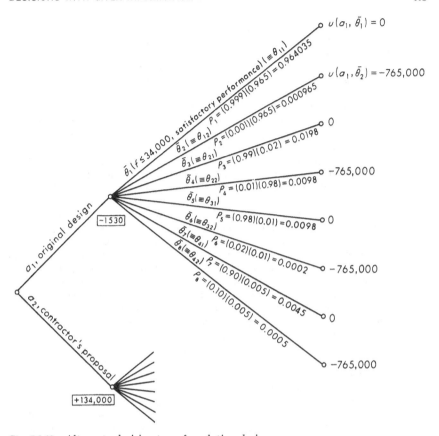

Fig. 5.1.13 Alternate decision tree: foundation design.

The reader can quickly convince himself that this is the same number obtained above, where that computation can be shown as:

$$E[u \mid a_j] = \sum_{i=1}^{4} E[u \mid \theta_i, a_j] P[\theta_i]$$

$$= \sum_{i=1}^{4} P[\theta_i] \sum_{n=1}^{2} u(\theta_{in}, a_j) P[\theta_{in} \mid \theta_i, a_j]$$

$$= \sum_{i=1}^{4} \sum_{n=1}^{2} u(\theta_{in}, a_j) P[\theta_{in} \mid \theta_i, a_j] P[\theta_i] \tag{5.1.7}$$

The equivalence of the two computations implies that the expected utilities in the original tree, Fig. 5.1.12, can be computed either by working to the left with sets of *conditional expectations* to obtain the expected utility of an action (as done) or by working to the right with sets of conditional probabilities to find the probability of each possible terminal point (and taking expectations only as a final step).

As a third alternative for the decision tree, the engineer might have preferred only two states, those prompted by observing that there is only a pair of utilities appearing in the larger decision trees (Figs. 5.1.12 or 5.1.13). These two states could be called "satisfactory performance" and "unsatisfactory performance." Their probabilities could be found by adding the probabilities of the mutually exclusive events which make them up. For example, the probability of satisfactory performance is the sum of the probabilities of $\bar{\theta}_1$, $\bar{\theta}_3$, $\bar{\theta}_5$, and $\bar{\theta}_7$ in Fig. 5.1.13. The purpose of mentioning this large number of alternative decision trees is simply to make the point that there is no single "proper" formulation. The engineer has freedom in how he chooses to view a problem, and the several possible views will lead to the same action.

A second major observation is that the decision may well have been altered had the engineer assessed his *personal* preferences in making the utility assignments. For illustration, let us simplify the decision tree to the elementary one shown in Fig. 5.1.14. (Note that the engineer has chosen now to assume that there is no probability of unsatisfactory performance with action a_1.) The engineer responsible for the decision might recognize that the consequences *to him* of each action-state pair are quite complicated. Failure to accept the immediate, sure construction savings (i.e., choosing action a_1) might brand him as inflexible and overconservative. If he takes this action, the possible poor performance that might have been (and was!) found under a_2 could never take place, and the engineer's conservatism (appropriate, after the fact) would never be appreciated. If the engineer takes action a_2, either he will obtain the benefits for himself and the people which he represents (θ_1) or the poor performance will be observed (θ_2). In the latter case, the consequences to the engineer will depend in part upon the review of his decision which is likely to follow. If the flood which causes the failure is truly a large one, larger at least than any on record, the engineer may escape personal criticism even though financial losses are quite large.

Viewed in this light, the professional decision situation, like most, is quite personal in nature. The methodology of decision theory remains applicable, however. Only a preference statement on the part of the engineer is needed. Assign, arbitrarily, $u(a_2,\theta_1) = 10$ and $u(a_2,\theta_2) = -10$. Then, through interro-

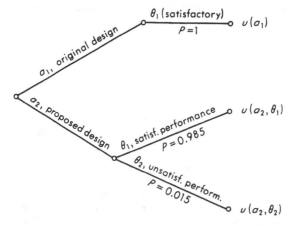

Fig. 5.1.14 Simplified tree: bridge-foundation design.

gation, an indifference probability p^* is sought. It is the value at which the engineer has no preference between (1) obtaining the consequences of a_1 for sure and (2) being given a chance in a lottery where the consequences of a_2 and θ_1 will be found with probability p^* and the consequences of a_2 and θ_2 will be his with probability $1 - p^*$. Then

$$u(a_1) = p^* u(a_2, \theta_1) + (1 - p^*) u(a_2, \theta_2)$$
$$= p^* 10 + (1 - p^*)(-10)$$

Assume that the engineer's preference statements imply that $p^* = 0.98$. (That is, $1 - p^*$ is about equal to the likelihood that the engineer can correctly name the top card on a deck of cards.) Then

$$u(a_1) = (0.98)(10) + (0.02)(-10)$$
$$= 9.8 - 0.2 = 9.6$$

With these preferences the engineer should still take action a_2, since

$$E[u \mid a_1] = 9.6$$

while

$$E[u \mid a_2] = 0.985(10) + (0.015)(-10) = 9.7$$

As a final observation, notice that the dollar consequences of an action-state pair may also be uncertain. In this case, still another level of branches can be established to represent the various levels of costs associated with a given action and outcome. Conditional probability assignments must be made to each new branch. The first step in the analysis will be to calculate the right-most expected value, which will now be the *conditional expectation* of the dollar consequences given the action and outcome. With this value the analysis proceeds exactly as above. In this sense the dollar estimates in the original tree (Fig. 5.1.12) can be thought of as *expected* dollar consequences given the action and outcome. Notice, however, that this interpretation is valid only if a utility and dollars are linear for the decision maker. If not, the dollar values must all be converted to utilities *before* taking expected values.

Illustration: General capacity-demand problem The preceding problem can be considered to be a special case of a more general capacity-demand type of problem. The flow rate is a measure of demand; the resistance of the foundation to this flow is a measure of capacity. The relationship between capacity and demand levels determines the performance of the system and hence the payoff or consequences: capacity greater than demand defines a more-or-less satisfactory behavior; demand greater than capacity implies a system failure of some magnitude.

Both capacity and demand may be uncertain. Commonly the probabilities of demand levels are largely independent of the action or design, whereas the specified design level or action determines the most likely capacity as well as the dispersion about that value. Performance and consequences depend upon the action and the value of the two-dimensional state, demand and capacity. If for a given action, 10 discrete demand levels and 10 discrete capacity levels are considered, 100 combinations of levels exist and each must be considered.

As mentioned in the bridge-foundation-design illustration, it may be possible to simplify the decision-tree analysis by lumping many of the demand-capacity

Table 5.1.3 Safety margin levels

Capacity	Demand	Safety margin	Probability of capacity-demand pair	Safety-margin characterization
100	0	100	$(0.1)(0.4) = 0.04$	Adequate
100	50	50	$(0.1)(0.3) = 0.03$	Adequate
100	100	0	$(0.1)(0.2) = 0.02$	Marginal
100	150	-50	$(0.1)(0.1) = 0.01$	Inadequate
120	0	120	$(0.2)(0.4) = 0.08$	Adequate
120	50	70	$(0.2)(0.3) = 0.06$	Adequate
120	100	20	$(0.2)(0.2) = 0.04$	Marginal
120	150	-30	$(0.2)(0.1) = 0.02$	Inadequate
140	0	140	$(0.4)(0.4) = 0.16$	Adequate
140	50	90	$(0.4)(0.3) = 0.12$	Adequate
140	100	40	$(0.4)(0.2) = 0.08$	Adequate
140	150	-10	$(0.4)(0.1) = 0.04$	Inadequate
160	0	160	$(0.3)(0.4) = 0.12$	Adequate
160	50	110	$(0.3)(0.3) = 0.09$	Adequate
160	100	60	$(0.3)(0.2) = 0.06$	Adequate
160	150	10	$(0.3)(0.1) = \underline{0.03}$	Marginal
			1.00	

pairs into a smaller number of common *performance* levels. This is possible if the consequences are equivalent for each action-state pair in a particular performance level.

For example, for a particular action, the decision maker may consider the four possible demand levels 0, 50, 100, and 150 and the four possible capacity levels 100, 120, 140, and 160. Assume that the probabilities of the demands are 0.4, 0.3, 0.2, and 0.1, respectively, whereas the probabilities of the capacities are 0.1, 0.2, 0.4, and 0.3. If performance can be related to simply the "safety margin" or the difference between capacity and demand, it is appropriate to calculate the possible values of the safety margin with their likelihoods. Assuming independence of demand and capacity, these values all appear in Table 5.1.3. The performance levels associated with each safety margin are shown. Negative values of the safety margin are termed "inadequate," non-negative values less than 30 are called "marginal," and larger values are considered "adequate." In these terms the probabilities of the three performances are 0.07, 0.09, and 0.84 for the inadequate, marginal, and adequate safety margins. Analysis of the tree could continue based on only these three outcomes, with their respective probabilities and utility values.

5.1.5 Summary

The simple decision model consists of a set of possible actions, a_1, a_2, . . . , a_n, and a set of possible states, $\theta_1, \theta_2, \ldots, \theta_m$. Decision theory prescribes how one should choose among the actions when the state is uncertain. The prescribed decision criterion is maximization of expected utility.

The criterion is appropriate if the decision maker's preferences among the various possible action-state pairs are expressed in terms of a particular utility measure $u(a_j, \theta_i)$ (or any positive linear transformation of that measure). The proper utility measure may or may not coincide with simple monetary units.

In decision analysis, probability assignments $p(\theta_i)$ are also interpreted with reference to the particular decision maker, and hence they are, in general, subjective quantities rather than, say, physical attributes of the uncertain states.

Analysis of the decisions involves only the computation of expected utilities $E[u \mid a_j]$ and choice of the action with the largest expected utility (or minimum expected loss).

5.2 TERMINAL ANALYSIS

In the previous section we discussed the analysis of a decision tree when the various probability assignments were available. In this section we consider *terminal analysis*, the analysis of the tree when new information about the states has become available. We shall learn in Sec. 5.2.1 to incorporate the new information with the old to yield new probability assignments. After this has been done, the decision-tree analysis is identical to that discussed in Sec. 5.1.

5.2.1 Decision Analysis Given New Information

Our interest turns now to the problem of reassessment of the probabilities of state in the light of new information. No new ideas are involved; we wish only to reconsider some familiar ideas in the context of a decision situation. For clarity we denote our initial or prior probabilities as $P'[\theta_i]$ and our revised or posterior probabilities as $P''[\theta_i]$. (For notational simplicity we suppress possible dependence of these state probabilities on the action a.)

Calculation of posterior probabilities The prior probabilities of state are known. As the result of an experiment e,† we have observed new information, namely, an outcome z_k in the space Z of all possible outcomes of that experiment. The problem is to combine this new information with the prior probabilities of the state $P'[\theta_i]$ to obtain posterior probabilities of state $P''[\theta_i]$. We do this through Bayes' rule, Eq. (2.1.13). This rule,

† This is an experiment in the general sense of Sec. 2.1.1.

recall, is derived as follows. Consider the two alternative products which equal the probability of z_k *and* θ_i:

$$P[\theta_i \mid z_k]P[z_k] = P[z_k \mid \theta_i]P'[\theta_i]$$

in which an obvious short-hand notation has been adopted, e.g., $P[z_k \mid \theta_i]$ is the probability that the outcome of the experiment would be z_k if θ_i is the true state of nature.

Solving for $P[\theta_i \mid z_k]$ or $P''[\theta_i]$,

$$P''[\theta_i] \equiv P[\theta_i \mid z_k] = \frac{P[z_k \mid \theta_i]P'[\theta_i]}{P[z_k]}$$

Substituting for $P[z_k]$ in the denominator,

$$P''[\theta_i] = \frac{P[z_k \mid \theta_i]P'[\theta_i]}{\sum_j P[z_k \mid \theta_j]P'[\theta_j]} \tag{5.2.1}$$

Bayes' rule states that the posterior probability of a state is the product of three factors:

$$\begin{pmatrix} \text{Posterior proba-} \\ \text{bility of } \theta \\ \text{given the} \\ \text{sample outcome} \end{pmatrix} = \begin{pmatrix} \text{normalizing} \\ \text{constant} \end{pmatrix} \begin{pmatrix} \text{sample likeli-} \\ \text{hood given } \theta \end{pmatrix} \begin{pmatrix} \text{prior proba-} \\ \text{bility of } \theta \end{pmatrix}$$

$$\tag{5.2.2}$$

$$P''[\theta_i] = P[\theta_i \mid z_k] = NP[z_k \mid \theta_i]P'[\theta_i] \tag{5.2.3}$$

in which

$$N = \frac{1}{\sum_j P[z_k \mid \theta_j]\, P'[\theta_j]} \tag{5.2.4}$$

The normalizing factor simply insures that the $P''[\theta_i]$ form a proper set of probabilities. The mixing of new information and old appears through the product of the *sample likelihood*, $P[z_k \mid \theta_i]$, and the prior probability $P'[\theta_i]$. The sample likelihood is the probability of obtaining the observed sample as a function of the true state of nature θ_i. Or (as discussed in Sec. 4.1.4) the sample likelihoods can also be interpreted as the relative likelihoods of the various states given the observation z_k. If this likelihood is relatively higher for θ_i than other states, the posterior probability of this state will be increased over its prior probability. If θ_i is a relatively unlikely state to be associated with observation z_k, $P''[\theta_i]$ will reflect this by being smaller than $P'[\theta_i]$. If either the sample likelihood or the prior is sharply defined, strongly favoring one or a small group of states, the subsequent posterior probabilities will be predominantly influenced by this function (unless the other is also strongly peaked).   (This notion is demonstrated in Chap. 6.)

Once the posterior probabilities have been computed, the decision analysis proceeds exactly as in Sec. 5.1.

A simple example We reconsider the pile decision discussed in Sec. 5.1. A construction engineer must choose a length of steel section to be driven to a firm layer below. The decision elements of that problem are as follows. The engineer has a choice between two actions:

a_0:   Select a 40-ft steel pile to drive
a_1:   Select a 50-ft steel pile to drive

The possible states of nature are assumed to be just two:

θ_0:   The depth to the firm stratum is 40 ft
θ_1:   The depth to the firm stratum is 50 ft

The utilities† are shown in Table 5.1.1 and in the decision tree shown in Fig. 5.2.1.

The prior probabilities of state, representing the engineer's assessment of such information as large-scale geological maps, depths of piles

† As discussed in Prob. 5.7, these utilities are, in fact, the negative of *opportunity losses*, relative to the best action available for each state.

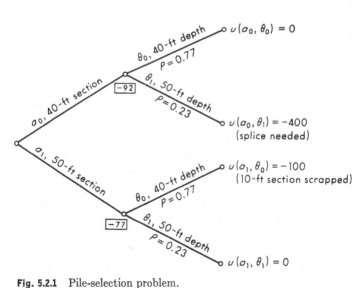

Fig. 5.2.1 Pile-selection problem.

driven several hundred feet away, etc., are assumed to be

$$P'[\theta_0] = 0.7$$

$$P'[\theta_1] = 0.3$$

The engineer plans, as the experiment or source of information, to use a simple sonic test to give a depth indication. An instrument records the time required for the sound wave created by a hammer blow on the surface to travel to the firm stratum and return. Owing to soil irregularities, measuring errors, etc., the indicated depth Z is not a wholly reliable estimate of the true depth. Assume that the sample space of possible experimental outcomes is discrete with only three values, 40, 45, and 50 ft. Two of the possible outcomes might be said to favor a particular state of nature, while the third possible outcome, 45 ft, is ambiguous. The engineer summarizes the test's properties in a table of *sample likelihoods* or conditional probabilities of observing particular test outcomes *given* each of the states of nature.

Sample likelihoods $P[z_k|\theta_i]$

Sample outcome z_k / True state θ_i	θ_0 40-ft depth	θ_1 50-ft depth
z_0, 40-ft indication	0.6	0.1
z_1, 50-ft indication	0.1	0.7
z_2, 45-ft indication	0.3	0.2
Sum	1.0	1.0

In words, the right-hand column of the table states that if the true depth is 50 ft, the engineer believes that one of the three indications—40, 45, or 50 ft—will result, with the probabilities shown. Given this true depth and the local conditions, the instrument is 70 percent reliable; it has a 10 percent chance of being flatly wrong (by indicating 40 ft); and it will give an ambiguous 45-ft reading with probability 20 percent. The total is, of course, unity. Comparing the two columns, the instrument is apparently somewhat more likely to give a too-deep indication than a too-shallow indication. These sample likelihoods may be in part subjective, since in addition to the manufacturer's calibration tests and stated "tolerances," they may depend on the engineer's judgment, based

on his own previous experience with similar tests on a number of different
soil types and depths using the same instrument.

The engineer makes the test and the instrument indicates 45 ft; i.e.,
z_2 is observed. The posterior probabilities of state, given the observation
of a 45-ft depth indication, are [Eq. (5.2.1)]:

$$P''[\theta_0] = P[\theta_0 \mid z_2] \propto P[z_2 \mid \theta_0]P[\theta_0]$$
$$\propto (0.3)(0.7)$$
$$\propto 0.21$$
$$P''[\theta_1] = P[\theta_1 \mid z_2] \propto P[z_2 \mid \theta_1]P[\theta_1]$$
$$\propto (0.2)(0.3)$$
$$\propto 0.06$$

The true state is either θ_0 or θ_1; therefore the sum of the posterior probabilities must equal unity. Normalizing these posterior likelihoods by
their sum,

$$P''[\theta_0] = P[\theta_0 \mid z_2] = \frac{0.21}{0.21 + 0.06} = \frac{0.21}{0.27} = 0.77$$

$$P''[\theta_1] = P[\theta_1 \mid z_2] = \frac{0.06}{0.21 + 0.06} = \frac{0.06}{0.27} = 0.23$$

Notice that the ambiguous result has, in fact, altered the probabilities of
state, owing to the asymmetry of the "error" probabilities, 0.3 versus 0.2.

The decision analysis uses these posterior probabilities of state as
indicated in Fig. 5.2.1. The expected utility of each action is found as
the sum of products of utilities and posterior probabilities. (In this
problem, state probabilities are independent of the action.) The better
decision is a_1, i.e., the longer pile. Although the engineer believes that
the true state is more likely to be θ_0, the shallower one, his prior information and the ambiguous sonic test result do not give sufficient confidence to offset the relatively large loss (-400) that he will absorb if the
action appropriate to θ_0 is taken and θ_1 is found to be the true state.

Note, in passing, that if the engineer must drive a number of piles
in a close area, the choice of piles, *after* driving the first, can be made
without risk since the true state will have been determined. In decision
terms, by driving one pile the engineer will have carried out a perfectly
reliable experiment for which the conditional probabilities are all zero
except two:

$P[$state θ_0 is indicated by "the sample" $\mid \theta_0] = 1$

$P[$state θ_1 is indicated by "the sample" $\mid \theta_1] = 1$

The implication is that Bayes' rule will lead to a posterior probability of zero for one state and unity for the other, depending on whether the first pile hits a firm stratum at 40 or 50 ft. A more detailed, but similar, illustration appears in Sec. 5.3.1.

Illustration: Remodeling decision with imperfect testing In an illustration of Sec. 2.1 an engineer is concerned with the quality of concrete in an old building which is being considered for a new purpose. The concrete qualities are classed as 2000 psi (θ_1), 3000 psi (θ_2), or 4000 psi (θ_3). The engineer assigns the states prior probabilities of 0.3, 0.6, and 0.1, respectively. His available actions are either to report that the floor slabs of this concrete should be replaced (a_1), to restrict the use of the rooms they support to a light office category (a_2), or to give permission that the rooms be used for all office-use types including file rooms (a_3). The performance of any slab will depend upon its concrete strength, plus a number of other random factors such as the loads to which it is subjected.

The engineer summarizes these factors in the decision tree (Fig. 5.2.2) (perhaps after an analysis of additional branches representing load levels, etc., given each concrete state and action pair).

To better his information he has two cores taken and tested. The reliability of any one of these core tests in defining in-place concrete quality is defined

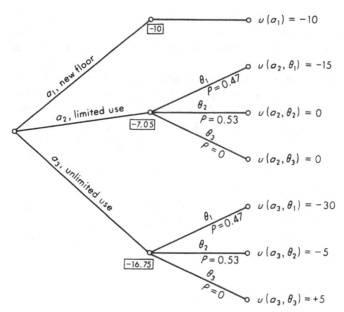

Fig. 5.2.2 Remodeling decision.

by the table of sample likelihoods or conditional probabilities:

Sample likelihoods $P[z_k \mid \theta_i]$

Core-quality indications	$P[core\text{-}quality\ indication \mid true\ quality] = P[z_k \mid \theta_i]$		
	2000 psi, θ_1	3000 psi, θ_2	4000 psi, θ_3
z_1, indicates 2000	0.7	0.2	0
z_2, indicates 3000	0.3	0.6	0.3
z_3, indicates 4000	0	0.2	0.7
Sum	1.0	1.0	1.0

The two cores are taken, one indicating 2000, one indicating 3000; that is, the outcome is $\{z_1,z_2\}$. As discussed in detail in Sec. 2.1, we can find the posterior probabilities of state in either of two ways: (1) by first finding the conditional probabilities of the $\{z_1,z_2\}$ outcome given the various possible states and then finding the posterior probabilities of state; or (2) by first finding the posterior probabilities of the states given the first outcome z_1 and then, using these posterior probabilities as priors, finding the posterior probabilities given the second outcome z_2. In either case, the results are the same. With the former approach, we find first (assuming independence of the samples) the conditional probabilities or sample likelihoods.

$$P[\{z_1,z_2\} \mid \theta_1] = P[z_1 \mid \theta_1]P[z_2 \mid \theta_1] = (0.7)(0.3) = 0.21$$
$$P[\{z_1,z_2\} \mid \theta_2] = P[z_1 \mid \theta_2]P[z_1 \mid \theta_2] = (0.2)(0.6) = 0.12$$
$$P[\{z_1,z_2\} \mid \theta_3] = P[z_1 \mid \theta_3]P[z_2 \mid \theta_3] = (0)(0.3) = 0$$

Then the posterior probabilities are

$$P''[\theta_1] = P[\theta_1 \mid \{z_1,z_2\}] = NP[\{z_1,z_2\} \mid \theta_1]P[\theta_1]$$
$$= N(0.21)(0.3) = N(0.063)$$
$$P''[\theta_2] = P[\theta_2 \mid \{z_1,z_2\}] = NP[\{z_1,z_2\} \mid \theta_2]P[\theta_2]$$
$$= N(0.12)(0.6) = N(.072)$$
$$P''[\theta_3] = P[\theta_3 \mid \{z_1,z_2\}] = NP[\{z_1,z_2\} \mid \theta_3]P[\theta_3]$$
$$= N(0)(0.1) = 0$$

In which N is the normalizing factor needed to make these posterior probabilities sum to 1.

$$N = \frac{1}{\sum_j P[\{z_1,z_2\} \mid \theta_i]P[\theta_i]} = \frac{1}{0.063 + 0.072 + 0} = \frac{1}{0.135}$$

Thus:

$$P''[\theta_1] = P[\theta_1 \mid \{z_1, z_2\}] = \frac{0.063}{0.135} = 0.47$$

$$P''[\theta_2] = P[\theta_2 \mid \{z_1, z_2\}] = \frac{0.072}{0.135} = 0.53$$

$$P''[\theta_0] = P[\theta_3 \mid \{z_1, z_2\}] = \frac{0}{0.135} = 0$$

The core outcomes have eliminated the possibility that the concrete is of high strength, and they suggest that the lower-strength state is more likely then the engineer had originally judged.

Using these probabilities in the decision tree, Fig. 5.2.2, the best decision seems to be a_2, limiting the use of rooms.

In fact the engineer might reconsider his position, which now finds the concrete quality quite uncertain. He may also recognize that he can, at some expense, carry out further experimentation to gain further information. In this case, he might want to order another or several more cores or even a load test. In Sec. 5.3, we shall treat such questions as, "When should he postpone the decision and pay for more information?" and, "Which experiment should he pay for?"

Illustration: A traffic-sampling problem   In a study of the dispersion of traffic throughout a city, one car in every r cars entering the city between 8 and 9 A.M. via a certain route, route A, was stopped and small, easily visible stickers were placed on its bumpers. During the day various points in the city were monitored by men who counted the total number of marked autos passing certain points during a certain time interval. A decision as to whether to restrict a particular street to one-way traffic is going to be based on the total number of cars that enter the city via route A and then use this street during the time interval. The performance of the total traffic system will be improved if the change is made and this number is large. If the decision to restrict the street is made, but the traffic on it from route A is now small, the inconvenience to others will outweigh the gain for route A commuters.

The engineer summarizes these potential consequences in a utility assignment of

$$u(a_1, \theta) = -c_1 + b_1 \theta + b_2 \theta^2$$

in which a_1 is the action to restrict the street to one-way flow. The action will incur cost c_1 to others, but it will bring benefits to route A users which the engineer assumes are an increasing function of θ, the total number of cars from route A using the street during the time interval (on the day of the study, for simplicity). To take action a_2 is to leave the street as is, with (relative) utility of zero:

$$u(a_2, \theta) = 0$$

We need information about the state of nature θ. The engineer summarizes the information contained in a number of related, but less specific, past traffic surveys and traffic-assignment analyses by assigning θ a prior distribution, $P'[\theta]$, $\theta = 0, 1, 2 \ldots$. His decision will be determined by an expected utility criterion based on the posterior distribution of θ.

The results of the day's survey are an observation that z_0 cars with stickers used the street during the specified time interval. What is the posterior distribution of θ?

We ask first for the sample likelihood $P[Z = z \mid \theta]$ in which Z is the random number of marked route A cars using the street. The probability that any one of the θ route A cars using the street has a sticker is $1/r$. Assuming independence of cars, Z, the total number of marked cars in the total of θ cars, has a binomial distribution:

$$P[Z = z \mid \theta] = \frac{\theta!}{z!(\theta - z)!} \left(\frac{1}{r}\right)^z \left(1 - \frac{1}{r}\right)^{\theta - z} \qquad z = 0, 1, \ldots, \theta \qquad (5.2.5)$$

Thus, given the observation that $Z = z_0$, the likelihood function, a function of θ, is

$$L(\theta \mid Z = z_0) = P[Z = z_0 \mid \theta] = \frac{\theta!}{z_0!(\theta - z_0)!} \left(\frac{1}{r}\right)^{z_0} \left(1 - \frac{1}{r}\right)^{\theta - z_0}$$
$$\theta = z_0, z_0 + 1, \ldots, \infty \qquad (5.2.6)$$

since any value of θ equal to or larger than z_0 could lead to the observation of $Z = z_0$. As a function now of θ, the likelihood function gives the sample-implied relative likelihoods of the various possible values of the state θ. It should be emphasized that in this case the sample likelihoods were derived from a theoretical analysis of the experiment, not from purely subjective assignments.

The posterior distribution of θ is thus

$$P''[\theta] = P[\theta \mid Z = z_0]$$
$$= \begin{cases} N \dfrac{\theta!}{z_0!(\theta - z_0)!} \left(\dfrac{1}{r}\right)^{z_0} \left(1 - \dfrac{1}{r}\right)^{\theta - z_0} P'[\theta] & \theta = z_0, z_0 + 1, \ldots \\ 0 & \theta < z_0 \end{cases} \qquad (5.2.7)$$

The interesting part ($\theta \geq z_0$) can be rewritten

$$P''[\theta] = N_1 \frac{\theta!}{(\theta - z_0)!} \left(1 - \frac{1}{r}\right)^{\theta} P'[\theta] \qquad \theta = z_0, z_0 + 1, \ldots \qquad (5.2.8)$$

in which the new normalizing constant is

$$N_1 = \frac{1}{\displaystyle\sum_{\theta = z_0}^{\infty} [\theta!/(\theta - z_0)!](1 - 1/r)^{\theta} P'[\theta]} \qquad (5.2.9)$$

It should be expected that the larger the proportion of cars that are stopped for stickers, the "better" the survey. In the limit, if every car is stopped, $r = 1$, the information on θ is perfect. The likelihood function [Eq. (5.2.6)] becomes simply

$$P[Z = z_0 \mid \theta] = \begin{cases} 0 & \theta \neq z_0 \\ 1 & \theta = z_0 \end{cases}$$

in which z_0 might equal, say, 64 cars. In this case Eq. (5.2.7) becomes

$$P''[\theta] = \begin{cases} 1 & \theta = 64 \\ 0 & \theta \neq 64 \end{cases}$$

This result is true no matter what the engineer's prior distribution is (as long as that distribution does not specifically exclude the possibility of $\theta = 64$, that is, as long as $P'[64] > 0$). The expected costs of the actions are

$E[u(a_1)] = -c_1 + b_1 64 + b_2 4096$

$E[u(a_2)] = 0$

implying that a_1 is the better decision only if the cost of the change c_1 is less then $64b_1 + 4096b_2$.

The cost of a survey that stops every car may be high. Had the engineer decided to stop only every other car, the sample likelihood function would, of course, not be concentrated solely at $\theta = z_0$. If every other car were stopped ($r = 2$) and the number of marked cars observed were $z_0 = 32$, the sample-likelihood function [Eq. (5.2.6)] would appear as shown in Fig. 5.2.3. Combined with the prior distribution [through Eq. (5.2.8)], this would yield the posterior distribution $P''[\theta]$ and finally $E[u(a_1)]$. The shapes of two posterior distributions are shown in Fig. 5.2.4. In both cases the engineer's prior distribution was triangular, either increasing from $\theta = 50$ to $\theta = 90$ (Fig. 5.2.4a) or decreasing from $\theta = 50$ to $\theta = 90$ (Fig. 5.2.4b). Compare these with the likelihood function, Fig. 5.2.3. Even for these diverse prior shapes, the posterior distribution and hence the decision are still influenced predominantly by the shape of the sample-likelihood function, that is, by the information in the experimental results.

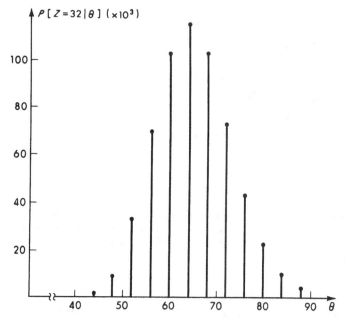

Fig. 5.2.3 Selected values of the sample-likelihood function given an observation of 32 marked cars; every second car marked.

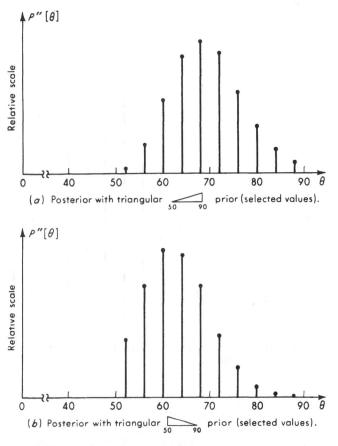

Fig. 5.2.4 Posterior distributions based on every second car marked.

If an even smaller proportion of cars were stopped and marked, say one in four, and 16 were observed† on the street, the likelihood function becomes much more diffuse (Fig. 5.2.5), implying that it contains less information about the value of θ. In this case the engineer's prior information will be of more relative significance. Figure 5.2.6, parts a and b, shows posterior distributions based on the two triangular prior distributions mentioned above. Clearly, these prior distributions have a strong influence on the final shape, and hence potentially on the decision.

The trend is clear. If only every eighth car is marked, we know that unless the value of Z observed is very large or very small, the information to be gained

† The numbers $z_0 = 16$ and $z_0 = 32$ were chosen for illustrative purposes to give sample likelihood functions centered on a value of $\theta = 64$. The numbers are kept small for ease of computation. For larger numbers, the binomial likelihood function could be replaced by an approximating normal distribution.

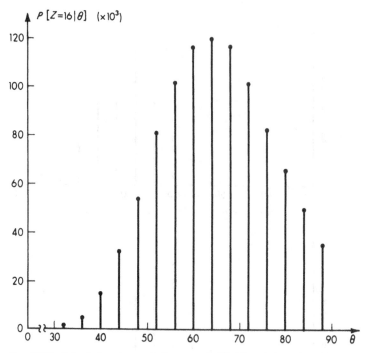

Fig. 5.2.5 Selected values of the sample-likelihood function given an observation of 16 marked cars; every fourth car marked.

will be of such small value relative to the engineer's triangular prior distribution as to have little influence on the posterior distribution of θ or the decision. This conclusion is influenced, of course, by the shape of the prior distribution.

In Sec. 5.3 we shall investigate the method by which the engineer here could have planned whether to mark every car, every second car, or every rth car, or whether even to dispense with the new survey entirely and base his decision on his prior distribution. We should anticipate that the decision to experiment and the level of experimentation will depend on the nature of the engineer's prior distribution, the costs of various experiments, and the quality of information we expect to gain from them.

Illustration: A flow-measurement problem   In this example we shall illustrate several factors commonly appearing in engineering problems. These include the use of inexact, empirical functional relationships, the observation of an easily measured variable to help predict the state of interest, and the presence of joint state variables, one or more of which may not be of direct interest, but whose influence must be accounted for.

In the decision as to the amount of pollutant that can be safely released into a river in any day, the flow rate θ for that day must be estimated. This flow rate is the integrated effect of the distribution of the downstream component of the velocity across the cross-sectional area of the channel on that day. Owing to the difficulty and expense of making all the velocity and depth measurements

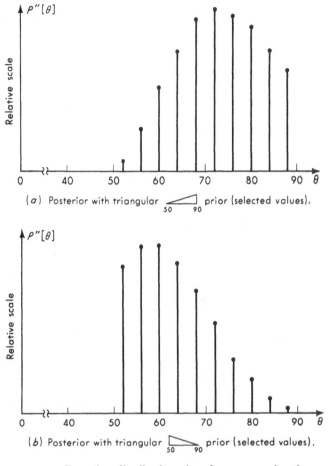

Fig. 5.2.6 Posterior distributions based on every fourth car marked.

necessary to calculate θ, however, it is predicted using only an easily measured depth Z at a convenient reference point.

To "calibrate" the measuring procedure, the river was carefully surveyed a number of times under different flow conditions to measure θ, the total flow rate. At the same time the reference depth Z was observed and a reference, midstream, velocity V was measured. A regression analysis (Sec. 4.3) of the observations of θ/V versus Z led to the conclusion that a straight line gave a reasonable fit in the region of interest, but the scatter of the observations about this line was significant.†

† Including V in the analysis avoids errors due to systematic seasonal changes in velocity. The choice of the form θ/V versus Z was based on its validity for a hypothetical, uniform flow in a rectangular river cross section.

The channel conditions are constantly changing. On a particular day, the engineer responsible for the decision on that day interprets historical data, seasonal information, and other indirect information (such as a careful measurement 5 miles upstream made earlier in the summer), by putting a joint prior distribution on θ and V. We shall see that a distribution on θ alone is insufficient in this problem. For simplicity we use the simple discrete distribution shown in Fig. 5.2.7.

The consequences of the decision are a function only of θ, the flow rate. The available actions, a_1, a_2, . . . , are various rates, f_1, f_2, . . . , of discharging the pollutant into the river. Assume that the utility function for any action-state pair is

$$u(a_i,\theta) = \begin{cases} cf_i & f_i \leq 0.1\theta \\ cf_i - be^{\alpha(f_i-0.1\theta)} & f_i > 0.1\theta \end{cases}$$

in which 0.1θ is the "permitted" discharge rate. The consequences of mistakenly using a discharge rate in excess of this value include a flat "fine" of b units and other, exponentially rising losses.

Without an experiment the decision could only be based on the prior on θ, now a marginal to the joint prior given in Fig. 5.2.7:

$$P'[\theta] = \begin{cases} 0.1 & \theta = 8,000;\ 12,000 \\ 0.2 & \theta = 9,000 \\ 0.3 & \theta = 10,000;\ 11,000 \end{cases}$$

Given an observation on the reference depth Z, however, we can compute a posterior distribution on θ which reflects the uncertainty in the calibration "curve" and the dependence of that curve upon the uncertain velocity V.

The sample-likelihood function, now necessarily a function of θ and V, $P[Z = z \mid \theta,V]$, must be deduced† from the plot of observations of the careful surveys. In discrete form, for simplicity, this is given as a probability of 0.5 that Z will equal its "most likely" value, which, is linearly related to flow rate over reference velocity; specifically, it is $2 + 3 \times 10^{-3}\ \theta/V$. The scatter about this empirical line is represented by discrete probabilities of 0.25 that Z will be 1.0 to 1.5 ft above or 1.0 to 1.5 ft below this most likely value.

† For example, by a regression analysis of Z on θ/V.

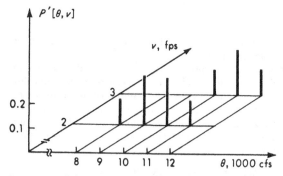

Fig. 5.2.7 Joint prior on flow rate and reference velocity.

Suppose that the engineer measures the reference depth and observes a value of $Z = 13$ ft. Any pair of values of θ and V such that $2 + 3 \times 10^{-3} \theta/V$ is within 1.5 ft of 13 feet *might* have given rise to this observation. The *nonzero* sample liklihoods (that also match possible state pairs $\{\theta, V\}$ indicated in the prior) are thus

$$P[Z = 13 \mid \theta = 11{,}000; \ V = 3] = 0.5$$

$$P[Z = 13 \mid \theta = 10{,}000; \ V = 3] = 0.25$$

$$P[Z = 13 \mid \theta = 12{,}000; \ V = 3] = 0.25$$

$$P[Z = 13 \mid \theta = \ \ 8{,}000; \ V = 2] = 0.25$$

The latter three cases arise from possible errors of 1.0 to 1.5 ft in the empirical relationship. The last, for example, arises from an error of 1 ft, since $2 + (3 \times 10^{-3})\,(8000/2)$ equals 14.

Based on an obvious extension of Bayes' theorem [Eq. (5.2.2)] to a two-dimensional state of nature, the posterior (joint) probability mass function on θ and V is zero except at four points:

$$P''[\theta = 11{,}000; \ V = 3]$$
$$= NP[Z = 13 \mid \theta = 11{,}000; \ V = 3]P'[\theta = 11{,}000; \ V = 3]$$
$$= N(0.5)(0.2) = 0.1N$$

$$P''[\theta = 10{,}000; \ V = 3]$$
$$= NP[Z = 13 \mid \theta \ 10{,}000; \ V = 3]P'[\theta = 10{,}000; \ V = 3]$$
$$= N(0.25)(0.1) = 0.025N$$

and, similarly,

$$P''[\theta = 12{,}000; \ V = 3] = N(0.25)(0.1) = 0.025N$$

$$P''[\theta = 8{,}000; \ V = 2] = N(0.25)(0.1) = 0.025N$$

Therefore

$$N = \frac{1}{0.1 + 0.025 + 0.025 + 0.025} = \frac{1}{0.175} = 5.7$$

The posterior joint distribution of θ and V is shown in Fig. 5.2.8. The

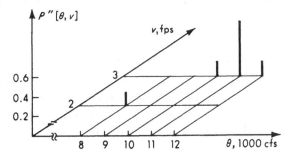

Fig. 5.2.8 Joint posterior on flow rate and reference velocity.

marginal distribution of θ is now

$$P''[\theta] = \begin{cases} 0.14 & \theta = 8,000; \ 10,000; \ 12,000 \\ 0.57 & \theta = 11,000 \end{cases}$$

This distribution should be used to compute the expected utility associated with each action. For example, for a_j with pollutant rate $f_j = 1000$:

$$E[u(a_j)] = \sum_k P''[\theta_k]u(a_j,\theta_k) = 1000c - 0.14\,be^{\alpha(1000-800)}$$

Reflection on this illustration will reveal the necessity of including the uncertainty in V in the analysis. It cannot, for example, be replaced by $E[V] = 2.4$ ft/sec if a correct marginal posterior distribution on θ is to be obtained.

The role of the dispersion in an empirical formula is to introduce the probability of values away from the "calculated" or "most likely" values of θ (or here θ/V) being possible states given an observation of Z. The width of this dispersion controls the number and relative weights of these values of θ, through the sample-likelihood function.

5.2.2 Summary

Terminal analysis or decision analysis given new information follows the same pattern as prior analysis, Sec. 5.1, once posterior probabilities of state $P''[\theta]$ have been calculated. These are based on Bayes' theorem:

$$P''[\theta_i] = NP[z_k \mid \theta_i]\,P'[\theta_i]$$

in which N is a normalizing constant, $P'(\theta_i)$ is the prior probability of state θ_i, and $P[z_k \mid \theta_i]$ is the sample likelihood or conditional probability of observing the experimental outcome $Z = z_k$, given that θ_i is the true state. Note that once z_k has been observed, only the sample likelihoods of that result, $P[z_k \mid \theta_i]$ for all θ_i, are needed for the decision analysis.

Posterior probabilities are influenced by the sample likelihoods and by the prior probabilities. If either set of probabilities is sharply concentrated at some particular θ_i's, relative to the other set, the former set will have the dominant influence of the posterior probabilities. In short, information from a "reliable" experiment will dominate a vague prior distribution, and vice versa.

5.3 PREPOSTERIOR ANALYSIS†

In the previous two sections we have dealt with decision making using given information and with processing of new information prior to the decision. In this section our concern is with deciding whether a price should be paid for new information and which source of information is the best choice.

† The material in this section will not be used subsequently and might be excluded on a first reading.

5.3.1 The Complete Decision Model

Often the decision maker has the option of paying to observe the outcome of an experiment before making his choice of actions. If the cost of this experiment is sufficiently small compared to the information on θ that it promises, the engineer should select to experiment. If several types or levels of experimentation are possible, he should choose among them in a manner which achieves the best balance between experiment cost and reduced risk in the action choice.

Identifying the possible experiments as e_1, e_2, . . . , and denoting, for convenience, the choice of no further experimentation as e_0, the complete, two-stage decision tree becomes that shown in Fig. 5.3.1. E is the set of available experiments, e_0, e_1, e_2, ... Having chosen an experiment e, an outcome z in the sample space Z of the experiment will be observed. The decision maker must then choose an action a from the set of actions A, after which the state of nature θ will be found in the space Θ. As a result of this sequence of consequences, value measured

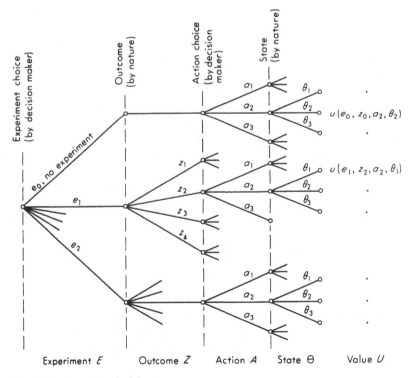

Fig. 5.3.1 Complete decision tree.

by a utility assignment, $u(e,z,a,\theta)$, will be received. Note in particular that these consequences include the cost of experimentation.

In practical situations the possible experiments are represented by a number of alternatives to collect more data on various quantities that might alter the present (prior) probabilities of the state of nature. Before deciding on the number of lanes in a highway, for example, the engineer might want to consider calling for an additional origin-destination survey to help predict the demand on the future road. The cost of this experiment may include delay of design and construction.

Before settling upon a foundation design, the engineer may want to consider drilling a series of test holes at the site, or if a preliminary survey is already available, he may consider a second set. Various kinds of experiments are possible, ranging from a single additional hole in a critical or questionable area, through a selective pattern over certain regions, to a fine-scale pattern over the entire building area. The costs and the potential information content vary from experiment to experiment.

By the analysis of such a decision tree we mean determining the best choice of experiment, *and*, in the process, determining the best choice of action given that a particular experiment has been chosen and a particular outcome of it has been observed. We shall outline the analysis procedure symbolically and then illustrate it.

The analysis of the tree proceeds from right to left in Fig. 5.3.1. We recognize that, *given* a particular experiment e and outcome z, the analysis of the *remainder* of a branch is precisely that which we studied in Sec. 5.2. It is a terminal or *posterior* analysis, that is, an analysis made *after* the experiment and outcome. Given the experiment and outcome, posterior probabilities can be computed and an optimal act can be selected on the basis of maximum expected utility. In the analysis of the complete tree, we must make *all* these posterior analyses for *every* possible experiment-outcome pair. The result of each pair is a corresponding optimal action and its associated expected utility, which we denote $u^*(e,z)$:

$$u^*(e,z) = \max_a \ [E[u(e,z,a) \mid e,z]] \tag{5.3.1}$$

in which the ("posterior") expected utilities are

$$E[u(e,z,a) \mid e,z] = \sum_i u(e,z,a,\theta_i) P[\theta_i \mid e,z] \tag{5.3.2}$$

The probabilities of state used in this equation are the posterior probabilities $P''[\theta_i]$ or $P[\theta_i \mid e,z]$, except in the case of the no-experiment or e_0 branch, when they remain the prior probabilities. Again for notational convenience we suppress the possible dependence of these probabilities of θ on the action a.

In order to chose from among the available experiments, we need only to calculate the expected value of each and select e to maximize expected value; that is, we must choose e to maximize:

$$E[u(e)] = \sum_k u^*(e,z_k)P[z_k \mid e] \tag{5.3.3}$$

To compute this expected value for each e we need, in addition to $u^*(e,z)$, only the probabilities of the various outcomes given that experiment e is taken. These are simply

$$P[z_k \mid e] = \sum_i P[z_k \mid e,\theta_i]P'[\theta_i] \tag{5.3.4}$$

The conditional probabilities of outcomes given the state, i.e., the sample likelihoods, are familiar to us. Here we only need multiply them by $P'[\theta_i]$ and sum to obtain the probability that a particular outcome will be observed.

Because a number of terminal or posterior analyses are made conditional upon, but actually *before* the experiment and the outcome, this entire analysis is called *preposterior*. The analysis procedure is illustrated numerically in the next example.

Illustration: Preposterior analysis   A single, simple example will be given, for the computational aspects are similar to those already presented.

Reconsider the engineer who was faced with selecting a steel pile to be driven to a firm stratum at an uncertain depth below. We put him now in a more realistic setting than originally found in Secs. 5.1.1 or 5.2. The engineer now must place an order for a large number of piles to be used at the same site. The stratum is known to be at a constant but unknown depth below the entire site. The relative utilities are -400 if the short piles are ordered (a_0) and the true depth is 50 ft (θ_1), and -100 if the long piles are ordered (a_1) and the depth is 40 ft (θ_0). Otherwise, the losses are 0. The engineer's prior probabilities are:

$$P'[\theta_0] = 0.7$$
$$P'[\theta_1] = 0.3$$

The losses are now associated with having to wait for more steel if insufficient lengths are obtained or having to scrap or return at a loss the excess steel if unnecessarily long piles are ordered.

The engineer has three options with regard to experimentation. He may choose to order on the basis of present information (e_0), to use the imperfect sonic device discussed in Sec. 5.2 (e_1), or to have a hole drilled to determine the true depth accurately (e_2). In the latter two cases he will postpone ordering until the outcome has been determined. With the use of e_1, the sonic tester, the experimental cost is small. It amounts to 20 units, which we assume can be subtracted from the relative utilities of 0, -100, and -400 associated with no experimentation to yield the utilities of the total consequences including experimentation.† The drilling operation is more reliable but more expensive, being represented by a cost of 50 units, subtractable, it is assumed, from other utilities. The elements of the decision tree, Fig. 5.3.2, are in their places, except for the probabilities.

For the e_0 branch, no experiment is made, nothing is observed, and the probabilities of state are the prior probabilities. As indicated in Fig. 5.3.3, the expected value of action a_0 is -120 and that of a_1 is -70. The latter is preferred, so -70 becomes $u^*(e_0,z)$; this figure is entered on the e_0 branch.

† In general, the utility of the total consequence, $u(e,z,a,\theta)$, must be assessed. Recall from Sec. 5.1.2 that owing to possible nonlinearities in the cost-utility relationships, $u(\$x + \$y)$ does not necessarily equal $u(\$x) + u(\$y)$.

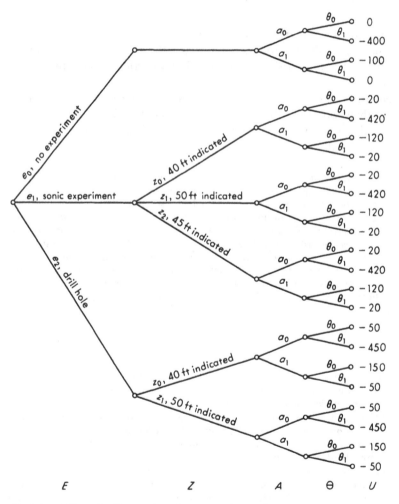

Fig. 5.3.2 Pile-order illustration of preposterior analysis.

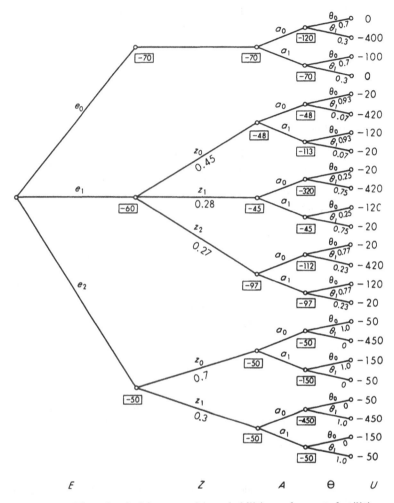

Fig. 5.3.3 Pile-order decision tree with probabilities and expected utilities.

Consider next the e_2 or drilling-experiment branch. In this case, there is no possibility that an incorrect or ambiguous state indication will be made. Thus the posterior probabilities given the outcome z_0 are:

$$P[\theta_0 \mid z_0, e_2] = 1$$
$$P[\theta_1 \mid z_0, e_2] = 0$$

and those given z_1 are:

$$P[\theta_0 \mid z_1, e_2] = 0$$
$$P[\theta_1 \mid z_1, e_2] = 1$$

These intuitively obvious results can be verified through formal application of Bayes' rule. The sample likelihoods needed for this formal analysis are:

$$P[z_0 \mid \theta_0, e_2] = 1$$
$$P[z_0 \mid \theta_1, e_2] = 0$$
$$P[z_1 \mid \theta_0, z_2] = 0$$
$$P[z_1 \mid \theta_1, z_2] = 1$$

Multiplying by prior probabilities and normalizing will yield the results deduced informally above.

Entering these probabilities, a posterior analysis is carried out on the $\{e, z_0\}$ branch and on the $\{e, z_1\}$ branch, as shown in Fig. 5.3.3. The results indicate, obviously, that a_0 is optimal if z_0 is observed and that a_1 is optimal if z_1 is observed. In both cases the maximum expected utilities are -50, the cost of obtaining this perfect information:

$$u^*(e_2, z_0) = u^*(e_2, z_1) = -50$$

Because both these utilities equal -50, we conclude that the expected utility of e_2 is -50 no matter what the probabilities of z_0 and z_1. This then is $E[u(e_2)]$. This intuitive result is verified formally as follows. From Eq. (5.3.4)

$$P[z_0 \mid e_2] = P[z_0 \mid e_2, \theta_0]P'[\theta_0] + P[z_0 \mid e_2, \theta_1]P'[\theta_1]$$
$$= (1.0)(0.7) + (0)(0.3) = 0.7$$

and

$$P[z_1 \mid e_2] = P[z_1 \mid e_2, \theta_0]P'[\theta_0] + P[z_1 \mid e_2, \theta_1]P'[\theta_1]$$
$$= (0)(0.7) + (1.0)(0.3) = 0.3$$

These are the probabilities that z_0 or z_1 will be found if experiment e_2 is taken. The expected utility of this experiment is [Eq. (5.3.3)]:

$$E[u(e_2)] = u^*(e_2, z_0)P[z_0 \mid e_2] + u^*(e_2, z_1)P[z_1 \mid e_2]$$
$$= (-50)(0.7) + (-50)(0.3) = -50$$

Finally, we consider again the sonic test e_1. The sample likelihoods have been given in Sec. 5.2 as:

Sample likelihoods $P[z_k \mid \theta_i, e_1]$

Sample outcome z_k / True state θ_i	θ_0 40-ft depth	θ_1 50-ft depth
z_0, 40-ft indication	0.6	0.1
z_1, 50-ft indication	0.1	0.7
z_2, 45-ft indication	0.3	0.2

The posterior analysis given the ambiguous outcome z_2 has been discussed in detail in Sec. 5.2. We found posterior probabilities† of state through Bayes' rule as

$$P[\theta_0 \mid z_2,e_1] = NP[z_2 \mid \theta_0,e_1]P'[\theta_0]$$
$$= N(0.3)(0.7) = N(0.21) = 0.77$$
$$P[\theta_1 \mid z_2,e_1] = NP[z_2 \mid \theta_1,e_1]P'[\theta_1]$$
$$= N(0.2)(0.3) = N(0.06) = 0.23$$

in which the normalizing constant N is $1/(0.21 + 0.06)$.

Multiplication of the posterior probabilities by corresponding utilities and summation leads to expected utilities. For example, for a_0,

$$E[u(e_1,z_2,a_0) \mid e_1,z_2] = u(e_1,z_2,a_0,\theta_0)P[\theta_0 \mid e_1,z_2] + u(e_1,z_2,a_0,\theta_1)P[\theta_1 \mid e_1,z_2]$$
$$= (-20)(0.77) + (-420)(0.23)$$
$$= -112$$

Similarly, the posterior expectation for a_1 given experiment e_1 and outcome z_2 is $(-120)(0.77) + (-20)(0.23) = -97$. The better action in the event of outcome z_2 to experiment e_1 is a_1; thus

$$u^*(e_1,z_2) = \max_a [-112, -97] = -97$$

The same steps must be repeated for experimental outcome z_0 and outcome z_1. The results are, for z_0,

$$P[\theta_0 \mid z_0,e_1] = 0.93$$
$$P[\theta_1 \mid z_0,e_1] = 0.07$$
$$E[u(e_1,z_0,a_0)] = (-20)(0.93) + (-420)(0.07) = -48$$
$$E[u(e_1,z_0,a_1)] = (-120)(0.93) + (-20)(0.07) = -113$$
$$u^*(e_1,z_0) = \max_a [-48, -113] = -48$$

with a_0 the better action. And, for z_1,

$$P[\theta_0 \mid z_1,e_1] = 0.25$$
$$P[\theta_1 \mid z_1,e_1] = 0.75$$
$$E[u(e_1,z_1,a_0)] = -320$$
$$E[u(e_1,z_1,a_1)] = -45$$
$$u^*(e_1,z_0) = \max_a [-320, -45] = -45$$

with a_1 the better action. These values appear in Fig. 5.3.3.

The expected value associated with choosing experiment e_1 is found by weighting the optimal utilities $u^*(e_1,z)$ by the probabilities that z will be the

---

† The state probabilities are independent of the action in this case and so the argument a is suppressed. By the same token, the argument z could have been suppressed in the utility function since the consequences do not depend on the outcome of the experiment.

outcome and summing. These probabilities are:

$$P[z_0 \mid e_1] = P[z_0 \mid e_1, \theta_0]P'[\theta_0] + P[z_0 \mid e_1, \theta_1]P'[\theta_1]$$
$$= (0.6)(0.7) + (0.1)(0.3) = 0.45$$
$$P[z_1 \mid e_1] = P[z_1 \mid e_1, \theta_0]P'[\theta_0] + P[z_1 \mid e_1, \theta_1]P'[\theta_1]$$
$$= (0.1)(0.7) + (0.7)(0.3) = 0.28$$
$$P[z_2 \mid e_1] = 1 - P[z_0 \mid e_1] - P[z_1 \mid e_1]$$
$$= 1 - 0.45 - 0.28 = 0.27$$

which are also shown in Fig. 5.3.3.

Thus the expected utility of e_1 is:

$$E[u(e_1)] = \sum_k u^*(e_1, z_k)P[z_k \mid e_1]$$
$$= (-48)(0.45) + (-45)(0.28) + (-97)(0.27)$$
$$= -60$$

Comparing the expected values of the three experiments, -70, -60, and -50, shown in Fig. 5.3.3, it is apparent that the engineer should not make a decision before obtaining more information from one of the available experiments. Of the two experiments, the reliable but expensive one e_2 is preferable (in this case) to the cheaper but less reliable sonic experiment e_1. The engineer should absorb the initial cost of the drilling and then order accordingly, rather than run the risk of making an incorrect order. This risk results from his present uncertainty, which might be reduced but not eliminated by the less expensive sonic test e_1.

Discussion on the value of information   Because we have assumed additivity of utilities in this example, we can associate the costs of experimenting directly with the e branches, as shown in dashed boxes in Fig. 5.3.4. In this case, the cost of experimentation need not be considered until after the various terminal analyses. Therefore, all other utilities on this tree are simply increased in value (over those utilities entered in Fig. 5.3.3) by the costs of experimentation associated with their branches. Interpretation of *the value of information* is much simpler on this tree. Note, for instance, that the expected cost (-70) of choosing e_0 and making a decision with present information is entirely risk-related; there is no experimentation cost. In contrast, with the perfect experiment (e_2), the cost (-50) is entirely that required to "buy information." Once the experiment is paid for, there is no *risk* (expected cost) involved. No matter what the outcome of the experiment, the decision appropriate to the true state of nature will be made with certainty. In this example, the perfect information is worth its price, 50 units. If this price were 75 units, or 50 percent higher, however, it would be more prudent for the engineer to run the risk on the basis of his present knowledge than to pay

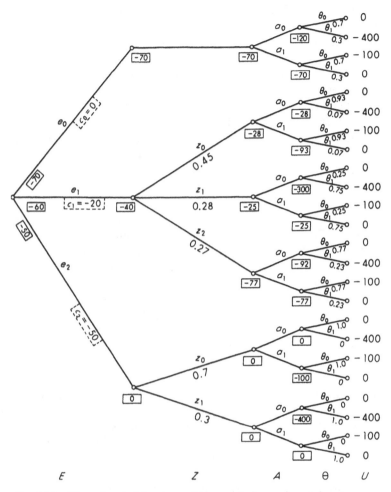

Fig. 5.3.4 Pile-order decision tree with separated experimental costs.

for this perfect information.† This is the decision which is consistent with his stated preference and uncertainty assignments. Recall that these preference statements may contain relatively subjective elements. If the engineer feels, for example, that there would be a serious loss of

† The *value of perfect information* is in general defined to be the difference between (1) the prior expectation $\Sigma u(\theta_i)P'(\theta_i)$, in which $u(\theta_i)$ is the utility associated with the best choice of action *given* that it is known *with certainty* that θ_i is the true state, and (2) the expected utility associated with no experiment e_0. In this example, this is simply $[(0)(0.7) + (0)(0.3)] - (-70) = 70$. No experiment need even be considered if it costs more than the value of perfect information.

personal reputation if he makes an incorrect order on the basis of his present information (which may rest strongly upon his personal professional judgment), then this fact should be reflected in the utilities (now -400 and -100) assigned to these "bad outcomes" and experiment e_0. His aversion to risk should be reflected in these utility or preference assignments, not in biased, "conservative" uncertainty assignments.

The e_1, or imperfect, inexpensive experiment, offers a compromise between the two extreme level-of-experiment decisions e_0 and e_2. Note that if the outcome of the experiment *favors* either θ_0 or θ_1 (that is, if either z_0 or z_1 is observed), the risk is reduced from the no experiment or present expected utility of -70 to either -28 or -25. Because the method is not wholly reliable, however, the risk has not been eliminated entirely, as it is in e_2. In particular, if the ambiguous result z_2 is observed, the engineer has in fact *increased* the risk† that he faces from -70 to -77. This ambiguous outcome is sufficiently unlikely, however, that the experiment choice e_1 as a whole represents a risk of only -40. That this risk reduction can be purchased for 20 units makes it more attractive than not experimenting at all, but less attractive than paying 50 units for perfect information. Notice that only a 25 percent increase in the cost of the drilling would make e_2 less attractive than e_1.

Again, professional factors can be reflected in the decision analysis. Suppose that the engineer chooses e_1; having observed outcome z_1, he takes action a_1, which is later revealed to be an incorrect decision (that is, θ_0 was in fact the true state). He may not be held responsible for his choice *if* he can point to the fact that it was "based" on a sonic test result which happened to be misleading. If this is the case, under e_1, the losses to the engineer given bad outcomes should be *less* than the no-experiment values of -100 and -400. In this case·utility is a function of e and z as well as θ and a.

Comment The authors look upon decision theory as an extremely accurate conceptual model of professional engineering decision making. It permits problems to be broken down, studied in detail (by different people, if appropriate), and then reassembled in a consistent manner. Quite apart from possible quantitative help, however, simple understanding of the concept of statistical decision theory may be a useful guide in a practical, professional decision situation. The model points out the need to balance cost and likelihood, to incorporate preference and judgment as well as data, and to consider the cost of obtaining more information in relation to its potential benefit in reducing uncertainty

† Some engineers might call for a drill hole in the event of this outcome to experiment e_1. The relative advantage of this "hedging," compound strategy can be investigated by considering it as a fourth type of experiment, e_3. See Prob. 5.11.

and risk. The theory deserves much wider understanding among civil engineers.

In the next chapter we shall consider some more advanced decision-theory topics. Conceptually we add nothing to the basic decision model; the emphasis will be on random variables and random processes rather than random events (or states) as in this chapter. Continuous sample spaces will be used more freely, with resulting enhancement of model accuracy and computational economy.

5.3.2 Summary

Preposterior analysis is designed to choose among available experiments or sources of information e_i (including the "do-nothing" experiment e_0). The analysis requires a terminal analysis of each experiment e and each possible outcome z of that experiment. For the various experiments, the resulting optimal-actions' expected utilities $u^*(e,z)$ are weighted by the probabilities of each possible experimental outcome to obtain the expected utility of carrying out that experiment. (The probability of a particular experimental outcome z is obtained by weighting the sample likelihoods of that outcome by the prior probabilities of state θ_i.) The experiment with the highest expected utility is the best choice consistent with the engineer's stated probabilities and preferences.

Preposterior analysis can reveal the value of information in the form of reductions in the risks (expected losses) associated with any particular experiment.

5.4 SUMMARY FOR CHAPTER 5

Bayesian decision theory is a prescriptive or normative† model for decision making under uncertainty. It demands that the decision maker state his (possibly subjective) prior probabilities of state $P'(\theta)$ and his preferences among different pairs of actions a and states of nature θ. If these preferences are properly assigned in the form of a utility function $u(a,\theta)$, then decisions should be based on a maximum-expected-utility criterion. Acceptance of the theory rests in principle on an acceptance of both subjective probabilities and utility functions, although these two factors may in practice be indistinguishable from "objective" relative frequencies and dollar values, respectively.

Decision analysis is simply the application of the fundamental probability theory of Chap. 2. It may include simple prior analysis, based on the prior probabilities of state $P'[\theta]$, or terminal analysis, based on the posterior probabilities of state $P''[\theta]$. The latter are found through the

---

† That is, it tells one how to make a decision; it does not describe how decisions are made, as a descriptive model would.

application of Bayes' theorem, given the outcome of an experiment $Z = z$. Required information includes, in addition to the prior probabilities, sample likelihoods $P[z \mid \theta]$. These may be obtained directly or indirectly through theoretical analysis of the experimental scheme. In addition, decision analysis may include choosing from among several possible experiments prior to choosing the action. This preposterior analysis involves making a terminal analysis of each possible experiment-outcome $\{e,z\}$ pair and combining this with prior probabilities of the possible outcomes to obtain the expected utility associated with each experiment.

REFERENCES

General

Bayes, T. [1958]: Essay towards Solving a Problem in the Doctrine of Chances, *Biometrika*, vol. 45, pp. 293–315. (Reproduction of 1763 paper.)

Benjamin, J. R. [1968]: Probabilistic Models for Seismic Force Design, *J. Structural Div. ASCE*, vol. 94, no. ST7, p. 1175, July.

Benjamin, J. R. [1968]: Probabilistic Structural Analysis and Design, *J. Structural Div. ASCE*, vol. 94, no. ST5, p. 1665, May.

Fishburn, P. C. [1965]: "Decision and Value Theory," John Wiley & Sons, Inc., New York.

Good, I. J. [1950]: "Probability and the Weighing of Evidence," Charles Griffin & Company, Ltd., London.

Good, I. J. [1965]: "The Estimation of Probabilities," Research monograph no. 30, Massachusetts Institute of Technology Press, Cambridge, Mass.

Grayson, C. J. Jr. [1960]: "Decisions Under Uncertainty, Drilling Decisions by Oil and Gas Operators," Harvard University Press, Cambridge, Mass.

Hadley, G. [1967]: "Introduction to Probability and Statistical Decision Theory," Holden-Day Inc., Publisher, San Francisco, Calif.

I.E.E.E. [1968]: "Special Issue on Decision Analysis," *IEEE Tran.* vol. SSC-4, No. 3, September.

Jeffreys, H. [1961]: "Theory of Probability," 3d ed., Clarendon Press, Oxford.

Jeffrey, R. C. [1965]: "The Logic of Decision," McGraw-Hill Book Company, New York.

Kaufman, G. M. [1963]: "Statistical Decision and Related Techniques in Oil and Gas Exploration," Prentice-Hall, Inc., Englewood Cliffs, N.J.

Lindley, D. V. [1965]: "Introduction to Probability and Statistics from a Bayesian Viewpoint," parts 1 and 2, Cambridge University Press, Cambridge, England.

Luce, R. D. and H. Raiffa [1958]: "Games and Decisions," John Wiley & Sons, Inc., New York.

Machol, R. and P. Gray [1962]: "Recent Developments in Information and Decision Processes," The Macmillan Company, New York.

Pratt, J. W., H. Raiffa and R. Schlaifer [1965]: "Introduction to Statistical Decision Theory," McGraw-Hill Book Company, New York.

Raiffa, H. [1968]: "Decision Analysis, Introductory Lectures on Choices under Uncertainty," Addison-Wesley Press, Inc., Reading, Mass.

Raiffa, H. and R. Schlaifer [1961]: "Applied Statistical Decision Theory," Harvard University Press, Cambridge, Mass.

Savage, L. J. [1954]: "The Foundations of Statistics," John Wiley & Sons, Inc., New York.

Savage, L. J. [1960]: The Foundations of Statistics Reconsidered, *4th Berkeley Symp. Math. Statist. Probability*, vol. 1, pp. 575–86.

Schlaifer, R. [1961]: "Introduction to Statistics for Business Decisions," McGraw-Hill Book Company, New York.

Schlaifer, R. [1959]: "Probability and Statistics for Business Decisions," McGraw-Hill Book Company, New York.

Schlaifer, R. [1969]: "Analysis of Decisions Under Uncertainty," McGraw-Hill Book Company, New York.

Tribus, M. [1969]: "Rational Descriptions, Decisions, and Designs," Pergamon Publishing Company, Elmsford, N.Y.

Turkstra, C. J. [1962]: A Formulation of Structural Design Decisions, Ph.D. dissertation, University of Waterloo, Waterloo, Ontario.

Von Neumann, J. and O. Morganstern [1953]: "Theory of Games and Economic Behavior," Princeton University Press, Princeton, N.J.

Weiss, L. [1961]: "Statistical Decision Theory," McGraw-Hill Book Company, New York.

Specific text references

Abraham, C. and J. Thedié [1960]: The Price of a Human Life in Economic Decisions, *Revue Français de Recherche Opérationnelle*, pp. 157–168.

Altouney, E. G. [1963]: The Role of Uncertainties in the Economic Evaluation of Water Resources Projects, *Eng. Economic Planning Program Report EEP-7*, Stanford Univ.

Howard, R. A. [1966]: Decision Analysis: Applied Decision Theory, *Proc. 4th Inter. Conf. on Operational Research*, Inter. Fed. Oper. Res. Soc., Boston.

Rascon, O. A. [1967]: Stochastic Model to Fatigue, *J. Eng. Mech. Div., ASCE*, vol. 93, no. EM3, pp. 147–156.

PROBLEMS

5.1. You are running a small (three-man) design office and suddenly the job you are working on is stopped pending the outcome of an election to occur 3 months from now. Office operation costs $3,000 per month.

(a) Set up the decision tree for the actions you can take now:

(i) Do nothing.

(ii) Close the office for one month. This reduces the losses to zero, but you cannot count on rehiring an adequate staff at that time. A small job may be received at the end of the month.

(iii) Same as (ii) except close office for 3 months.

(iv) Make the rounds of prospective clients and offer to work for a reduced fee. Note that it is very difficult to raise fee levels once they are dropped. You can expect to receive sufficient jobs to continue to operate your office. Be very careful in defining the possible states. Note that probability measures exist for each possible state—undesirable as well as desirable. It is a fact that many offices are totally unrealistic in such circumstances.

(b) What additional information do you need to make the outlined study in (a) useful. Note that the election may prove negative and the job may not resume.

5.2. Develop a utility function to represent the utility of your grade-point average. Assume a numerical grade scale from 0 to 4 with grades assigned numbers according to:

Grade	grade points
Failure	0
D	1
C	2
B	3
A	4

(a) First, assume that a grade-point average of 2.0 is required for graduation and your *only* concern is with obtaining a degree.

(b) Second, assume you are a graduate student and you need a grade-point average of 2.75 to receive your degree. Consider too the influence of grades on your social status.

(c) Third, assume that you require financial aid to continue your education. All other factors being equal, the level of aid you receive depends entirely on your grade-point average and no aid is likely for less than a 3.0 average.

(d) Comment on the difficulty in comparison of utilities for this problem and those for other situations involving economic decisions under uncertainty.

5.3. Construct your personal utility function for money in the range of −$1,000 to +$1,000. The decision situation is related to possible added income and/or expenses associated with traveling to Europe to interview for a possible job. How would this function change if you had no cash on hand? How would it change if you had $10,000 in personal savings?

How would the utility function change if you needed $500 to make the trip? Assume that a major loss is received if the trip is not made.

5.4. You are setting up your own consulting office with very limited financial backing. One of the first problems is to set a fee-per-hour level. If your fee is too large, you will lose clients, and if your fee is too small, you will not receive adequate payment for your services and, in addition, will find it difficult to increase your income from this set of clients. You are considering charging $20/hr or $25/hr. These are the actions.

If your fee level is $25/hr ($a_1$) and the business response (θ_1) is adequate, you assign the gain to be +100. If you charge the same rate and the response is unsatisfactory (θ_2), the loss is 50. If you ask for $20/hr ($a_2$) and the response is adequate (θ_1), the gain is 50, while if the response is unsatisfactory (θ_2), your loss is 100. Assume:

(a) $P[\theta_1 \mid a_1] = 0.5$ $P[\theta_2 \mid a_2] = 0.5$
(b) $P[\theta_1 \mid a_1] = 0.3$ $P[\theta_2 \mid a_2] = 0.4$

What is the optimum action under (a) and (b)?

5.5. Construct a decision tree and determine the optimum decision for an engineer in business for himself. The potential losses involved are small compared to his working capital.

The engineer either may accept the beam design his employee has selected and shown on the plans or he may ask for a revision in the design involving a drafting cost of $100. There is no gain or loss associated with a design which performs satisfactorily. If, however, the long-term creep deflection of the concrete beam is excessive, the engineer estimates a (discounted) future loss of $500 in potential fees and inconvenience to his client (as the client's losses are reflected through professional responsibility in dollar loss to the engineer).

The sustained loads are quite well known for this structural type, but uncertainty exists in the long-term effective modulus of elasticity of a reinforced-concrete beam. The engineer is concerned because the likelihood of this modulus being greater than 4000 ksi, i.e., the value needed for satisfactory performance of the present design, is only 0.3. With redesign a value of the modulus of 3000 ksi would be satisfactory. This value will be exceeded, the engineer feels, with odds 3 in 4.

5.6. Reconsider the first example in Sec. 5.2. Based on the ambiguous depth indication and the utilities, the engineer used a long pile even though the shallow depth was more likely. Would this still be the correct action if the instrument had indicated a shallow depth? How large would the "error probability" $P[z_0 \mid \theta_1]$ have to be before the long pile would be the better decision no matter what depth the test indicated? Would the test be justified at any cost in this circumstance?

5.7. Opportunity loss or regret. Consider a simple decision problem with a general payoff table of utilities:

Payoff table of utilities

	Actions	
States	a_1	a_2
θ_1	u_{11}	u_{12}
θ_2	u_{21}	u_{22}

Assume that the actions are numbered such that $u_{11} > u_{12}$ and $u_{22} > u_{21}$. Note that if the engineer knew that θ_i was the true state he would choose action a_i and obtain utility u_{ii} as the best he could do given that state.

If he had instead taken the other action, he would not receive this much utility but something less, namely, u_{12} or u_{21}. We define the "lost" amount, $u_{11} - u_{12}$ or $u_{22} - u_{21}$, as the *regret* or *opportunity loss* associated with having not taken the best action given the state. We can construct, in general, an opportunity loss table for a decision problem by *subtracting the utility values in any row from the largest utility value in that row*. In this case:

Opportunity-loss table

	Actions	
States	a_1	a_2
θ_1	0	$l_{12} = u_{11} - u_{12}$
θ_2	$l_{21} = u_{22} - u_{21}$	0

Show for this simple case that one will obtain the same choice of decisions if he maximizes expected utilities or *minimizes expected opportunity losses.* *Hint:* Compare the values of these expectations for the two actions for arbitrary $p = P[\theta_1]$, showing that $E[u(a_1)] > E[u(a_2)]$ if and only if $E[l(a_1)] < E[l(a_2)]$.

This result holds in general for any number of states or actions. It is often easier to think about and evaluate the *relative* losses associated with several actions for a *fixed* state θ_i than to deal with the absolute utilities. The negative of opportunity losses were in fact used in Table 5.1.1 and in Fig. 5.2.1 in the pile-selection illustration. Of course, to minimize expected opportunity losses is to maximize expected negative opportunity losses. The point remains that the engineer can concentrate on each state of nature *separately* and look at utilities or losses *relative* to the best action available, given that state.

5.8. Specifications for the shop welding of certain structural assemblies say that no welds with internal flaws greater than a critical volume can be permitted in order to guarantee adequate strength. An expensive x-ray inspection system is presently being used to inspect all welds. The test is considered to be absolutely accurate, and history shows that 15 percent of all welds are rejected and rewelded. The fabricator is considering introducing a new, less-expensive, but not perfectly reliable ultrasonic inspection device. By retesting known good and bad welds, the fabricator has found that the new device will, if a flaw is present, indicate this fact with probability 0.80 and erroneously pass the weld with probability 0.20. If no flaw is present, however, the device will pass the specimen with probability 0.90 and incorrectly indicate a critical flaw with probability 0.10.

(a) What is the probability of finding an indication of a bad weld using the sonic device? What are the probabilities of passing a poor weld and of a good weld being wrongly indicated as bad?

(b) If perfect inspection costs $1.00 per weld while the new method costs $0.50, and it costs $5.00 to redo a weld and $30.00 to have a defective weld leave the shop, should the new device be accepted?

5.9. Reconsider the flow-rate-measurement illustration in Sec. 5.2. Assume that the engineer incorrectly neglects the uncertainty in V and carries out a terminal analysis based on θ alone, using his point estimate of $V = 2.5$ ft/sec. Recompute the marginal posterior distribution of θ and compare it with that found in the illustration.

Discuss the influence of neglecting the possible uncertainty in variables such as V which are embedded within a problem (in the sense that utility does not depend on them nor are they observed in an experiment).

What is gained if the flow rate is estimated using a daily measurement of V as well as Z? How do these results differ for observations $Z = 13$ and $V = 2$?

5.10. Recompute the marginal posterior distribution of θ in the flow-rate illustration in Sec. 5.2, assuming now that there is negligible scatter about the relationship between Z and θ/V.

Consider two cases. An observation of:

(a) $Z = 13$ ft

(b) $Z = 14$ ft

Compute also the results for an observation of $Z = 14$ ft when equation uncertainty is retained. Then discuss such uncertainty in this type of problem.

5.11. Investigate the relative advantage of a "hedging" strategy in the pile-order-decision illustration in Sec. 5.3. This new strategy calls for a sonic measurement to be followed by a drill hole if and only if the sonic test is "ambiguous," i.e., if z_2 is observed. Keep the same experiment costs and assume additivity of utilities. *Hint:* Add a fourth alternative experiment to Fig. 5.3.5 with this form:

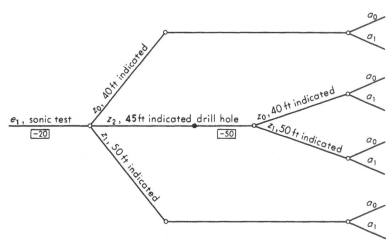

Fig. P5.11

Is the conclusion changed if it is recognized that the "double" experiment, sonic test *and* drilling, will lead to a delay in ordering with a loss of 50 units? *Comment:* This is a simple example of an n-stage sequential decision problem, where here $n = 2$. See Hadley [1967] for an elementary discussion of such problems.

5.12.† Discuss the construction of a decision tree for the engineer in the following situation. Which is the best action-state pair and how might utilities and probabilities be estimated? Make some reasonable guesses for numerical values.

A reinforced-concrete school building is under construction, and questions about concrete strength have arisen during construction. The contract cost of the entire structure is $2,161,713, and it is estimated that replacement of the concrete in question will cost $300,000 to $400,000. The superstructure of the two- and three-story building is specified to contain 4 percent of entrained air with a design strength of 3500 psi. Tests show up to 10 percent air, and tests made on concrete cylinders and 90 cores show a mean strength of 2100 psi when projected to a 28-day strength.

The situation is complicated by the results of load tests. In 16 tests using a superimposed load of 1.5 times the design live load plus half the dead load on the beams and slabs, the structure showed no signs of distress.

Replacement involves perhaps a year's delay in the completion of the badly needed structure. Acceptance of the present structure involves assumption of liability in the event of poor future performance. The decision is complicated by the unknown relationships between beam capacity and concrete cylinder strength, particularly with underreinforced sections, and by the likelihood that the entire specified design strength is not needed in most of the structure.

Note that the decisive value is to be received at some unknown future time. Thus value to a well-established firm may differ from the value to an office that will cease operations in the near future.

5.13. A number of probabilistic, physical models of the process of fatigue damage have been proposed in the literature. Each leads to a different distribution of the number of load applications to failure. These include the Weibull distribution, the lognormal

† *Reference: Engineering News-Record*, August, 29, 1963, pp. 20 and 72.

distribution, and the gamma distribution. Other distributions exist (see, for example, Rascon [1967]), but several can be excluded on physical grounds, and others can be excluded because of their lack of ability to "fit" statistical observations. Remaining contending models have better or poorer physical arguments, and better or poorer "fits" to data. An expert in the field has summarized the situation at a particular point in time by saying, "Neither the Weibull nor gamma distributions can be excluded. If I had to give them relative weights now, I would say perhaps 60 to 40 Weibull over gamma."

Two designs for a structural member are under consideration. If alternative *a* is analyzed assuming that the Weibull distribution holds, it has a reliability with respect to fatigue distress during its lifetime of 0.98, but the probability is 0.99 under the gamma model. On the other hand, the parameters are such that for the second design the reverse is true. The probability of failure is 0.01 under the Weibull hypothesis and 0.02 under the gamma hypothesis. (It is not known that these precise numbers could be constructed with these two particular distributions, but the situation is conceptually important even if there is only a change in the relative values of the probabilities.)

If the designs are of comparable initial cost and if the cost of failure is the same in both cases, which design should the designer use (to reflect best the information available to him)?

5.14. A large reinforced-concrete building is to be constructed on very poor foundation material. The building is 60 ft wide, 200 ft long, and 6 stories in height. Two basic foundation alternatives are friction piles or some type of raft foundation. A differential settlement on the order of 3 in. can be expected to involve legal problems with the client, since you have already indicated that the site is adequate for the proposed construction.

Routine tests and studies indicate that a pile foundation can be expected to show long-term differential settlements of up to 3 in. Odds cannot be quoted that the settlement will not be larger, but it is unlikely. The odds are approximately even that the differential settlement will be less than 1 in., approximately 4 to 1 that it will be less than 2 in., and perhaps approximately 1 in 100 that it will be approximately 3 in. The cost of a pile foundation is so large that it is "even money" that the client will abandon the job if this design is suggested. Use discrete states and note that the measures of likelihood must be normalized.

Studies for a raft foundation indicate that a relatively flexible slab coupled with rock surcharge around the building will be within the cost estimates. The odds are more or less even that the differential settlement will be less than 2 in. and approximately 10 to 1 that it will be less than 3 in.

If the settlement is too large, the client will take legal action against you; you expect the cost to you will be $1,000 or $10,000 with equal likelihood. Your profit on the job will probably be $3,000, providing that it goes to completion.

(*a*) Set up the decision tree for this decision and determine the optimum action.

(*b*) If the client now says he will not pay your usual fee, determine the offered fee level below which you would sue him for the original fee assuming that the cost of collection is $500. Assume that the client will be required to pay you $3,000 and normal engineering costs are not influenced by the legal problems. Note that this type of operation does not imply that you will not be offered future jobs. In fact, if you reduce your fee (under pressure), you can expect to receive future jobs (with attendant fee difficulties). Note that you could hedge against such problems by asking for more than you expect in the initial negotiations.

5.15. A large apartment complex for low-income renters is being considered. The choice must be made between conventional steel construction and using an innovative scheme of "pushup" construction that promises to cost less. There is uncertainty in the cost estimates in both of the proposals, but clearly it is greater in the case of the latter, untried, system of construction.

Assume that the best estimate of the conventional cost is x_0, with uncertainty in the estimate being summarized by saying there is a 50–50 chance that the true cost will be within $x_0 \pm 0.05x_0$. Assume that the best estimate of the pushup cost is $0.8x_0$, but that the "probable error" is $0.10x_0$; that is, there is a 50–50 chance that the true cost will be within $0.8x_0 \pm 0.1x_0$. Assume (for simplicity in your computations) that in both cases the engineer is prepared to state that his uncertainty in the costs can be described by symmetrical triangular distributions, centered on the "best estimates."

(a) As the technical representative of the nonprofit developer, decide which alternative to take if the project is being carried out with funding under government regulations which place a maximum rent on the units and which require that the rent be set so that *no* profit is made by the developer. The implication to the (charitable) developer is that he will make no money no matter what the construction cost, but he will lose money from his own "pocket" if the construction cost exceeds that value c which the maximum rents (minus interest, etc.) will just cover. This value can be accurately calculated to be $c = 1.10x_0$. Sketch the function of utility to the developer versus the cost.

(b) As the technical representative of an association representing the steel industry, decide whether your industry should underwrite the pushup project by guaranteeing that it will pay any cost in excess of c. Your industry gains nothing if the pushup project is not undertaken, but gains information on a potentially profitable new development, no matter what its final cost. Assume that this *benefit* to the industry is judged to be worth $0.02x_0$, independently of the final cost, plus an amount which decreases linearly from $0.05x_0$ at 40 percent of conventional construction cost to zero at the cost of conventional construction x_0. Sketch the utility to the industry versus true cost.

5.16. The Boston Port Authority (BPA) is trying to decide whether to begin immediate design and construction of a major airport in the Boston Harbor to supplement Logan Airport (a_1) or to wait a number of years to make the decision (a_2). The following information has been provided by their consultants, MADCO, Inc.:

(a) Passenger demand for conventional flights will do one of these things in the next decade:

θ_1: Increase very rapidly (Logan overtaxed)

θ_2: Increase moderately (Logan still satisfactory)

θ_3: Remain about fixed owing to the introduction of new systems (high-speed ground transportation or vertical-takeoff aircraft)

(b) The losses to the BPA under the possible action-state pairs are

	θ_1	θ_2	θ_3
a_1	0	-2	-5
a_2	-10	0	0

(c) Available information on demand and new system development projections can best be summarized by saying

$P[\theta_1] = 0.4$

$P[\theta_2] = 0.2$

$P[\theta_3] = 0.4$

(i) Based on the expected-cost criterion, determine the optimal choice of action for BPA.

(ii) Assume now that reliable information says there will be no new system (therefore $P[\theta_1] = 0.4/(0.4 + 0.2) = \frac{2}{3}$ and $P[\theta_2] = \frac{1}{3}$), but the demand remains uncertain. A major portion of the demand may be from New York–Boston commuters. Therefore, instead of choosing a_1 or a_2 at once, the BPA hired a second consultant, RIA, to conduct interviews with local businessmen. The "indication" of the survey was that demand would increase very rapidly. RIA pointed out, however, that such polls are not wholly reliable and that there is a 30 percent chance that this indication would be obtained even if the increase would be only moderate. On the other hand, if the true state is θ_1, the polling procedure will "indicate" θ_1 with 90 percent reliability. Reevaluate the BPA decision in the light of this new information.

5.17. Equivalent lotteries. The following decision-tree formulation and interpretation is often illuminating and useful. Return to the problem of assigning utilities to possible outcomes A, B, and C, which was discussed in Sec. 5.1.2. A and C were the most favorable and least favorable outcomes, respectively. Through interrogation, the decision analyst finds, as before, a probability p^* of obtaining outcome A rather than C such that the decision maker is indifferent between obtaining outcome B and this "A versus C" lottery. Rather than calculating $u(B)$ in terms of assigned values $u(A)$ and $u(C)$, we now *replace* outcome B by the "equivalent" lottery, namely, one with a probability p^* of obtaining A and $1 - p^*$ of obtaining C. Therefore, a decision tree which was of the form

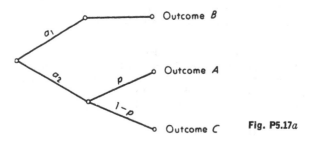

Fig. P5.17a

is replaced by a tree of the form

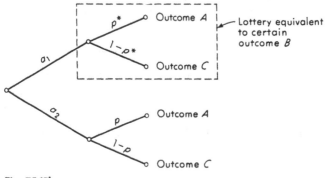

Fig. P5.17b

Or, a more complicated tree (still with A and C as the extreme outcomes)

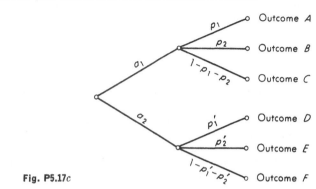

Fig. P5.17c

would become

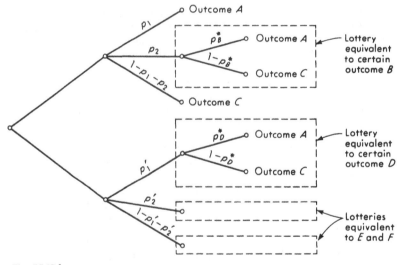

Fig. P5.17d

(a) Demonstrate that the original criterion, "choose a to maximize the expected utility," can now be replaced by the criterion "choose a to maximize the probability of obtaining the most favorable outcome A." The reader can appreciate that this interpretation is more appealing to many users of decision theory.

(b) Apply this approach to the pavement-design illustration in Sec. 5.1.4.

5.18. A stream flows through a new residential area. An engineer must decide how much temporary flood protection should be provided during the 12-month period prior to completion of an upstream dam which will remove the threat. The channel has been reshaped and lined, and so the engineer has no historical data on stream heights at the site during floods. Based on limited data on flood flow rates and based on rough analysis of the channel's hydraulic behavior, the engineer assigns to the coming year's peak flood height an exponential distribution with mean 20 ft.

Find the expected utility of a design for the temporary protection which provides complete protection up to 24 ft. Assume that above 24 ft the damage

(a) Varies linearly:

$$u(\theta) = -100(\theta - 24) \qquad \theta \geq 24$$

(b) Varies as

$$u(\theta) = -10(\theta - 24) - 1000 \left(\frac{\theta - 24}{24}\right)^2 \qquad \theta \geq 24$$

5.19. Consider the aggregate-borrow-pit illustration in Sec. 5.1. It is convenient to fit an analytical expression to the distribution on θ. The beta distribution with limits 3000 and 5250 yd^3 is a reasonable choice.

Define and sketch a utility function over the same limits, a to b or 3000 to 5250 yd^3, in a "beta form":

$$u(\theta) = u_b - (u_b - u_a) \left(\frac{b - \theta}{b - a}\right)^m \qquad a \leq \theta \leq b$$

Find the simple closed-form solution for the expected utility in terms of the state distribution parameters (a,b,r,t) and the utility parameters (u_a,u_b,m). Fit a reasonable PDF to the CDF defined by Fig. 5.1.9. Evaluate $u(\theta)$ for $u_a = -10$, $u_b = +1$, $m = 3$.

5.20. The optimum design of a sewage-treatment plant involves minimizing the losses from possible over- and underdesign. Altouney [1963] has reviewed 71 public-works population projections after a 15-year period. The mean ratio of actual to predicted population was found to be 0.88 and the standard deviation of the ratio was found to be 0.12, with the normal distribution providing a satisfactory fit.

Size, million gal/day	Cost per million gal/day	Total cost
7	\$111,000	\$ 777,000
8	108,000	864,000
9	109,000	945,000
10	102,000	1,020,000
11	100,000	1,100,000
12	98,000	1,175,000

Use the accompanying cost figures.† Assume that the population is estimated to be 100,000 in 15 years and waste generation is 100 gal/(capita)(day). Design is to be based only on conditions 15 years from now. Study alternate actions of design for 7, 8, 9, 10, 11, 12 million gal/day. Assume that the cost of adding capacity is \$150,000 per million gal/day needed.

5.21. An engineer has been called upon to judge the adequacy of an existing structure. Complete plans are available and preliminary computations show that the steel reinforcing is ample. Site inspection raises questions of quality of concrete and construction of a number of long-span slabs because of large deflections.

† C. J. Velz [1948], How Much Should Sewage Treatment Cost?, *Engineering News-Record*, October 14.

The engineer's fee of $1,000 is guaranteed independent of his report conclusions. The owner has stated that he desires one of three reports based on strength according to code.

(i)   Structure is unsafe.
(ii) Structure is safe for office loading, 50 psf.
(iii) Structure is safe for light storage load, 125 psf.

The engineer's prior probabilities of capacity are

State θ; live load capacity, psf	$P'[\theta]$
20	0.20
50	0.30
80	0.20
110	0.20
140	0.10

The owner believes very strongly that the structure is safe for light storage. If the engineer reports that the structure is unsafe, he faces loss of future jobs for the owner with a present worth of $5,000, whether he is correct or not. If he reports that the structure can be used for offices only, he has a probability of only 0.20 of receiving the desired jobs. If the engineer reports that the structure is safe for light storage, the future jobs are assured, provided that the performance is adequate. The total cost of reporting "adequate for office loading" and finding $\theta < 50$ is $5,000; the cost of being wrong ($\theta < 125$) with a report "adequate light storage" is $10,000.

Construct the decision tree for this situation. Is the fee of $1,000 important to the decision? Why?

5.22. As technical coordinator of a proposed harbor-bottom exploration device, you must report on the field of vision of a television camera which will be *dropped* without control to rest on the irregular floor of the harbor. One factor in your analysis is the slope of the floor at the point (a 1-ft circle) where the camera container comes to rest. In the Boston Harbor, which is to be explored, no appropriate data is available, but you must decide on certain dimensions of the device based in part on a probability distribution of the extent of the field of vision, and therefore you must obtain some probability distribution on the floor slope.

(a) Choose some engineering friend and assume (for all his probable ignorance of the subject) that he is the most knowledgeable consultant available on this topic. Find through a sequence of simple hypothetical lotteries the three slope values which have probabilities of 0.25, 0.50, and 0.75 of being exceeded. Sketch a CDF through these points.

(b) Repeat this with a second friend. Have the two compare their CDF's. Can the differences be reconciled through discussion?

5.23. There is circumstantial evidence that a fraction of the cement was inadvertently left out of one batch of concrete now in place in a tall retaining wall. Core data has been obtained. If the cement has been left out, the mean core strength will now be

3000 psi; if not, it will be 3500 psi.   In either case, the standard deviation of the (normally distributed) core strengths would be 450 psi.

(a) What are the relative values of the two sample likelihoods if the sample of four cores yielded values of 3000, 3300, 2800, and 3500?

(b) If the prior probabilities on the two mean strengths were 0.6 and 0.4 on 3000 psi and 3500 psi, respectively, what are the posterior probabilities?

(c) If only two actions are available—a_0, "do nothing," and a_1, "order the concrete removed and replaced"—what decision should be made if the payoff table is

$u(a_i, \theta_j)$

	θ_0 3500 *psi*	θ_1 3000 *psi*
a_0	0	-10
a_1	-100	0

(d) At what value of $u(a_1, \theta_0)$ would the engineer just be indifferent between a_0 and a_1?

5.24. Decide whether the engineer in Prob. 5.23 should test an additional four cores at a cost of eight (additive) units before making a decision?   If cores cost two units each, plot the (preposterior) expected utility versus number of additional cores.   How many should the engineer order?   What is the value of perfect information?

6

Decision Analysis of Independent Random Processes

In this chapter we discuss the analysis of decision problems where information is made available through *past* observations of a random process upon whose *future* values the consequences of a decision will depend. The process considered is simply a sequence of independent random variables. Past observations of maximum annual floods at a site, for example, give information about the probabilities of future flood levels; the performance of a proposed flood-control project will depend on these levels.

Our situation is in part like that in Chap. 4 where a sample is made available and where our concern is to extract information from these observations of a random variable. As in Sec. 4.1, *we shall assume in this chapter that we know with certainty the form of the underlying probability distribution, but that the parameters of this distribution are not precisely known.* Whereas our concern in Chap. 4 was solely with making estimates or confidence statements about these parameter values, our purpose is now the more practical one of making engineering decisions in which these parameter uncertainties play a central role.

Our assumption that the type of distribution is known implies that we are prepared to act as if this level of uncertainty has a negligible influence on the choice of decisions.   The engineer should recognize that this may not be the case in important decisions sensitive to the tails of distributions, where the form of the distribution may be critical.†

We shall use the same basic decision model discussed in Chap. 5. We shall now find it appropriate, however, to place prior distributions on the fixed, but unknown parameters of the process, rather than directly on the value-"producing" state (i.e., the as-yet-unknown future value or values of the random process).   After discussing briefly the prior analysis of this decision situation, we shall consider the more important terminal analysis.   In the latter context, we are answering the question, "How should I modify or update the distribution on the parameters in the light of new information given to me in the form of past observations of the random process?"   We shall find that the conditional probabilities needed for this updating are simply the likelihood functions encountered in Chap. 4.   Just as in Chap. 5, once this terminal-analysis modification in the parameter distribution is made, the actual choice of a decision follows the same procedure used in the prior analysis.

Prior and terminal analysis make up Secs. 6.1 and 6.2.   In Sec. 6.3 we discuss an alternate view of the analysis that gives a useful engineering interpretation of the procedure.

6.1 THE MODEL AND ITS PRIOR ANALYSIS

Formally, we want to analyze decision situations where the stochastic component can be modeled by a simple random process.   This simple process is one generating a sequence, $X_1, X_2, X_3, \ldots$, of *independent, identically distributed* random variables.   Although we assume that we know the form of the common distribution (e.g., exponential, normal, etc.), we are *not* certain about the common parameter value (or values, if the distribution form has more than one parameter).

We choose to deal with this uncertainty by *treating the unknown parameters as random variables*.   This is totally consistent with the subjective or operational interpretation of probability in decision making that we have discussed in Chap. 5.   It is comparable to our saying that the fixed but unknown depth to the firm stratum below a site is a random quantity.   With sufficient information the uncertainty in this depth

---

† Although it has not as yet been practiced or even well studied, it is possible to use the decision model of Chaps. 5 and 6 with the *type of distribution* as one of the dimensions of θ.   Thus, the engineer would assign prior probabilities and calculate sample likelihoods of the true distribution being either Weibull or gamma, say.   Prob. 5.13 is a simple example.

or in the parameter of a random variable will be nonexistent or negligible. This is in contrast to the inherent or "natural" randomness of the process generating the X_i's themselves. For example, after only one inspection of an unbalanced coin we would be uncertain as to the probability p that it would come up heads, whereas after observing 1,000,000 tosses of that coin, we would have a virtually perfectly reliable estimate of p, but we would be left unable to say for certain whether the *next* flip would be heads or tails. The proposed treatment of unknown parameters as random variables is, however, contrary to the more traditional approach which was described in Chap. 4. There, the parameters were assumed to be fixed, unknown *constants* about which statistical inferences were made on the basis of a sample. It was this restriction which forced us to the use of "confidence intervals" and "significance levels," which, recall, are not in principle to be either referred to or operated upon as probabilities.

In this section we will work only with *prior* probability distributions on the parameters. In Sec. 6.2 we shall assume that information becomes available to us in the form of a *random sample*, that is, as a set of independent observations $x_1, x_2, x_3, . . , x_n$. Then we will deal with posterior distributions.

The consequences of any decision will, as in Chap. 5, depend upon the action chosen and the true state of nature. The state of nature of interest is, by *our* definition, that state upon which value depends. Its relationship to the process may be simple or complex. For example, the X_i might be the sequence of total annual flows in a river. The flows are assumed to be independent normally distributed random variables with the same, but unknown, parameters m and σ. If the decision concerns the design of a temporary cofferdam which will impound all water during a single year only, the consequences will depend upon the value of the total flow in year r. Our definition of the state of nature is thus the value of X_r. On the other hand, if the decision is related to the design of a small storage dam for an irrigation system, utility may depend upon the size of the dam chosen and on the *average flow* over the dam's lifetime, a period of 10 years from year r to year $r + 9$. In this case, the state of nature would conveniently be defined as

$$Y = \frac{1}{10}(X_r + X_{r+1} + \cdot \cdot \cdot + X_{r+9})$$

In yet another situation, the performance of a proposed operating policy for the irrigation system might depend solely on the annual flow in the year when the flow is smallest, the state of nature now being defined as

$$Z = \min [X_r, X_{r+1}, ... , X_{r+9}]$$

Finally, the engineer may define the consequences of some decision to depend only on the true mean annual flow m. The state of nature of interest in this last case coincides with a *parameter* of the process, as opposed to the value of a particular variable X_r or the value of a function of one or more *observed* quantities, such as $\frac{1}{10}(X_r + X_{r+1} + \cdots + X_{r+9})$ or min $[X_r, X_{r+1}, \ldots, X_{r+9}]$.

We shall refer to the simplest situation, one (like the 1-year cofferdam) where the state of nature is some particular *single* future value of the process, as "the special case $u(a,X)$." As a simple and very common case, it will be considered in detail in Sec. 6.1.1 and again in Sec. 6.3.1.

We shall see in Sec. 6.1.2 that no matter what the relationship between the process and the state of nature, the decision analysis can be reduced to the same form. The analysis of this common form, being directly parallel to Chap. 5, prompts the choice to denote the *unknown value of the process's parameter* (or parameters) as θ (rather than to denote the state of nature upon which the consequences directly depend as θ). It is θ to which the engineer will assign a prior distribution, and it is θ whose distribution will be updated when new information becomes available (Sec. 6.2).†

6.1.1   Prior Analysis of the Special Problem, $u(a,X)$

In this section we treat the decision analysis of that situation in which the consequences of the decision depend on the chosen action a and on the value x of a random variable X.‡ The distribution of X is assumed

---

† This notational choice is consistent, too, with the theoretical literature in the field. In this literature, for historical reasons associated with classical problems in parameter estimation, major interest lies in situations where it is logical to treat the unknown parameter and the state of nature as equivalent. In such estimation problems, the consequences of an action (i.e., saying the parameter equals some number t) are usually assumed to be a direct function of the true parameter (e.g., $u(a,\theta) = (\theta - t)^2$). In engineering decisions, however, utility is more frequently a direct function of the future value of some random variable in the process X_r, or of some function of several of these values, $Y = \frac{1}{10}(X_r + X_{r+1} + \cdots + X_{r+9})$, for example. We still maintain a distinction, calling X_r or Y the state of nature and θ the unknown parameter value. Notice that they may coincide, as in the last example above, where the engineer stated that utility depended directly on one of the unknown parameters, the mean value m of the X_i. The reader should be cautioned, then, that most literature defines θ as the process parameter *and* as the state of nature. Actually it is possible to maintain this convention for all situations, since, as will be seen, all decision problems reduce finally to consideration of an expectation of a function of θ. That function of θ may be directly a utility function if value depends immediately upon the process parameter, or the function of θ may already be an expectation if utility depends on some future value (or values) of the process, say X_r. In short, any distinctions are only in the semantics of the definition of the state of nature. There are no notational differences. Nonetheless the definition of terms needs discussion.

‡ In the simple prior analysis of this special case no association need be made to a process, $X_1, X_2, \ldots$. It becomes a necessity, however, in Secs. 6.1.2 and 6.2.

to be known, but there is uncertainty about the distribution's parameter(s) θ. For this reason we show the decision tree, as in Fig. 6.1.1, with an additional set of branches, representing the various possible values of the parameter(s), θ_1, θ_2, . . . , with their assigned probabilities. (As indicated in Fig. 6.2.1, we assume throughout Chap. 6 that the utility u depends on the action a and the state of nature x but not *directly* on the parameter θ. The extension necessary to include this uncommon dependence is obvious, but it is notationally cumbersome.)

Examples of the situations represented by this model include the 1-year cofferdam mentioned on page 597. As another example, the long-term deflection performance of a particular beam design for a building may depend upon the compressive strength of the concrete in the beam. That strength may be assumed to be normally distributed with unknown mean and coefficient of variation. The coefficient of variation may depend, however, only on the contractor who is low bidder on the job. Prior to bid opening, the contractor and the value of the coefficient of variation are uncertain, but this uncertainty can be expressed through discrete probability assignments on the contractors and thence on the values of the coefficient of variation. The mean may be uncertain because inadequate data are available to estimate it accurately. The engineer might express this uncertainty by assigning to the unknown

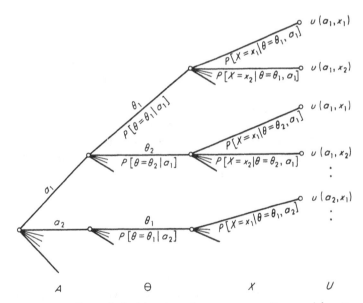

Fig. 6.1.1 General decision tree for a process: the special case $u(a,x)$, showing branches asociated with parameter uncertainty and uncertainty in the future value of the process.

mean a smooth distribution, centered on the specified mix design, 4500 psi, and dispersed to reflect the degree of his uncertainty. (At a finer level of the modeling of this concrete-strength-dependent decision, the engineer may recognize that even after the contractor is known and a significant amount of on-site data collected, the mean and coefficient of variation are not wholly certain because they depend to a degree upon daily conditions such as the weather, the cement supplier, the aggregate source, and numerous other factors which the engineer is uncertain about, but which he may be prepared to make probability assignments on.)

Decision analysis The first step of the analysis of a decision tree like that in Fig. 6.1.1 is to compute the expected utilities given the parameter value $E[u \mid a,\theta]$. This value follows directly from the definition of the expectation of the function of a random variable:

$$E[u \mid a,\theta] = \sum_{\text{all } x_i} u(a,x_i)p_X(x_i \mid a,\theta) \tag{6.1.1}$$

or

$$E[u \mid a,\theta] = \int_{-\infty}^{\infty} u(a,x)f_X(x \mid a,\theta) \, dx \tag{6.1.2}$$

in which the known probability distribution of X is written as:

$$p_X(x \mid a,\theta)$$

or

$$f_X(x \mid a,\theta)$$

simply to emphasize that the distribution may depend on the action and that in this computation the values of the parameters are given. This expected utility will depend, of course, on θ, the value of the parameter(s) of the distribution of X. For each value of θ there will, in general, be a different value $E[u \mid a,\theta]$.

Placing these results on the reduced or "pared" decision tree, Fig. 6.1.2, the second step of the analysis is to weigh these expectations by the discrete or continuous probability assignment on the parameter(s) $P[\theta = \theta_i \mid a]$ or $f_\Theta(\theta \mid a)$† to obtain the expected utility for each action,

$$E[u \mid a] = \sum_i E[u \mid \theta_i,a]P[\theta = \theta_i \mid a] \tag{6.1.3}$$

or

$$E[u \mid a] = \int_{-\infty}^{\infty} E[u \mid \theta,a]f_\Theta(\theta \mid a) \, d\theta \tag{6.1.4}$$

† We are treating the unknown parameter in every way as a random variable, but, to avoid awkward references, in this chapter we shall use the upper case notation for a parameter treated as random variable only when subscripting. So, for example, we refer here to a distribution $f_\Theta(\theta)$ on the parameter θ, not on Θ, as would be more nearly consistent with Chap. 2.

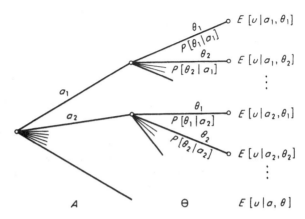

Fig. 6.1.2 The reduced decision tree, showing expected utility values given the parameter value.

The distribution of X may be continuous or discrete (or mixed); the probability assignment on the parameter(s) θ may be continuous or discrete (or mixed); and *any* combination of these situations may occur in a particular problem. Notationally, the distribution's possible dependence on the action a will usually be suppressed.

As in Chap. 5 the optimal choice of action is that with the largest expected utility $E[u \mid a]$. It is important to notice that the reduced tree, Fig. 6.1.2, and the calculations in Eqs. (6.1.3) or (6.1.4) are identical to those in the simple model of Sec. 5.1 [Fig. 5.1.1 and Eq. (5.1.1)]. The only difference is the replacement of an actual utility $u(\theta,a)$ by an expected utility $E[u \mid \theta,a]$.[†] After numerical illustrations of this special problem, $u(a,x)$, we show in Sec. 6.1.2 that even the analysis of more complicated relationships between the process and the state of nature of interest reduce to this familiar form.

Illustration: Hurricane-barrier design Suppose that two alternate designs are being considered for a hurricane barrier protecting a small boat harbor. One design a_1 will provide certain protection at the expense of a constricted opening. The other design a_2 restricts passage to and from the harbor much less, but it would permit large damaging waves to enter in the event that a severe hurricane approaches directly from the east. The performance of this design depends, in part, on the number of such hurricanes that occur in the design lifetime of 50 years. We consider, in general, a process $X_1, X_2, \ldots$, in which X_i is the number of damaging hurricanes in the ith 50-year interval. The consequences here depend only on the number in the next 50-year interval; call that

[†] Even this replacement is not new. A similar substitution was used in the second step of the bridge-foundation illustration, Eq. (5.1.5), where the state of nature was two-dimensional—flow and performance.

number simply X. As rare events the hurricanes will be treated here as Poisson arrivals with an average number of occurrences of ν per 50 years. Thus, the PMF of X is

$$p_X(x \mid \nu) = \frac{\nu^x e^{-\nu}}{x!} \qquad x = 0, 1, 2, \ldots$$

In the absence of historical records at the site the engineer must base his estimate of the unknown parameter ν on any related information and professional judgment. In this case, on the basis of generalized meteorological analyses and gross regional data, the meteorological consultant suggests that ν might lie somewhere between 0.25 and 1.0 with smaller values being somewhat more strongly indicated than larger ones. The engineer responsible for the decision chooses to summarize his available information by a probability statement:

$$P[\nu = 0.25] = 0.2$$
$$P[\nu = 0.50] = 0.4$$
$$P[\nu = 0.75] = 0.3$$
$$P[\nu = 1.0] = 0.1$$

These four discrete values represent, in the engineer's judgment, a sufficiently refined representation of his information.

Assume that differences in initial costs and in losses due to the restrictions imposed by the first design are summarized by the engineer in the following utility statements:

$$u(a_1) = 0$$
$$u(a_2, X) = 4 - 8X$$

In words, the difference in initial cost and the freer access of the second design are worth four units of value relative to the first design. The damage to boats in the harbor due to *each* critical hurricane, should one or more occur, is estimated at eight units.

The decision tree is shown in Fig. 6.1.3. The analysis is begun by calculation of the expected utilities given the parameter values, Eq. (6.1.1),

$$E[u \mid a_2, \nu] = \sum_{x=0}^{\infty} u(a_2, x) p_X(x \mid \nu) \tag{6.1.5}$$

Substituting for the PMF of X, the expectation in Eq. (6.1.5) becomes

$$E[u \mid a_2, \nu] = \sum_{x=0}^{\infty} (4 - 8x) \frac{\nu^x e^{-\nu}}{x!} = 4 \sum \frac{\nu^x e^{-\nu}}{x!} - 8 \sum x \frac{\nu^x e^{-\nu}}{x!}$$

Since the first sum is unity,

$$E[u \mid a_2, \nu] = 4 - 8 \sum x \frac{\nu^x e^{-\nu}}{x!}$$

$$= 4 - 8E[X]$$

$$= 4 - 8\nu$$

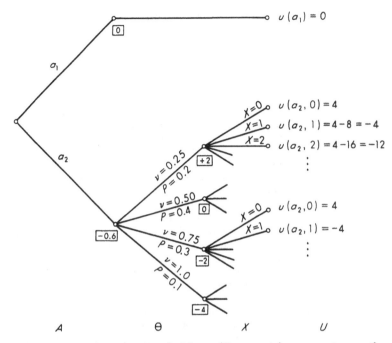

Fig. 6.1.3 Hurricane-barrier decision with uncertain parameter ν, the expected number of critical hurricanes per 50 years.

Or, more directly,

$$E[u \mid a_2, \nu] = E[4 - 8X] = 4 - 8E[X] = 4 - 8\nu$$

In particular,

$$E[u \mid a_2, \nu = 0.25] = 4 - 8(0.25) = +2$$
$$E[u \mid a_2, \nu = 0.5] = 4 - 8(0.5) = 0$$
$$E[u \mid a_2, \nu = 0.75] = 4 - 8(0.75) = -2$$
$$E[u \mid a_2, \nu = 1.0] = 4 - 8(1.0) = -4$$

These numbers are entered in boxes at the appropriate nodes of the tree, Fig. 6.1.3.

The second step of the analysis requires computation of $E[u \mid a]$, Eq. (6.1.3). Here, for a_2,†

$$E[u \mid a_2] = \sum_i E[u \mid \nu_i, a_2] P[\nu = \nu_i \mid a_2] \qquad (6.1.6)$$

$$E[u \mid a_2] = (2)P[\nu = 0.25] + (0)P[\nu = 0.50] + (-2)P[\nu = .75] \\ + (-4)P[\nu = 1.0]$$

$$= (2)(0.2) + (0)(0.4) + (-2)(0.3) + (-4)(0.1)$$

$$= -0.6$$

† Note that with a set of several alternate designs with varying degrees of exposure and with X defined as the number of damaging hurricanes associated with each, the true value of ν and the prior distribution on ν would depend on the design a_i.

This value and $E[u \mid a_1] = 0$ are entered on the tree also. Comparison suggests that a_1 is the preferred action.

In Sec. 6.2 we shall reconsider this decision in the light of new information, i.e., historical data that can be interpreted as observations of random variables distributed like X.

Illustration: Transportation mode choice   In planning a transportation system, it is important to determine how users will decide among the alternative modes available to them. It is apparent that in addition to total travel time, variability or unpredictability of the travel time is a factor that influences users' decisions. Let us consider how one particular user might (best) make his personal decision.

A business man has an appointment in another city at a particular hour t. He has a choice of planning to travel the major distance there by train, a_1, or by air, a_2. In either case, he estimates that his loss due to arriving at $t + X$ hr is X^2 units. Since an early arrival implies wasted time, this loss relationship is assumed to hold for negative values of X as well. In short, these losses are assumed to be associated *not* with total travel time but with the unpredictability of arrival time.

If he travels by train, action a_1, the mean travel time, door-to-door, will be $m = 5$ hr and the standard deviation $\sigma = 0.25$ hr. These are numbers based on extensive observations. As a (suboptimization) problem, the reader can easily show that the man should leave at $t - 5$, or 5 hr before his appointment to minimize his expected loss if he takes the train. (We assume throughout that the scheduling is flexible enough that he can find a train or plane leaving at exactly the hour he prefers. More realistic choices do not alter the problem-solution method.) If he takes the train, the mean of X is thus zero and its standard deviation is $\sigma_X = 0.25$. (As the sum of small delays, X might be assumed normally distributed, but we shall find that, in fact, no distribution assumption is needed in this problem.)

If the man chooses air, a_2, and if the weather is good, the mean travel time, door-to-door, will be $m = 2.5$ hr with an unknown standard deviation σ. Based on his limited air-travel experience, the man estimates that σ for this route is somewhere between about 0.50 and 0.80 hr. He expresses his uncertainty about σ by assigning three possible discrete values—0.55, 0.65, and 0.75—equal weight.

On the other hand, if bad weather is encountered on the day of the scheduled trip, the mean travel time by air increases to $m = 3.0$ hr, with the standard deviation σ lying somewhere in the range 0.75 to 1.25. In this case, the traveler gives the values $\sigma = 0.8$, 1.0, and 1.2 equal weights. Almanac data suggest that such weather conditions occur on 10 percent of all days.

If the man travels by air, he will leave 2.5 hr before his appointment.† Thus if the weather is good, $m_X = 0$ and σ_X is uncertain, lying between 0.5 and 0.8. If the weather is bad, $m_X = 3.0 - 2.5 = 0.5$, and σ_X lies between 0.75 and 1.25. Relative to the train choice, the later scheduled departure time of the airplane flight with the additional $(5 - 2.5)$ or 2.5 hr in his office is worth 0.4 unit of utility to the traveler. We assume that we can add this to the "unpredictability" cost X^2. Thus the total utility for a_2 given $X = x$ is $0.4 - x^2$.

---

† This is *not* quite a suboptimal strategy, given that he goes by air. What is the best planned departure time? See Prob. 6.4.

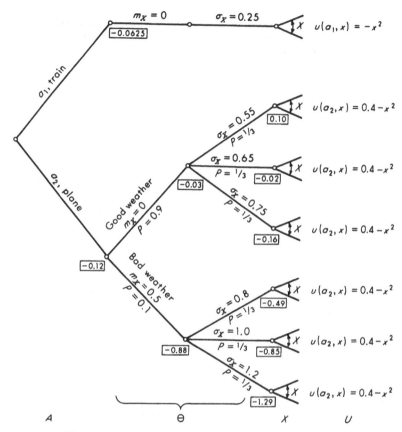

Fig. 6.1.4 Transportation-mode-selection illustration, train versus plane.

The decision tree for this problem is shown in Fig. 6.1.4. A separate set of branches is shown for each of the unknown parameters, m_X and σ_X. (A single set of six branches could also be used.) The analysis of the tree is straightforward. It is facilitated by recalling that, since $\sigma^2 = E[X^2] - m^2$,

$$E[a + bX^2] = a + bE[X^2] = a + b(m_X^2 + \sigma_X^2) \qquad (6.1.7)$$

This is true no matter what the distribution of X. Thus this particular, quadratic structure of the utility versus X relationship simplifies the analysis and eliminates the need to make a specific assumption as to the form of the distribution of X. In particular, the distribution of X might well have a large positive or negative skewness, but this need not be ascertained, as it will not influence the decision.

For action a_1, the train, $a = 0$ and $b = -1$, while $m_X = 0$ and $\sigma_X^2 = (0.25)^2$. Thus:

$$E[u \mid a_1] = 0 - 1[0 + (0.25)^2] = -0.0625$$

For action a_2, the plane, $a = 0.4$ and $b = -1$, but the values of the parameters of X are uncertain. Thus a two-stage analysis is required, the first step being to find the expected utility given the parameters $E[u \mid a_1,\theta]$. Here θ is a pair of values m_X and σ_X.

$$E[u \mid a_1,\theta] = 0.4 - 1(m_X{}^2 + \sigma_X{}^2)$$

For example, if the weather is bad, $m_X = 0.5$; if, also, $\sigma_X = 0.8$,

$$E[u \mid a_2, m_X = 0.5, \sigma_X = 0.8] = 0.4 - [(0.5)^2 + (0.8)^2]$$
$$= 0.4 - (0.25 + 0.64)$$
$$= -0.49$$

Six pairs of mean and standard deviation values exist. The pared tree is shown in Fig. 6.1.5.

The second stage of the analysis requires weighting the expected utilities which are conditional on the parameter values by the probabilities of these

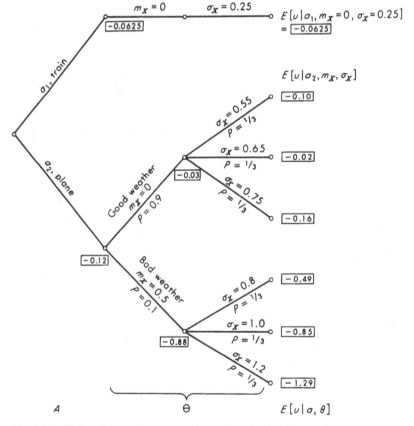

Fig. 6.1.5 Reduced tree: transportation-mode-selection illustration.

parameter values and summing. In this case, this is conveniently done in two steps: the first finds the expected utility given the mean value by summing over values of the standard deviation; the second finds the marginal expected utility (conditional on neither parameter) by summing over values of the mean. For example, given $m_X = 0$,

$$E[u \mid a_2, m_X = 0] = \sum_i E[u \mid a_2, m_X = 0, \sigma_X = \sigma_i] P[\sigma_X = \sigma_i \mid m_X = 0]$$

$$= (0.10)(\tfrac{1}{3}) + (-0.02)(\tfrac{1}{3}) + (-0.16)(\tfrac{1}{3})$$

$$= -0.03$$

Similarly, given $m_X = 0.5$, the expected utility is -0.88. Notice that m_X and σ_X are *not* independent random variables. The marginal expected utility for the plane is

$$E[u \mid a_2] = \sum_i E[u \mid a_2, m_X = m_i] P[m_X = m_i]$$

$$= (-0.03)(0.9) + (-0.88)(0.1)$$

$$= -0.12$$

For this traveler's stated preferences, the train appears to be a better choice. Notice that if air travel were not influenced by bad weather, the benefits of the shorter scheduled travel time would offset the larger variability and make air a better choice (-0.03 versus -0.0625). But inclusion of the possible influence of bad weather swings the decision to the train.

6.1.2 More General Relationships between the Process and the State of Interest

In Sec. 6.1.1, we assumed that a future value of a particular single variable in the process $X_1, X_2, \ldots$ was the state of nature of interest, that is, the factor to which the consequences of an action would be directly related. In this section we shall find that under more general relationships between the state of nature and the process, the prior analysis reduces to the same form. Only the computation of the conditional utility $E[u \mid a,\theta]$ will differ. Recall that θ represents the common, but unknown parameter(s) of the process of independent identically distributed random variables, $X_1, X_2, \ldots$.

The analysis To carry out the decision analysis, it is necessary to know only how the conditional expectation given the parameters (and the action) depends on θ, that is, to know the function $g(\theta)$ in

$$E[u \mid a,\theta] = g(\theta) \tag{6.1.8}$$

For notational convenience, we have suppressed the indication of the dependence of $g(\theta)$ on the action a. With knowledge of $g(\theta)$, the analysis proceeds as before, finding next $E[u \mid a]$ as [Eqs. (6.1.3) or (6.1.4)]:

$$E[u \mid a] = \sum_j E[u \mid a,\theta_j]P[\theta = \theta_j \mid a] \tag{6.1.9}$$

or

$$E[u \mid a] = \int_{-\infty}^{\infty} E[u \mid a,\theta]f_\Theta(\theta \mid a) \, d\theta \tag{6.1.10}$$

In terms of $g(\theta)$,

$$E[u \mid a] = \sum_j g(\theta_j)P[\theta = \theta_j \mid a] \tag{6.1.11}$$

or

$$E[u \mid a] = \int g(\theta)f_\Theta(\theta \mid a) \, d\theta \tag{6.1.12}$$

Therefore it is clear that the analysis proceeds as before when $g(\theta)$ is known.   Let us consider several examples of finding $g(\theta)$.

Examples of $g(\theta)$ Suppose that for a particular action a the consequences of a decision (say the design of a small water-storage dam) are quadratically related to the sum of β values of the process $X_1, X_2, \ldots + X_\beta$ (the total annual inflows):

$$u(a,X_1, \ldots, X_\beta) = c_1 + c_2(X_1 + X_2 + \cdots + X_\beta)$$
$$+ c_3(X_1 + X_2 + \cdots + X_\beta)^2$$

Then given the parameter(s) of the distribution of the X_i's,

$$g(\theta) = E[u \mid a,\theta] = E[\{c_1 + c_2(X_1 + X_2 + \cdots + X_\beta)$$
$$+ c_3(X_1 + X_2 + \cdots + X_\beta)^2\} \mid a,\theta]$$
$$= c_1 + c_2 E[X_1 + X_2 + \cdots + X_\beta \mid a,\theta]$$
$$+ c_3 E[(X_1 + X_2 + \cdots + X_\beta)^2 \mid a,\theta]$$
$$= c_1 + \beta c_2 m_X + c_3(\beta^2 m_X^2 + \beta \sigma_X^2) \tag{6.1.13}$$

If, for example, the X's are all exponentially distributed with parameter λ, then

$$g(\theta) = g(\lambda) = c_1 + \beta c_2 \frac{1}{\lambda} + c_3 \left(\frac{\beta^2}{\lambda^2} + \frac{\beta}{\lambda^2}\right) \tag{6.1.14}$$

With a distribution on λ, the decision analysis for this action is completed by finding

$$E[u \mid a] = \int_0^{\infty} g(\lambda)f_\Lambda(\lambda) \, d\lambda \tag{6.1.15}$$

As a second example of finding $g(\theta)$, assume that the evaluation of the performance of a design is related solely to its response to Y, the *largest* of the demands during its lifetime of β years.   The annual demands

X_i are independent, with a common type II extreme-value distribution
(Sec. 3.3.3)

$$F_X(x) = e^{-(x/v)^{-k}} \qquad x \geq 0 \tag{6.1.16}$$

with unknown parameters v and k. Now

$$Y = \max [X_1, X_2, \ldots, X_\beta] \tag{6.1.17}$$

Then the CDF of Y is

$$F_Y(y) = P[X_1, X_2, \ldots, X_\beta \text{ are all} \leq y]$$
$$= [e^{-(y/v)^{-k}}]^\beta$$
$$= e^{-\beta(y/v)^{-k}} \tag{6.1.18}$$

The density function of Y can be found by differentiation of $F_Y(y)$; and
for a prescribed lifetime β it too depends only on the parameters k and v
of the process.

Knowing the dependence of utility on the value of Y, the function
$E[u \mid a, \theta] = g(\theta)$ follows directly. For example, if

$$u(a, y) = cy^j \tag{6.1.19}$$

(in which constants c and j might depend upon the action a), then

$$E[u \mid a, \theta] = g(\theta) = g(k, v) = \int_0^\infty u(a, y) f_Y(y) \, dy$$
$$= \int_0^\infty \left(\frac{y}{r}\right)^j \frac{k}{v'} \left(\frac{y}{v'}\right)^{-k-1} e^{-(y/v')^{-k}} \, dy \tag{6.1.20}$$

in which

$$v' = v\beta^{1/k}$$

and

$$r = c^{-1/j}$$

The integral is easily evaluated after the change of variable $z = (y/v')^{-k}$.
The result is

$$E[u \mid a, \theta] = g(k, v) = \left(\frac{v'}{r}\right)^{-j} \Gamma\left(1 - \frac{j}{k}\right)$$
$$= c(v\beta^{1/k})^{-j} \Gamma\left(1 - \frac{j}{k}\right) \tag{6.1.21}$$

With a distribution $f_{K,V}$ expressing the uncertainty in parameters k and v,
the decision analysis is again completed by weighting:

$$E[u \mid a] = \int E[u \mid a, \theta] f_\Theta(\theta \mid a) \, d\theta = \iint g(k, v) f_{K,V}(k, v \mid a) \, dk \, dv \tag{6.1.22}$$

It has been pointed out that in some cases the consequences of a decision may be most easily expressed as being related directly to the parameters of the process rather than to specific values of random variables. For example, consider the choice of the size of a storage lane for left-turning vehicles at an intersection. The number of vehicles delayed each day owing to the insufficiently large lane of a particular design a is a sequence $X_1, X_2, \ldots$ of assumed independent, identically distributed random variables. The performance of the design might be stated by the engineer to be proportional to the (unknown) average number of vehicles delayed per day m. Then immediately one has

$$E[u \mid a,\theta] = g(\theta) = g(m) = -cm \tag{6.1.23}$$

Weighting by the distribution on the mean $f_M(m \mid a)$ completes the decision analysis. One can argue, however, that this situation is no different from the previous examples, i.e., that the utility is in fact dependent on the X_i's. The engineer has assumed implicitly that utility is proportional to the (random number) of cars delayed in some arbitrary number of days, β. Choosing the proportionality constant to be $-c/\beta$,

$$u = -\frac{c}{\beta}(X_1 + X_2 + \cdots + X_\beta)$$

implying, as before, that

$$E[u \mid a,\theta] = -\frac{c}{\beta} E[X_1 + X_2 + \cdots + X_\beta] = \frac{-c\beta m}{\beta} = -cm \tag{6.1.24}$$

It is helpful, nonetheless, to be free to think in terms of the consequences depending directly on process parameters rather than values of the random variables.

Illustration: A sea-wall design As a numerical example of more general relationships between the simple process of independent, identically distributed random variables and the state of nature upon which value depends directly, we consider this case. The process is a sequence $X_1, X_2, \ldots$ of independent, identically distributed maximum annual storm intensities. For each different design of a sea wall, there will be damage only in those years when the maximum storm† exceeds a different intensity x_0. *In the upper tail* (i.e., at least for x greater than some value x' which is less than the smallest x_0 of interest), the distribution of the X_i's will be treated as exponential in form:

$(1 - F_X)$ is unknown for $x < x'$

$1 - F_X(x) = be^{-\lambda x}$ $\quad x \geq x'$

† We ignore here the probability of two or more damaging storms in a year.

Thus the probability of damage in any year is

$$p = 1 - F_X(x_0) = be^{-\lambda x_0}$$

Obtaining $g(\theta)$ Given the parameter values b and λ, the expectation of the present worth† of the total damages ΣC_i over β years is found as follows:

$$E[u \mid a,\theta] = g(\theta) = E\left[-\sum_{i=1}^{\beta} C_i\right]$$

$$= -\sum_{i=1}^{\beta} E[C_i]$$

in which we assume that the (random) present worth of the damage in year i is

$$C_i = 0 \qquad\qquad \text{if } X_i \leq x_0$$

$$C_i = c\,\frac{1}{(1+\epsilon)^i} \qquad \text{if } X_i > x_0$$

In these costs, c is the cost of each damaging storm (assumed independent of storm size and of sea-wall design), and $1/(1+\epsilon)^i$ is the capitalization factor if the cost is incurred in year i.   The interest rate is ϵ.   Clearly,

$$E[C_i] = 0(1-p) + \frac{c}{(1+\epsilon)^i}\,p$$

$$= \frac{cp}{(1+\epsilon)^i} \tag{6.1.25}$$

Thus

$$g(\theta) = -\sum_{i=1}^{\beta} \frac{cp}{(1+\epsilon)^i}$$

$$= -cp\,\frac{1}{\epsilon}\left[1 - \frac{1}{(1+\epsilon)^{\beta}}\right]$$

Substituting for p and letting the discount factor be

$$k_\epsilon = \frac{1}{\epsilon}\left[1 - \frac{1}{(1+\epsilon)^{\beta}}\right]$$

$$g(b,\lambda) = -ck_\epsilon be^{-\lambda x_0} \tag{6.1.26}$$

Assume that the design life β is 20 years and that the interest rate ϵ is 5 percent.   Then $k_\epsilon = 12.4$.   Let the value of c be \$50,000 (which might actually be the expected cost *given* that $X \geq x_0$ (see Prob. 6.5).   Three designs are under consideration, a_1, a_2, and a_3.   They have initial costs \$200,000, \$215,000, and \$230,000, respectively, but they will withstand progressively higher storm intensities—$x_0 = 75$, 90, and 100 mph, respectively.   (It should be clear to the reader how, following the procedure in illustrations in Chap. 5, he could also treat the case where the capacity x_0 of a particular design is also a random

† Many of the previous illustrations could have been more realistic by including, as this example does, the influence of interest.

variable.)   With these assumptions we have

$$g(b,\lambda) = -50,000(12.4)be^{-\lambda x_0}$$

$$= -620,000be^{-\lambda x_0} \tag{6.1.27}$$

For the particular three actions available, the expected utilities due to damages are

$$E[u \mid a_1,\theta] = g_1(b,\lambda) = -620,000be^{-75\lambda}$$

$$E[u \mid a_2,\theta] = g_2(b,\lambda) = -620,000be^{-90\lambda}$$

$$E[u \mid a_3,\theta] = g_3(b,\lambda) = -620,000be^{-100\lambda}$$

The initial costs must also be subtracted from these values.

From this point on, the analysis is exactly like that in the previous section.
Parameter Distribution The choice among actions requires a distribution on the unknown parameters b and λ. In this case, looking at data (observed values of the X_i) plotted versus the natural logarithm of their complementary cumulative frequency polygon (i.e., using exponential probability paper), the engineer considered several straight-line fits to the upper tail. He decided that they all passed through the value 0.2 at $x' = 50$ mph but that the slope was not certain. He expressed his uncertainty in this slope λ by stating that it had a gamma distribution with mean 0.02 and coefficient of variation of 0.3. (We shall discuss more formal ways to treat the information in observed values of the X_i in Sec. 6.2.)

The implication of the cumulative distribution passing through 0.2 at $x' = 50$ is that

$$1 - F_X(50) = be^{-50\lambda} = 0.2$$

or

$$b = 0.2e^{50\lambda}$$

Thus any of the conditional expected values $E[u \mid a,\theta]$ can be written in terms of the parameter λ only:

$$E[u \mid a,\theta] = g(\lambda) = -620,000(0.2)e^{50\lambda}e^{-x_0\lambda}$$

$$= -124,000e^{-(x_0-50)\lambda} \tag{6.1.28}$$

The implications of the engineer's mean and coefficient of variation choices are that λ is gamma-distributed, $G(r,\nu)$, with parameters (Sec. 3.2.3):

$$\frac{1}{\sqrt{r}} = 0.3 \quad \text{or} \quad r = 11$$

$$\frac{r}{\nu} = 0.02 \quad \text{or} \quad \nu = \frac{11}{0.02} = 550$$

Decision The remaining step in the decision analysis is the computation of the expected utility for each action. These are [Eq. (6.1.4)],

$$E[u \mid a] = \int E[u \mid a,\theta]f_\Theta(\theta \mid a)\,d\theta$$

$$= -\int_0^\infty 124,000e^{-(x_0-50)\lambda}\frac{\lambda(\nu\lambda)^{r-1}e^{-\nu\lambda}}{\Gamma(r)}\,d\lambda$$

This integral can be rearranged to a constant times an integral over a gamma PDF, leaving

$$E[u \mid a] = -124{,}000 \frac{\nu^r}{(\nu + x_0 - 50)^r}$$

Thus

$$E[u \mid a] = -124{,}000 \left(\frac{550}{500 + x_0} \right)^{11} \tag{6.1.29}$$

Evaluating this expression for $x_0 = 75$, 90, and 100, and subtracting the initial costs, we have

$$E[u \mid a_1] = -124{,}000(^{550}\!/_{575})^{11} - 200{,}000$$
$$= -75{,}000 - 200{,}000 = -275{,}000$$
$$E[u \mid a_2] = -124{,}000(^{550}\!/_{590})^{11} - 215{,}000$$
$$= -271{,}000$$
$$E[u \mid a_3] = -124{,}000(^{550}\!/_{600})^{11} - 230{,}000$$
$$= -277{,}000$$

The best design consistent with the engineer's value and probability statements is a_2, which will withstand a 90-mph storm.

6.1.3 Summary

Given a process of independent, identically distributed random variables X_1, X_2, . . . , with unknown parameter(s) θ, the utility of any action may depend on a particular X_i, on some function of several of the X_i, or directly on the value of θ. In any case we can find for each action the expected utility *given* the parameter(s):

$$g(\theta) = E[u \mid a, \theta]$$

If the engineer is prepared to state his information on the unknown process parameters in the form of a distribution $f_\Theta(\theta)$ on θ, then the analysis of a decision is carried out by comparing expected utilities

$$E[u \mid a] = \int g(\theta) f_\Theta(\theta) \, d\theta$$

6.2 TERMINAL ANALYSIS GIVEN OBSERVATIONS OF THE PROCESS

If observations of the process become available to the engineer, they provide additional information regarding the true value(s) of the unknown parameter(s) of the process. As in Sec. 4.1, we are concerned here with the second type of uncertainty, that which is associated with unknown parameters. This form of uncertainty tends, recall, to decrease as the sample size grows. Unlike Chap. 4, in Chap. 6 our interest in the possible values of the parameters rests explicitly on their influence upon choices among actions. In Chap. 4 only information in the sample was analyzed. Now, however, we have some distribution on the parameters *prior* to taking the sample. The data-analysis problem is one of modifying the prior distribution of the parameter(s) θ given a sample of observations of the process. The end product is a new or *posterior distribution* on the parameters. The components of this procedure—Bayes' theorem and sample-likelihood functions—are already familiar to us.

Section 6.2.1 treats the general case; Sec. 6.2.2 deals with situations where the prior information is diffuse; and Sec. 6.2.3 introduces important computation-saving choices for prior distributions. Since, given the posterior distribution on θ, the decision analysis reduces to the same form as Sec. 6.1, we emphasize throughout Sec. 6.2 the process of obtaining the posterior distributions from the prior distribution of θ and the sample, rather than the subsequent decision analysis. We shall not discuss the problem of preposterior analysis of processes; this, recall, is the problem of whether to pay for more information, i.e., more observations of the process. As discussed in Sec. 5.3, this is in principle a straightforward extension involving a number of terminal analyses.

6.2.1 The General Case

The problem to be analyzed is the following. We have a prior distribution, $p_\theta'(\theta)$ or $f_\theta'(\theta)$, on the unknown parameter(s) θ of the process. This distribution is assigned by the engineer to reflect his professional assessments of all available related information. There is made available to the decision maker the outcome z of an experiment which consists of n independent observations of the process

$$X_1 = x_1, X_2 = x_2, \ldots, X_n = x_n$$

As we have encountered several times, Bayes' theorem relates the posterior distribution, $p_\theta''(\theta)$ or $f_\theta''(\theta)$, to the prior distribution and the sample-likelihood function [Eq. (5.2.2)]

$$\begin{pmatrix} \text{Posterior prob-} \\ \text{ability of } \theta \\ \text{given the} \\ \text{sample} \end{pmatrix} = \begin{pmatrix} \text{normalizing} \\ \text{constant} \end{pmatrix} \begin{pmatrix} \text{likelihood of the} \\ \text{sample given } \theta \end{pmatrix} \begin{pmatrix} \text{prior prob-} \\ \text{ability of } \theta \end{pmatrix}$$

$$(6.2.1)$$

For a discrete distribution on θ, the posterior distribution $p_\theta''(\theta)$ becomes

$$p_\theta''(\theta) = P[\theta \mid z] = P[\theta \mid X_1 = x_1, X_2 = x_2, \ldots, X_n = x_n]$$
$$= NP[X = x_1, X = x_2, \ldots, X_n = x_n \mid \theta]p_\theta'(\theta)$$

or

$$p_\theta''(\theta) = NL(\theta \mid x_1, x_2, \ldots, x_n)p_\theta'(\theta) \qquad (6.2.2)$$

in which N is a normalizing constant and in which the sample likelihood function L (Sec. 4.1.4) is the probability of the observed sample given θ:

$$L(\theta \mid x_1, x_2, \ldots, x_n) = P[(X_1 = x_1) \cap (X_2 = x_2) \cap \cdots$$
$$\cap (X_n = x_n) \mid \theta] \quad (6.2.3)$$

If the X_i's are discrete random variables,

$$L(\theta \mid x_1, x_2, \ldots, x_n) = p_{X_1, X_2, \ldots, X_n}(x_1, x_2, \ldots, x_n \mid \theta)$$

For this process of independent X_i's, this equation becomes

$$L(\theta \mid x_1, x_2, \ldots, x_n) = \prod_{i=1}^{n} p_X(x_i \mid \theta) \quad (6.2.4)$$

in which $p_X(x \mid \theta)$ is the common PMF of the random variables of the process. If the X_i's are continuous random variables,

$$L(\theta \mid x_1, x_2, \ldots, x_n) = \prod_{i=1}^{n} f_X(x_i \mid \theta) \quad (6.2.5)$$

If the distribution (prior and posterior) on θ is continuous, Eq. (6.2.2) becomes

$$f_\Theta''(\theta) = NL(\theta \mid x_1, x_2, \ldots, x_n) f_\Theta'(\theta) \quad (6.2.6)$$

We shall see examples shortly.

The decision analysis described in Sec. 6.1 is unchanged. It is only necessary that the most up-to-date distribution of the parameter(s) be used in the expression for expected utility, Eq. (6.1.11) or Eq. (6.1.12). For example,

$$E[u \mid a] = \int E[u \mid a, \theta] f_\Theta''(\theta) \, d\theta \quad (6.2.7)$$

Illustration: Hurricane-barrier design extended Reconsider the hurricane-barrier-design illustration of Sec. 6.1.1. The process is a sequence of random variables $X_1, X_2, \ldots$, representing the number of critical hurricanes in each of a sequence of (nonoverlapping) 50-year intervals. Based on information other than direct observations, the engineer's prior distribution on the parameter of the process was

$$p_N'(\nu) = \begin{cases} 0.2 & \nu = 0.25 \\ 0.4 & \nu = 0.50 \\ 0.3 & \nu = 0.75 \\ 0.1 & \nu = 1.0 \end{cases}$$

Seeking more information on this mean number per 50-year interval, the engineer searched weather-station data. It indicated that no hurricanes had traveled in the critical direction in the past 100 years. Newspapers and historical records prior to the organization of the weather stations revealed one such hurricane 125 years before. This source of information was considered reliable for 150 years prior to the weather-station data. In summary, the engineer has obtained observations of the random variables $X_1, X_2, \ldots, X_5$:

$x_1 = 0 \qquad x_4 = 0$

$x_2 = 0 \qquad x_5 = 0$

$x_3 = 1$

The consequences of his decision depend upon the value of X_6, the number during the *next* 50 years.

The sample-likelihood function [Eq. (6.2.4)] is, for these Poisson random variables,

$$L(\nu \mid x_1, x_2, \ldots, x_n) = \prod_{i=1}^{n} e^{-\nu} \frac{\nu^{x_i}}{x_i!}$$

$$= \frac{e^{-n\nu} \nu^{\Sigma x_i}}{\Pi x_i!}$$

Any factors not depending on ν can be dropped from the likelihood function and included later, more easily, in the normalizing constant. This leaves only

$$L(\nu \mid x_1, x_2, \ldots, x_n) = \nu^{\Sigma x_i} e^{-n\nu} \qquad (6.2.8)$$

In this case $n = 5$ and

$\Sigma x_i = 0 + 0 + 1 + 0 + 0 = 1$

Therefore

$L(\nu \mid 0,0,1,0,0) = \nu e^{-5\nu}$

Thus, the posterior distribution of ν is, Eq. (6.2.2),

$$p_N''(\nu) = N p_N'(\nu) L(\nu \mid 0,0,1,0,0) = \begin{cases} N(0.2)(0.25)e^{-1.25} & \nu = 0.25 \\ N(0.4)(0.50)e^{-2.50} & \nu = 0.50 \\ N(0.3)(0.75)e^{-3.75} & \nu = 0.75 \\ N(0.1)(1.0)e^{-5.0} & \nu = 1.0 \end{cases}$$

or

$$p_N''(\nu) = \begin{cases} 1.43N \times 10^{-2} & \nu = 0.25 \\ 1.64N \times 10^{-2} & \nu = 0.50 \\ 0.52N \times 10^{-2} & \nu = 0.75 \\ 0.06N \times 10^{-2} & \nu = 1.0 \end{cases}$$

The normalizing constant N must be

$$N = \frac{1}{1.43 + 1.64 + 0.52 + 0.06} \times 10^2$$

$$= 27.4$$

Finally, then,

$$
p_N''(\nu) = \begin{cases} 0.392 & \nu = 0.25 \\ 0.450 & \nu = 0.50 \\ 0.142 & \nu = 0.75 \\ 0.016 & \nu = 1.0 \end{cases}
$$

The historical data has made the smaller values of ν more probable.

With these new probabilities of ν, the expected utility given action a_2 becomes, Eq. (6.1.6),

$$
\begin{aligned}
E[u \mid a_2] &= \sum_i E[u \mid \nu_i, a_2] P[\nu = \nu_i \mid a_2] \\
&= (2)(0.392) + (0)(0.450) + (-2)(0.142) + (-4)(0.016) \\
&= +0.436
\end{aligned}
$$

Compared to $E[u \mid a_1] = 0$, this expected utility implies that a_2 is the better design. Notice that in this case the observations have altered the distribution on ν to such an extent that the choice of actions has been changed from that based on prior information alone (Sec. 6.1).

Illustration: Soil strength　A soils engineering firm with long experience in the same local area is willing to assume that they know precisely the standard deviation σ_X of unconfined compressive strength tests in a local grey-clay layer. It is 300 psf. The tests are normally distributed.† The mean strength of the layer varies from site to site, however. Past experience leads the soils engineer for a new building site to assign this mean strength m a prior normal distribution with prior mean $m_M' = 1600$ psf and prior standard deviation $\sigma_M' = 200$ psf. A crew returns with the results of five tests: 1200, 1400, 1600, 1400, 1100, and 1700 psf.

The sample-likelihood function is

$$
L(m \mid z) = L(m \mid x_1, x_2, \ldots, x_n) = \prod_{i=1}^{n} \left(\frac{1}{\sqrt{2\pi}\sigma_X} \right)^n \exp\left[-\frac{1}{2}\left(\frac{x_i - m}{\sigma_X} \right)^2 \right]
$$

$$
= \left(\frac{1}{\sqrt{2\pi}\sigma_X} \right)^n \exp\left[-\frac{1}{2} \sum_{i=1}^{n} \left(\frac{x_i - m}{\sigma_X} \right)^2 \right]
$$

Dropping the term which does not depend on m and rearranging,

$$
L(m \mid z) = \exp\left[-\frac{1}{2}\left(\frac{m - \bar{x}}{\sigma_X/\sqrt{n}} \right)^2 \right]
$$

in which $\bar{x}$ is the average of the $n = 5$ observations.

The posterior distribution on the mean M is

$$
\begin{aligned}
f_M''(m) &= N L(m \mid z) f_M'(m) \\
&= N \exp\left[-\frac{1}{2}\left(\frac{m - \bar{x}}{\sigma_X/\sqrt{n}} \right)^2 \right] \frac{1}{\sqrt{2\pi}\sigma_M'} \exp\left[-\frac{1}{2}\left(\frac{m - m_M'}{\sigma_M'} \right)^2 \right]
\end{aligned}
$$

† The scatter observed includes both small-scale spatial variation in the soil property and testing errors.

Some rearranging of terms will yield

$$f''_M(m) = N \exp\left[-\frac{1}{2}\left(\frac{m - m''_M}{\sigma''_M} \right)^2 \right]$$

(6.2.9)

in which

$$m''_M = \frac{m'_M/(\sigma'_M)^2 + \bar{x}/(\sigma_X^2/n)}{1/(\sigma'_M)^2 + 1/(\sigma_X^2/n)}$$

and

$$\sigma''_M = \left[\frac{1}{1/(\sigma'_M)^2 + 1/(\sigma_X^2/n)} \right]^{\frac{1}{2}}$$

In short, the posterior distribution of m is normal, as the prior distribution was. Its posterior mean is a weighted average of the prior mean m'_M and the sample mean $\bar{x}$. The weighting factors are reciprocal variances. The reciprocal of the posterior variance $(\sigma''_M)^2$ is the sum of the reciprocals of the prior variance $(\sigma'_M)^2$ and the sample mean variance σ_X^2/n. The normalizing factor N is, by inspection, $1/\sqrt{2\pi}\,\sigma''_M$. These simple relationships among prior and posterior parameters, and this maintenance of the posterior distribution in the same form as the prior distribution, is not to be expected in general (Sec. 6.2.3).

The numerical values are

$$m''_M = \frac{1600/(200)^2 + (1400)/[(300)^2/5]}{1/(200)^2 + 1/[(300)^2/5]} = \frac{0.040 + 0.078}{0.25 \times 10^{-4} + 0.55 \times 10^{-4}}$$

$$= 1490 \text{ psf}$$

$$\sigma''_M = \left(\frac{1}{0.25 \times 10^{-4} + 0.55 \times 10^{-4}} \right)^{\frac{1}{2}} = 112 \text{ psf}$$

Since embankment or footing stability depends on average soil strength rather than on its local strength, the soils engineer assigns utility directly to the parameter value m. He must report a single "recommended soil strength" m_0 for use by the building designers. The decision is what value of m_0 to select. He assigns the utility curve shown in Fig. 6.2.1. It reflects the facts that an increase in m_0 will benefit the client through a cheaper design, that a true strength significantly less than the recommended value may lead to a major loss to the soils engineer, and that true strength in excess of m_0 will involve only a small relative benefit in the form of an additional margin of safety.

The expected utility of any action (any choice of m_0) is

$$E[u \mid a] = bm_0 + c - ce^{+dm_0} \int_{-\infty}^{\infty} e^{-dm} \frac{1}{\sqrt{2\pi}\sigma''_M} \exp\left[-\frac{1}{2}\left(\frac{m - m''_M}{\sigma''_M} \right)^2 \right] dm$$

$$= bm_0 + c - ce^{+dm_0} \exp\left[-m''_M d + \frac{1}{2}(\sigma''_M d)^2 \right]$$

$$= bm_0 + c[1 - e^{-d(m_M'' - m_0)}e^{\frac{1}{2}(\sigma_M''d)^2}]$$

The best choice of m_0 maximizes the expected utility. Setting the derivative of $E[u \mid a]$ with respect to m_0 equal to zero yields

$$(m_0)_{\text{optimal}} = m''_M - \frac{d}{2}(\sigma''_M)^2 + \frac{1}{d}\ln\frac{b}{cd}$$

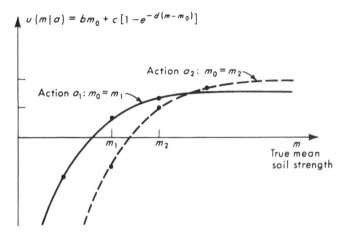

Fig. 6.2.1 Example of design-recommendation utility curves; an action a_i is to recommend a design strength $m_0 = m_i$.

The greater the engineer's uncertainty σ_M'' in the mean strength m, the more conservative will be his recommended value m_0.

Discussion In the process of discussing further topics there will be several additional illustrations of this procedure of "updating" the distribution on unknown parameters. Recall that any of the four combinations of discrete and continuous distributions on θ and X may arise. A discrete prior distribution on a number of values of θ, like that in the first illustration, may simplify computation in many complicated cases. It has the disadvantage of restricting the possible values of θ to the set used in the prior distribution, no matter what the sample outcome. It is preferable, when appropriate and when computationally feasible, to use continuous prior distributions on the parameters.

The posterior distribution of θ should be recognized as the product of two functions, one representing prior information, the other carrying new sample information as to the relative likelihoods of values of the unknown parameter. In Fig. 6.2.2 we see a number of sketches representing situations ranging from that where the prior information is well defined and the sample information minimal to that where the prior information is vague and the sample information, owing to the large sample size, is dominant. Notice that the sampling virtually always leads to a sharpened posterior distribution, indicating a reduction in the variance of the parameter and hence in the uncertainty in the value of the unknown parameter. In those cases where one source of information or the other is relatively weak, the shape of the posterior distribution is very nearly

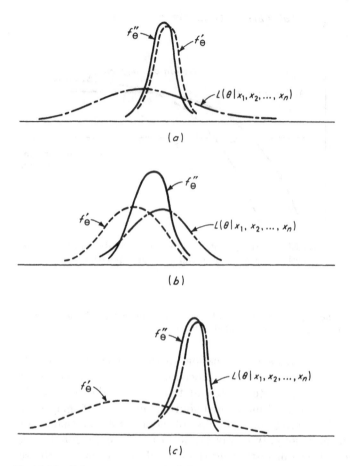

Fig. 6.2.2 Prior versus sample information relationships. (*a*) Strong prior and relatively small sample. (*b*) Prior and sample information of comparable weight. (*c*) Vague prior and relatively large sample.

simply that of the other (or dominant) function. We will consider one of these cases in more detail next.

6.2.2 Data-based Decisions: Diffuse Priors

The decision depends upon the posterior distribution of θ. This posterior distribution is the product of a constant and of two functions: the prior distribution $f'_\Theta(\theta)$ and the sample-likelihood function $L(\theta \mid x_1, x_2, \ldots, x_n)$. If, as shown in Fig. 6.2.2c, the prior distribution has a nearly constant value c throughout that region where the sample-likelihood function is significantly large in magnitude, the shape of the posterior will be virtually identical to that of the likelihood function:

$$f_\Theta''(\theta) = NL(\theta \mid z)f_\Theta'(\theta) \approx cNL(\theta \mid z) \qquad (6.2.10)$$

The influence of the prior distribution is then indistinguishable from that of the normalizing constant, and *the prior may be ignored in the determination of the posterior*. The decisions will then depend solely on the information contained in the sample of direct observations. The prior is said to be "flat" or "diffuse" or "vague" relative to the sample-likelihood function. (These priors are sometimes said to be uniform distributions, although this cannot be strictly true for a parameter free to take on values over an infinitely large range, for example, 0 to ∞.)

Example If the X_i have the Type II extreme-value distribution (Sec. 3.3.3),

$$F_X(x) = e^{-(x/u)^{-k}} \qquad x \geq 0 \qquad (6.2.11)$$

Assume that k is known, but u is not. For example, with maximum annual wind velocities, it is known that the coefficient of variation (and hence k) varies little from site to site, whereas the mode u does. It is simpler in this case to work with $w = u^k$ as the unknown parameter. Then

$$F_X(x) = e^{(-wx^{-k})}$$
$$f_X(x) = kwx^{-(k+1)}e^{(-wx^{-k})}$$

Assuming that the engineer has no firm prior distribution on the parameter w at his site, i.e., assuming that he adopts a very broadly spread prior on w (the parameter treated as a random variable), the posterior distribution on w is

$$f_W''(w) = cNL(w \mid z)$$

$$= cNk^n w^n e^{-w\Sigma x_i^{-k}} \prod_{i=1}^{n} (x_i^{-k-1})$$

$$= \left[cNk^n \prod (x_i^{-k-1}) \right] w^n e^{-w\Sigma x_i^{-k}}$$

Absorbing terms independent of w in a new normalizing constant N',

$$f_W''(w) = N'w^n e^{-\overline{x^{-k}}nw} \qquad w \geq 0 \qquad (6.2.12)$$

in which $\overline{x^{-k}}$ is the observed $-k$th moment of the sample $\Sigma x_i^{-k}/n$. By inspection, the distribution of the parameter w is gamma, $G(r,\nu)$, with

$$r = n + 1$$
$$\nu = n\overline{x^{-k}}$$

Decisions dependent on utilities or expected utilities that are functions of the parameter w (or, alternatively, u) can be made in the now familiar manner.

Data-based decisions   Prior distributions which may be treated as diffuse will arise in particular when the data are relatively large in number such that the information contained in the sample is large compared to that upon which the prior is based. The resulting posterior distribution can be said to be *data-based*. Being virtually free of subjective judgment in the prior, the statements or decisions based on such a posterior distribution might be called *objective*. (This is true at least with respect to the probability assignments; utilities remain subjective, although in some cases all involved might agree to a common measure, such as dollars.) In this context, it also is possible to make such statements as "The data indicate that the probability that the true mean lies between a and b is 95 percent," or "The data indicate that the hypothesis that $\sigma > \sigma_0$ is true with probability 90 percent." This type of statement is, recall, *not* permitted in classical estimation or hypothesis testing, Secs. 4.1 and 4.2.

Bayesian estimation   In fact, all parameter estimation (Sec. 4.1) could be revisited from a Bayesian point of view. Given the (prior or posterior) distribution on parameter θ, one can ask for the single number $\hat{\theta}$ (i.e., a *point estimate* of the unknown parameter) which maximizes some utility function. This is known as a *Bayesian estimator*. For example, if the loss associated with an estimation error, $\hat{\theta} - \theta$, is assumed proportional to the squared error $(\hat{\theta} - \theta)^2$, the decision will be to use the (prior or posterior) mean value of θ as an estimate. For example, if the posterior distribution happens to be data-based, i.e., simply the normalized sample-likelihood function, then for many distributions, the mean of the posterior distribution on the true mean is simply the familiar sample average. This is the same estimator suggested by apparently quite different criteria in Sec. 4.1.†

Consider again the Type II extreme-value wind velocity distribution we were just studying. A local committee of engineers must recommend wind-load code provisions in terms of the distributions of maximum loads over building lifetimes. These depend on the parameter u (or on $w = u^k$).

---

† The reader should compare this formulation with that in Sec. 4.1. Notice in particular that in Sec. 4.1 the estimator was a random variable with a distribution centered on the true parameter. Here the true parameter is treated as a random variable with a distribution centered on the posterior expected value. In the gaussian case, we had in Sec. 4.1 the sample average distributed about the true mean (before taking the sample); now we will have the true mean distributed about the observed sample average. [See Eq. (6.2.14).]

Not knowing the details of all the individual future decisions which the distributions will influence and lacking professional knowledge beyond the data, the committee chooses to report an "objective, data-based" estimate $\hat{w}$ of w. They choose to select the estimate to minimize the expected square error $(w - \hat{w})^2$, when the distribution on w is the normalized likelihood function, Eq. (6.2.10). This estimate is the posterior mean of the gamma distribution, Eq. (6.2.12), or

$$\hat{w} = \frac{r}{\nu} = \frac{n+1}{n} \, (\overline{x^{-k}})^{-1} \tag{6.2.13}$$

In short, the point estimate should be found by finding the reciprocal of the $(-k)$th moment of the sample and multiplying it by $(n+1)/n$. The committee could also report objectively the uncertainty in the unknown parameter $w = u^k$ by using the data-based distribution on w to calculate the probability that u^k lies in any particular interval.† These statements correspond closely to confidence-interval estimates, but the "confidence values" may now be properly interpreted and operated upon as true probabilities. This interpretation is clearly closer in spirit to the way in which engineers commonly use and discuss confidence limits.

Bayesian hypothesis testing   In a similar vein we could reconsider hypothesis testing (Sec. 4.2) in a Bayesian light. This would permit removal of the conventional interpretation of significance levels, which many engineers seem to find "nonintuitive." The conventional interpretation can be replaced by simple probabilities. If the posterior distributions on the parameters are data-based normalized likelihood functions, the conclusions would be just as "objective" as those arrived at by conventional tests.

Illustration: Compaction-acceptance test   Suppose, for example, that an engineer is to be responsible for accepting a contractor's work done on compacting a soil embankment for a highway. The contract specifies that a value of "95 percent of the optimum density" should be obtained. Owing to acceptable spatial variations in placement and owing to testing techniques, the results of tests on compaction are random variables X_i. Assume that they are normal with unknown mean m, but known‡ standard deviation $\sigma_X = 1$ percent. To be conservative, the engineer must assume that the mean compaction is less than the contract specified value of 95 *unless* the data are sufficient to convince him otherwise. His hypothesis, then, is that the unknown mean compaction m is *less* than 95. Wishing to be objective, he adopts a diffuse prior distribution;

---

† Such data-based Bayesian probabilities correspond to the so-called "fiducial probabilities," introduced years ago without Bayesian arguments by a famous statistician, R. A. Fisher. The concept of fiducial probabilities is not widely used.

‡ Treatment of normal variables with unknown σ will be domonstrated in the next section.

that is, he assumes that the posterior distribution of m is proportional to the sample-likelihood function.   The posterior distribution on m, given a sample $x_1, x_2, \ldots, x_n$, is therefore

$$f_M''(m) = N \prod_{i=1}^{n} \frac{1}{\sqrt{2\pi}\,\sigma_X} \exp\left[-\frac{1}{2}\left(\frac{x_i - m}{\sigma_X}\right)^2 \right]$$

which can be rearranged and reduced to

$$f_M''(m) = \frac{\sqrt{n}}{\sqrt{2\pi}\,\sigma_X} \exp\left[-\frac{1}{2\sigma_X^2/n}(m - \bar{x})^2 \right] \qquad -\infty \leq m \leq \infty \qquad (6.2.14)$$

This result indicates that the (data-based) posterior distribution of m is normal with mean $\bar{x}$ and standard deviation $\sigma_X/\sqrt{n}$.

If the soil-density observations were 25 in number with observed sample average $\bar{x} = 95.5$, then the probability that the hypothesis is incorrect is found using the normal tables of the standardized variate U;

$$P[m < 95] = F_M(95)$$
$$= F_U\left(\frac{95.0 - 95.5}{1/\sqrt{25}}\right)$$
$$= F_U(-2.5)$$
$$= 0.01$$

Thus, the engineer might state, "The data indicate that there is only a probability of 1 in 100 that the mean compaction is less than the specified value." Depending on the relative importance of incorrectly accepting below-standard work or incorrectly refusing adequate work, the engineer must next decide how to act.   Clearly if the former loss is less than 99 times the latter, the engineer should accept the work.   (A more accurate decision analysis would recognize that the losses or utilities are probably a function of the true mean.   It is this entire function $u(a,m)$ which would be weighted by $f_M''(m)$ in a complete decision analysis.)   In this case, 0.01 equals the p-level of a conventional test.

Alternatively, the engineer might simply follow a department directive.   The directive might state that "compaction work should be accepted if the data indicate a probability greater than 90 percent that the mean compaction exceeds specifications."†   The directive is quite general.   It can be stated independently of the values of σ_X and n which apply at any particular construction site.   Such an inspection directive, notice, puts the decision in the hands of the directive writer, who is normally a more experienced engineer than the inspector.   It does so, however, at the expense of not letting the on-the-job inspector incorporate quantitatively (through a prior distribution) pertinent information other than the 25 tests.

Discussion   Complete "objectivity" in probability statements about unknown parameters is not, in fact, gained by basing the posterior dis-

---

† This directive should be compared with one written using conventional hypothesis testing.   It would say, "Test the hypothesis that m is less than specified at the 10 percent significance level; if the hypothesis is rejected, accept the work."

tribution solely on the sample-likelihood function.   It is a simple matter
to demonstrate that these probabilities may depend to a degree on the
choice of the form of the parameterization of the problem.   For example,
a prior distribution assumed smooth on the mean, $\beta = 1/\lambda$, of an exponential distribution is *not* equivalent to a prior assumed smooth on λ
itself.   A smooth prior on the standard deviation σ of a normal random
variable X will give one result for, say, the posterior expected utility
$E[u(X)] = E[(a + bX^2)]$, and a smooth prior on the variance σ^2 will give
a somewhat different one.   The answers will not be significantly different
in most typical cases if sample sizes are not small and if potential difficulties such as those around the origin (e.g., consider $\beta = 1/\lambda$ versus λ)
are avoided.   In short, the apparent objectivity of data-based probability statements and decisions is not strictly true.   In choosing the form
of the parameter to be used, the engineer has already made the equivalent of a diffuse prior distribution assignment on that parameter to the
exclusion of a diffuse prior assignment on some alternate parameter choice.

6.2.3 Use of Conjugate Priors

It is frequently possible with common probability laws to make significant computational savings in the terminal analyses of simple processes
if the decision maker is willing to use a particular form for his prior distribution.   This form of the prior distribution is the one that is compatible with the sample-likelihood function, or, technically, a *conjugate*
of this function.   The convenient result of this choice of a prior is that
the posterior distribution is of the same form as the prior.   Commonly
the parameters of the posterior distribution are simply related to the
parameters of the prior distribution and to simple statistics of the sample.
We saw one example in the second illustration in Sec. 6.2.1.   The normal
distribution is a conjugate prior for the mean of a normal distribution,
with σ known.

Binomial case   If the decision maker is prepared to summarize his prior
information about the unknown parameter p of a process of Bernoulli
random variables X_i, not by some arbitrary distribution on the parameter, but by a beta distribution, then he is using the conjugate prior
for this process.   The process might involve a succession of concrete
cylinders. $X_i = 1$ if the ith cylinder is below specified strength.   The
unknown fraction of low-strength cylinders is p.   We will treat this fraction as a random variable p.   Suppose the engineer states that his prior
information (previous experience with similar mix designs, etc.) suggests
a point estimate (here, a prior mean value) of p equal to some number
m'_p and that this uncertainty in the fraction is expressed by a standard
deviation of σ'_p.   The corresponding values of the parameters r' and t' of

the beta distribution can be found by solving Eqs. (3.4.11) and (3.4.12). Then [Eq. (3.4.8)]

$$f'_P(p) = \frac{1}{B'} p^{r'-1}(1 - p)^{t'-r'-1} \qquad 0 \leq p \leq 1 \tag{6.2.15}$$

in which

$$B' = \frac{\Gamma(r')\Gamma(t' - r')}{\Gamma(t')} \tag{6.2.16}$$

Assume that this shape is judged by the engineer to be a reasonable reflection of his prior information.

Suppose that a sample of the process is now observed. Of the n observations, s are found to be successes (i.e., below specified strength). Then the sample-likelihood function† is

$$L(p \mid z) = p^s(1 - p)^{n-s} \qquad 0 \leq p \leq 1 \tag{6.2.17}$$

Thus the posterior distribution of p is, after dropping uninteresting factors,

$$\begin{aligned} f''_P(p) &= NL(p \mid z)f'_P(p) \\ &= Np^s(1 - p)^{n-s}p^{r'-1}(1 - p)^{t'-r'-1} \\ &= Np^{s+r'-1}(1 - p)^{n-s+t'-r'-1} \qquad 0 \leq p \leq 1 \end{aligned} \tag{6.2.18}$$

This distribution is clearly also of the beta family, with "posterior" parameters

$$r'' = s + r' \tag{6.2.19}$$
$$t'' = n + t' \tag{6.2.20}$$

That is,

$$f''_P(p) = \frac{1}{B''} p^{r''-1}(1 - p)^{t''-r''-1} \qquad 0 \leq p \leq 1 \tag{6.2.21}$$

in which

$$B'' = \frac{\Gamma(r'')\Gamma(t'' - r'')}{\Gamma(t'')} \tag{6.2.22}$$

One action available to the engineer, that of keeping the present mix design, might involve a loss proportional to the total number of

---

† This likelihood function is valid if the data are made available in the complete, ordered form, e.g., "success, success, failure, . . . , etc." If the data were made available in the form "s out of n were successes" the likelihood function would have a binomial coefficient in front. The final results will be identical. Since the coefficient is independent of p, it would be absorbed in the normalizing constant.

understrength cylinders in the next v tests. Then, if p were known,

$$u = c - b \sum_{i=1}^{v} X_i$$

$$g(\theta) = E[u \mid p] = c - b\Sigma E[X_i]$$

$$= c - bvp$$

With the information available on p, the expected value of this action is

$$E[u] = \int_0^1 E[u \mid p] f_P''(p) \, dp$$

$$= \int_0^1 (c - bvp) f_P''(p) \, dp$$

$$= E[c - bvp]$$

$$= c - bvm_P$$

$$= c - bv \frac{r''}{t''}$$

$$= c - bv \frac{s + r'}{n + t'} \tag{6.2.23}$$

It is interesting to note in Eqs. (6.2.19) and (6.2.20) the simple relationships of the posterior parameters r'' and t'' to the prior parameters r' and t' and to the sample statistics s and n. It is clear that the total information in the sample, i.e., the sequence of observations—success, failure, etc.—is summarized sufficiently by the two numbers, sample size n and total number of successes s. These are called the *sufficient statistics* of a Bernoulli-process sample. Similarly, in the normal illustration with known σ (Sec. 6.2.1), the sample average, $\bar{x}$ and n were found to be sufficient statistics.

One can also observe (Eq. 6.2.18) that the prior parameters t' and r', no matter what their source or how determined, appear in the analysis as an "equivalent" prior sample size $t' - 2$ and as an "equivalent" prior observed number of successes $r' - 1$. (This does not mean, however, that t' and r' need be integer-valued.) The relative weight of the prior information and the sample information can be measured by the relative magnitudes of the sample sizes $t' - 2$ and n. In particular, note that if $t' - 2$ and $r' - 1$ are zero, the posterior distribution with $r'' = s + 1$ and $t'' = n + 2$ is equivalent to the data-based distribution of p obtained by simply normalizing the sample-likelihood function, since then $f_P'(p)$ is a constant, independent of p.

Poisson and exponential case Inspection of the likelihood function, Eq. (6.2.8), will reveal that the gamma distribution is the natural or conjugate prior distribution for the parameter v of a process of Poisson-distributed random variables. This fact will be used in Sec. 6.3.1. It should not be unexpected if one recalls the central role of the distribution in a Poisson process (Sec. 3.2), and that the gamma distribution is also the conjugate distribution for the parameter λ of an exponential distribution.

Normal case: m and σ unknown As a final example of conjugate prior distributions we shall consider the general case of a normal process, that where neither parameter is known. The likelihood function was found in Sec. 4.1.4 to be [Eq. (4.1.65)]

$$L(m,\sigma \mid x_1,x_2, \ldots ,x_n) = \frac{1}{\sigma^n} \exp\left[-\frac{1}{2\sigma^2} \sum_{i=1}^{n} (x_i - m)^2 \right]$$

$$-\infty \leq m \leq \infty, \sigma \geq 0 \quad (6.2.24)$$

Some rearrangement is desirable. The sum in the exponent can be expanded and rewritten in terms of sample statistics $\bar{x}$, s^2, and n:

$$L(m,\sigma \mid x_1,x_2, \ldots ,x_n) = \frac{1}{\sigma^n} \exp\left\{ -\frac{1}{2\sigma^2} [(n-1)s^2 + n(m - \bar{x})^2] \right\}$$

$$= \left\{ \frac{1}{\sigma} \exp\left[-\frac{1}{2}\left(\frac{m - \bar{x}}{\sigma/\sqrt{n}} \right)^2 \right] \right\} \left[\frac{1}{\sigma^{n-1}} \exp\left(-\frac{n-1}{2}\frac{s^2}{\sigma^2} \right) \right]$$

in which the sample variance s^2 is defined here as

$$s^2 = \frac{1}{n-1} \sum (x_i - \bar{x})^2 \quad (6.2.25)$$

The form of the joint-likelihood function of m and σ appears like the product of a normal density and a gamma density. The exponential forms of these distributions suggest why it is possible to find a (joint) conjugate prior distribution on m and σ. It is

$$f'_{M,\Sigma}(m,\sigma) = \left\{ \frac{1}{\sqrt{2\pi}\,\sigma/\sqrt{n'}} \exp\left[-\frac{1}{2}\left(\frac{m - \bar{x}'}{\sigma/\sqrt{n'}} \right)^2 \right] \right\}$$

$$\times \left[\frac{\left(\dfrac{n'-1}{2} \right)^{(n'-2)/2}}{\Gamma\left(\dfrac{n'-2}{2} \right)} \frac{2}{s'}\left(\frac{s'^2}{\sigma^2} \right)^{(n'-1)/2} \exp\left(-\frac{n'-1}{2}\frac{s'^2}{\sigma^2} \right) \right]$$

$$-\infty \leq m \leq \infty, \sigma > 0 \quad (6.2.26)$$

with parameters n', $\bar{x}'$, s'^2. It is a straightforward matter to demonstrate that this is a proper joint density function. It is simpler still to show that the *posterior* joint density function

$$f''_{M,\Sigma}(m,\sigma) = N f'_{M,\Sigma}(m,\sigma)L(m,\sigma \mid x_1,x_2, \ldots ,x_n)$$

is of the same form as the prior, Eq. (6.2.26), with parameters found easily from the relationships:

$$n'' = n + n' \tag{6.2.27}$$

$$n''\bar{x}'' = n\bar{x} + n'\bar{x}' \tag{6.2.28}$$

$$(n'' - 1)s''^2 + n''\bar{x}''^2 = [(n - 1)s^2 + n\bar{x}^2]$$
$$+ [(n' - 1)s'^2 + n'\bar{x}'^2] \tag{6.2.29}$$

Again, we can interpret n' as an equivalent prior sample size. Note the weightings by sample sizes n' and n, which appear in Eqs. (6.2.28) and (6.2.29) for $\bar{x}''$ and s''^2. A sufficiently large amount of information (as measured by $n'' = n + n'$) will, of course, reduce to a negligible level the uncertainty in the unknown parameters, leaving their distribution concentrated on the true values of m and σ. Small prior information ($n' \approx 0$ and $(n' - 1)s'^2 = 0$) reduces the posterior distribution to a constant times the likelihood function, or a data-based distribution.

We can find the marginal distributions on the unknown mean m and standard deviation σ by integration:

$$f_M''(m) = \int_0^\infty f_{M,\,\Sigma}''(m,\sigma)\, d\sigma$$

$$= \frac{1}{\sqrt{\pi}} \frac{1}{s''} \sqrt{\frac{n''}{n'' - 1}} \frac{\Gamma(n'' - 1/2)}{\Gamma[(n'' - 2)/2]}$$

$$\times \left[1 + \frac{n''}{n'' - 1} \left(\frac{m - \bar{x}''}{s''} \right)^2 \right]^{-(n''-1)/2} \tag{6.2.30}$$

By comparison with Eq. (3.4.33), the reader will see that the factor $(M - x'') \sqrt{n''}/s''$ has a t distribution with $n'' - 2$ degrees of freedom. The same distribution arose, Sec. 4.1, when we dealt with the distribution of the sample mean $\bar{X}$ of a sample of normal random variables with unknown variance. The mean and variance of M are

$$m_M = \bar{x}'' \tag{6.2.31}$$

$$\text{Var}\,[M] = s''^2 \frac{n'' - 1}{n''(n'' - 2)} \tag{6.2.32}$$

Integration over m reveals that the marginal distribution of the variance σ^2 is related to the χ^2 distribution. More precisely, $[(n'' - 1)s''^2](1/\sigma^2)$ is CH$[n'' - 2]$. The expected value of σ^2 is $s''^2(n'' - 1)/(n'' - 4)$. The posterior mean and variance of the standard deviation are

$$m_\Sigma = s'' \sqrt{\frac{n'' - 1}{2}} \frac{\Gamma[(n'' - 3)/2]}{\Gamma[(n'' - 2)/2]} \qquad n'' > 3 \tag{6.2.33}$$

$$\text{Var}\,[\sigma] = s''^2 \frac{n'' - 1}{n'' - 4} - m_\Sigma^2 \qquad n'' > 4 \tag{6.2.34}$$

It is clear from Eq. (6.2.30) that even if utility depends upon only the unknown mean, the second unknown parameter, i.e., variance, cannot be ignored in the terminal decision analysis unless n'' is large. Such a parameter is called a *nuisance* parameter.

The reader is referred to Raiffa and Schlaifer [1961] for a systematic treatment and cataloging of conjugate distributions in decision theory.

Illustration: Sustained office loads Suppose that an engineer is concerned with the long-term deflection of a prestressed-concrete beam design. The critical source of uncertainty is assumed to be the value of the sustained, long-term load on the beam. For file-room occupancies of the type this beam will support, the sustained load is assumed to be normally distributed since it is the sum of many individual loads. The engineer knows neither the mean m nor the standard deviation σ of X, the (equivalent, uniform) sustained load in such file rooms. Judgment suggests point estimates of m and σ of 60 and 20 psf, respectively. Thus he takes parameters of the conjugate prior distribution (whose form he is willing to accept) as

$\bar{x}' = 60$

$s' = 20$

The engineer's prior belief that the mean is about twice as likely to lie inside the range 60 ± 15 as it is to lie out of it suggests that n' be taken as about 4. [This is based on Eq. (6.2.32) and Table A.3.]

Following this, the engineer requests measurements to be made in five similar, existing file rooms. The results are

$$x_1 = 25 \qquad x_2 = 37 \qquad x_3 = 59 \qquad x_4 = 54 \qquad x_5 = 75$$

The relevant sample statistics, *the sufficient statistics*, are

$n = 5$

$\bar{x} = 50.0$

$$s^2 = \frac{1}{5-1}(1516) = 379 \qquad s = 19.5$$

The joint posterior density function of m and σ is of the same form as the prior, but with parameters

$n'' = n' + n = 4 + 5 = 9$

$$\bar{x}'' = \frac{nx + n'\bar{x}'}{n''} = \frac{(5)(50.0) + (4)(60)}{9} = 54.4 \text{ psf}$$

$$s''^2 = \frac{(n-1)(s^2) + n\bar{x}^2 + (n'-1)s'^2 + n'\bar{x}'^2 - n''\bar{x}''^2}{n''-1}$$

$$= \frac{(4)(379) + 5(2500) + (3)(400) + 4(3600) - (9)(2960)}{8} = 372$$

$s'' = 19.3 \text{ psf}$

By choice of dimensions and prestressing force the engineer can design a beam which will have no long-term deflection under a particular sustained load a. If the true load is X, there will be a deflection (sag positive, camber negative) proportional to the difference, or $X - a$. The loss associated with poor performance of the design would be some function of the sag or camber and hence of $X - a$. Assume, for example, that the loss is proportional to the square of the difference:

$$u(a,X) = -(X - a)^2$$

Then, the expected utility given m and σ is, for any action or design a,

$$g(\theta) = E[u \mid a,m,\sigma] = E[-(X - a)^2] = -(E[X^2] - 2aE[X] + a^2)$$
$$= -(m^2 + \sigma^2 - 2am + a^2)$$

With the posterior distribution on m and σ, the expected loss $-E[u \mid a]$ is

$$-E[u \mid a] = \int_0^\infty \int_{-\infty}^\infty (m^2 + \sigma^2 - 2am + a^2) f''_{M,\Sigma}(m,\sigma) \, dm \, d\sigma$$
$$= E[m^2] + E[\sigma^2] - 2aE[m] + a^2$$

Using results above for the posterior means and variances of m and σ,

$$-E[u \mid a] = \bar{x}''^2 + \frac{s''^2(n'' - 1)}{n''(n'' - 2)} + \frac{s''^2(n'' - 1)}{n'' - 3} - 2a\bar{x}'' + a^2$$

Clearly the expected loss is minimized by the intuitively obvious choice of $a = \bar{x}'' = 54.4$ psf. Notice, however, that some expected loss, proportional to s''^2, is unavoidable.

6.2.4 Summary

Given new information about the process in the form of a sample z of n observations $x_1, x_2, \ldots, x_n$, a posterior distribution on the process parameter(s) θ can be found through Bayes' theorem

$$f''_\theta(\theta \mid z) = NL(\theta \mid z)f'_\theta(\theta)$$

in which N is a normalizing constant and $L(\theta \mid z)$ is the sample-likelihood function

$$L(\theta \mid z) = \prod_{i=1}^n f_X(x_i \mid \theta)$$

Decision analysis follows as in Sec. 6.1 with the posterior distribution in place of the prior.

If the prior distribution is relatively flat compared to the sample-likelihood function, as it well may be when prior information is vague and the sample size is large, the posterior distribution reduces effectively to a data-based normalized likelihood function.

In certain cases a natural or conjugate prior distribution can be found such that the posterior distribution and prior distribution are of the same analytical form, with subsequent computational savings.

6.3 THE BAYESIAN DISTRIBUTION OF A RANDOM VARIABLE

In this section we shall deal briefly with an extended interpretation of the distribution of a random variable. This new interpretation recognizes that in virtually all practical problems at least two of the three levels of uncertainty exist (Chap. 4). These are (1) the inherent randomness as described by the underlying (or model) distribution of the X_i, and (2) the (statistical) uncertainty as to the true values of the parameters of this model. Our Bayesian treatment of the latter problem permits us to deal with both types of uncertainty concurrently and in a parallel manner.

By combining the model distribution of a random variable X with the prior or posterior distribution on the model's parameters, we can obtain (through the method of Sec. 3.5.3) a *compound distribution* on X which incorporates both sources of uncertainty simultaneously.

Although the derivations will be general and although we expect most ultimate uses to be in decision situations, for brevity we shall illustrate only cases in which the prior distributions are chosen to be natural conjugate distributions (Sec. 6.2.3) and in which our interest ceases with statements (distributions, moments, etc.) about X (rather than about the expected utility). Derivations will assume that the random variables of interest are continuously distributed, but illustrations will include discrete cases.

6.3.1 The Simple Case; X Only

Formally we should like to deal with a description of the random variable X which incorporates both the model $f_X(x)$ and the uncertainty in the parameter(s) θ of that model. Let us write the model distribution as $f_X(x \mid \theta)$ to emphasize that it is a function of its parameters. In this chapter we have assumed that the information on the parameters can be contained in a distribution $f_\Theta(\theta)$. This may be a prior or posterior distribution, but in general it will be based at least in part upon a sequence of observations, $x_1, x_2, \ldots, x_n$, of the process $X_1, X_2, \ldots$. We discuss in this section the case where interest lies only in a particular single future value X of the process. The performance of a temporary exhibition structure, for example, might depend on the maximum wind velocity X in the next year only. X might be assumed to have an extreme value distribution with unknown parameters. Information on the parameters' values is available through past observations of annual maximum wind velocities.

Bayesian distribution We note that in this chapter we are treating the parameters in every way as random variables. Recall the definition of a compound distribution (Sec. 3.5.3) as the distribution of a random variable which possesses a distribution with parameters which are in turn random variables. Armed with this notion, we define a new distribution $\tilde{f}_X(x)$ on X as [Eq. (3.5.10)]

$$\tilde{f}_X(x) = \int f_X(x \mid \theta) f_\Theta(\theta) \, d\theta \tag{6.3.1}$$

Let us call this new distribution the *Bayesian distribution* of X as distinct from the *model distribution* of X, $f_X(x \mid \theta)$. The former name is prompted by the fact that the treatment of unknown parameters as random variables is a central concept in Bayesian statistics. More specifically the new distribution might be referred to as the prior (Bayesian) distribution or the posterior (Bayesian) distribution on X, depending on whether a prior or posterior distribution of θ is used to determine $\tilde{f}_X(x)$.

The new distribution can be interpreted as a weighted average of all possible distributions $f_X(x \mid \theta)$ which are associated with different values of θ. In this sense,† Eq. (6.3.1) can be interpreted as an application of the total probability theorem, Eq. (2.1.12). In any event we note that the unknown parameter will *not* appear in $\tilde{f}_X(x)$, as it has been "integrated out" of the equation. We also note that as more and more data become available, the distribution $\tilde{f}_X(x)$ will approach the true distribution of X, since the distribution of θ will be becoming more and more concentrated about the true value of the parameter. We should generally expect the distribution $\tilde{f}_X(x)$ to be wider, e.g., to have a larger variance, than the true $f_X(x)$, since the former incorporates *both* inherent *and* statistical uncertainty.

Illustration: Earthquake prediction   Consider an example of predicting X, the number of occurrences of large earthquakes in the next t years in a given region. The region surrounds a new dam, whose clay foundation is in a relatively weak condition prior to total consolidation. Assuming that major earthquakes are Poisson arrivals with average rate λ, X is Poisson-distributed with mean λt:

$$p_X(x \mid \lambda) = \frac{e^{-\lambda t}(\lambda t)^x}{x!} \qquad x = 0, 1, \ldots$$

The (prior or posterior) distribution on λ is taken to be gamma in form, $G(r_0 + 1, t_0)$ (Sec. 3.2.3):

$$f_\Lambda(\lambda) = \frac{e^{-\lambda t_0} \lambda^{r_0} t_0^{r_0+1}}{\Gamma(r_0 + 1)} \qquad \lambda \geq 0 \tag{6.3.2}$$

in which t_0 and r_0 are given numbers. (The gamma distribution is a convenient choice; it is the natural or conjugate prior distribution‡ and it leads to an expression for $\tilde{p}_X(x)$ which can be integrated easily.)

---

† A less interesting interpretation is that $\tilde{f}_X(x)$ is the Bayesian estimator of $f_X(x)$ which minimizes the expected value of a quadratic estimation loss function (Sec. 6.2.2).

‡ With prior parameters r' and t' and an observation of r occurrences in t time units, the posterior distribution parameters are $r'' = r' + r$ and $t'' = t' + t$.

The Bayesian distribution on X becomes [Eq. (6.3.1)]

$$\tilde{p}_X(x) = \int_0^\infty \frac{e^{-\lambda t}(\lambda t)^x}{x!} \frac{e^{-\lambda t_0}\lambda^{r_0}t_0^{r_0+1}}{\Gamma(r_0+1)} \, d\lambda$$

After collection of terms, the integral over a (constant times a) gamma density will be recognized. By inspection, the result will be seen to be

$$\tilde{p}_X(x) = \frac{t^x t_0^{r_0+1}}{(t_0+t)^{r_0+x+1}} \frac{(x+r_0)!}{x!\Gamma(r_0+1)} \qquad x = 0, 1, 2, \ldots \tag{6.3.3}$$

Note that λ no longer occurs in the distribution of X.

As a numerical illustration assume that the engineer assigns to the distribution on λ the parameters $t_0 = 60$ and $r_0 = 4$. These values would result, for example, if the engineer assumed a diffuse prior† on λ and then obtained data showing in 60 years a total of four occurrences, that is, if the engineer was using a data-based distribution (Sec. 6.2.2) on λ. For these values, and a future time period of interest of $t = 1$ year,

$$\tilde{p}_X(x) = \frac{60^5(x+4)!}{(61)^{5+x}4! \, x!} \qquad x = 0, 1, 2, \ldots$$

The (Bayesian) probability of no occurrences during the construction is

$$\tilde{p}_X(0) = \frac{60^5}{61^5} = 0.921$$

This result is somewhat different from that which would be obtained if a point estimate, $\hat{\lambda} = 4/60 = 0.067$, were used with $p_X(x \mid \theta)$. This number is

$$p_X(0 \mid 0.067) = e^{-0.067} = 0.935$$

But note that if this same point estimate (0.067) were based on, say, 600 years of observations and 40 observed occurrences, the Bayesian result would be ($r_0 = 40$, $t_0 = 600$):

$$\tilde{p}_X(0) = \left(\frac{600}{601}\right)^{41} = \left(1 - \frac{1}{601}\right)^{41} = 1 - \frac{41}{601} + \frac{(41)(40)}{2(601)^2} - \cdots$$

$$\approx 1 - \frac{40}{600} + \frac{1}{2}\left(\frac{40}{600}\right)^2 - \cdots$$

$$\approx e^{-0.067} = 0.935$$

Thus we observe the intuitively satisfying conclusion that if there were a large amount of data to support the point estimate of 0.067, the two approaches would give the same result. The Bayesian result for the smaller amount of data, namely 0.921, is consistent with a larger point estimate of $\hat{\lambda} = 0.082$. The gamma distribution on λ is, of course, skewed to the right or toward the larger values of λ. Thus these values have larger weights in the integral, Eq. (6.3.1).

Bayesian moments Just as we have discussed above a Bayesian distribution of X, we can now consider a (prior or posterior) *Bayesian mean* and

---

† For a discussion of *prior* distributions on earthquake occurrence rates see the work of L. Esteva, e.g., Esteva [1969].

variance of X. Their natural definitions are

$$\tilde{m}_X = \int x \tilde{f}_X(x)\, dx \tag{6.3.4}$$

and

$$\tilde{\sigma}_X{}^2 = \int (x - \tilde{m}_X)^2 \tilde{f}_X(x)\, dx \tag{6.3.5}$$

One might also quite naturally think of the Bayesian mean of X as the mean of X given the parameter value $m_{X|\Theta}$ weighted and summed over all values of the parameter:

$$\tilde{m}_X = \int m_{X|\Theta} f_\Theta(\theta)\, d\theta \tag{6.3.6}$$

That the two interpretations are identical is easily shown by the substitution of $m_{X|\Theta}$ by its definition

$$\tilde{m}_X = \int [\int x f_X(x \mid \theta)\, dx] f_\Theta(\theta)\, d\theta \tag{6.3.7}$$

and by an interchange of the order of integration

$$\tilde{m}_X = \int x [\int f_X(x \mid \theta) f_\Theta(\theta)\, d\theta]\, dx \tag{6.3.8}$$

The inner brackets are $\tilde{f}_X(x)$ by its definition [Eq. (6.3.1)], yielding Eq. (6.3.4).

For example, the Bayesian mean of X in the earthquake-occurrence example above is given either by

$$\tilde{m}_X = \sum_{x=0,1,\ldots} x \tilde{p}_X(x) = \sum x \frac{t^x t_0{}^{r_0+1}}{(t_0 + t)^{r_0+x+1}} \frac{(x + r_0)!}{x!\Gamma(r_0 + 1)}$$

or by

$$\tilde{m}_X = \int_0^\infty \lambda t f_\Lambda(\lambda)\, d\lambda = \int_0^\infty \frac{\lambda t e^{-\lambda t_0} \lambda^{r_0} t_0{}^{r_0+1}}{\Gamma(r_0 + 1)}\, d\lambda = t E[\lambda] \tag{6.3.10}$$

In either case the answer is $t(r_0 + 1)/t_0$ [Eq. (3.2.16)]. The numerical result, for $r_0 = 4$ and $t_0 = 60$, is $\tilde{m}_X = 5t/60 = 0.083t$.

The Bayesian variance of X is likewise open to one of two interpretations. Since it reflects model *and* parameter uncertainty, we would expect this Bayesian variance to be larger than the true (unknown) variance of X unless the amount of information available (e.g., the sample size) is large.

Expected utilities   In general, we can consider the expected value of any function of X based on its Bayesian distribution. Of particular interest in decision problems is the expected value of the utility of an action, when utility is a function of only a single future value X (Sec. 6.1.1):

$$E[u \mid a] = \int u(a,x) \tilde{f}_X(x)\, dx \tag{6.3.11}$$

We should expect, and indeed an argument parallel to that above for the mean will prove, that this expectation also equals

$$E[u \mid a] = \int E[u \mid a, \theta] f_\Theta(\theta) \, d\theta \tag{6.3.12}$$

in which [Eq. (6.1.2)]

$$E[u \mid a, \theta] = \int u(a,x) f_X(x \mid \theta) \, dx \tag{6.3.13}$$

This latter approach [Eq. (6.3.12) rather than Eq. (6.3.11)] has been used throughout Chap. 6 until this point, but either interpretation is valid. In fact, the introduction of the concept of the Bayesian distribution of X can best be justified on the basis that its use in a decision analysis will give the same result as the more conventional approach used in the previous sections of this chapter.

For example, if the total earthquake damage to the dam undergoing consolidation is proportional to the number of large earthquakes in the region during the next year, then for a particular embankment slope design, action a, the expected loss is

$$u(a,x) = -c_a x$$

and

$$
\begin{aligned}
E[u(a)] &= \Sigma u(a,x) \tilde{p}_X(x) \\
&= -c_a \Sigma x \tilde{p}_X(x) \\
&= -c_a \tilde{m}_X
\end{aligned}
$$

which under the assumptions above is

$$E[u(a)] = -\frac{c_a(r_0 + 1)}{t_0} = -0.083 c_a$$

Updating distributions It is clear that if new data become available, we can talk meaningfully of the prior *and* the posterior Bayesian distribution of X. But it is important to note that we cannot apply Bayes' theorem directly to the prior and posterior distribution of X. Rather we must work indirectly, first updating the prior distribution of the parameter(s) θ to the posterior distribution of θ, and only then calculating a new, posterior Bayesian distribution on X. Thus given the experimental outcome z (say, observations $x_1, x_2, \ldots, x_n$) one *cannot* say

$$\tilde{f}_X''(x) = NL(x \mid z)\tilde{f}_X'(x) \qquad (invalid)$$

since the likelihood function $L(x \mid z)$ is not meaningful. Instead the posterior Bayesian distribution on X is

$$
\begin{aligned}
\tilde{f}_X''(x) &= \int f_X(x \mid \theta) f_\Theta''(\theta) \, d\theta \\
&= \int f_X(x \mid \theta) NL(\theta \mid z) f_\Theta'(\theta) \, d\theta
\end{aligned}
\tag{6.3.14}
$$

The important implication is that it is not sufficient to retain and report only the Bayesian distribution of X; this information alone does not, in general, permit the updating of this distribution when new data becomes available.

For example, it is not appropriate to report simply that the distribution of the number of earthquakes in any year is

$$\tilde{p}'_X(x) = \frac{60^5(x + 4)!}{61^{5+x}4!x!}$$

Rather one should state, in addition, that this is based on a Poisson model *and* on a prior distribution on λ of

$$f'_\Lambda(\lambda) = \frac{e^{-60\lambda}\lambda^4 60^5}{4!}$$

One year later there will be new data upon which to base the Bayesian distribution for X for the subsequent year. The new data are an observation of X in the intervening year. Suppose that no earthquake occurred in that year; that is, $X = 0$. Then the posterior distribution on λ has parameters $t_0 = 60 + 1$ and $r_0 = 4 + 0$, and

$$f''_\Lambda(\lambda) = \frac{e^{-61\lambda}\lambda^r 61^5}{4!}$$

while the new distribution on X in any one year is

$$\tilde{p}''_X(x) = \frac{61^5(x + 4)!}{62^{5+x}4!x!}$$

6.3.2 The General Case $Y = h(X_1, X_2, \ldots, X_\beta)$

More generally interest may lie in some random variable Y which is a function h of one or more future values of the process, X_1, X_2, ... A number of examples were discussed in Sec. 6.1. As other examples, Y might be the largest of the next 15 maximum annual wind velocities, X_1, X_2, ... , X_{15}. Another common form for h is

$$Y = h(X_1, X_2, \ldots, X_\beta) = \sum_{i=1}^{\beta} X_i$$

In any such case, if the (common) model distribution $f_X(x \mid \theta)$ of the X_i is known, the model distribution $f_Y(y \mid \theta)$ of $Y = h(X_1, X_2, \ldots, X_\beta)$ can be determined by the methods of derived distributions outlined in Sec. 2.3. If, however, the parameters are not known, but are only available as random variables with a distribution $f_\Theta(\theta)$, then it becomes appropriate to discuss the Bayesian distribution of Y, defined as

$$\tilde{f}_Y(y) = \int f_Y(y \mid \theta) f_\Theta(\theta) \, d\theta \qquad (6.3.15)$$

Thus the variable $Y = h(X_1, X_2, \ldots, X_\beta)$ can be treated as X was in Sec. 6.3.1, with a Bayesian mean, with prior and posterior Bayesian distributions, and, in particular, with expected utility values

$$E[u \mid a] = \int u(a,y) \hat{f}_Y(y) \, dy \tag{6.3.16}$$

This calculation arises when, as in Sec. 6.1.2, the utility is associated with some function of the X_i's. In that section, of course, the alternate approach was taken; the expected utility given θ was computed first as [Eqs. (6.1.13) and (6.1.14), for example]

$$g(\theta) = E[u \mid a,\theta] = E[u(a,X_1,X_2, \ldots, X_\beta \mid \theta]$$
$$= \int\int \cdots \int u(a,x_1,x_2, \ldots, x_\beta) f_X(x_1 \mid \theta) f_X(x_2 \mid \theta) \cdots$$
$$f_X(x_\beta \mid \theta) \, dx_1 \, dx_2 \cdots dx_\beta$$

or as [Eq. (6.1.20), for example]

$$g(\theta) = E[u \mid a,\theta] = E[u(a,Y) \mid \theta]$$
$$= \int u(a,y) f_Y(y \mid \theta) \, dy$$

in which $f_Y(y \mid \theta)$ had first been derived from the distribution of the X_i's. After computation of $g(\theta) = E[u \mid a,\theta]$, the (prior or posterior) distribution of θ, $f_\Theta(\theta)$, was introduced to determine $E[u \mid a]$ [Eqs. (6.1.10) or (6.1.12)].

Again, the approach of either section is valid. The approach of Sec. 6.1 can usually be expected to be more expedient computationally in the solution of decision problems than that in this section. But both interpretations are of value.

Illustration: Port planning   When probability statements are desired prior to or in addition to a decision analysis, it is particularly useful to consider the Bayesian distribution of Y. Consider a port-planning example in which the X_i are a sequence of independent exponentially distributed interarrival times:

$$f_X(x \mid \lambda) = e^{-\lambda x} \qquad x \geq 0 \tag{6.3.17}$$

The times are those between the arrivals of members of a certain class of cargo ships. The ships require a special unloading facility, and the decision is among various levels of automation (and expense) in unloading equipment. For one particular system, loading requires l days. Therefore, a ship will be delayed in unloading if it arrives within l days of the one before. Interest lies then in Y, the shortest of the interarrival times of a fixed number β of ships.

$$Y = \min [X_1, X_2, \ldots, X_\beta] \tag{6.3.18}$$

The distribution of Y is found as

$$1 - F_Y(y) = P[\text{all } \beta \text{ of the } X_i\text{'s} \geq y]$$
$$= [1 - F_X(y)]^\beta$$
$$= (e^{-\lambda y})^\beta = e^{-\beta \lambda y} \tag{6.3.19}$$

That is, Y is also exponential, with parameter $\beta\lambda$.

Assume that, based on rough predictions of future cargo activity, a (conjugate) gamma prior has been assigned to λ,

$$f_\Lambda(\lambda) = \frac{e^{-\lambda t_0}\lambda^{r_0}t_0^{r_0+1}}{\Gamma(r_0 + 1)} \qquad \lambda \geq 0 \qquad\qquad (6.3.20)$$

with given values t_0 and r_0. Then the Bayesian distribution of Y is

$$\tilde{f}_Y(y) = \int_0^\infty f_Y(y \mid \lambda)f_\Lambda(\lambda)\, d\lambda$$

$$= \int_0^\infty \beta\lambda e^{-\beta\lambda y}\frac{e^{-\lambda t_0}\lambda^{r_0}t_0^{r_0+1}}{\Gamma(r_0 + 1)}\, d\lambda$$

Combination of terms reveals a gammalike form. By rearrangement and inspection one can obtain

$$\tilde{f}_Y(y) = \beta(r_0 + 1)t_0^{r+1}(\beta y + t_0)^{-(r+2)} \qquad y \geq 0 \qquad\qquad (6.3.21)$$

The Bayesian CDF of Y is

$$\tilde{F}_Y(y) = \int_0^y f_Y(u)\, du = 1 - \left(\frac{t_0}{\beta y + t_0}\right)^{r+1} \qquad y \geq 0 \qquad\qquad (6.3.22)$$

Thus the (Bayesian) probability that at least one of the next β ships will be delayed is

$$\tilde{F}_Y(l) = 1 - \left(\frac{t_0}{\beta l + t_0}\right)^{r+1} \qquad\qquad (6.3.23)$$

Let us consider some numerical values, $r_0 = 10$ and $t_0 = 50$. If it requires ($l =$) 3 days to unload a ship with the proposed system, the probability that at least one of the next $\beta = 10$ ships will have to wait is

$$F_Y(3) = 1 - \left(\frac{50}{(3)(10) + 50}\right)^{11}$$

$$= 1 - (\tfrac{5}{8})^{11} \approx 1$$

Even if the unloading time can be reduced to $l = 1$ day, this probability is high:

$$\tilde{F}_Y(1) = 1 - \left(\frac{50}{10 + 50}\right)^{11} = 0.87$$

The expected minimum of 10 arrival times is

$$\tilde{m}_Y = \int_0^\infty y\tilde{f}_Y(y)\, dy$$

For the gamma distribution on λ,

$$\tilde{m}_Y = \frac{t_0}{\beta r_0} \qquad\qquad (6.3.24)$$

For the numerical values above

$$\tilde{m}_Y = \frac{50}{(10)(5)} = 1 \text{ day}$$

Illustration: Flood flows As a final illustration we consider an example prompted by an important civil-engineering paper by Thomas [1948]. The problem is to determine the likelihoods of future events solely on the basis of past data, without assumption about the form of the underlying distribution. In particular, what is the probability that in the next m years (trials) there will be y occurrences of floods in excess of some given magnitude if from available data of n years, s such excessive magnitudes have been observed?

Let $X_i = 1$ denote an occurrence in year i. Then we know that the distribution of

$$Y = X_{n+1} + X_{n+2} + \cdots + X_{n+m} \tag{6.3.25}$$

is binomial, $B(m,p)$, if the parameter p is known. That is,

$$p_Y(y \mid p) = \binom{m}{y} p^y (1 - p)^{m-y} \qquad y = 0,1,2, \ldots \tag{6.3.26}$$

Treating the parameter p as a random variable, we adopt the data-based beta distribution [Eq. (6.2.18)] with $r'' = s + 1$ and $t'' = n + 2$

$$f_P(p) = \frac{(n + 1)!}{(s)!(n - s)!} p^s (1 - p)^{n-s} \qquad 0 \le p \le 1 \tag{6.3.27}$$

The "Bayesian" distribution of Y becomes

$$\begin{aligned}
\tilde{p}_Y(y) &= \int_0^1 \frac{m!}{m!(m - y)!} p^y (1 - p)^{m-y} \frac{(n + 1)!}{(s)!(n - s)!} p^s (1 - p)^{n-s} \, dp \\
&= \frac{n + 1}{s + y + 1} \binom{m}{y} \binom{n}{s} \left[\binom{m + n + 1}{s + y + 1} \right]^{-1}
\end{aligned} \tag{6.3.28}$$

We shall use the same numerical data considered by Thomas. During a 25-year period the annual maximum (24-hr average) runoffs (floods) in each year in a small stream at Good Heights, Massachusetts, ranged from 629 to 4720 cfs. Interest lies in floods in excess of 1870 cfs, the magnitude which a cofferdam protecting a construction site is designed to withstand. The data show $s = 5$ floods in excess of 1870 cfs in $n = 25$ years.

The Bayesian distribution of Y is therefore

$$\tilde{p}_Y(y) = \frac{26}{6 + y} \binom{m}{y} \binom{25}{5} \left[\binom{m + 26}{6 + y} \right]^{-1} \qquad y = 0,1,2, \ldots$$

The Bayesian probability that the cofferdam will not be overtopped in 1 year ($m = 1$, $y = 0$) is

$$\begin{aligned}
\tilde{p}_Y(0) &= \binom{26}{6} \binom{25}{5} \left[\binom{27}{6} \right]^{-1} \\
&= 0.78
\end{aligned}$$

whereas the probability of no "excessive" floods in 5 years is

$$\begin{aligned}
\tilde{p}_Y(0) &= \binom{26}{6} \binom{5}{0} \binom{25}{5} \left[\binom{31}{6} \right]^{-1} \\
&= 0.312
\end{aligned}$$

6.3.3   Summary

Putting together our treatment of unknown parameters as random variables and the notion of compound distributions, we defined a compound, Bayesian distribution of a random variable X as

$$\tilde{f}_X(x) = \int f_X(x \mid \theta) f_\Theta(\theta) \, d\theta$$

in which $f_X(x \mid \theta)$ is the true, underlying, "model" distribution with unknown parameter(s) θ, and $f_\Theta(\theta)$ is a prior or posterior distribution on Θ. The Bayesian distribution of X couples the randomness inherent in the model with the statistical uncertainty in its parameters.

One can define, too, a Bayesian expectation of X (or, generally, any function of X) as

$$\tilde{m}_X = \int x \tilde{f}_X(x) \, dx = \int E[X \mid \theta] \, f_\Theta(\theta) \, d\theta$$

Similar definitions hold for variables that are functions of the X_i of the process.

6.4   SUMMARY OF CHAPTER 6

This chapter deals with the decision analysis of problems involving a sequence of independent, identically distributed random variables, X_1, $X_2, \ldots$, with unknown parameter(s) θ. In general, the engineer has assigned a prior distribution to θ, $f'_\Theta(\theta)$, and then incorporates the information in a sample, $x_1, x_2, \ldots, x_n$, through Bayes' theorem, to find a posterior distribution on θ, $f''_\Theta(\theta)$. The decision analysis requires finding

$$E[u \mid a] = \int g(\theta) f''_\Theta(\theta) \, d\theta$$

in which $g(\theta)$ is the expected utility given the value of θ. The utility itself depends in general on some function of a set of future values of the process.

Data-based distributions on θ arise from neglecting all but the sample likelihood to determine the shape of the posterior. The use of a conjugate prior distribution simplifies calculations. The Bayesian distribution of a random variable is defined as its model distribution $f_X(x \mid \theta)$, weighted by the available distribution on the parameter(s) $f_\Theta(\theta)$.

REFERENCES

General

See Chap. 5 references. In particular, Pratt, Raiffa, and Schlaifer [1965] and Raiffa and Schlaifer [1961] will be of value. The following papers may also prove interesting.

Blake, R. E. [1966]: On Predicting Structural Reliability, *Proc. AIAA 4th Aerospace Sci. Meeting*, Los Angeles, June.

Cornell, C. A. [1969]: Bayesian Statistical Decision Theory and Reliability-Based Structural Design, *Proc. Intern. Conf. Structural Safety and Reliability, Washington, D.C.* (Proceedings to be published by Pergamon Press, New York.)

DeHardt, J. H., and H. D. McLaughlin [1966]: Using Bayesian Methods to Select a Design with Known Reliability without a Confidence Coefficient, *AIAA Ann. Reliability Maintainability*, vol. 5, p. 611.

Esteva, L. [1968]: *Bases Para la Formulacion de Decisiones de Deseño Sismico*, Ph.d. thesis, National University of Mexico, August.

Shinozuka, M. and J-N. Yang [1969]: Optimum Structural Design Based on Reliability and Proof-Load Test, *AIAA Ann. Reliability and Maintainability*, vol. 8.

Specific text references

Esteva, L. [1969]: Seismicity Prediction: A Bayesian Approach, *Proc. 4th World Conf. Earthquake Engineering*, Santiago, Chile.

Thomas, H. A. [1948]: Frequency of Minor Floods, *J. Boston Soc. Civil Eng.*, vol. 34, pp. 425–442.

PROBLEMS

6.1. In Sec. 6.2.2 a professional committee chose a semiconventional method to select an estimate of u^k for a wind-design code to be used with a Type II extreme-value distribution in calculating a design wind velocity which would have probability of, say, 0.02 of being exceeded during the lifetime β of the structure.

(a) For $\beta = 15$ years, calculate the design velocity using their point estimate of u^k. There are 10 years of data: 62, 85, 71, 69, 95, 83, 78, 89, 73, and 80 mph, for which

$$\overline{x^{-k}} = \frac{1}{n} \sum x^{-k} = 1.73 \times 10^{-16}$$

when k equals its known value of 9.

(b) Alternatively, determine the Bayesian distribution of $Y = \max [X_1, X_2, \dots, X_\beta]$ using a posterior distribution of the normalized likelihood form [Eq. (6.2.12)], namely, gamma, with vague prior distribution. Note that the distribution of the parameter $w = u^k$ is gamma, $f_W(w) = N_0 w^n e^{[-\overline{x^{-k}} n w]}$, in which

$$N_0 = \frac{(n\overline{x^{-k}})^{n-1}}{\Gamma(n+1)}$$

(c) For $\beta = 15$ years, calculate the design wind velocity y_0 using this Bayesian distribution, i.e., such that $1 - \bar{F}_Y(y_0) = 0.02$.

6.2. Value of perfect information.† In routine estimating of the average daily flow on a rural highway, traffic counts are taken on n days. Assuming that the daily traffic flows are normal, independent, and homogeneous in time (unknown mean flow m and the daily variance σ^2 are the same every day). Assume that σ is known to be 2000. In absence of knowledge as to exactly how his traffic counts will be used in future decision about new traffic links, etc., the state engineer responsible for maintaining

† Adapted from W. F. Johnson [1969], Decision Theoretic Techniques for Information Acquisition in Transportation Network Planning, Sc.D. Thesis, Massachusetts Institute of Technology.

a data pool of traffic counts assumes that the cost of his providing designers an estimate $\bar{x}$ of the mean in error by an amount $\bar{x} - m$ is proportional to $(\bar{x} - m)^2$ and equals $c(\bar{x} - m)^2$. The cost of providing a counter and its maintenance is d per day. Consider the following preposterior analysis problems:

(a) Assuming that the engineer has a very vague prior distribution on m with a mean of 10,000, what is the number of days n he should count to minimize his total expected cost?

(b) If the engineer, basing his judgment on out-of-date traffic counts and on growth projections, has a normal prior distribution on the present value of m with mean 10,000 and standard deviation 500, what should n be?

(c) What in case (b) is the value of perfect information (see Sec. 5.3)? What then is the largest number of counts that should even be considered?

(d) Now, owing to equipment and crew limitations, the engineer has a fixed available counting capacity of $n_0 = 20$ days, which he must allocate to two independent highways. For the first, more important highway, the constant c is twice that on the other. Also, σ is 4000 on the first highway, compared with 2000 on the other. The engineer's prior distributions on both means are vague. How should he allocate the 20 days?

(e) *Extra credit:* Instead of assuming known σ in part (d), assume known coefficients of variation, $V = 15$ percent and $V = 20$ percent, respectively. Now determine the allocation scheme. Assume normal prior distributions on the two means with prior means and standard deviations of 20,000, 10,000, 700, and 500, respectively. Compare these results with those in Prob. 4.18 where a scheme for soil testing on highways was proposed.

6.3. Modify the hurricane-barrier illustration of Sec. 6.2.1 by attempting to account for the possible lack of reliability of the old historical records as follows. The direction of the hurricane observed during the third interval could not be precisely estimated from the records. The engineer assigns a degree-of-belief probability of 0.80 to the fact that this hurricane did travel in the critical direction, and 0.20 to the fact that it did not. In the first time period, on the other hand, there were two hurricanes whose directions might have been critical ones. The limited historical data are assessed by the engineer to suggest that either of these hurricanes, each with probability 0.1, might have been critical. *Hint:* Add a level of branches to the decision tree reflecting the possibility that any number of total observations from zero to three might have taken place. Assign each branch a probability.

Comment: Such historical data, for all its questionable reliability, is invaluable in assessing the risk of rare natural events such as earthquakes.

6.4. In the second illustration of Sec. 6.1.1, the passenger neglected to make an optimal choice of departure time given that he was going to go by air.

(a) Find his best departure time given this choice of action.

(b) Reanalyze his decision problem with this departure time and determine whether he should go by air or train. Assume that the utility associated with the additional time spent in his office (relative to the choice of a train) can be linearly interpolated. It is 0.4 if the added time is $5 - 2.5 = 2.5$ hr, and it is 0 if the time is $5 - 5 = 0$ hr.

6.5. Reconsider the sea-wall illustration of Sec. 6.1.2 under the assumption that the damage C given an occurrence of a storm intensity greater than x_0 is not a constant c but rather depends on the intensity x in an exponential way:

$$C = \begin{cases} 0 & x \leq x_0 \\ de^{\alpha(x-x_0)} & x \geq x_0,\, \alpha < \lambda \end{cases}$$

(a) Show that (for interest rate ϵ equal to zero)

$$E[u \mid a,b,\lambda] = g(b,\lambda) = -\frac{\beta\lambda \, db}{\lambda - \alpha} e^{-(\lambda-\alpha)x_0} \qquad \alpha < \lambda$$

(b) Show that the expected loss, given that $x > x_0$, is $\lambda d/(\lambda - \alpha)$, which is independent of b. Assume that $d = \$12,500$ and $\alpha = 0.015$. Show that for $\lambda = r/\nu = 0.02$ (the prior expected value of λ), these numbers give the same conditional expected cost, given that $x > x_0$, as the fixed cost used in the original illustration, that is, c.

(c) Using the values of d and α in part (b), reanalyze the decision situation in the illustration. (Unless a closed-form solution can be found for the integral $E[u \mid a]$, this step may require approximate numerical integration by hand or computer.) It is not necessarily true that the same decision will result, since $E[\lambda d/(\lambda - \alpha)]$ is not necessarily equal to $E[\lambda]d/(E[\lambda] - \alpha)$. For $\lambda \le \alpha$, assume $g(b,\lambda) = \$500,000$.

6.6. Assume that, based on the long observation of a contractor's previous satisfactory production of other types of concrete, an engineer assigns a conjugate joint prior distribution on the unknown mean and variance of the (normally distributed) concrete cylinder strength of a new mix. He adopts parameters

$n' = 20$

$\bar{x}' = 4200$ psi

$s' = 600$ psi

(a) What is his prior probability that the true mean is greater than 3500, the "design strength?"

(b) After observing five values of cylinder strengths, 4080, 4310, 4820, 3610, 3930, what are the posterior parameters of the marginal distribution on m?

(c) About what value m_{90} is the engineer prepared to say that the true mean exceeds m_{90} with probability 90 percent?

(d) Assume now that the engineer chooses to use a diffuse prior distribution on the unknown parameters. What is his posterior value of m_{90} now? Compare with the one-sided, lower, 90 percent confidence limit on m, as defined in Sec. 4.1.3.

6.7. In the (upper) tail of interest to an engineer, demands (e.g., earthquake and wind loads) follow a distribution of the form

$$1 - F(x) = kx^{-\beta} \qquad x \ge x_0$$

in which the parameter β must be greater than zero. k can be found in terms of β and x_0:

$$k = x_0{}^\beta$$

(a) Find a conjugate prior distribution for the (unknown) parameter β.

(b) Find a simple relationship between the prior and posterior parameters of the (conjugate) distribution on β, in terms of some simple statistics of the sample.

(c) For $x_0 = 10$ and for a prior distribution of β of

$$f_B(\beta) = k' x_0{}^{k''\beta} k^{(-\beta-1)} \beta^{k''} \qquad \beta \ge 0$$

$$k' = \text{normalizing constant}$$

$$k = 226{,}512$$

$$k'' = 5$$

find the posterior distribution on β if the following observations are obtained: $x_1 = 12$, $x_2 = 11$, and $x_3 = 13$.

(d) If the demand exceeds 13, the system fails with loss of $1,000. What is the expected loss if a single maximum value is to be found next year? Note that numerical integration may be required to evaluate the integral.

6.8. Consider Eq. (6.3.3). Calculate the values of $\tilde{p}_X(0)$ and $\tilde{p}_X(1)$ as $t_0 \to \infty$ but r_0/t_0 remains constant. Compare with $p_X(0)$ and $p_X(1)$ with $\nu = r_0/t_0$. *Hint:* A series expansion of $e^{-\nu}$ will help. Are these results consistent with your intuition? Explain.

6.9. In 1955 the existing sewage-treatment facilities of a large oceanside city were operating at full capacity. Two alternate plans were considered for expansion. The first plan involved construction of two large ocean outfalls. One of these was to discharge digested sludge at a point 7 miles out at sea at a minimum depth of 300 ft and the other was to transport unchlorinated effluent from primary treatment 5 miles into the ocean at a minimum depth of 200 ft. This plan involved some modifications of the existing complete treatment plant to a primary plant of greater capacity.

The alternate plan called for expansion of the existing treatment plant to allow for continued complete high-rate activated sludge treatment of all the waste including disinfection by chlorination and disposal through an existing 1-mile outfall.

The designs were based on a dry-weather flow of 420 million gal/day in the year 2000. The first alternative was estimated to cost $27,000,000. Operation cost was estimated at $15 per million gals. The second plan was estimated to involve a cost of $19,500,000 with an operating cost of $18 per million gal treated. A demand of 100 gal per capita-day was assumed.

(a) Make a decision assuming that the demand is normally distributed, with parameter:

Mean demand, million gal/day	$p'_M(m)$
380 md	0.4
420 md	0.5
460 md	0.1

Use an interest rate of 4 percent and assume a current mean demand of approximately 250 million gal/day with the demand increasing linearly to full value in 45 years.

(b) Assume that possible pollution of the adjacent ocean beaches may occur with the two new outfalls. In this case an added cost of $10 per million gal treated is estimated. The probability of unsatisfactory performance is estimated at 0.20 with plan 1. Repeat the decision analysis.

(c) Would it be advisable to consider the construction of plan 2 in two equal steps assuming a 2 percent annual increase in construction cost?

(d) A small-scale experiment can be performed to better define the possibility of pollution. If the sampling reliability is 0.80 with both satisfactory and unsatisfactory performance, is an experiment desirable? If so, what cost of experiment can be justified?

6.10. The wind velocity has been measured for the past 5 years at an airport with the following results.

Time	Velocity, mph	Number of events
1st year	40–50	2
	50–60	1
	Over 60	0
2nd year	40–50	1
	50–60	0
	Over 60	0
3d year	40–50	0
	50–60	0
	Over 60	0
4th year	40–50	3
	50–60	1
	Over 60	1
5th year	40–50	1
	50–60	1
	Over 60	0

A large construction project is proceeding adjacent to the site and you are asked to determine the probabilities of large winds in the first two months of construction and in the first year of construction. Use a Poisson model and a diffuse prior distribution on the mean rate of occurrence.

6.11. An economic study is being made on the design of a small bridge. The traffic load has been studied with the following loads and occurrences noted during the last 5 years. Use a Poisson model for the occurrence of large loads.

Load	Events
T-1	1
T-2	6
T-3	80

In the first design under consideration, if a load T-1 occurs, a contribution to cumulative fatigue damage is expected that implies an eventual loss of useful life worth $500 to the state. Vehicle T-2 will involve an average damage of $100 each occurrence, while damage under T-3 "costs" $0.50 per occurrence.

An alternate design costing more than the first design is being considered. This design is expected to have an average damage of $100 for each T-1 occurrence and no damage with the other vehicles.

Determine the expected annual cost due to fatigue damage for each design. Use data-based distributions for each of the three Poisson-process parameters.

6.12. The tabulated concrete cylinder compressive test results were furnished by a commercial testing agency for the concrete used in a major remodeling job. The specification 28-day strength was 3000 psi. The engineer must accept or reject the concrete.

Make a statistical study of the data. Assume that value to the engineer is a function of true mean strength. The engineer's values are linear with mean strength. Acceptance loss is 100 at 3000 psi (28-day strength) and zero at 3500 psi. Rejection loss is zero below 3000 psi, increasing linearly to 400 at 3500 psi.

The two mixes are similar except for maximum aggregate size. You must decide which model and which data you will use.

Mix 600-2		Mix 700-1	
7 days	28 days	7 days	28 days
2190 psi	4120 psi	1310 psi	3290 psi
1870	3160	2120	3080
2190		1840	3130
1870		1860	3250
2190		1890	3110
		1860	2970
		1840	3280
		1610	3550
		1310	3800
		1890	3040
		1680	3990
		2120	
		1610	
		1680	
		2240	
		1890	
		1910	
		2240	
		1910	
		1890	

6.13. Most physical measurements involve some degree of random error. The problem is to infer the true value from the observations. Suppose that it is desired to estimate the depth d to basement rock below a proposed power plant using a sonic device. Assume that any measurement gives a reading

$$X_i = d + Y_i$$

in which the Y_i's are normally distributed errors with zero mean (implying that $m_X = d$). Assume that repeated measurements are independent.

(a) Assuming that $\sigma_{Y_i} = 100$ ft, what is the (data based) 90 percent confidence interval on d if the five readings are $x_1 = 2000$ ft, $x_2 = 2300$ ft, $x_3 = 2100$ ft, $x_4 = 2200$ ft, $x_5 = 2100$ ft?

(b) Assuming that $V_{X_i} = 0.05$, what is this interval, given the same readings?

(c) What is the interval if neither σ_{Y_i} nor V_{X_i} is known?

(d) If the procedure has a known bias of $+100$ ft, that is, if

$$X_i = d + 100 \text{ ft} + Y_i$$

what is the confidence interval? Note that if a procedure has a known bias or other systematic error it can be removed and an unbiased procedure obtained. For example, in the case above let $X_i' = X_i - 100$. Then the X_i' represent an unbiased procedure, since $E[X_i'] = d$.

(e) To calibrate the procedure (i.e., to determine a correction factor or curve, say, that maps or translates an instrument reading into an unbiased depth estimate X_i), the device is run on a variety of site conditions which have a precisely known depth to rock. Explain how you would obtain an estimate of V_{X_i}.

(f) If a particular site at which the procedure was to be used to infer the depth had some peculiar factor (e.g., some unusual combination of soil properties) which caused a systematic error in all readings at that site, is there any way to detect it, eliminate it, or reduce it (by, say, increasing the number of tests)? What does this imply about the confidence intervals above? That is, are they not in fact conditional on site conditions "not dissimilar" from those at the calibration sites?

6.14. Consider the Poisson process earthquake occurrences in Sec. 6.3.1.

(a) Is the probability of an occurrence in the next t years zero if no occurrences have been observed in the past t_0 years? Assume a diffuse prior on λ. Compare to a forecast using the maximum-likelihood point estimate of λ.

(b) An earthquake of intensity MM (modified Mercalli) IX can be expected to disrupt a particular irrigation system (slides and fracture of canals). Intensities up to VIII have been recorded in the vicinity, but none of MM IX has been included in the record for the past 150 years. If each IX occurrence involves a random amount of damage with an expected value of $100,000, using the results above, what is the expected cost from earthquakes of intensity IX in the next 20 years? Assume that the system is fully repaired between earthquakes and that the amount of damage given an earthquake is independent of such repairs. (See, also, Cornell [1969].)

6.15. Exponential case. (a) If the occurrence of earthquakes is a Poisson process, the time between major earthquakes is exponentially distributed. In a particular region two such events have been found in the last 120 years. What is the chance of finding a time of less than 2 years to the next event? Use the maximum-likelihood estimator of the mean rate of occurrence.

(b) Repeat (a) using a Bayesian data-based approach. Show that the Bayesian PDF of the time to the next event T, given r such events in a total time of t, is given by:

$$\tilde{f}_T(t_0 \mid \Sigma t, r) = \frac{r+1}{\Sigma t} \frac{1}{(t_0/\Sigma t + 1)^{(r+2)}} \qquad t_0 \geq 0$$

Compare the estimate of $P[T \leq 2]$ based on this Bayesian distribution with that in part (a).

6.16. Multinomial case. A water-supply system includes four reservoirs so separated that a small storm will yield important runoff either to only one or to none of the reservoirs. Thus one of five mutually exclusive events will occur at each storm. A multinomial model can be used to represent the X_i, the number of occurrences of type i ($i = 1, 2, \ldots, k$; here $k = 5$) in n storms:

$$p_{X_1, X_2, \ldots, X_k}(x_1, x_2, \ldots, x_k) = \frac{n!}{x_1! x_2! \cdots x_k!} p_1{}^{x_1} p_2{}^{x_2} \cdots p_k{}^{x_k}$$

$$p_1 + p_2 + \cdots + p_k = 1$$

$$x_1 + x_2 + \cdots + x_k = n$$

The Bayesian data-based forecast relationship for $x_1', x_2', \ldots, x_k'$ future events in n_0 future trials given the occurrence of $x_1, x_2, \ldots, x_k$ events in n past trials can be found to be:

$$\tilde{p}(x_1', x_2', \ldots, x_k' \mid x_1, x_2, \ldots, x_k) = \frac{n_0! (n + k - 1)! (x_1 + x_1')! \cdots (x_k + x_k')!}{x_1'! \cdots x_k'! x_1! \cdots x_k! (n + n_0 + k - 1)!}$$

$$x_1' + x_2' + \cdots + x_k' = n_0$$

(a) A single small storm is observed, and no reservoir runoff is observed. What is the probability of observing this same event with the next storm? Compare this result with one based on a maximum-likelihood estimate of p_i.

(b) Two more small storms yield important runoffs to reservoir 1. Combine the information with that in (a) and forecast the probability that the next two storms will influence reservoir 2 once and reservoir 1 once.

(c) A typical year includes 10 small storms of the type under study. During the last year the following results were found.

Reservoir	Events last year
1	2
2	1
3	1
4	1
None	5

A cloud-seeding operation is being considered to increase the fraction of storms which drop their precipitation into the watersheds of the reservoirs. It will be run for 1 year on an experimental basis. How would you judge whether or not the experiment has been successful? If the results of the experiment are as follows, is the change a chance result or the result of the experiment?

Reservoir	Events
1	2
2	2
3	2
4	2
None	2

Note that the analysis can be made on the basis of a binomial model.

6.17. If X is lognormally distributed, determine the sample likelihood of observing $x_1, x_2, \ldots, x_n$. What problems arise in developing a conjugate distribution analysis?

6.18. Bayesian distribution of a normal random variable. It can be shown after suitable arithmetic that if the X_i's are a process of independent normal random variables with two unknown parameters, and if the conjugate distribution [Eq. (6.2.26)] on m and σ is used, with parameters $\bar{x}$, s, and n, then the Bayesian distribution on X is the t distribution. Specifically, the variate $(X - \bar{x})/ks$ has the t distribution with $n - 2$ degrees of freedom where k is

$$\sqrt{\frac{(n-1)(n+1)}{n(n-2)}} \quad \text{if} \quad s^2 = \frac{1}{n-1} \sum_{i=1}^{n} (x_i - \bar{x})^2$$

(a) Use tables of t to find the value of X which has 5 percent chance of being exceeded given a diffuse prior and three observations with $\bar{x} = 30$ and $s = 5$.

(b) Compare the result in (a) with that found with the normal distribution and the given point estimates of the parameters. Sketch this PDF and the Bayesian PDF of X found in (a).

(c) Decide whether to use system design a_1 or a_2 if X is the peak demand on the system. Use the distribution of X found in part (a). The utilities are

Design a_1: $u(a_1,x) = 1000 - x - \frac{1}{10}x^2$
Design a_2: $u(a_2,x) = 1200 - x - \frac{1}{8}x^2$

Note that the expected utilities can be found in terms of the mean and variance of X which are given in Sec. 3.4.

6.19. In the search for a new quality-control specification for concrete more definitive than just a lower limit, a confidence-interval type of specification has been suggested. It is argued that if the 28-day strength of standard cylinders is to be well controlled, the specification should state that, say, 90 percent of the test results shall lie in a given range. Questions arise in the use of such a decision rule owing to limitations on sample size. For example, it is desired that the concrete have a mean strength of 4000 psi. Assume that concrete cylinder strengths are normally distributed and that the standard deviation is stable and known to equal 600 psi. The specification states that 90 percent of the test results must lie in the range 3000 to 5000 psi (i.e., about $m \pm 1.65\sigma$).

Unfortunately, the actual sample size may be as small as three. Assume that three strengths—3600, 4600, and 4400 psi—have been observed. Assuming a "perfectly vague" prior distribution on the mix mean, what is the probability that the true mean is equal to or greater than 4000 psi? If the true mean is in the range 3500 to 4500 psi, no significant loss is involved, but a loss of $1,000 is assumed if the true mean is below 3500 psi or above 4500 psi, and it is not reported. Compute the expected loss of accepting (based solely on the three observations) the mix as having a true mean in this range.

6.20. A manufacturer wants to have an existing column removed from a work area. It will be necessary to remove two existing beams and replace them with a single large beam. The contractor will have to remove the roof adjacent to the two beams in order to take them out of the building and install the new beam. The owner desires that the change be made as soon as possible and that operations in the balance of the building not be influenced by the occurrence of bad weather while the roof is off for the 3 days required to make the necessary changes. In fact, he will pay $2,000 extra to the contractor if the change can be made right away. However, the contractor must pay any cost of damage from bad weather, estimated at $1,000 per day.

Records indicate that "bad weather" in this month occurred on 10, 12, 15, and 6 days in the last 4 years. If the occurrence of such events is assumed to be Poisson, should the contractor accept the deal in so wet a month? Base the decision on the data available. Comment on the model chosen. What would the influence of positive correlation among rainy-day occurrences be?

6.21. A large housing project will have major broken window glass from a variety of sources. These repairs are beyond the resources of routine maintenance. A sample of size 12 of these large costs has been obtained. The data on cost per repair call are:

$33.10	58.40	201.80
33.10	67.70	228.10
50.50	106.30	278.40
50.50	116.70	513.90

The 12 calls listed occurred in a period of 3 months. They did not appear to occur "systematically" in that period.

It is desired to set up a fund (or purchase insurance) to cover this expense for the next 3 months. What fund or insurance cost should be accepted to cover this expense assuming that the factors governing such costs will not change in that period?

An analysis requires assumptions about the distributions of both the number of events N and the size of damage X_i at each occurrence. Make reasonable assumptions. Compare the latter assumption versus the available data. Determine data-based distributions on the *means* of N and X_i. Note that only these two parameters are involved, since the total cost is

$$C = \sum_{i=1}^{N} X_i$$

and [Eq. (2.4.103)]

$$E[C] = E[N]E[X_i] = m_N m_X$$

6.22. Show that the following useful expected utility values hold:

Utility function $u(X)$	Probability distribution on X	Expected value $E[u]$
(a) $u = k_0 + k_1(x - m)$ in which $k_0 =$ utility at m, mean value of X	Any	k_0
(b) $u = k_2(x - x_0)^2$	Any	$k_2 \operatorname{Var}[X] + k_2(m - x_0)^2$
(c) Gamma: $u = u_0(\lambda_u x)^{r_u-1} e^{-\lambda_u x}$	Gamma: $f_X(x) = \dfrac{\lambda}{\Gamma(r)}(\lambda x)^{r-1} e^{-\lambda x}$	$\dfrac{u_0 \lambda \lambda_u^{r_u-1}\Gamma(r_0)}{\lambda_0^{r_0}\Gamma(r)}$; $r_0 - 1 = r + r_u - 2$ $\lambda_0 = \lambda + \lambda_u$
(d) Beta: $u = \dfrac{u_0(x-a)^{r_u-1}(b-x)^{t_u-r_u-1}}{(b-a)^{t_u-1}}$	Beta: $f_X(x) = \dfrac{1}{B}\dfrac{(x-a)^{r-1}(b-x)^{t-r-1}}{(b-a)^{t-1}}$	$\dfrac{u_0 B_0}{B(b-a)}$; $r_0 - 1 = r_u + r - 2$ $t_0 - r_0 = t_u + t - r_u - r - 1$ $B_0 = \dfrac{(r_0-1)!\,(t_0 - r_0 - 1)!}{(t_0 - 1)!}$
(e) Exponential: $u = \alpha + \beta e^{-\gamma x}$	Gamma: $f_X(x) = \dfrac{\lambda}{\Gamma(r)}(\lambda x)^{r-1} e^{-\lambda x}$	$\alpha + \beta\left(\dfrac{\lambda}{\lambda + \gamma}\right)^r$
(f) Exponential: $u = \alpha + \beta e^{-\gamma x}$	Normal: $f_X(x) = \dfrac{1}{\sigma\sqrt{2\pi}} e^{-\frac{1}{2}[(x-m)/\sigma]^2}$	$\alpha + \beta e^{(-m\gamma + \frac{1}{2}\sigma^2\gamma^2)}$
(g) Discontinuous linear: $u = \begin{cases} \alpha + \beta x & x_0 \le x \le x_1 \\ 0 & x_1 \le x < x_0 \end{cases}$	Normal	$(\alpha + \beta m)[F_X(x_1) - F_X(x_0)] + \beta\sigma^2[f_X(x_0) - f_X(x_1)]$

Tables

Table A.1† Values of the standardized normal distribution

Ordinates of the probability density function, $f_U(u) = (1/\sqrt{2\pi})e^{-\frac{1}{2}u^2}$

u	0.00	0.01	0.02	0.03	0.04	0.05	0.06	0.07	0.08	0.09
0.0	0.3989	0.3989	0.3989	0.3988	0.3986	0.3984	0.3982	0.3980	0.3977	0.3973
0.1	0.3970	0.3965	0.3961	0.3956	0.3951	0.3945	0.3939	0.3932	0.3925	0.3918
0.2	0.3910	0.3902	0.3894	0.3885	0.3876	0.3867	0.3857	0.3847	0.3836	0.3825
0.3	0.3814	0.3802	0.3790	0.3778	0.3765	0.3752	0.3739	0.3725	0.3712	0.3697
0.4	0.3683	0.3668	0.3653	0.3637	0.3621	0.3605	0.3589	0.3572	0.3555	0.3538
0.5	0.3521	0.3503	0.3485	0.3467	0.3448	0.3429	0.3410	0.3391	0.3372	0.3352
0.6	0.3332	0.3312	0.3292	0.3271	0.3251	0.3230	0.3209	0.3187	0.3166	0.3144
0.7	0.3123	0.3101	0.3079	0.3056	0.3034	0.3011	0.2989	0.2966	0.2943	0.2920
0.8	0.2897	0.2874	0.2850	0.2827	0.2803	0.2780	0.2756	0.2732	0.2709	0.2685
0.9	0.2661	0.2637	0.2613	0.2589	0.2565	0.2541	0.2516	0.2492	0.2468	0.2444
1.0	0.2420	0.2396	0.2371	0.2347	0.2323	0.2299	0.2275	0.2251	0.2227	0.2203
1.1	0.2179	0.2155	0.2131	0.2107	0.2083	0.2059	0.2036	0.2012	0.1989	0.1965
1.2	0.1942	0.1919	0.1895	0.1872	0.1849	0.1826	0.1804	0.1781	0.1758	0.1736
1.3	0.1714	0.1691	0.1669	0.1647	0.1626	0.1604	0.1582	0.1561	0.1539	0.1518
1.4	0.1497	0.1476	0.1456	0.1435	0.1415	0.1394	0.1374	0.1354	0.1334	0.1315
1.5	0.1295	0.1276	0.1257	0.1238	0.1219	0.1200	0.1182	0.1163	0.1145	0.1127
1.6	0.1109	0.1092	0.1074	0.1057	0.1040	0.1023	0.1006	0.0983	0.09728	0.09566
1.7	0.09405	0.09246	0.09089	0.08933	0.08780	0.08628	0.08478	0.08329	0.08183	0.08038
1.8	0.07895	0.07754	0.07614	0.07477	0.07341	0.07206	0.07074	0.06943	0.06814	0.06687
1.9	0.06562	0.06438	0.06316	0.06195	0.06077	0.05959	0.05844	0.05730	0.05618	0.05508
2.0	0.05399									
2.5	0.01753									
3.0	0.00443									
3.5	0.000873									
4.0	0.000134									
5.0	0.00000149									

† Table A.1 is taken in part from Hald [1952] and National Bureau of Standards [1953] with respective permissions of the authors and publishers. See Chap. 3 for full reference details of all appendix references.

Table A.1 Values of the standardized normal distribution (Continued)

The cumulative distribution function, $F_U(u) = \int_{-\infty}^{u} f_U(u)\, du$

u	0.00	0.01	0.02	0.03	0.04	0.05	0.06	0.07	0.08	0.09
0.0	0.5000	0.5040	0.5080	0.5120	0.5160	0.5199	0.5239	0.5279	0.5319	0.5359
0.1	0.5398	0.5438	0.5478	0.5517	0.5557	0.5596	0.5636	0.5675	0.5714	0.5753
0.2	0.5793	0.5832	0.5871	0.5910	0.5948	0.5987	0.6026	0.6064	0.6103	0.6141
0.3	0.6179	0.6217	0.6255	0.6293	0.6331	0.6368	0.6406	0.6443	0.6480	0.6517
0.4	0.6554	0.6591	0.6628	0.6664	0.6700	0.6736	0.6772	0.6808	0.6844	0.6879
0.5	0.6915	0.6950	0.6985	0.7019	0.7054	0.7088	0.7123	0.7157	0.7190	0.7224
0.6	0.7257	0.7291	0.7324	0.7357	0.7389	0.7422	0.7454	0.7486	0.7517	0.7549
0.7	0.7580	0.7611	0.7642	0.7673	0.7703	0.7734	0.7764	0.7794	0.7823	0.7852
0.8	0.7881	0.7910	0.7939	0.7967	0.7995	0.8023	0.8051	0.8078	0.8106	0.8133
0.9	0.8159	0.8186	0.8212	0.8238	0.8264	0.8289	0.8315	0.8340	0.8365	0.8389
1.0	0.8413	0.8438	0.8461	0.8485	0.8508	0.8531	0.8554	0.8577	0.8599	0.8621
1.1	0.8643	0.8665	0.8686	0.8708	0.8729	0.8749	0.8770	0.8790	0.8810	0.8830
1.2	0.8849	0.8869	0.8888	0.8907	0.8925	0.8944	0.8962	0.8980	0.8997	0.90147
1.3	0.90320	0.90490	0.90658	0.90824	0.90988	0.91149	0.91309	0.91466	0.91621	0.91774
1.4	0.91924	0.92073	0.92220	0.92364	0.92507	0.92647	0.92785	0.92922	0.93056	0.93189
1.5	0.93319	0.93448	0.93574	0.93699	0.93822	0.93943	0.94062	0.94179	0.94295	0.94408
1.6	0.94520	0.94630	0.94738	0.94845	0.94950	0.95053	0.95154	0.95254	0.95352	0.95449
1.7	0.95543	0.95637	0.95728	0.95818	0.95907	0.95994	0.96080	0.96164	0.96246	0.96327
1.8	0.96407	0.96485	0.96562	0.96638	0.96712	0.96784	0.96856	0.96926	0.96995	0.97062
1.9	0.97128	0.97193	0.97257	0.97320	0.97381	0.97441	0.97500	0.97558	0.97615	0.97670
2.0	0.97725									
2.1	0.98214									
2.2	0.98610									
2.3	0.98928									
2.4	0.99180									
2.5	0.99379									
3.0	0.99865									
3.5	0.999767									
4.0	0.9999683									
4.5	0.9999966									
5.0	0.99999971									
5.5	0.999999981									

u	2.32	3.09	3.72	4.27	4.75	5.20	5.61	6.00	6.36	6.71
$1 - F(u)$	10^{-2}	10^{-3}	10^{-4}	10^{-5}	10^{-6}	10^{-7}	10^{-8}	10^{-9}	10^{-10}	10^{-11}

Table A.2† Tables for evaluation of the CDF of the χ^2, gamma, and Poisson distributions

ν	$\chi^2 = 0.001$ / $m = 0.0005$	0.002 / 0.0010	0.003 / 0.0015	0.004 / 0.0020	0.005 / 0.0025	0.006 / 0.0030	0.007 / 0.0035	0.008 / 0.0010	0.009 / 0.0045	0.010 / 0.0050
1	0.97477	0.96433	0.95632	0.94957	0.94363	0.93826	0.93332	0.92873	0.92442	0.92034
2	0.99950	0.99900	0.99850	0.99800	0.99750	0.99700	0.99651	0.99601	0.99551	0.99501
3	0.99999	0.99998	0.99996	0.99993	0.99991	0.99988	0.99984	0.99981	0.99977	0.99973
4							0.99999	0.99999	0.99999	0.99999

ν	$\chi^2 = 0.01$ / $m = 0.005$	0.02 / 0.010	0.03 / 0.015	0.04 / 0.020	0.05 / 0.025	0.06 / 0.030	0.07 / 0.035	0.08 / 0.040	0.09 / 0.045	0.10 / 0.050
1	0.92034	0.88754	0.86249	0.84148	0.82306	0.80650	0.79134	0.77730	0.76418	0.75183
2	0.99501	0.99005	0.98511	0.98020	0.97531	0.97045	0.96561	0.96079	0.95600	0.95123
3	0.99973	0.99925	0.99863	0.99790	0.99707	0.99616	0.99518	0.99412	0.99301	0.99184
4	0.99999	0.99995	0.99989	0.99980	0.99969	0.99956	0.99940	0.99922	0.99902	0.99879
5			0.99999	0.99998	0.99997	0.99995	0.99993	0.99991	0.99987	0.99984
6							0.99999	0.99999	0.99999	0.99998

ν	$\chi^2 = 0.1$ / $m = 0.05$	0.2 / 0.10	0.3 / 0.15	0.4 / 0.20	0.5 / 0.25	0.6 / 0.30	0.7 / 0.35	0.8 / 0.40	0.9 / 0.45	1.0 / 0.50
1	0.75183	0.65472	0.58388	0.52709	0.47950	0.43858	0.40278	0.37109	0.34278	0.31731
2	0.95123	0.90484	0.86071	0.81873	0.77880	0.74082	0.70469	0.67032	0.63763	0.60653
3	0.99184	0.97759	0.96003	0.94024	0.91889	0.89643	0.87320	0.84947	0.82543	0.80125
4	0.99879	0.99532	0.98981	0.98248	0.97350	0.96306	0.95133	0.93845	0.92456	0.90980
5	0.99984	0.99911	0.99764	0.99533	0.99212	0.98800	0.98297	0.97703	0.97022	0.96257
6	0.99998	0.99985	0.99950	0.99885	0.99784	0.99640	0.99449	0.99207	0.98912	0.98561
7		0.99997	0.99990	0.99974	0.99945	0.99889	0.99834	0.99744	0.99628	0.99483
8			0.99998	0.99994	0.99987	0.99973	0.99953	0.99922	0.99880	0.99825
9				0.99999	0.99997	0.99993	0.99987	0.99978	0.99964	0.99944
10					0.99999	0.99998	0.99997	0.99994	0.99989	0.99983
11							0.99999	0.99998	0.99997	0.99995
12									0.99999	0.99999

ν	$\chi^2 = 1.1$ / $m = 0.55$	1.2 / 0.60	1.3 / 0.65	1.4 / 0.70	1.5 / 0.75	1.6 / 0.80	1.7 / 0.85	1.8 / 0.90	1.9 / 0.95	2.0 / 1.00
1	0.29427	0.27332	0.25421	0.23672	0.22067	0.20590	0.19229	0.17971	0.16808	0.15730
2	0.57695	0.54881	0.52205	0.49659	0.47237	0.44933	0.42741	0.40657	0.38674	0.36788
3	0.77707	0.75300	0.72913	0.70553	0.68227	0.65939	0.63693	0.61493	0.59342	0.57241
4	0.89427	0.87810	0.86138	0.84420	0.82664	0.80879	0.79072	0.77248	0.75414	0.73576
5	0.95410	0.94488	0.93493	0.92431	0.91307	0.90125	0.88890	0.87607	0.86280	0.84915
6	0.98154	0.97689	0.97166	0.96586	0.95949	0.95258	0.94512	0.93714	0.92866	0.91970
7	0.99305	0.99093	0.98844	0.98557	0.98231	0.97864	0.97457	0.97008	0.96517	0.95984
8	0.99753	0.99664	0.99555	0.99425	0.99271	0.99092	0.98887	0.98654	0.98393	0.98101
9	0.99917	0.99882	0.99838	0.99782	0.99715	0.99633	0.99537	0.99425	0.99295	0.99147
10	0.99973	0.99961	0.99944	0.99921	0.99894	0.99859	0.99817	0.99766	0.99705	0.99634
11	0.99992	0.99987	0.99981	0.99973	0.99962	0.99948	0.99930	0.99908	0.99882	0.99850
12	0.99998	0.99996	0.99994	0.99991	0.99987	0.99982	0.99975	0.99966	0.99954	0.99941
13	0.99999	0.99999	0.99998	0.99997	0.99996	0.99994	0.99991	0.99988	0.99983	0.99977
14			0.99999	0.99999	0.99999	0.99998	0.99997	0.99996	0.99994	0.99992
15						0.99999	0.99999	0.99999	0.99998	0.99997
16									0.99999	0.99990

Table A.2† Tables for evaluation of the CDF of the χ^2, gamma, and Poisson distributions (*Continued*)

ν	$\chi^2 = 2.2$ $m = 1.1$	2.4 1.2	2.6 1.3	2.8 1.4	3.0 1.5	3.2 1.6	3.4 1.7	3.6 1.8	3.8 1.9	4.0 2.0
1	0.13801	0.12134	0.10686	0.09426	0.08327	0.07364	0.06520	0.05778	0.05125	0.04550
2	0.33287	0.30119	0.27253	0.24660	0.22313	0.20190	0.18268	0.16530	0.14957	0.13534
3	0.53195	0.49363	0.45749	0.42350	0.39163	0.36181	0.33397	0.30802	0.28389	0.26146
4	0.69903	0.66263	0.62682	0.59183	0.55783	0.52493	0.49325	0.46284	0.43375	0.40601
5	0.82084	0.79147	0.76137	0.73079	0.69999	0.66918	0.63857	0.60831	0.57856	0.54942
6	0.90042	0.87949	0.85711	0.83350	0.80885	0.78336	0.75722	0.73062	0.70372	0.67668
7	0.94795	0.93444	0.91938	0.90287	0.88500	0.86590	0.84570	0.82452	0.80250	0.77978
8	0.97426	0.96623	0.95691	0.94628	0.93436	0.92119	0.90681	0.89129	0.87470	0.85712
9	0.98790	0.98345	0.97807	0.97170	0.96430	0.95583	0.94631	0.93572	0.92408	0.91141
10	0.99457	0.99225	0.98934	0.98575	0.98142	0.97632	0.97039	0.96359	0.95592	0.94735
11	0.99766	0.99652	0.99503	0.99311	0.99073	0.98781	0.98431	0.98019	0.97541	0.96992
12	0.99903	0.99850	0.99777	0.99680	0.99554	0.99396	0.99200	0.98962	0.98678	0.98344
13	0.99961	0.99938	0.99903	0.99856	0.99793	0.99711	0.99606	0.99475	0.99314	0.99119
14	0.99985	0.99975	0.99960	0.99938	0.99907	0.99866	0.99813	0.99743	0.99655	0.99547
15	0.99994	0.99990	0.99984	0.99974	0.99960	0.99940	0.99913	0.99878	0.99832	0.99774
16	0.99998	0.99996	0.99994	0.99989	0.99983	0.99974	0.99961	0.99944	0.99921	0.99890
17	0.99999	0.99999	0.99998	0.99996	0.99993	0.99989	0.99983	0.99975	0.99964	0.99948
18			0.99999	0.99998	0.99997	0.99995	0.99993	0.99989	0.99984	0.99976
19				0.99999	0.99999	0.99998	0.99997	0.99995	0.99993	0.99989
20						0.99999	0.99999	0.99998	0.99997	0.99995
21								0.99999	0.99999	0.99998
22										0.99999

ν	$\chi^2 = 4.2$ $m = 2.1$	4.4 2.2	4.6 2.3	4.8 2.4	5.0 2.5	5.2 2.6	5.4 2.7	5.6 2.8	5.8 2.9	6.0 3.0
1	0.04042	0.03594	0.03197	0.02846	0.02535	0.02259	0.02014	0.01796	0.01603	0.01431
2	0.12246	0.11080	0.10026	0.09072	0.08209	0.07427	0.06721	0.06081	0.05502	0.04979
3	0.24066	0.22139	0.20354	0.18704	0.17180	0.15772	0.14474	0.13278	0.12176	0.11161
4	0.37962	0.35457	0.33085	0.30844	0.28730	0.26739	0.24866	0.23108	0.21459	0.19915
5	0.52099	0.49337	0.46662	0.44077	0.41588	0.39196	0.36904	0.34711	0.32617	0.30622
6	0.64963	0.62271	0.59604	0.56971	0.54381	0.51843	0.49363	0.46945	0.44596	0.42319
7	0.75647	0.73272	0.70864	0.68435	0.65996	0.63557	0.61127	0.58715	0.56329	0.53975
8	0.83864	0.81935	0.79935	0.77872	0.75758	0.73600	0.71409	0.69194	0.66962	0.64723
9	0.89776	0.88317	0.86769	0.85138	0.83431	0.81654	0.79814	0.77919	0.75976	0.73992
10	0.93787	0.92750	0.91625	0.90413	0.89118	0.87742	0.86291	0.84768	0.83178	0.81526
11	0.96370	0.95672	0.94898	0.94046	0.93117	0.92109	0.91026	0.89868	0.88637	0.87337
12	0.97955	0.97509	0.97002	0.96433	0.95798	0.95096	0.94327	0.93489	0.92583	0.91608
13	0.98887	0.98614	0.98298	0.97934	0.97519	0.97052	0.96530	0.95951	0.95313	0.94615
14	0.99414	0.99254	0.99064	0.98841	0.98581	0.98283	0.97943	0.97559	0.97128	0.96649
15	0.99701	0.99610	0.99501	0.99369	0.99213	0.99029	0.98816	0.98571	0.98291	0.97975
16	0.99851	0.99802	0.99741	0.99666	0.99575	0.99467	0.99338	0.99187	0.99012	0.98810
17	0.99928	0.99902	0.99869	0.99828	0.99777	0.99715	0.99639	0.99550	0.99443	0.99319
18	0.99966	0.99953	0.99936	0.99914	0.99886	0.99851	0.99809	0.99757	0.99694	0.99620
19	0.99985	0.99978	0.99969	0.99958	0.99943	0.99924	0.99901	0.99872	0.99836	0.99793
20	0.99993	0.99990	0.99986	0.99980	0.99972	0.99962	0.99950	0.99934	0.99914	0.99890
21	0.99997	0.99995	0.99993	0.99991	0.99987	0.99982	0.99975	0.99967	0.99956	0.99943
22	0.99999	0.99998	0.99997	0.99996	0.99994	0.99991	0.99988	0.99984	0.99978	0.99971
23	0.99999	0.99999	0.99999	0.99998	0.99997	0.99996	0.99994	0.99992	0.99989	0.99986
24			0.99999	0.99999	0.99999	0.99998	0.99997	0.99996	0.99995	0.99993
25					0.99999	0.99999	0.99999	0.99998	0.99998	0.99997
26								0.99999	0.99999	0.99998
27									0.99999	0.99999

Table A.2† Tables for evaluation of the CDF of the χ^2, gamma, and Poisson distributions (*Continued*)

ν	$\chi^2 = 6.2$ $m = 3.1$	6.4 3.2	6.6 3.3	6.8 3.4	7.0 3.5	7.2 3.6	7.4 3.7	7.6 3.8	7.8 3.9	8.0 4.0
1	0.01278	0.01141	0.01020	0.00912	0.00815	0.00729	0.00652	0.00584	0.00522	0.00468
2	0.04505	0.04076	0.03688	0.03337	0.03020	0.02732	0.02472	0.02237	0.02024	0.01832
3	0.10228	0.09369	0.08580	0.07855	0.07190	0.06579	0.06018	0.05504	0.05033	0.04601
4	0.18470	0.17120	0.15860	0.14684	0.13589	0.12569	0.11620	0.10738	0.09919	0.09158
5	0.28724	0.26922	0.25213	0.23595	0.22064	0.20619	0.19255	0.17970	0.16761	0.15624
6	0.40116	0.37990	0.35943	0.33974	0.32085	0.30275	0.28543	0.26890	0.25313	0.23810
7	0.51660	0.49390	0.47168	0.45000	0.42888	0.40836	0.38845	0.36918	0.35056	0.33259
8	0.62484	0.60252	0.58034	0.55836	0.53663	0.51522	0.49415	0.47349	0.45325	0.43347
9	0.71975	0.69931	0.67869	0.65793	0.63712	0.61631	0.59555	0.57490	0.55442	0.53415
10	0.79819	0.78061	0.76259	0.74418	0.72544	0.70644	0.68722	0.66784	0.64837	0.62884
11	0.85969	0.84539	0.83049	0.81504	0.79908	0.78266	0.76583	0.74862	0.73110	0.71330
12	0.90567	0.89459	0.88288	0.87054	0.85761	0.84412	0.83009	0.81556	0.80056	0.78513
13	0.93857	0.93038	0.92157	0.91216	0.90215	0.89155	0.88038	0.86865	0.85638	0.84360
14	0.96120	0.95538	0.94903	0.94215	0.93471	0.92673	0.91819	0.90911	0.89948	0.88933
15	0.97619	0.97222	0.96782	0.96269	0.95765	0.95186	0.94559	0.93882	0.93155	0.92378
16	0.98579	0.98317	0.98022	0.97693	0.97326	0.96921	0.96476	0.95989	0.95460	0.94887
17	0.99174	0.99007	0.98816	0.98599	0.98355	0.98081	0.97775	0.97437	0.97064	0.96655
18	0.99532	0.99429	0.99309	0.99171	0.99013	0.98833	0.98630	0.98402	0.98147	0.97864
19	0.99741	0.99679	0.99606	0.99521	0.99421	0.99307	0.99176	0.99026	0.98857	0.98667
20	0.99860	0.99824	0.99781	0.99729	0.99669	0.99598	0.99515	0.99420	0.99311	0.99187
21	0.99926	0.99905	0.99880	0.99850	0.99814	0.99771	0.99721	0.99662	0.99594	0.99514
22	0.99962	0.99950	0.99936	0.99919	0.99898	0.99873	0.99843	0.99807	0.99765	0.99716
23	0.99981	0.99974	0.99967	0.99957	0.99945	0.99931	0.99913	0.99892	0.99867	0.99837
24	0.99990	0.99987	0.99983	0.99978	0.99971	0.99963	0.99953	0.99941	0.99926	0.99908
25	0.99995	0.99994	0.99991	0.99989	0.99985	0.99981	0.99975	0.99968	0.99960	0.99949
26	0.99998	0.99997	0.99996	0.99994	0.99992	0.99990	0.99987	0.99983	0.99978	0.99973
27	0.99999	0.99999	0.99998	0.99997	0.99996	0.99995	0.99993	0.99991	0.99989	0.99985
28		0.99999	0.99999	0.99999	0.99998	0.99998	0.99997	0.99996	0.99994	0.99992
29					0.99999	0.99999	0.99999	0.99998	0.99998	0.99997
30						0.99999	0.99999	0.99999	0.99999	0.99998

ν	$\chi^2 = 8.2$ $m = 4.1$	8.4 4.2	8.6 4.3	8.8 4.4	9.0 4.5	9.2 4.6	9.4 4.7	9.6 4.8	9.8 4.9	10.0 5.0
1	0.00419	0.00375	0.00336	0.00301	0.00270	0.00242	0.00217	0.00195	0.00175	0.00157
2	0.01657	0.01500	0.01357	0.01228	0.01111	0.01005	0.00910	0.00823	0.00745	0.00674
3	0.04205	0.03843	0.03511	0.03207	0.02929	0.02675	0.02442	0.02229	0.02034	0.01857
4	0.08452	0.07798	0.07191	0.06630	0.06110	0.05629	0.05184	0.04773	0.04394	0.04043
5	0.14555	0.13553	0.12612	0.11731	0.10906	0.10135	0.09413	0.08740	0.08110	0.07524
6	0.22381	0.21024	0.19736	0.18514	0.17358	0.16264	0.15230	0.14254	0.13333	0.12465
7	0.31529	0.29865	0.28266	0.26734	0.25266	0.23861	0.22520	0.21240	0.20019	0.18857
8	0.41418	0.39540	0.37715	0.35945	0.34230	0.32571	0.30968	0.29423	0.27935	0.26503
9	0.51412	0.49439	0.47499	0.45594	0.43727	0.41902	0.40120	0.38383	0.36692	0.35049
10	0.60931	0.58983	0.57044	0.55118	0.53210	0.51323	0.49461	0.47626	0.45821	0.44049
11	0.69528	0.67709	0.65876	0.64035	0.62189	0.60344	0.58502	0.56669	0.54846	0.53039
12	0.76931	0.75314	0.73666	0.71991	0.70293	0.68576	0.66844	0.65101	0.63350	0.61596
13	0.83033	0.81660	0.80244	0.78788	0.77294	0.75768	0.74211	0.72627	0.71020	0.69393
14	0.87865	0.86746	0.85579	0.84365	0.83105	0.81803	0.80461	0.79081	0.77666	0.76218
15	0.91551	0.90675	0.89749	0.88774	0.87752	0.86683	0.85569	0.84412	0.83213	0.81974
16	0.94269	0.93606	0.92897	0.92142	0.91341	0.90495	0.89603	0.88667	0.87686	0.86663
17	0.96208	0.95723	0.95198	0.94633	0.94026	0.93378	0.92687	0.91954	0.91179	0.90361
18	0.97551	0.97207	0.96830	0.96420	0.95974	0.95493	0.94974	0.94418	0.93824	0.93191
19	0.98454	0.98217	0.97955	0.97666	0.97348	0.97001	0.96623	0.96213	0.95771	0.95295
20	0.99046	0.98887	0.98709	0.98511	0.98291	0.98047	0.97779	0.97486	0.97166	0.96817
21	0.99424	0.99320	0.99203	0.99070	0.98921	0.98755	0.98570	0.98365	0.98139	0.97891
22	0.99659	0.99593	0.99518	0.99431	0.99333	0.99222	0.99098	0.98958	0.98803	0.98630
23	0.99802	0.99761	0.99714	0.99659	0.99596	0.99524	0.99442	0.99349	0.99245	0.99128
24	0.99888	0.99863	0.99833	0.99799	0.99760	0.99714	0.99661	0.99601	0.99532	0.99455
25	0.99937	0.99922	0.99905	0.99884	0.99860	0.99831	0.99798	0.99760	0.99716	0.99665
26	0.99966	0.99957	0.99947	0.99934	0.99919	0.99902	0.99882	0.99858	0.99830	0.99798
27	0.99981	0.99977	0.99971	0.99963	0.99955	0.99944	0.99932	0.99917	0.99900	0.99880
28	0.99990	0.99987	0.99984	0.99980	0.99975	0.99969	0.99962	0.99953	0.99942	0.99930
29	0.99995	0.99993	0.99991	0.99989	0.99986	0.99983	0.99979	0.99973	0.99967	0.99960
30	0.99997	0.99997	0.99996	0.99994	0.99993	0.99991	0.99988	0.99985	0.99982	0.99977

Table A.2† Tables for evaluation of the CDF of the χ^2, gamma, and Poisson distributions (*Continued*)

ν	$\chi^2 = 10.5$ $m = 5.25$	11.0 5.5	11.5 5.75	12.0 6.0	12.5 6.25	13.0 6.5	13.5 6.75	14.0 7.0	14.5 7.25	15.0 7.5
1	0.00119	0.00091	0.00070	0.00053	0.00041	0.00031	0.00024	0.00018	0.00014	0.00011
2	0.00525	0.00409	0.00318	0.00248	0.00193	0.00150	0.00117	0.00091	0.00071	0.00055
3	0.01476	0.01173	0.00931	0.00738	0.00585	0.00464	0.00367	0.00291	0.00230	0.00182
4	0.03280	0.02656	0.02148	0.01735	0.01400	0.01128	0.00907	0.00730	0.00586	0.00470
5	0.06225	0.05138	0.04232	0.03479	0.02854	0.02338	0.01912	0.01561	0.01273	0.01036
6	0.10511	0.08838	0.07410	0.06197	0.05170	0.04304	0.03575	0.02964	0.02452	0.02026
7	0.16196	0.13862	0.11825	0.10056	0.08527	0.07211	0.06082	0.05118	0.04297	0.03600
8	0.23167	0.20170	0.17495	0.15120	0.13025	0.11185	0.09577	0.08177	0.06963	0.05915
9	0.31154	0.27571	0.24299	0.21331	0.18657	0.16261	0.14126	0.12233	0.10562	0.09094
10	0.39777	0.35752	0.31991	0.28506	0.25299	0.22367	0.19704	0.17299	0.15138	0.13206
11	0.48605	0.44326	0.40237	0.36364	0.32726	0.29333	0.26190	0.23299	0.20655	0.18250
12	0.57218	0.52892	0.48662	0.45468	0.40640	0.36904	0.33377	0.30071	0.26992	0.24144
13	0.65263	0.61082	0.56901	0.52764	0.48713	0.44781	0.40997	0.37384	0.33960	0.30735
14	0.72479	0.68604	0.64639	0.60630	0.56622	0.52652	0.48759	0.44971	0.41316	0.37815
15	0.78717	0.75259	0.71641	0.67903	0.64086	0.60230	0.56374	0.52553	0.48800	0.45142
16	0.83925	0.80949	0.77762	0.74398	0.70890	0.67276	0.63591	0.59871	0.56152	0.52464
17	0.88135	0.85656	0.82942	0.80014	0.76896	0.73619	0.70212	0.66710	0.63145	0.59548
18	0.91436	0.89436	0.87195	0.84724	0.82038	0.79157	0.76106	0.72909	0.69596	0.66197
19	0.93952	0.92384	0.90587	0.88562	0.86316	0.83857	0.81202	0.78369	0.75380	0.72260
20	0.95817	0.94622	0.93221	0.91608	0.89779	0.87738	0.85492	0.83050	0.80427	0.77641
21	0.97166	0.96279	0.95214	0.93962	0.92513	0.90862	0.89010	0.86960	0.84718	0.82295
22	0.98118	0.97475	0.96686	0.95738	0.94618	0.93316	0.91827	0.90148	0.88279	0.86224
23	0.98773	0.98319	0.97748	0.97047	0.96201	0.95199	0.94030	0.92687	0.91165	0.89463
24	0.99216	0.98901	0.98498	0.97991	0.97367	0.96612	0.95715	0.94665	0.93454	0.92076
25	0.99507	0.99295	0.99015	0.98657	0.98206	0.97650	0.96976	0.96173	0.95230	0.94138
26	0.99696	0.99555	0.99366	0.99117	0.98798	0.98397	0.97902	0.97300	0.96581	0.95733
27	0.99815	0.99724	0.99598	0.99429	0.99208	0.98925	0.98567	0.98125	0.97588	0.96943
28	0.99890	0.99831	0.99749	0.99637	0.99487	0.99290	0.99037	0.98719	0.98324	0.97844
29	0.99935	0.99899	0.99846	0.99773	0.99672	0.99538	0.99363	0.99138	0.98854	0.98502
30	0.99963	0.99940	0.99907	0.99860	0.99794	0.99704	0.99585	0.99428	0.99227	0.98974

ν	$\chi^2 = 15.5$ $m = 7.75$	16.0 8.0	16.5 8.25	17.0 8.5	17.5 8.75	18.0 9.0	18.5 9.25	19.0 9.5	19.5 9.75	20.0 10.0
1	0.00008	0.00006	0.00005	0.00004	0.00003	0.00002	0.00002	0.00001	0.00001	0.00001
2	0.00043	0.00034	0.00026	0.00020	0.00016	0.00012	0.00010	0.00008	0.00006	0.00005
3	0.00144	0.00113	0.00090	0.00071	0.00056	0.00044	0.00035	0.00027	0.00022	0.00017
4	0.00377	0.00302	0.00242	0.00193	0.00154	0.00123	0.00099	0.00079	0.00063	0.00050
5	0.00843	0.00684	0.00555	0.00450	0.00364	0.00295	0.00238	0.00192	0.00155	0.00125
6	0.01670	0.01375	0.01131	0.00928	0.00761	0.00623	0.00510	0.00416	0.00340	0.00277
7	0.03010	0.02512	0.02092	0.01740	0.01444	0.01197	0.00991	0.00819	0.00676	0.00557
8	0.05012	0.04238	0.03576	0.03011	0.02530	0.02123	0.01777	0.01486	0.01240	0.01034
9	0.07809	0.06688	0.05715	0.04872	0.04144	0.03517	0.02980	0.02519	0.02126	0.01791
10	0.11487	0.09963	0.08619	0.07436	0.06401	0.05496	0.04709	0.04026	0.03435	0.02925
11	0.16073	0.14113	0.12356	0.10788	0.09393	0.08158	0.07068	0.06109	0.05269	0.04534
12	0.21522	0.19124	0.16939	0.14960	0.13174	0.11569	0.10133	0.08853	0.07716	0.06709
13	0.27719	0.24913	0.22318	0.19930	0.17744	0.15752	0.13944	0.12310	0.10840	0.09521
14	0.34485	0.31337	0.28380	0.25618	0.23051	0.20678	0.18495	0.16495	0.14671	0.13014
15	0.41604	0.38205	0.34962	0.31886	0.28986	0.26267	0.23729	0.21373	0.19196	0.17193
16	0.48837	0.45296	0.41864	0.38560	0.35398	0.32390	0.29544	0.26866	0.24359	0.22022
17	0.55951	0.52383	0.48871	0.45437	0.42102	0.38884	0.35797	0.32853	0.30060	0.27423
18	0.62740	0.59255	0.55770	0.52311	0.48902	0.45565	0.42320	0.39182	0.36166	0.33282
19	0.69033	0.65728	0.62370	0.58987	0.55603	0.52244	0.48931	0.45684	0.42521	0.39458
20	0.74712	0.71662	0.68516	0.65297	0.62031	0.58741	0.55451	0.52183	0.48957	0.45793
21	0.79705	0.76965	0.74093	0.71111	0.68039	0.64900	0.61718	0.58514	0.55310	0.52126
22	0.83990	0.81589	0.79032	0.76336	0.73519	0.70599	0.67597	0.64533	0.61428	0.58304
23	0.87582	0.85527	0.83304	0.80925	0.78402	0.75749	0.72983	0.70122	0.67185	0.64191
24	0.90527	0.88808	0.86919	0.84866	0.82657	0.80301	0.77810	0.75199	0.72483	0.69678
25	0.92891	0.91483	0.89912	0.88179	0.86287	0.84239	0.82044	0.79712	0.77254	0.74683
26	0.94749	0.93620	0.92341	0.90908	0.89320	0.87577	0.85683	0.83643	0.81464	0.79156
27	0.96182	0.95295	0.94274	0.93112	0.91806	0.90352	0.88750	0.87000	0.85107	0.83076
28	0.97266	0.96582	0.95782	0.94859	0.93805	0.92615	0.91285	0.89814	0.88200	0.86446
29	0.98071	0.97554	0.96939	0.96218	0.95383	0.94427	0.93344	0.92129	0.90779	0.89293
30	0.98659	0.98274	0.97810	0.97258	0.96608	0.95853	0.94986	0.94001	0.92891	0.91654

Table A.2†   Tables for evaluation of the CDF of the χ^2, gamma, and Poisson distributions (*Continued*)

ν	$\chi^2 = 21$ $m = 10.5$	22 11.0	23 11.5	24 12.0	25 12.5	26 13.0	27 13.5	28 14.0	29 14.5	30 15.0
1	0.00001									
2	0.00003	0.00002	0.00001	0.00001						
3	0.00011	0.00007	0.00004	0.00003	0.00002	0.00001	0.00001			
4	0.00032	0.00020	0.00013	0.00008	0.00005	0.00003	0.00002	0.00001	0.00001	0.00001
5	0.00081	0.00052	0.00034	0.00022	0.00014	0.00009	0.00006	0.00004	0.00002	0.00002
6	0.00184	0.00121	0.00080	0.00052	0.00034	0.00022	0.00015	0.00009	0.00006	0.00004
7	0.00377	0.00254	0.00171	0.00114	0.00076	0.00050	0.00033	0.00022	0.00015	0.00010
8	0.00715	0.00492	0.00336	0.00229	0.00155	0.00105	0.00071	0.00047	0.00032	0.00021
9	0.01265	0.00888	0.00620	0.00430	0.00297	0.00204	0.00140	0.00095	0.00065	0.00044
10	0.02109	0.01511	0.01075	0.00760	0.00535	0.00374	0.00260	0.00181	0.00125	0.00086
11	0.03337	0.02437	0.01768	0.01273	0.00912	0.00649	0.00460	0.00324	0.00227	0.00159
12	0.05038	0.03752	0.02773	0.02034	0.01482	0.01073	0.00773	0.00553	0.00394	0.00279
13	0.07293	0.05536	0.04168	0.03113	0.02308	0.01700	0.01244	0.00905	0.00655	0.00471
14	0.10163	0.07861	0.06027	0.04582	0.03457	0.02589	0.01925	0.01423	0.01045	0.00763
15	0.13683	0.10780	0.08414	0.06509	0.04994	0.03802	0.02874	0.02157	0.01609	0.01192
16	0.17851	0.14319	0.11374	0.08950	0.06982	0.05403	0.04148	0.03162	0.02394	0.01800
17	0.22629	0.18472	0.14925	0.11944	0.09471	0.07446	0.05807	0.04494	0.03453	0.02635
18	0.27941	0.23199	0.19059	0.15503	0.12492	0.09976	0.07900	0.06206	0.04838	0.03745
19	0.33680	0.28426	0.23734	0.19615	0.16054	0.13019	0.10465	0.08343	0.06599	0.05180
20	0.39713	0.34051	0.28880	0.24239	0.20143	0.16581	0.13526	0.10940	0.08776	0.06985
21	0.45894	0.39951	0.34398	0.29306	0 24716	0.20645	0.17085	0.14015	0.11400	0.09199
22	0.52074	0.45989	0.40173	0.34723	0.29707	0.25168	0.21123	0.17568	0.14486	0.11846
23	0.58109	0.52025	0.46077	0.40381	0.35029	0.30087	0.25597	0.21578	0.18031	0.14940
24	0.63873	0.57927	0.51980	0.46160	0.40576	0.35317	0.30445	0.26004	0.22013	0.18475
25	0.69261	0.63574	0.57756	0.51937	0.46237	0.40760	0.35588	0.30785	0.26392	0.22429
26	0.74196	0.68870	0.63295	0.57597	0.51898	0.46311	0.40933	0.35846	0.31108	0.26761
27	0.78629	0.73738	0.68501	0.63032	0.57446	0.51860	0.46379	0.41097	0.36090	0.31415
28	0.82535	0.78129	0.73304	0.68154	0.62784	0.57305	0.51825	0.46445	0.41253	0.36322
29	0.85915	0.82019	0.77654	0.72893	0.67825	0.62549	0.57171	0.51791	0.46507	0.41400
30	0.88789	0.85404	0.81526	0.77203	0.72503	0.67513	0.62327	0.57044	0.51760	0.46565

ν	$\chi^2 = 31$ $m = 15.5$	32 16.0	33 16.5	34 17.0	35 17.5	36 18.0	37 18.5	38 19.0	39 19.5	40 20.0
5	0.00001	0.00001								
6	0.00003	0.00002	0.00001	0.00001						
7	0.00006	0.00004	0.00003	0.00002	0.00001	0.00001				
8	0.00014	0.00009	0.00006	0.00004	0.00003	0.00002	0.00001	0.00001		
9	0.00030	0.00020	0.00013	0.00009	0.00006	0.00004	0.00003	0.00002	0.00001	0.00001
10	0.00059	0.00040	0.00027	0.00019	0.00012	0.00008	0.00006	0.00004	0.00003	0.00002
11	0.00110	0.00076	0.00053	0.00036	0.00025	0.00017	0.00012	0.00008	0.00005	0.00004
12	0.00197	0.00138	0.00097	0.00068	0.00047	0.00032	0.00022	0.00015	0.00011	0.00007
13	0.00337	0.00240	0.00170	0.00120	0.00085	0.00059	0.00041	0.00029	0.00020	0.00014
14	0.00554	0.00401	0.00288	0.00206	0.00147	0.00104	0.00074	0.00052	0.00036	0.00026
15	0.00878	0.00644	0.00469	0.00341	0.00246	0.00177	0.00127	0.00090	0.00064	0.00045
16	0.01346	0.01000	0.00739	0.00543	0.00397	0.00289	0.00210	0.00151	0.00109	0.00078
17	0.01997	0.01505	0.01127	0.00840	0.00622	0.00459	0.00337	0.00246	0.00179	0.00129
18	0.02879	0.02199	0.01669	0.01260	0.00945	0.00706	0.00524	0.00387	0.00285	0.00209
19	0.04037	0.03125	0.02404	0.01838	0.01397	0.01056	0.00793	0.00593	0.00442	0.00327
20	0.05519	0.04330	0.03374	0.02613	0.02010	0.01538	0.01170	0.00886	0.00667	0.00500
21	0.07366	0.05855	0.04622	0.03624	0.02824	0.02187	0.01683	0.01289	0.00981	0.00744
22	0.09612	0.07740	0.06187	0.04912	0.03875	0.03037	0.02366	0.01832	0.01411	0.01081
23	0.12279	0.10014	0.08107	0.06516	0.05202	0.04125	0.03251	0.02547	0.01984	0.01537
24	0.15378	0.12699	0.10407	0.08467	0.06840	0.05489	0.04376	0.03467	0.02731	0.02139
25	0.18902	0.15801	0.13107	0.10791	0.08820	0.07160	0.05774	0.04626	0.03684	0.02916
26	0.22827	0.19312	0.16210	0.13502	0.11165	0.09167	0.07475	0.06056	0.04875	0.03901
27	0.27114	0.23208	0.19707	0.16605	0.13887	0.11530	0.09507	0.07786	0.06336	0.05124
28	0.31708	0.27451	0.23574	0.20087	0.16987	0.14260	0.11886	0.09840	0.08092	0.06613
29	0.36542	0.31987	0.27774	0.23926	0.20454	0.17356	0.14622	0.12234	0.10166	0.08394
30	0.41541	0.36753	0.32254	0.28083	0.24264	0.20808	0.17714	0.14975	0.12573	0.10486

Table A.2† Tables for evaluation of the CDF of the χ^2, gamma, and Poisson distributions (*Continued*)

ν	$\chi^2 = 42$ $m = 21$	44 22	46 23	48 24	50 25	52 26	54 27	56 28	58 29	60 30
10	0.00001									
11	0.00002	0.00001								
12	0.00003	0.00002	0.00001							
13	0.00006	0.00003	0.00001	0.00001						
14	0.00012	0.00006	0.00003	0.00001	0.00001					
15	0.00023	0.00011	0.00005	0.00003	0.00001	0.00001				
16	0.00040	0.00020	0.00010	0.00005	0.00002	0.00001	0.00001			
17	0.00067	0.00034	0.00017	0.00009	0.00004	0.00002	0.00001	0.00001		
18	0.00111	0.00058	0.00030	0.00015	0.00008	0.00004	0.00002	0.00001		
19	0.00177	0.00094	0.00050	0.00026	0.00013	0.00007	0.00003	0.00002	0.00001	
20	0.00277	0.00151	0.00081	0.00043	0.00022	0.00011	0.00006	0.00003	0.00001	0.00001
21	0.00421	0.00234	0.00128	0.00069	0.00036	0.00019	0.00010	0.00005	0.00003	0.00001
22	0.00625	0.00355	0.00198	0.00109	0.00059	0.00031	0.00016	0.00009	0.00004	0.00002
23	0.00908	0.00526	0.00299	0.00167	0.00092	0.00050	0.00027	0.00014	0.00007	0.00004
24	0.01291	0.00763	0.00443	0.00252	0.00142	0.00078	0.00043	0.00023	0.00012	0.00006
25	0.01797	0.01085	0.00642	0.00373	0.00213	0.00120	0.00066	0.00036	0.00020	0.00011
26	0.02455	0.01512	0.00912	0.00540	0.00314	0.00180	0.00102	0.00056	0.00031	0.00017
27	0.03292	0.02068	0.01272	0.00768	0.00455	0.00265	0.00152	0.00086	0.00048	0.00026
28	0.04336	0.02779	0.01743	0.01072	0.00647	0.00384	0.00224	0.00129	0.00073	0.00041
29	0.05616	0.03670	0.02346	0.01470	0.00903	0.00545	0.00324	0.00189	0.00109	0.00062
30	0.07157	0.04769	0.03107	0.01983	0.01240	0.00762	0.00460	0.00273	0.00160	0.00092

ν	$\chi^2 = 62$ $m = 31$	64 32	66 33	68 34	70 35	72 36	74 37	76 38
21	0.00001							
22	0.00001	0.00001						
23	0.00002	0.00001	0.00001					
24	0.00003	0.00002	0.00001					
25	0.00006	0.00003	0.00002	0.00001				
26	0.00009	0.00005	0.00003	0.00001	0.00001			
27	0.00014	0.00008	0.00004	0.00002	0.00001	0.00001		
28	0.00023	0.00012	0.00007	0.00004	0.00002	0.00001	0.00001	
29	0.00035	0.00019	0.00011	0.00006	0.00003	0.00002	0.00001	
30	0.00052	0.00029	0.00016	0.00009	0.00005	0.00003	0.00001	0.00001

† Compiled from E. S. Pearson and H. O. Hartley (eds.) [1954], "Biometrika Tables for Statisticians," vol. 1, Cambridge University Press, Cambridge, England (by permission).
Note: See Sec. 3.4.2 for illustrations of application. Tables yield:

1. $1 - F(\chi^2)$ for a χ^2 distribution with ν degrees of freedom.
2. $1 - F(y)$, where $y = \chi^2/2\lambda$, for a gamma distribution with parameters $k = \nu/2$ and λ. Enter table with $\nu = 2k$ and $\chi^2 = 2\lambda y$.
3. $F(u)$, where $u = (\frac{1}{2})\nu - 1$, for a Poisson distribution with mean $m = \frac{1}{2}\chi^2$. Enter table with m and $\nu = 2(u + 1)$.

For $\nu \geq 30$, the χ^2 distribution is approximately normal, with mean ν and variance 2ν. Somewhat more accurately (Hald [1952]) for $\nu \geq 30$, the value $\chi_p{}^2$ such that $F(\chi_p{}^2) = p$ is approximately $\chi_p{}^2 = \frac{1}{2}(\sqrt{2\nu - 1} + u_p)^2$, in which u_p is the value from the standardized normal Table A.1 such that $F(u_p) = p$.

Table A.3† Cumulative distribution of Student's t distribution

t \ ν	1	2	3	4	5	6	7	8	9	10
0.0	0.50000	0.50000	0.50000	0.50000	0.50000	0.50000	0.50000	0.50000	0.50000	0.50000
0.1	0.53173	0.53527	0.53667	0.53742	0.53788	0.53820	0.53843	0.53860	0.53873	0.53884
0.2	0.56283	0.57002	0.57286	0.57438	0.57532	0.57596	0.57642	0.57676	0.57704	0.57726
0.3	0.59277	0.60376	0.60812	0.61044	0.61188	0.61285	0.61356	0.61409	0.61450	0.61484
0.4	0.62112	0.63608	0.64203	0.64520	0.64716	0.64850	0.64946	0.65019	0.65076	0.65122
0.5	0.64758	0.66667	0.67428	0.67834	0.68085	0.68256	0.68380	0.68473	0.68546	0.68605
0.6	0.67202	0.69529	0.70460	0.70958	0.71267	0.71477	0.71629	0.71745	0.71835	0.71907
0.7	0.69440	0.72181	0.73284	0.73875	0.74243	0.74493	0.74674	0.74811	0.74919	0.75006
0.8	0.71478	0.74618	0.75890	0.76574	0.76999	0.77289	0.77500	0.77659	0.77784	0.77885
0.9	0.73326	0.76845	0.78277	0.79050	0.79531	0.79860	0.80099	0.80280	0.80422	0.80536
1.0	0.75000	0.78868	0.80450	0.81305	0.81839	0.82204	0.82469	0.82670	0.82828	0.82955
1.1	0.76515	0.80698	0.82416	0.83346	0.83927	0.84325	0.84614	0.84834	0.85006	0.85145
1.2	0.77886	0.82349	0.84187	0.85182	0.85805	0.86232	0.86541	0.86777	0.86961	0.87110
1.3	0.79129	0.83838	0.85777	0.86827	0.87485	0.87935	0.88262	0.88510	0.88705	0.88862
1.4	0.80257	0.85177	0.87200	0.88295	0.88980	0.89448	0.89788	0.90046	0.90249	0.90412
1.5	0.81283	0.86380	0.88471	0.89600	0.90305	0.90786	0.91135	0.91400	0.91608	0.91775
1.6	0.82219	0.87464	0.89605	0.90758	0.91475	0.91964	0.92318	0.92587	0.92797	0.92966
1.7	0.83075	0.88439	0.90615	0.91782	0.92506	0.92998	0.93354	0.93622	0.93833	0.94002
1.8	0.83859	0.89317	0.91516	0.92688	0.93412	0.93902	0.94256	0.94522	0.94731	0.94897
1.9	0.84579	0.90109	0.92318	0.93488	0.94207	0.94691	0.95040	0.95302	0.95506	0.95669
2.0	0.85242	0.90825	0.93034	0.94194	0.94903	0.95379	0.95719	0.95974	0.96172	0.96331
2.1	0.85854	0.91473	0.93672	0.94817	0.95512	0.95976	0.96306	0.96553	0.96744	0.96896
2.2	0.86420	0.92060	0.94241	0.95367	0.96045	0.96495	0.96813	0.97050	0.97233	0.97378
2.3	0.86945	0.92593	0.94751	0.95853	0.96511	0.96945	0.97250	0.97476	0.97650	0.97787
2.4	0.87433	0.93077	0.95206	0.96282	0.96919	0.97335	0.97627	0.97841	0.98005	0.98134
2.5	0.87888	0.93519	0.95615	0.96662	0.97275	0.97674	0.97950	0.98153	0.98307	0.98428
2.6	0.88313	0.93923	0.95981	0.96998	0.97587	0.97967	0.98229	0.98419	0.98563	0.98675
2.7	0.88709	0.94292	0.96311	0.97295	0.97861	0.98221	0.98468	0.98646	0.98780	0.98884
2.8	0.89081	0.94630	0.96607	0.97559	0.98100	0.98442	0.98674	0.98840	0.98964	0.99060
2.9	0.89430	0.94941	0.96875	0.97794	0.98310	0.98633	0.98851	0.99005	0.99120	0.99208
3.0	0.89758	0.95227	0.97116	0.98003	0.98495	0.98800	0.99003	0.99146	0.99252	0.99333
3.1	0.90067	0.95490	0.97335	0.98189	0.98657	0.98944	0.99134	0.99267	0.99364	0.99437
3.2	0.90359	0.95733	0.97533	0.98355	0.98800	0.99070	0.99247	0.99369	0.99459	0.99525
3.3	0.90634	0.95958	0.97713	0.98503	0.98926	0.99180	0.99344	0.99457	0.99539	0.99599
3.4	0.90895	0.96166	0.97877	0.98636	0.99037	0.99275	0.99428	0.99532	0.99606	0.99661
3.5	0.91141	0.96358	0.98026	0.98755	0.99136	0.99359	0.99500	0.99596	0.99664	0.99714
3.6	0.91376	0.96538	0.98162	0.98862	0.99223	0.99432	0.99563	0.99651	0.99713	0.99758
3.7	0.91598	0.96705	0.98286	0.98958	0.99300	0.99496	0.99617	0.99698	0.99754	0.99795
3.8	0.91809	0.96860	0.98400	0.99045	0.99369	0.99552	0.99664	0.99738	0.99789	0.99826
3.9	0.92010	0.97005	0.98504	0.99123	0.99430	0.99601	0.99705	0.99773	0.99819	0.99852
4.0	0.92202	0.97141	0.98600	0.99193	0.99484	0.99644	0.99741	0.99803	0.99845	0.99874
4.2	0.92560	0.97386	0.98768	0.99315	0.99575	0.99716	0.99798	0.99850	0.99885	0.99909
4.4	0.92887	0.97602	0.98912	0.99415	0.99649	0.99772	0.99842	0.99886	0.99914	0.99933
4.6	0.93186	0.97792	0.99034	0.99498	0.99708	0.99815	0.99876	0.99912	0.99936	0.99951
4.8	0.93462	0.97962	0.99140	0.99568	0.99756	0.99850	0.99902	0.99932	0.99951	0.99964
5.0	0.93717	0.98113	0.99230	0.99625	0.99795	0.99877	0.99922	0.99947	0.99963	0.99973
5.2	0.93952	0.98248	0.99309	0.99674	0.99827	0.99899	0.99937	0.99959	0.99972	0.99980
5.4	0.94171	0.98369	0.99378	0.99715	0.99853	0.99917	0.99950	0.99968	0.99978	0.99985
5.6	0.94375	0.98478	0.99437	0.99750	0.99875	0.99931	0.99959	0.99975	0.99983	0.99989
5.8	0.94565	0.98577	0.99490	0.99780	0.99893	0.99942	0.99967	0.99980	0.99987	0.99991
6.0	0.94743	0.98666	0.99536	0.99806	0.99908	0.99952	0.99973	0.99984	0.99990	.099993
6.2	0.94910	0.98748	0.99577	0.99828	0.99920	0.99959	0.99978	0.99987	0.99992	0.99995
6.4	0.95066	0.98822	0.99614	0.99847	0.99931	0.99966	0.99982	0.99990	0.99994	0.99996
6.6	0.95214	0.98890	0.99646	0.99863	0.99940	0.99971	0.99985	0.99992	0.99995	0.99997
6.8	0.95352	0.98953	0.99675	0.99878	0.99948	0.99975	0.99987	0.99993	0.99996	0.99998
7.0	0.95483	0.99010	0.99701	0.99890	0.99954	0.99979	0.99990	0.99994	0.99997	0.99998
7.2	0.95607	0.99063	0.99724	0.99901	0.99960	0.99982	0.99991	0.99995	0.99997	0.99999
7.4	0.95724	0.99111	0.99745	0.99911	0.99964	0.99984	0.99993	0.99996	0.99998	0.99999
7.6	0.95836	0.99156	0.99764	0.99920	0.99969	0.99986	0.99994	0.99997	0.99998	0.99999
7.8	0.95941	0.99198	0.99781	0.99927	0.99972	0.99988	0.99995	0.99997	0.99999	0.99999
8.0	0.96042	0.99237	0.99796	0.99934	0.99975	0.99990	0.99996	0.99998	0.99999	0.99999

† Compiled from E. S. Pearson and H. O. Hartley (eds.) [1954], "Biometrika Tables for Statisticians," vol. 1, Cambridge University Press, Cambridge, England (by permission).

Table A.3 Cumulative distribution of Student's t distribution (*Continued*)

t \ ν	11	12	13	14	15	16	17	18	19	20
0.0	0.50000	0.50000	0.50000	0.50000	0.50000	0.50000	0.50000	0.50000	0.50000	0.50000
0.1	0.53893	0.53900	0.53907	0.53912	0.53917	0.53921	0.53924	0.53928	0.53930	0.53933
0.2	0.57744	0.57759	0.57771	0.57782	0.57792	0.57800	0.57807	0.57814	0.57820	0.57825
0.3	0.61511	0.61534	0.61554	0.61571	0.61585	0.61598	0.61609	0.61619	0.61628	0.61636
0.4	0.65159	0.65191	0.65217	0.65240	0.65260	0.65278	0.65293	0.65307	0.65319	0.65330
0.5	0.68654	0.68694	0.68728	0.68758	0.68783	0.68806	0.68826	0.68843	0.68859	0.68873
0.6	0.71967	0.72017	0.72059	0.72095	0.72127	0.72155	0.72179	0.72201	0.72220	0.72238
0.7	0.75077	0.75136	0.75187	0.75230	0.75268	0.75301	0.75330	0.75356	0.75380	0.75400
0.8	0.77968	0.78037	0.78096	0.78146	0.78190	0.78229	0.78263	0.78293	0.78320	0.78344
0.9	0.80630	0.80709	0.80776	0.80883	0.80883	0.80927	0.80965	0.81000	0.81031	0.81058
1.0	0.83060	0.83148	0.83222	0.83286	0.83341	0.83390	0.83433	0.83472	0.83506	0.83537
1.1	0.85259	0.85355	0.85436	0.85506	0.85566	0.85620	0.85667	0.85709	0.85746	0.85780
1.2	0.87233	0.87335	0.87422	0.87497	0.87562	0.87620	0.87670	0.87715	0.87756	0.87792
1.3	0.88991	0.89099	0.89191	0.89270	0.89339	0.89399	0.89452	0.89500	0.89542	0.89581
1.4	0.90546	0.90658	0.90754	0.90836	0.90907	0.90970	0.91025	0.91074	0.91118	0.91158
1.5	0.91912	0.92027	0.92125	0.92209	0.92282	0.92346	0.92402	0.92452	0.92498	0.92538
1.6	0.93105	0.93221	0.93320	0.93404	0.93478	0.93542	0.93599	0.93650	0.93695	0.93736
1.7	0.94140	0.94256	0.94354	0.94439	0.94512	0.94576	0.94632	0.94683	0.94728	0.94768
1.8	0.95034	0.95148	0.95245	0.95328	0.95400	0.95463	0.95518	0.95568	0.95612	0.95652
1.9	0.95802	0.95914	0.96008	0.96089	0.96158	0.96220	0.96273	0.96321	0.96364	0.96403
2.0	0.96460	0.96567	0.96658	0.96736	0.96803	0.96861	0.96913	0.96959	0.97000	0.97037
2.1	0.97020	0.97123	0.97209	0.97283	0.97347	0.97403	0.97452	0.97495	0.97534	0.97569
2.2	0.97496	0.97593	0.97675	0.97745	0.97805	0.97858	0.97904	0.97945	0.97981	0.98014
2.3	0.97898	0.97990	0.98067	0.98132	0.98189	0.98238	0.98281	0.98319	0.98352	0.98383
2.4	0.98238	0.98324	0.98396	0.98457	0.98509	0.98554	0.98594	0.98629	0.98660	0.98688
2.5	0.98525	0.98604	0.98671	0.98727	0.98775	0.98816	0.98853	0.98885	0.98913	0.98938
2.6	0.98765	0.98839	0.98900	0.98951	0.98995	0.99033	0.99066	0.99095	0.99121	0.99144
2.7	0.98967	0.99035	0.99090	0.99137	0.99177	0.99211	0.99241	0.99267	0.99290	0.99311
2.8	0.99136	0.99198	0.99249	0.99291	0.99327	0.99358	0.99385	0.99408	0.99429	0.99447
2.9	0.99278	0.99334	0.99380	0.99418	0.99450	0.99478	0.99502	0.99523	0.99541	0.99557
3.0	0.99396	0.99447	0.99488	0.99522	0.99551	0.99576	0.99597	0.99616	0.99632	0.99646
3.1	0.99495	0.99541	0.99578	0.99608	0.99634	0.99656	0.99675	0.99691	0.99705	0.99718
3.2	0.99577	0.99618	0.99652	0.99679	0.99702	0.99721	0.99738	0.99752	0.99764	0.99775
3.3	0.99646	0.99683	0.99713	0.99737	0.99757	0.99774	0.99789	0.99801	0.99812	0.99821
3.4	0.99703	0.99737	0.99763	0.99784	0.99802	0.99817	0.99830	0.99840	0.99850	0.99858
3.5	0.99751	0.99781	0.99804	0.99823	0.99839	0.99852	0.99863	0.99872	0.99880	0.99887
3.6	0.99791	0.99818	0.99838	0.99855	0.99869	0.99880	0.99890	0.99898	0.99905	0.99911
3.7	0.99825	0.99848	0.99867	0.99881	0.99893	0.99903	0.99911	0.99918	0.99924	0.99929
3.8	0.99853	0.99874	0.99890	0.99902	0.99913	0.99921	0.99928	0.99934	0.99939	0.99944
3.9	0.99876	0.99895	0.99909	0.99920	0.99929	0.99936	0.99942	0.99948	0.99952	0.99956
4.0	0.99896	0.99912	0.99924	0.99934	0.99942	0.99948	0.99954	0.99958	0.99962	0.99965
4.2	0.99926	0.99938	0.99948	0.99955	0.99961	0.99966	0.99970	0.99973	0.99976	0.99978
4.4	0.99947	0.99957	0.99964	0.99970	0.99974	0.99978	0.99980	0.99983	0.99985	0.99986
4.6	0.99962	0.99969	0.99975	0.99979	0.99983	0.99985	0.99987	0.99989	0.99990	0.99991
4.8	0.99972	0.99978	0.99983	0.99986	0.99988	0.99990	0.99992	0.99993	0.99994	0.99995
5.0	0.99980	0.99985	0.99988	0.99990	0.99992	0.99993	0.99995	0.99995	0.99996	0.99997
5.2	0.99985	0.99989	0.99992	0.99993	0.99995	0.99996	0.99996	0.99996	0.99997	0.99998
5.4	0.99989	0.99992	0.99994	0.99995	0.99996	0.99997	0.99997	0.99998	0.99998	0.99999
5.6	0.99992	0.99994	0.99996	0.99997	0.99997	0.99998	0.99998	0.99999	0.99999	0.99999
5.8	0.99994	0.99996	0.99997	0.99998	0.99998	0.99999	0.99999	0.99999	0.99999	0.99999
6.0	0.99995	0.99997	0.99998	0.99998	0.99999	0.99999	0.99999	0.99999		
6.2	0.99997	0.99998	0.99998	0.99999	0.99999	0.99999	0.99999			
6.4	0.99997	0.99998	0.99999	0.99999	0.99999					
6.6	0.99998	0.99999	0.99999	0.99999						
6.8	0.99998	0.99999	0.99999							
7.0	0.99999	0.99999								

Table A.3 Cumulative distribution of Student's t distribution (*Continued*)

t \ ν	20	21	22	23	24	30	40	60	120	∞
0.00	0.50000	0.50000	0.50000	0.50000	0.50000	0.50000	0.50000	0.50000	0.50000	0.50000
0.05	0.51969	0.51970	0.51971	0.51972	0.51973	0.51977	0.51981	0.51986	0.51990	0.51994
0.10	0.53933	0.53935	0.53938	0.53939	0.53941	0.53950	0.53958	0.53966	0.53974	0.53983
0.15	0.55887	0.55890	0.55893	0.55896	0.55899	0.55912	0.55924	0.55937	0.55949	0.55962
0.20	0.57825	0.57830	0.57834	0.57838	0.57842	0.57858	0.57875	0.57892	0.57909	0.57926
0.25	0.59743	0.59749	0.59755	0.59760	0.59764	0.59785	0.59807	0.59828	0.59849	0.59871
0.30	0.61636	0.61644	0.61650	0.61656	0.61662	0.61688	0.61713	0.61739	0.61765	0.61791
0.35	0.63500	0.63509	0.63517	0.63524	0.63530	0.63561	0.63591	0.63622	0.63652	0.63683
0.40	0.65330	0.65340	0.65349	0.65358	0.65365	0.65400	0.65436	0.65471	0.65507	0.65542
0.45	0.67122	0.67134	0.67144	0.67154	0.67163	0.67203	0.67243	0.67283	0.67324	0.67364
0.50	0.68873	0.68886	0.68898	0.68909	0.68919	0.68964	0.69009	0.69055	0.69100	0.69146
0.55	0.70579	0.70594	0.70607	0.70619	0.70630	0.70680	0.70731	0.70782	0.70833	0.70884
0.60	0.72238	0.72254	0.72268	0.72281	0.72294	0.72349	0.72405	0.72462	0.72518	0.72575
0.65	0.73846	0.73863	0.73879	0.73893	0.73907	0.73968	0.74030	0.74091	0.74153	0.74215
0.70	0.75400	0.75419	0.75437	0.75453	0.75467	0.75534	0.75601	0.75668	0.75736	0.75804
0.75	0.76901	0.76921	0.76940	0.76957	0.76973	0.77045	0.77118	0.77191	0.77264	0.77337
0.80	0.78344	0.78367	0.78387	0.78405	0.78422	0.78500	0.78578	0.78657	0.78735	0.78814
0.85	0.79731	0.79754	0.79776	0.79796	0.79814	0.79897	0.79981	0.80065	0.80149	0.80234
0.90	0.81058	0.81084	0.81107	0.81128	0.81147	0.81236	0.81325	0.81414	0.81504	0.81594
0.95	0.82327	0.82354	0.82378	0.82401	0.82421	0.82515	0.82609	0.82704	0.82799	0.82894
1.00	0.83537	0.83565	0.83591	0.83614	0.83636	0.83735	0.83834	0.83934	0.84034	0.84134
1.05	0.84688	0.84717	0.84744	0.84769	0.84791	0.84895	0.84999	0.85104	0.85209	0.85314
1.10	0.85780	0.85811	0.85839	0.85864	0.85888	0.85996	0.86105	0.86214	0.86323	0.86433
1.15	0.86814	0.86846	0.86875	0.86902	0.86926	0.87039	0.87151	0.87265	0.87378	0.87493
1.20	0.87792	0.87825	0.87855	0.87882	0.87907	0.88023	0.88140	0.88257	0.88375	0.88493
1.25	0.88714	0.88747	0.88778	0.88807	0.88832	0.88952	0.89072	0.89192	0.89313	0.89435
1.30	0.89581	0.89616	0.89647	0.89676	0.89703	0.89825	0.89948	0.90071	0.90195	0.90320
1.35	0.90395	0.90431	0.90463	0.90492	0.90519	0.90648	0.90770	0.90896	0.91022	0.91149
1.40	0.91158	0.91194	0.91227	0.91257	0.91285	0.91411	0.91539	0.91667	0.91795	0.91924
1.45	0.91872	0.91908	0.91942	0.91972	0.92000	0.92128	0.92257	0.92387	0.92517	0.92647
1.50	0.92538	0.92575	0.92608	0.92639	0.92667	0.92797	0.92927	0.93057	0.93188	0.93319
1.55	0.93159	0.93196	0.93230	0.93260	0.93289	0.93419	0.93549	0.93680	0.93811	0.93943
1.60	0.93736	0.93773	0.93807	0.93838	0.93866	0.93996	0.94127	0.94257	0.94389	0.94520
1.65	0.94272	0.94309	0.94342	0.94373	0.94401	0.94531	0.94661	0.94792	0.94922	0.95053
1.70	0.94768	0.94805	0.94839	0.94869	0.94897	0.95026	0.95155	0.95284	0.95414	0.95543
1.75	0.95228	0.95264	0.95297	0.95327	0.95355	0.95483	0.95611	0.95738	0.95866	0.95994
1.80	0.95652	0.95688	0.95720	0.95750	0.95778	0.95904	0.96030	0.96156	0.96281	0.96407
1.85	0.96043	0.96078	0.96110	0.96140	0.96167	0.96291	0.96414	0.96538	0.96661	0.96784
1.90	0.96403	0.96437	0.96469	0.96498	0.96524	0.96646	0.96767	0.96888	0.97008	0.97128
1.95	0.96733	0.96767	0.96798	0.96827	0.96852	0.96971	0.97089	0.97207	0.97325	0.97441
2.0	0.97037	0.97070	0.97100	0.97128	0.97153	0.97269	0.97384	0.97498	0.97612	0.97725
2.1	0.97569	0.97601	0.97629	0.97655	0.97679	0.97788	0.97896	0.98003	0.98109	0.98214
2.2	0.98014	0.98043	0.98070	0.98094	0.98116	0.98218	0.98318	0.98416	0.98514	0.98610
2.3	0.98383	0.98410	0.98435	0.98457	0.98478	0.98571	0.98663	0.98753	0.98841	9.98928
2.4	0.98688	0.98712	0.98735	0.98756	0.98774	0.98860	0.98943	0.99024	0.99103	0.99180
2.5	0.98938	0.98961	0.98982	0.99000	0.99017	0.99094	0.99169	0.99241	0.99312	0.99379
2.6	0.99144	0.99164	0.99183	0.99200	0.99215	0.99284	0.99350	0.99414	0.99475	0.99534
2.7	0.99311	0.99329	0.99346	0.99361	0.99375	0.99436	0.99494	0.99550	0.99603	0.99653
2.8	0.99447	0.99463	0.99478	0.99492	0.99504	0.99557	0.99608	0.99657	0.99702	0.99744
2.9	0.99557	0.99572	0.99585	0.99596	0.99607	0.99654	0.99698	0.99740	0.99778	0.99813
3.0	0.99646	0.99659	0.99670	0.99681	0.99690	0.99730	0.99768	0.99804	0.99836	0.99865
3.1	0.99718	0.99729	0.99739	0.99748	0.99756	0.99791	0.99823	0.99853	0.99879	0.99903
3.2	0.99775	0.99785	0.99793	0.99801	0.99808	0.99838	0.99865	0.99890	0.99912	0.99931
3.3	0.99821	0.99829	0.99837	0.99844	0.99849	0.99875	0.99898	0.99918	0.99936	0.99952
3.4	0.99858	0.99865	0.99871	0.99877	0.99882	0.99904	0.99923	0.99940	0.99954	0.99966
3.5	0.99887	0.99893	0.99899	0.99904	0.99908	0.99926	0.99942	0.99956	0.99967	0.99977
3.6	0.99911	0.99916	0.99920	0.99925	0.99928	0.99943	0.99957	0.99968	0.99977	0.99984
3.7	0.99929	0.99933	0.99937	0.99941	0.99944	0.99957	0.99967	0.99976	0.99984	0.99989
3.8	0.99944	0.99948	0.99951	0.99954	0.99956	0.99967	0.99976	0.99983	0.99989	0.99993
3.9	0.99956	0.99959	0.99961	0.99964	0.99966	0.99975	0.99982	0.99988	0.99992	0.99995
4.0	0.99965	0.99967	0.99970	0.99972	0.99974	0.99981	0.99987	0.99991	0.99995	0.99997
5.0	0.99997	0.99997	0.99998	0.99998	0.99998	0.99999	0.99999			

Table A.4† Properties of some standardized beta distributions

r	0.1		0.2		0.2		0.4		0.4	
t	0.5		0.5		1.0		1.0		2.0	
m	0.20		0.40		0.20		0.40		0.20	
σ^2	0.327		0.400		0.283		0.346		0.231	
σ^2	0.107		0.160		0.080		0.120		0.053	
V	1.63		1.00		1.41		0.866		1.155	
γ_1	1.47		0.40		1.41		0.385		1.300	
x	PDF	CDF	PDF	CDF	PDF	CDF	PDF	CDF	PDF	CDF
0.05	1.28	0.624	1.47	0.357	2.08	0.515	1.86	0.230	2.95	0.377
0.10	0.71	0.671	0.88	0.412	1.21	0.592	1.26	0.305	1.89	0.493
0.15	0.51	0.701	0.66	0.450	0.88	0.643	1.01	0.361	1.43	0.575
0.20	0.41	0.724	0.55	0.480	0.71	0.683	0.87	0.407	1.16	0.639
0.25	0.35	0.742	0.48	0.505	0.60	0.715	0.78	0.449	0.98	0.692
0.30	0.31	0.759	0.43	0.528	0.53	0.744	0.72	0.486	0.84	0.738
0.35	0.28	0.773	0.40	0.549	0.47	0.768	0.68	0.521	0.73	0.777
0.40	0.26	0.787	0.38	0.569	0.43	0.791	0.64	0.554	0.64	0.811
0.45	0.25	0.800	0.37	0.587	0.40	0.812	0.62	0.585	0.57	0.841
0.50	0.24	0.812	0.37	0.606	0.37	0.831	0.61	0.616	0.50	0.868
0.55	0.23	0.823	0.36	0.624	0.35	0.849	0.60	0.646	0.45	0.892
0.60	0.23	0.835	0.37	0.642	0.34	0.866	0.59	0.676	0.40	0.913
0.65	0.23	0.846	0.38	0.661	0.33	0.883	0.60	0.705	0.35	0.932
0.70	0.24	0.858	0.40	0.680	0.32	0.899	0.61	0.735	0.30	0.948
0.75	0.25	0.870	0.43	0.701	0.31	0.915	0.63	0.766	0.26	0.962
0.80	0.27	0.883	0.48	0.724	0.31	0.930	0.66	0.798	0.22	0.974
0.85	0.30	0.898	0.55	0.749	0.31	0.946	0.71	0.832	0.18	0.984
0.90	0.37	0.914	0.70	0.780	0.32	0.962	0.81	0.870	0.14	0.992
0.95	0.53	0.936	1.09	0.823	0.35	0.978	1.03	0.915	0.09	0.997

r	0.8		0.6		1.2		0.4		1.6	
t	2.0		3.0		3.0		4.0		4.0	
m	0.40		0.20		0.40		0.20		0.40	
σ	0.283		0.200		0.245		0.18		0.22	
σ^2	0.080		0.040		0.060		0.03		0.05	
V	0.707		1.00		0.612		0.89		0.55	
γ_1	0.354		1.20		0.327		1.12		0.30	
x	PDF	CDF	PDF	CDF	PDF	CDF	PDF	CDF	PDF	CDF
0.05	1.69	0.106	3.34	0.291	1.23	0.052	3.46	0.230	0.83	0.027
0.10	1.45	0.184	2.34	0.429	1.36	0.118	2.67	0.382	1.17	0.077
0.15	1.32	0.253	1.84	0.533	1.41	0.187	2.17	0.502	1.38	0.142
0.20	1.23	0.317	1.51	0.616	1.42	0.257	1.80	0.601	1.51	0.214
0.25	1.17	0.377	1.26	0.685	1.41	0.328	1.49	0.683	1.57	0.291
0.30	1.11	0.433	1.06	0.743	1.38	0.398	1.23	0.751	1.59	0.371
0.35	1.06	0.488	0.90	0.791	1.34	0.466	1.02	0.807	1.58	0.450
0.40	1.01	0.539	0.76	0.833	1.29	0.532	0.83	0.853	1.53	0.528
0.45	0.97	0.589	0.64	0.868	1.24	0.595	0.67	0.890	1.45	0.602
0.50	0.94	0.637	0.54	0.900	1.17	0.656	0.53	0.920	1.35	0.672
0.55	0.90	0.683	0.45	0.922	1.10	0.712	0.41	0.944	1.23	0.737
0.60	0.86	0.727	0.37	0.943	1.01	0.765	0.31	0.962	1.10	0.796
0.65	0.83	0.769	0.30	0.959	0.93	0.814	0.23	0.975	0.96	0.847
0.70	0.79	0.809	0.23	0.972	0.83	0.858	0.16	0.985	0.81	0.891
0.75	0.75	0.848	0.17	0.983	0.73	0.897	0.11	0.992	0.65	0.928
0.80	0.71	0.884	0.12	0.990	0.62	0.930	0.06	0.996	0.50	0.957
0.85	0.66	0.919	0.08	0.995	0.50	0.958	0.03	0.998	0.34	0.978
0.90	0.60	0.950	0.04	0.998	0.36	0.980	0.01	0.999	0.20	0.991
0.95	0.52	0.978	0.02	1.000	0.21	0.994	0.00	1.000	0.08	0.998

† See text for integer r and t.

Table A.5† Values of the standardized Type I extreme-value distribution (largest value) $F_{II}(w) = e^{-e^{-w}}$

w	CDF	PDF	w	CDF	PDF	w	CDF	PDF
−3.0		0.00000 00	−0.50	0.19229 56	0.31704 19	3.5	0.97025 40	0.02929 91
−2.9	0.00000 00	0.00000 02	−0.45	0.20839 66	0.32638 10	3.6	0.97304 62	0.02658 72
−2.8	0.00000 01	0.00000 12	−0.40	0.22496 18	0.33560 36	3.7	0.97557 96	0.02411 98
−2.7	0.00000 03	0.00000 51	−0.35	0.24193 95	0.34332 85	3.8	0.97787 76	0.02187 59
−2.6	0.00000 14	0.00001 91	−0.30	0.25927 69	0.34998 72	3.9	0.97996 16	0.01983 63
−2.5	0.00000 51	0.00006 24	−0.25	0.27692 03	0.35557 27	4.0	0.98185 11	0.01798 32
−2.40	0.00001 63	0.00017 99	−0.20	0.29481 63	0.36008 95	4.2	0.98511 63	0.01477 24
−2.35	0.00002 79	0.00029 29	−0.15	0.31291 17	0.36355 15	4.4	0.98779 77	0.01212 75
−2.30	0.00004 66	0.00046 47	−0.10	0.33115 43	0.36598 21	4.6	0.98999 85	0.00995 13
			−0.05	0.34949 32	0.36741 21	4.8	0.99180 40	0.00816 23
−2.25	0.00007 58	0.00071 89	0.0	0.36787 94	0.36787 94			
−2.20	0.00012 04	0.00108 63	0.1	0.40460 77	0.36610 42	5.0	0.99328 47	0.00669 27
−2.15	0.00018 69	0.00160 46	0.2	0.44099 10	0.36105 29	5.2	0.99449 86	0.00548 62
−2.10	0.00028 41	0.00232 00	0.3	0.47672 37	0.35316 56	5.4	0.99549 36	0.00449 62
−2.05	0.00042 31	0.00328 66	0.4	0.51154 48	0.34289 88	5.6	0.99630 90	0.00368 42
						5.8	0.99697 70	0.00301 84
−2.00	0.00061 80	0.00456 63	0.5	0.54523 92	0.33070 43			
−1.95	0.00088 61	0.00622 81	0.6	0.57763 58	0.31701 33	6.0	0.99752 43	0.00247 26
−1.90	0.00124 84	0.00834 67	0.7	0.60860 53	0.30222 45	6.2	0.99797 26	0.00202 53
−1.85	0.00172 97	0.01100 04	0.8	0.63805 62	0.28669 71	6.4	0.99833 98	0.00165 88
−1.80	0.00235 87	0.01426 93	0.9	0.66593 07	0.27074 72	6.6	0.99864 06	0.00135 85
						6.8	0.99888 68	0.00111 25
−1.75	0.00316 82	0.01823 15	1.0	0.69220 06	0.25464 64			
−1.70	0.00419 46	0.02296 12	1.1	0.71686 26	0.23862 28	7.0	0.99908 85	0.00091 11
−1.65	0.00547 82	0.02852 48	1.2	0.73993 41	0.22286 39	7.2	0.99925 37	0.00074 60
−1.60	0.00706 20	0.03497 81	1.3	0.76144 92	0.20751 91	7.4	0.99938 89	0.00061 09
−1.55	0.00899 15	0.04236 34	1.4	0.78145 56	0.19270 46	7.6	0.99949 97	0.00050 02
						7.8	0.99959 03	0.00040 96
−1.50	0.01131 43	0.05070 71	1.5	0.80001 07	0.17850 65			
−1.45	0.01407 84	0.06001 78	1.6	0.81717 95	0.16498 57	8.0	0.99966 46	0.00033 54
−1.40	0.01733 20	0.07028 48	1.7	0.83303 17	0.15218 12	8.5	0.99979 66	0.00020 34
−1.35	0.02112 23	0.08147 77	1.8	0.84764 03	0.14011 40	9.0	0.99987 66	0.00012 34
−1.30	0.02549 44	0.09354 65	1.9	0.86107 93	0.12879 04	9.5	0.99992 51	0.00007 48
−1.25	0.03049 04	0.10642 20	2.0	0.87342 30	0.11820 50	10.0	0.99995 46	0.00004 54
−1.20	0.03614 86	0.12001 76	2.1	0.88474 45	0.10834 26	10.5	0.99997 25	0.00002 75
−1.15	0.04250 25	0.13423 10	2.2	0.89511 49	0.09918 16	11.0	0.99998 33	0.00001 67
−1.10	0.04958 01	0.14894 68	2.3	0.90460 32	0.09069 45	11.5	0.99998 99	0.00001 01
−1.05	0.05740 34	0.16403 90	2.4	0.91327 53	0.08285 05	12.0	0.99999 39	0.00000 61
−1.00	0.06598 80	0.17937 41	2.5	0.92119 37	0.07561 62	12.5	0.99999 63	0.00000 37
−0.95	0.07534 26	0.19481 41	2.6	0.92841 77	0.06895 69	13.0	0.99999 77	0.00000 23
−0.90	0.08546 89	0.21021 95	2.7	0.93500 30	0.06283 74	13.5	0.99999 86	0.00000 14
−0.85	0.09636 17	0.22545 23	2.8	0.94100 20	0.05722 24	14.0	0.99999 92	0.00000 08
−0.80	0.10800 90	0.24037 84	2.9	0.94646 32	0.05207 75	14.5	0.99999 95	0.00000 05
−0.75	0.12039 23	0.25487 04	3.0	0.95143 20	0.04736 90	15.0	0.99999 97	0.00000 03
−0.70	0.13348 68	0.26880 94	3.1	0.95595 04	0.04306 48	15.5	0.99999 98	0.00000 02
−0.65	0.14726 22	0.28208 67	3.2	0.96005 74	0.03913 41	16.0	0.99999 99	0.00000 01
−0.60	0.16168 28	0.29460 53	3.3	0.96378 87	0.03554 76	16.5	0.99999 99	0.00000 01
−0.55	0.17670 86	0.30628 08	3.4	0.96717 75	0.03227 79	17.0	1.00000 00	0.00000 00

† *Source:* National Bureau of Standards [1953], "Probability Tables for the Analysis of Extreme Value Data," Applied Math Series 22, Washington, D.C.

Table A.6† *F* distribution; value of z such that $F_Z(z) = 0.95$

ν_2	ν_1												
	1	2	3	4	5	6	7	8	9	10	11	12	∞
1	161	200	216	225	230	234	237	239	241	242	243	244	254
2	18.51	19.00	19.16	19.25	19.30	19.33	19.36	19.37	19.38	19.39	19.40	19.41	19.50
3	10.13	9.55	9.28	9.12	9.01	8.94	8.88	8.84	8.81	8.78	8.76	8.74	8.53
4	7.71	6.94	6.59	6.39	6.26	6.16	6.09	6.04	6.00	5.96	5.93	5.91	5.63
5	6.61	5.79	5.41	5.19	5.05	4.95	4.88	4.82	4.78	4.74	4.70	4.68	4.36
6	5.99	5.14	4.76	4.53	4.39	4.28	4.21	4.15	4.10	4.06	4.03	4.00	3.67
7	5.59	4.74	4.35	4.12	3.97	3.87	3.79	3.72	3.68	3.63	3.60	3.57	3.23
8	5.32	4.46	4.07	3.84	3.69	3.58	3.50	3.44	3.39	3.34	3.31	3.28	2.93
9	5.12	4.26	3.86	3.63	3.48	3.37	3.29	3.23	3.18	3.13	3.10	3.07	2.71
10	4.96	4.10	3.71	3.48	3.33	3.22	3.14	3.07	3.02	2.97	2.94	2.91	2.54
11	4.84	3.98	3.59	3.36	3.20	3.09	3.01	2.95	2.90	2.86	2.82	2.79	2.40
12	4.75	3.88	3.49	3.26	3.11	3.00	2.92	2.85	2.80	2.76	2.72	2.69	2.30
∞	3.84	3.00	2.60	2.37	2.21	2.10	2.01	1.94	1.88	1.83	1.79	1.75	1.00

† Compiled with permission from R. S. Burington and D. C. May [1953], "Handbook of Probability and Statistics with Tables," McGraw-Hill Book Company, New York.

Table A.7 Critical statistic for the Kolmogorov-Smirnov goodness-of-fit test†

Sample size	$\alpha = 0.10$	$\alpha = 0.05$	$\alpha = 0.01$
5	0.51	0.56	0.67
10	0.37	0.41	0.49
15	0.30	0.34	0.40
20	0.26	0.29	0.35
25	0.24	0.26	0.32
30	0.22	0.24	0.29
40	0.19	0.21	0.25
Large n	$1.22/\sqrt{n}$	$1.36/\sqrt{n}$	$1.63/\sqrt{n}$

† Table abstracted from Lindgren [1962], "Statistical Theory", The Macmillan Company, New York, (with the permission of the author and publisher).

Table A.8†   Table of random digits

52478	22835	33307	73842	67277	32880	76457	94489	82597	04836
80249	16089	01964	21414	72117	91712	11487	67479	13649	94539
94132	15190	08425	70298	02202	80519	23516	86294	32871	89573
56605	86696	37707	90117	17511	27701	35764	88217	70505	75300
58815	01919	22225	38562	45731	91743	99315	70350	78240	22015
69379	89366	50240	49343	31867	81661	41037	59120	44282	66605
75228	79546	65528	48794	73980	87645	22604	49290	08068	54935
14327	93484	49875	12103	77984	97966	08644	07089	18809	33738
90625	98430	03639	76657	26389	99093	51145	59343	22488	67026
06070	44497	21962	48270	68632	68338	39325	35105	42348	14412
33415	72559	19902	40024	74215	93857	04988	24389	22094	89237
41999	12790	87990	77646	33177	62684	34119	09212	89973	39638
75908	62356	27342	93069	60284	69329	83998	15037	96165	62149
91323	56853	08468	69550	90860	57946	70370	23114	67185	04633
03428	01736	91578	09165	67708	36704	59481	28243	71395	38607
02333	25192	93932	65485	73266	95972	72606	89242	91968	25721
55696	67106	73369	20689	27707	10432	53118	23692	21450	67362
74838	46105	29798	05504	62588	12700	46093	58754	15780	00361
25833	46204	42441	14284	07858	94467	64358	84445	86230	54172
87260	93170	35494	54207	82683	22976	12257	94522	61364	34228
73595	29104	59346	21213	30923	15747	67104	90389	75901	45606
66224	21746	81973	43832	55932	81707	89193	01511	83257	89931
48078	26348	33935	08981	44947	78208	94370	82235	34382	18908
14168	38881	02968	71715	10814	96338	09439	53864	51951	15691
47813	96995	19524	17227	73490	09448	13156	41802	28217	32658
39404	88593	71327	08978	41241	88350	34760	19507	39102	17168
84131	64236	26803	09167	39695	98995	22498	49489	16808	10807
65097	63684	50298	98391	93703	55438	22718	78013	64409	97879
44552	13101	96263	88862	32977	22191	32112	41046	50771	86355
11997	06462	80215	16900	75972	76712	14861	97496	18986	66671
89716	28633	77208	34231	79158	12531	31612	23543	57480	75667
85258	16576	27023	25722	44809	61284	07636	67054	26665	73238
45790	04380	06893	83032	91230	36690	39612	65695	94966	21734
92386	86028	01737	24812	45158	40744	95550	79951	05457	30445
80321	92435	23677	33356	76405	93136	60668	43458	08562	70311
16964	90116	77618	38200	45273	20442	80655	13676	41471	59063
03060	35414	80332	87759	13961	07849	08970	67354	16026	23225
46517	83209	83758	25428	07686	24628	95824	11554	01428	80580
41481	83999	09645	04406	13666	79199	59323	59115	41436	33185
11580	04688	11925	57414	56554	94938	18151	93058	26924	16181
92862	25355	38189	68819	61797	70112	84563	54657	21490	52086
27419	80915	50829	23146	11641	29047	45806	98176	75455	09782
01450	54579	23503	31250	56057	44450	55982	73182	23666	66578
61200	38309	29934	09351	17290	61419	39377	01770	48134	58599
66047	26430	22415	98215	10413	54380	10492	59665	42368	15138
71899	68860	08150	39941	60556	23386	92449	31012	41277	18925
36567	46306	69777	56251	20007	74448	75234	58915	64903	24311
86135	49654	63467	35906	50560	24921	21109	18652	39797	19964
82155	74998	68901	12964	65056	61967	08628	88194	26741	52840
75099	37473	98759	91653	76447	34010	86452	82362	25185	90842

† Table taken from RAND Corporation [1955], "A Million Random Digits with 100,000 Normal Deviates." The Free Press, Glencoe, Illinois (with the permission of the authors and publishers).

Table A.8   Table of random digits (*Continued*)

21283	43853	74865	73583	08876	67087	02525	03241	94350	04172
87917	97855	67348	20071	47344	90896	21679	94989	29591	37269
83862	26696	37392	03522	23203	08017	06790	67414	24561	12391
09471	85453	46931	40235	14176	34682	49192	99215	94701	81290
19759	48530	89733	32755	75692	71623	13795	90951	63440	44317
16244	57326	37848	66295	82312	77146	86703	61460	19394	53128
42460	46080	49116	17433	45753	46109	29109	10557	21508	11672
77034	52025	22004	86528	11113	90506	49771	32389	31743	75796
37750	03775	16934	50263	76022	12272	29174	13940	90468	89937
23949	49103	19843	88116	91538	08006	40005	76885	28744	41298
92294	96947	21833	37337	76138	26976	94045	17783	82313	06104
23371	17744	65796	85630	95253	48648	34753	23276	51600	73434
95917	24532	87339	14657	56532	19527	77319	47725	54536	12407
14049	00135	26916	49664	29364	77030	55504	98664	86582	92431
83273	62602	43860	04833	22429	21551	02298	68898	45197	20922
24205	72566	98820	06910	66819	35138	48382	90086	29851	92567
78857	76428	12777	50752	91240	21613	32001	54640	77537	89732
53171	34519	71235	82901	68766	39363	72408	35750	80331	22283
12044	43643	87828	29887	33370	56571	03655	99138	31425	59530
79443	97092	28610	75860	63375	48731	24410	11826	57240	26900
94510	20559	32692	10386	36002	30580	38869	53629	81677	54490
84271	47356	89034	95245	59812	18944	49435	02286	25393	09383
72220	18172	72317	24066	64958	87199	93564	38065	04548	51704
53059	47406	92312	77434	75372	93964	27030	69912	90407	70914
41835	02824	80725	86801	94677	06155	89349	02628	78079	27452
16940	31449	38981	24369	84086	98903	76385	64748	94077	74336
71127	57384	49660	61768	70085	06753	02517	92561	46648	41940
80965	02675	29466	50947	93438	53384	90732	31692	57615	61966
82087	29525	16108	93557	46002	04366	29731	48977	19853	04401
62935	59318	91999	65466	34319	83399	69708	72401	80732	52881
95379	47593	32196	95840	78499	86791	85631	85357	65270	06944
89452	05535	13573	80754	23743	20597	62292	35622	39526	01207
39456	49121	10159	09738	03562	42543	01315	80424	97945	03254
69537	18565	84307	00536	76220	26963	07795	64452	80839	71782
60594	00768	60848	35294	69980	20621	82557	73408	48201	35286
54675	61230	05929	03357	72937	16060	61474	10339	37174	90604
13680	68070	82342	18344	57481	60623	29449	75443	71486	88515
49826	58982	34936	98539	50746	99037	09234	17877	08364	36406
98045	65470	28792	31410	85092	14152	24717	36055	98868	79301
64593	88267	55920	84779	33728	90193	41664	19664	57268	52512
82470	36461	73573	42203	16381	29358	29665	08375	64965	97629
19527	18609	65436	54080	84406	19026	33173	51952	01367	33401
20792	38399	15718	79168	80543	09631	13507	03465	59044	91011
34507	94148	00184	54286	33729	70116	07980	16425	72211	47359
04160	76529	28271	63534	83397	40563	86611	80097	16246	00830
26102	37271	13794	72520	67022	08354	49589	68282	28640	56437
27718	94415	18969	51737	35639	13411	78331	90292	64469	24048
49816	02376	62684	16967	31540	11060	30737	89305	10431	79165
21884	52940	65576	61409	24223	10564	35762	26233	15569	92549
35477	68993	00738	03377	94401	01244	66478	81914	86146	76346

Derivation of the Asymptotic Extreme-value Distribution

We present here a simplified, not wholly rigorous, argument leading to the asymptotic Type I extreme-value distribution. The reader is referred to Gumbel [1958] for an entire book devoted to the subject of extreme value distributions. Our interest is in underlying distributions F_X having an upper tail of the exponential type, Eq. (3.3.47). Y_n is the largest of n independent, identically distributed random variables, $X_1, X_2, \ldots, X_n$.

For convenience, we introduce what is known as the characteristic value u_n of Y_n. It is defined to equal that value x for which $F_X(x) = 1 - 1/n$. For example, the characteristic value of u_2 of Y_2 is the median of the X_i, while u_{100}, the characteristic value of Y_{100}, is the value which any X_i exceeds with probability $\frac{1}{100}$. For the exponential-type case under consideration here, we can write

$$F_X(u_n) = 1 - \frac{1}{n} = 1 - e^{-g(u_n)}$$

Thus

$$e^{-g(u_n)} = \frac{1}{n}$$

and

$$\frac{1}{n} e^{g(u_n)} = 1 \tag{B.1}$$

We shall use this result below.

Consider now the distribution of Y_n, Eq. (3.3.45):

$$F_{Y_n}(y) = [F_X(y)]^n$$

If n is large, Y_n will almost certainly take on values only in the upper tail of the distribution of X. Thus in the region of interest

$$F_{Y_n}(y) = [1 - e^{-g(y)}]^n$$

Introducing a factor equal to unity [Eq. (B.1)],

$$F_{Y_n}(y) = \left[1 - \frac{1}{n} e^{g(u_n)} e^{-g(y)}\right]^n = \left\{1 - \frac{1}{n} e^{-[g(y)-g(u_n)]}\right\}^n \tag{B.2}$$

If y is "not far" from u_n (and u_n was seen in Sec. 3.3.3. to be the "most likely" value or mode of Y_n), then in the spirit of Sec. 2.4.4 we can replace $g(y)$ by a linear approximation. An expansion of $g(y)$ about $y = u_n$ gives:

$$g(y) \approx g(u_n) + \left[\frac{dg(y)}{dy}\bigg|_{u_n}\right](y - u_n) = g(u_n) + \alpha_n(y - u_n)$$

in which α_n depends only on n and not on y. Note, for example, that if $g(y) = \lambda y$, α_n is simply λ. Therefore,

$$g(y) - g(u_n) \approx \alpha_n(y - u_n)$$

Substituting into Eq. (B.2),

$$F_{Y_n}(y) \approx \left[1 - \frac{1}{n} e^{-\alpha_n(y-u_n)}\right]^n = \left(1 - \frac{\beta}{n}\right)^n$$

in which β has temporarily replaced $e^{-\alpha_n(y-u_n)}$. As n gets large, if $g(y)$ and hence α_n are reasonably well-behaved functions,[†]

$$\begin{aligned}
F_{Y_n}(y) &= \lim_{n \to \infty} \left(1 - \frac{\beta}{n}\right)^n \\
&= \lim_{n \to \infty} \left[1 - \frac{n}{n}\beta + \frac{n(n-1)}{2!n}\beta^2 - \cdots\right] \\
&= 1 - \beta + \frac{1}{2!}\beta^2 - \frac{1}{3!}\beta^3 \cdots \\
&= e^{-\beta}
\end{aligned}$$

[†] Rigorously we should formalize the assumption that we are always (i.e., for all n) interested in values of y near to the mode u_n [in terms of $1/\alpha_n$, which is proportional to σ_{Y_n}, Eq. (3.3.52)]. Then β can be considered a "constant" as n grows.

Substituting for β,

$$F_{Y_n}(y) = \exp\left[-e^{-\alpha_n(y-u_n)}\right]$$

This is the form of the asymptotic Type I extreme-value distribution introduced in Eq. (3.3.48).

Note that α and u depend in general on the number n of variables of which Y_n is a maximum. The most likely value of Y_n is the mode u_n. We should expect this to increase as n increases. In the text the subscript n is dropped because its value is usually not known. It is simply assumed that it is large enough that the functional form, Eq. (3.3.48), is an appropriate model for Y, the numerical estimates of α and u being obtained directly from observations of Y.

The dependence of u_n and α_n on n may be important, however. It is used, for example, in the theory of the strength of materials to explain *statistical size effect*, that is, the physical phenomenon of larger specimens of brittle material failing at smaller stresses than smaller specimens. Here the argument is that the mode of Y_n, the largest flaw (which determines the strength of a brittle specimen) grows as the size of the specimen grows (i.e., as the number n of unit volumes of material in which flaws occur grows).

The functional relationship between u_n and n depends on the functional form of $g(x)$, since, from Eq. (B.1),

$$g(u_n) = -\ln\frac{1}{n} = \ln n$$

or

$$u_n = g^{-1}(\ln n) \tag{B.3}$$

Thus if $g(x) = \lambda x$, $g^{-1}(\ln n) = (1/\lambda)\ln n$; while if $g(x) = \lambda x^2$,

$$u_n = g^{-1}(\ln n) = \frac{1}{\lambda}\sqrt{\ln n}$$

Thus from the dependence of the mode of specimen strength on the specimen or sample size n, one can infer the functional form of the upper tail of the distribution of the underlying X_i's. Consideration of the observed influence of n on the dispersion of Y_n can support these studies, particularly in distinguishing between underlying distributions of the exponential type [Eq. (3.3.47)] and that displayed in Eq. (3.3.65). This second type leads to another asymptotic extreme-value form, Eq. (3.3.66), whose central value and dispersion also depend on n, but in a different manner. A derivation parallel to that in this appendix will show these distinctions.

Name Index

Subject Index